Pitman Research Notes in Mathematics Series

Submission of proposals for consideration
Suggestions for publication, in the form of outlines and representative samples, are invited by the Editorial Board for assessment. Intending authors should approach one of the main editors or another member of the Editorial Board, citing the relevant AMS subject classifications. Alternatively, outlines may be sent directly to the publisher's offices. Refereeing is by members of the board and other mathematical authorities in the topic concerned, throughout the world.

Preparation of accepted manuscripts
On acceptance of a proposal, the publisher will supply full instructions for the preparation of manuscripts in a form suitable for direct photo-lithographic reproduction. Specially printed grid sheets are provided and a contribution is offered by the publisher towards the cost of typing. Word processor output, subject to the publisher's approval, is also acceptable.

Illustrations should be prepared by the authors, ready for direct reproduction without further improvement. The use of hand-drawn symbols should be avoided wherever possible, in order to maintain maximum clarity of the text.

The publisher will be pleased to give any guidance necessary during the preparation of a typescript, and will be happy to answer any queries.

Important note
In order to avoid later retyping, intending authors are strongly urged not to begin final preparation of a typescript before receiving the publisher's guidelines and special paper. In this way it is hoped to preserve the uniform appearance of the series.

Longman Scientific & Technical
Longman House
Burnt Mill
Harlow, Essex, UK
(tel (0279) 26721)

Titles in this series

1 Improperly posed boundary value problems
A Carasso and A P Stone
2 Lie algebras generated by finite dimensional ideals
I N Stewart
3 Bifurcation problems in nonlinear elasticity
R W Dickey
4 Partial differential equations in the complex domain
D L Colton
5 Quasilinear hyperbolic systems and waves
A Jeffrey
6 Solution of boundary value problems by the method of integral operators
D L Colton
7 Taylor expansions and catastrophes
T Poston and I N Stewart
8 Function theoretic methods in differential equations
R P Gilbert and R J Weinacht
9 Differential topology with a view to applications
D R J Chillingworth
10 Characteristic classes of foliations
H V Pittie
11 Stochastic integration and generalized martingales
A U Kussmaul
12 Zeta-functions: An introduction to algebraic geometry
A D Thomas
13 Explicit *a priori* inequalities with applications to boundary value problems
V G Sigillito
14 Nonlinear diffusion
W E Fitzgibbon III and H F Walker
15 Unsolved problems concerning lattice points
J Hammer
16 Edge-colourings of graphs
S Fiorini and R J Wilson
17 Nonlinear analysis and mechanics: Heriot-Watt Symposium Volume I
R J Knops
18 Actions of fine abelian groups
C Kosniowski
19 Closed graph theorems and webbed spaces
M De Wilde
20 Singular perturbation techniques applied to integro-differential equations
H Grabmüller
21 Retarded functional differential equations: A global point of view
S E A Mohammed
22 Multiparameter spectral theory in Hilbert space
B D Sleeman
24 Mathematical modelling techniques
R Aris
25 Singular points of smooth mappings
C G Gibson
26 Nonlinear evolution equations solvable by the spectral transform
F Calogero
27 Nonlinear analysis and mechanics: Heriot-Watt Symposium Volume II
R J Knops
28 Constructive functional analysis
D S Bridges
29 Elongational flows: Aspects of the behaviour of model elasticoviscous fluids
C J S Petrie
30 Nonlinear analysis and mechanics: Heriot-Watt Symposium Volume III
R J Knops
31 Fractional calculus and integral transforms of generalized functions
A C McBride
32 Complex manifold techniques in theoretical physics
D E Lerner and P D Sommers
33 Hilbert's third problem: scissors congruence
C-H Sah
34 Graph theory and combinatorics
R J Wilson
35 The Tricomi equation with applications to the theory of plane transonic flow
A R Manwell
36 Abstract differential equations
S D Zaidman
37 Advances in twistor theory
L P Hughston and R S Ward
38 Operator theory and functional analysis
I Erdelyi
39 Nonlinear analysis and mechanics: Heriot-Watt Symposium Volume IV
R J Knops
40 Singular systems of differential equations
S L Campbell
41 N-dimensional crystallography
R L E Schwarzenberger
42 Nonlinear partial differential equations in physical problems
D Graffi
43 Shifts and periodicity for right invertible operators
D Przeworska-Rolewicz
44 Rings with chain conditions
A W Chatters and C R Hajarnavis
45 Moduli, deformations and classifications of compact complex manifolds
D Sundararaman
46 Nonlinear problems of analysis in geometry and mechanics
M Atteia, D Bancel and I Gumowski
47 Algorithmic methods in optimal control
W A Gruver and E Sachs
48 Abstract Cauchy problems and functional differential equations
F Kappel and W Schappacher
49 Sequence spaces
W H Ruckle
50 Recent contributions to nonlinear partial differential equations
H Berestycki and H Brezis
51 Subnormal operators
J B Conway

52 Wave propagation in viscoelastic media
F Mainardi

53 Nonlinear partial differential equations and their applications: Collège de France Seminar. Volume I
H Brezis and J L Lions

54 Geometry of Coxeter groups
H Hiller

55 Cusps of Gauss mappings
T Banchoff, T Gaffney and C McCrory

56 An approach to algebraic K-theory
A J Berrick

57 Convex analysis and optimization
J-P Aubin and R B Vintner

58 Convex analysis with applications in the differentiation of convex functions
J R Giles

59 Weak and variational methods for moving boundary problems
C M Elliott and J R Ockendon

60 Nonlinear partial differential equations and their applications: Collège de France Seminar. Volume II
H Brezis and J L Lions

61 Singular systems of differential equations II
S L Campbell

62 Rates of convergence in the central limit theorem
Peter Hall

63 Solution of differential equations by means of one-parameter groups
J M Hill

64 Hankel operators on Hilbert space
S C Power

65 Schrödinger-type operators with continuous spectra
M S P Eastham and H Kalf

66 Recent applications of generalized inverses
S L Campbell

67 Riesz and Fredholm theory in Banach algebra
B A Barnes, G J Murphy, M R F Smyth and T T West

68 Evolution equations and their applications
F Kappel and W Schappacher

69 Generalized solutions of Hamilton-Jacobi equations
P L Lions

70 Nonlinear partial differential equations and their applications: Collège de France Seminar. Volume III
H Brezis and J L Lions

71 Spectral theory and wave operators for the Schrödinger equation
A M Berthier

72 Approximation of Hilbert space operators I
D A Herrero

73 Vector valued Nevanlinna Theory
H J W Ziegler

74 Instability, nonexistence and weighted energy methods in fluid dynamics and related theories
B Straughan

75 Local bifurcation and symmetry
A Vanderbauwhede

76 Clifford analysis
F Brackx, R Delanghe and F Sommen

77 Nonlinear equivalence, reduction of PDEs to ODEs and fast convergent numerical methods
E E Rosinger

78 Free boundary problems, theory and applications. Volume I
A Fasano and M Primicerio

79 Free boundary problems, theory and applications. Volume II
A Fasano and M Primicerio

80 Symplectic geometry
A Crumeyrolle and J Grifone

81 An algorithmic analysis of a communication model with retransmission of flawed message
D M Lucantoni

82 Geometric games and their applications
W H Ruckle

83 Additive groups of rings
S Feigelstock

84 Nonlinear partial differential equations and their applications: Collège de France Seminar. Volume IV
H Brezis and J L Lions

85 Multiplicative functionals on topological algebras
T Husain

86 Hamilton-Jacobi equations in Hilbert spaces
V Barbu and G Da Prato

87 Harmonic maps with symmetry, harmonic morphisms and deformations of metrics
P Baird

88 Similarity solutions of nonlinear partial differential equations
L Dresner

89 Contributions to nonlinear partial differential equations
C Bardos, A Damlamian, J I Díaz and J Hernández

90 Banach and Hilbert spaces of vector-valued functions
J Burbea and P Masani

91 Control and observation of neutral systems
D Salamon

92 Banach bundles, Banach modules and automorphisms of C*-algebras
M J Dupré and R M Gillette

93 Nonlinear partial differential equations and their applications: Collège de France Seminar. Volume V
H Brezis and J L Lions

94 Computer algebra in applied mathematics: an introduction to MACSYMA
R H Rand

95 Advances in nonlinear waves. Volume I
L Debnath

96 FC-groups
M J Tomkinson

97 Topics in relaxation and ellipsoidal methods
M Akgül

98 Analogue of the group algebra for topological semigroups
H Dzinotyiweyi

99 Stochastic functional differential equations
S E A Mohammed

100 Optimal control of variational inequalities
V Barbu
101 Partial differential equations and dynamical systems
W E Fitzgibbon III
102 Approximation of Hilbert space operators. Volume II
C Apostol, L A Fialkow, D A Herrero and D Voiculescu
103 Nondiscrete induction and iterative processes
V Ptak and F-A Potra
104 Analytic functions – growth aspects
O P Juneja and G P Kapoor
105 Theory of Tikhonov regularization for Fredholm equations of the first kind
C W Groetsch
106 Nonlinear partial differential equations and free boundaries. Volume I
J I Díaz
107 Tight and taut immersions of manifolds
T E Cecil and P J Ryan
108 A layering method for viscous, incompressible L_p flows occupying R^n
A Douglis and E B Fabes
109 Nonlinear partial differential equations and their applications: Collège de France Seminar. Volume VI
H Brezis and J L Lions
110 Finite generalized quadrangles
S E Payne and J A Thas
111 Advances in nonlinear waves. Volume II
L Debnath
112 Topics in several complex variables
E Ramírez de Arellano and D Sundararaman
113 Differential equations, flow invariance and applications
N H Pavel
114 Geometrical combinatorics
F C Holroyd and R J Wilson
115 Generators of strongly continuous semigroups
J A van Casteren
116 Growth of algebras and Gelfand–Kirillov dimension
G R Krause and T H Lenagan
117 Theory of bases and cones
P K Kamthan and M Gupta
118 Linear groups and permutations
A R Camina and E A Whelan
119 General Wiener–Hopf factorization methods
F-O Speck
120 Free boundary problems: applications and theory, Volume III
A Bossavit, A Damlamian and M Fremond
121 Free boundary problems: applications and theory, Volume IV
A Bossavit, A Damlamian and M Fremond
122 Nonlinear partial differential equations and their applications: Collège de France Seminar. Volume VII
H Brezis and J L Lions
123 Geometric methods in operator algebras
H Araki and E G Effros
124 Infinite dimensional analysis–stochastic processes
S Albeverio
125 Ennio de Giorgi Colloquium
P Krée
126 Almost-periodic functions in abstract spaces
S Zaidman
127 Nonlinear variational problems
A Marino, L Modica, S Spagnolo and M Degiovanni
128 Second-order systems of partial differential equations in the plane
L K Hua, W Lin and C-Q Wu
129 Asymptotics of high-order ordinary differential equations
R B Paris and A D Wood
130 Stochastic differential equations
R Wu
131 Differential geometry
L A Cordero
132 Nonlinear differential equations
J K Hale and P Martinez-Amores
133 Approximation theory and applications
S P Singh
134 Near-rings and their links with groups
J D P Meldrum
135 Estimating eigenvalues with *a posteriori/a priori* inequalities
J R Kuttler and V G Sigillito
136 Regular semigroups as extensions
F J Pastijn and M Petrich
137 Representations of rank one Lie groups
D H Collingwood
138 Fractional calculus
G F Roach and A C McBride
139 Hamilton's principle in continuum mechanics
A Bedford
140 Numerical analysis
D F Griffiths and G A Watson
141 Semigroups, theory and applications. Volume I
H Brezis, M G Crandall and F Kappel
142 Distribution theorems of L-functions
D Joyner
143 Recent developments in structured continua
D De Kee and P Kaloni
144 Functional analysis and two-point differential operators
J Locker
145 Numerical methods for partial differential equations
S I Hariharan and T H Moulden
146 Completely bounded maps and dilations
V I Paulsen
147 Harmonic analysis on the Heisenberg nilpotent Lie group
W Schempp
148 Contributions to modern calculus of variations
L Cesari
149 Nonlinear parabolic equations: qualitative properties of solutions
L Boccardo and A Tesei
150 From local times to global geometry, control an[d] physics
K D Elworthy

151 A stochastic maximum principle for optimal control of diffusions
U G Haussmann
152 Semigroups, theory and applications. Volume II
H Brezis, M G Crandall and F Kappel
153 A general theory of integration in function spaces
P Muldowney
154 Oakland Conference on partial differential equations and applied mathematics
L R Bragg and J W Dettman
155 Contributions to nonlinear partial differential equations. Volume II
J I Díaz and P L Lions
156 Semigroups of linear operators: an introduction
A C McBride
157 Ordinary and partial differential equations
B D Sleeman and R J Jarvis
158 Hyperbolic equations
F Colombini and M K V Murthy
159 Linear topologies on a ring: an overview
J S Golan
160 Dynamical systems and bifurcation theory
M I Camacho, M J Pacifico and F Takens
161 Branched coverings and algebraic functions
M Namba
162 Perturbation bounds for matrix eigenvalues
R Bhatia
163 Defect minimization in operator equations: theory and applications
R Reemtsen
164 Multidimensional Brownian excursions and potential theory
K Burdzy
165 Viscosity solutions and optimal control
R J Elliott
166 Nonlinear partial differential equations and their applications. Collège de France Seminar. Volume VIII
H Brezis and J L Lions
167 Theory and applications of inverse problems
H Haario
168 Energy stability and convection
G P Galdi and B Straughan
169 Additive groups of rings. Volume II
S Feigelstock
170 Numerical analysis 1987
D F Griffiths and G A Watson
171 Surveys of some recent results in operator theory. Volume I
J B Conway and B B Morrel
172 Amenable Banach algebras
J-P Pier
173 Pseudo-orbits of contact forms
A Bahri
174 Poisson algebras and Poisson manifolds
K H Bhaskara and K Viswanath
175 Maximum principles and eigenvalue problems in partial differential equations
P W Schaefer
176 Mathematical analysis of nonlinear, dynamic processes
K U Grusa
177 Cordes' two-parameter spectral representation theory
D F McGhee and R H Picard
178 Equivariant K-theory for proper actions
N C Phillips
179 Elliptic operators, topology and asymptotic methods
J Roe
180 Nonlinear evolution equations
J K Engelbrecht, V E Fridman and E N Pelinovski
181 Nonlinear partial differential equations and their applications. Collège de France Seminar. Volume IX
H Brezis and J L Lions
182 Critical points at infinity in some variational problems
A Bahri
183 Recent developments in hyperbolic equations
L Cattabriga, F Colombini, M K V Murthy and S Spagnolo
184 Optimization and identification of systems governed by evolution equations on Banach space
N U Ahmed
185 Free boundary problems: theory and applications. Volume I
K H Hoffmann and J Sprekels
186 Free boundary problems: theory and applications. Volume II
K H Hoffmann and J Sprekels
187 An introduction to intersection homology theory
F Kirwan
188 Derivatives, nuclei and dimensions on the frame of torsion theories
J S Golan and H Simmons
189 Theory of reproducing kernels and its applications
S Saitoh
190 Volterra integrodifferential equations in Banach spaces and applications
G Da Prato and M Iannelli
191 Nest algebras
K R Davidson
192 Surveys of some recent results in operator theory. Volume II
J B Conway and B B Morrel
193 Nonlinear variational problems. Volume II
A Marino and M K Murthy
194 Stochastic processes with multidimensional parameter
M E Dozzi
195 Prestressed bodies
D Iesan
196 A Hilbert space approach to some classical transforms
R H Picard
197 Stochastic calculus in application
J R Norris

Mathematical analysis of nonlinear dynamic processes

K-U Grusa

Centre for Interdisciplinary Research (ZiF), Bielefeld

Mathematical analysis of nonlinear dynamic processes

An introduction to processes governed by partial differential equations

Copublished in the United States with
John Wiley & Sons, Inc., New York

Longman Scientific & Technical
Longman Group UK Limited
Longman House, Burnt Mill, Harlow
Essex CM20 2JE, England
and Associated Companies throughout the world.

Copublished in the United States with
John Wiley & Sons, Inc., 605 Third Avenue, New York, NY 10158

First published 1988

AMS Subject Classifications: 34-XX, 35-XX, 49A22, 49C10, 90A16

ISSN 0269-3674

British Library Cataloguing in Publication Data
Grusa, K.U.
Mathematical analysis of nonlinear dynamic processes
1. Partial differential equations
I. Title
515.3'53

ISBN 0-582-02880-9

Library of Congress Cataloging-in-Publication Data
Grusa, Karl-Ulrich
Mathematical analysis of nonlinear dynamic processes.
(Pitman research notes in mathematics series,
ISSN 0269-3674;176)
Bibliography: p.
Includes index.
1. Differential equations, partial.
2. Nonlinear theories. 3. Differentiable dynamic systems.
I. Title. II. Series.
QA374.G78 1988 515.3'53 88-9490
ISBN 0-470-21146-6 (USA only)

Printed and bound in Great Britain by
Biddles Ltd, Guildford and King's Lynn

To the memory
of my beloved parents
Arthur and Magdalena Grusa

psalm 148

Contents

Preface

In recent years, nonlinear diffusion equations and nonlinear wave equations have attracted increasing interest within the mathematical community, a fact which reflects growth in the range and importance of their applications as well as recent contributions to the understanding of the mathematical behaviour of their solutions.

The study of the solutions of such equations has proved to be a rich and rewarding area of mathematical endeavour, yielding valuable insights into the phenomena being modelled.

In this volume an attempt will be made to describe and gain a preliminary insight into the subject of nonlinear, dynamic processes governed by nonlinear diffusion, reaction-diffusion and wave equations. The nonlinear, dynamic processes are represented with the help of the mathematical methods of multi-connected domains and controlled by differential games and von Stackelberg strategies.

Chapter III deals with reaction-diffusion processes and some applications in biology and chemistry. Chapter IV offers new modelling methods of influence-phenomena in economics. Chapter VI devotes itself to image transmission and reconstruction processes in physics. Chapter VII deals with new methods for modelling the genesis of cardiac arrhythmias in medicine.

The field of applications of the theory of nonlinear, dynamic processes is extremely wide. The methods of model-building we have used are quite general and hence more general nonlinear, dynamic processes can be described and solved in an analytic form using the same methods.

K-U Grusa

I Model-building in multiconnected domains

I.1 Variational formulation of initial-boundary-value problems

The optimal control theory described in the works of Pontryagin-Boltyanskii-Gamkrelidze-Mischenko (282) and Hestenes (156) is concerned with the study of a family of ordinary differential (or integro-differential) operators.

Due to the complexity of the system to be controlled, it is often advantageous to describe it by a family of partial differential operators (see Butkowskii (50) and Wang (353,f) with bibliography). Lions (220) considers, in his work, systems, whose state is given by the solution of a partial differential equation to which we must add appropriate boundary conditions, and in the case of evolution equations initial conditions (214-219, 220a,b,c,d).

We are considering systems, which are governed by partial differential equations with respect to their boundary and initial value problems in multiconnected regions.

In order not to overburden this work, we have restricted ourselves to comparatively simple models and model-building-methods. However our modelling-methods are quite general and hence more general problems can be solved using our model-building-processes.

Amongst other things we are studying now the minimization of positive definite quadratic forms, defined on a closed, convex subset of a Hilbert space.

Let V and H be two Hilbert spaces, with dense and continuous imbedding $V \subset H$ and norms $|. .|$, $\|. .\|$. Let V' be the dual space of V, and by identifying H with its dual, we have

$$V \subset H \simeq H' \subset V'.$$

Denoting by (f,v) the scalar product of $f \in V'$ with $v \in V$ and let $\| \|_*$ be the norm of V'.

Introducing another Hilbert space $\mathcal{H}$ with norm $\| \|_{\mathcal{H}}$, then we denote by

$L^2((0,T),\mathcal{H})$ the space of equivalence class of functions, where the mapping

$$[0,T] \to \mathcal{H}$$

$$t \mapsto v(t)$$

is measurable, such that

$$\left(\int_0^T \|v(t)\|_{\mathcal{H}}^2 \, dt\right)^{\frac{1}{2}} < \infty.$$

It can be shown that $L^2(0,T),\mathcal{H})$ is a Hilbert space. Let

$$a: V \times V \to \mathbf{R}$$

$$(u,v) \mapsto a(u,v)$$

be a continuous bilinear form on V, then there exists a well determinated operator $A \in \mathcal{L}(V,V')$ in the space of all continuous operators from V to V', such that

$$a(u,v) = (Au,v).$$

Now, we can formulate the evolution problem

$$\frac{du}{dt} + Au = f \text{ in } (0,T),$$

with the initial condition $u(0) = u_0$.

Introducing the bilinear form $a(u,v)$, the evolution problem can be written in the equivalent form

$$\left(\frac{du}{dt},v\right) + a(u(t),v) = (f(t),v) \text{ for } v \in V.$$

We are able to define for $u \in L^2(0,T),\mathcal{H})$ the derivative with respect to t by a distribution with values in $\mathcal{H}$.

Let $\mathcal{D}(\Omega)$ be the space of infinitely differentiable functions in Ω with compact support in Ω, endowed with the inductive limit topology of L. Schwartz (308), and $\mathcal{D}'(\Omega)$, the dual of $\mathcal{D}(\Omega)$, is the space of distributions on (0, T).

If $\mathcal{B}$ is a Banach space, then $\mathcal{D}'((\theta,T),\mathcal{B})$ denotes the space of distributions on $(0,T)$ with values in $\mathcal{B}$, L. Schwartz (310, 311).

We say f is a distribution on $(0,T)$ with values in $\mathcal{H}$, if

$$f \in L(\mathcal{D}((0,T));\mathcal{H}) = \mathcal{D}'((0,T),\mathcal{H}).$$

Now, if $f \in L^2((0,T),\mathcal{H})$, then $f \in \mathcal{D}'((0,T),\mathcal{H})$. For $\varphi \in \mathcal{D}((0,T))$ we can write

$$f(\varphi) = \int_0^T f(t)\cdot\varphi(t)dt \in \mathcal{H},$$

and this defines a linear and continuous mapping:

$$\mathcal{D}((0,T)) \to \mathcal{H}$$

$$\varphi \mapsto f(\varphi).$$

Introducing the differential operator $\frac{d}{dt}$,

$$\frac{d}{dt} \in L(\mathcal{D}'((0,T),\mathcal{H});\ \mathcal{D}'(0,T);\mathcal{H})) \text{ by } \frac{d}{dt} f(\varphi) = -f \frac{\partial\varphi}{\partial t} \text{ for all } \varphi \in \mathcal{D}((0,T)),$$

thus, for $f \in L^2(0,T);\mathcal{H})$ we have $f \in \mathcal{D}'(0,T);\mathcal{H})$ and define $\frac{d}{dt} f \in \mathcal{D}'((0,T);\mathcal{H})$. So we can introduce the Sobolev space $H^1(\cdot)$ of the first order with values in $\mathcal{H}$ by

$$f \in L^2(0,T);\mathcal{H});\ \frac{df}{dt} \in L^2(0,T);\mathcal{H}),$$

this is a Hilbert space, as

$$\left(\int_0^T \left(\|f(t)\|_{\mathcal{H}}^2 + \left\|\frac{df}{dt}\right\|_{\mathcal{H}}^2\right)dt\right)^{\frac{1}{2}} < \infty.$$

For $v \in L^2(0,T);V)$ we have $\frac{dv}{dt} \in \mathcal{D}'((0,T);V) \subset \mathcal{D}'((0,T);V')$, then

$$\left(\int_0^T \left(\|v(t)\|^2 + \left\|\frac{dv}{dt}\right\|_*^2\right)dt\right)^{\frac{1}{2}}$$

is a norm of the space, where $\|..\|_*$ is the norm of V'. Introducing

$$W = \{v | v \in L^2((0,T);V), \frac{dv}{dt} \in L^2((0,T);V'),$$

Lions-Magenes (216) have demonstrated that $v \in W$ is continuous on $[0,T] \to H$, after a modification of a set of measure 0.

So we define the trace (in the origin) by the linear, continuous and surjective (216) mapping:

$$W \to H$$

$$v \mapsto v(0).$$

Now we are able to formulate the evolution problem exactly:
Let $f \in L^2((0,T);V')$, $u_o \in H$, we are searching for $u \in W$, that is a solution of

$$\frac{du}{dt} + Au = f \text{ in } (0,T),$$

with the initial condition

$$u(0) = u_o;$$

and the variational formulation of the evolution problem can be given by

$$(\frac{du}{dt},v) + a(u(t),v) = (f(t),v).$$

Assuming, that there exists $\lambda \in \mathbf{R}$, such that

$$a(u,v) + \lambda(v,v) \geqq \alpha(v,v) \text{ for } \alpha > 0,\ v \in V',$$

we have by Lions-Magenes the

<u>Theorem</u>:

The evolution problem has a unique solution $u \in W$, if $f \in L^2(0,T),V')$, $u_o \in H$, and satisfies the inequality

$$\|u\|_W \leqq c\ \|f\|_{L^2(0,T),V')} + \|u_o\|_H \ .$$

Generalizing this theorem, we introduce the continuous, time-dependent bilinear form

$$a : [0,T] \times V \times V \to \mathbb{R}$$

$$(t,u,v) \mapsto a(t,u,v),$$

such that for u, $v \in V$

$t \mapsto (a(t,u,v))$ is measurable, and there exists a constant M independent of u and v

$$|a(t,u,v)| \leq M \cdot \|u\| \cdot \|v\|,$$

let there exist $\lambda \in \mathbb{R}$ such that

$$a(t,v,v) + \lambda(v,v) \geq \alpha \|v\|^2 \text{ for } \alpha > 0 \text{ and } v \in V,\ t \in (0,T),$$

defining

$A(t) \in L(V,V')$ by

$$a(t,u,v) = (A(t)u,v) \text{ for } u,v \in V.$$

Then we obtain the generalized

Theorem:

There exists a unique solution $u \in W$ of the evolution problem, if $f \in L^2((0,T);V')$, and $u_0 \in H$, such that

$$\frac{du(t)}{dt} + A(t) \cdot u(t) = f(t), \quad u(0) = u_0$$

in equivalent variational form

$$\left(\frac{du(t)}{dt},v\right) + a(t,u(t),v) = (f(t),v) \text{ for } v \in V$$

and the initial condition $u(0) = u_0$ fixed.

Remark:

We consider the following example to understand the variational formulation of boundary value problems.

Let there be a doubly-connected domain with interdependent diffusion processes governed by the system

$$-\Delta u_o = f_o \text{ in } \Omega_o$$

$$-\varepsilon\Delta u_1 = f_1 \text{ in } \Omega_1, \ \varepsilon = \text{small parameter},$$

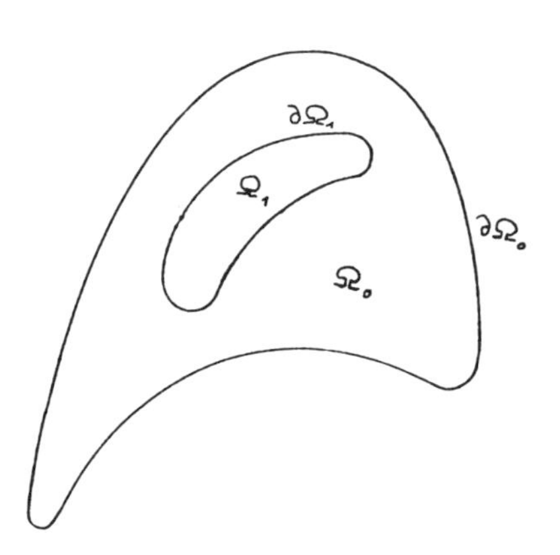

with the boundary condition

$$u_o = 0 \text{ on } \partial\Omega ,$$

the transmission condition

$$u_o = u_1, \text{ and } \frac{\partial u_o}{\partial n} = \frac{\partial u_1}{\partial n} \text{ on } \partial\Omega_1.$$

Denoting $\Omega = \bar{\Omega}_1 \cup \Omega_o$ and introducing the bilinear forms

$$a_i(u,v) = \int_{\Omega_i} \text{grad } u \cdot \text{grad } v \, dx \text{ for } i=0,1,$$

and the space

$$V = \{v | v \in H^1(\Omega), \ v = 0 \text{ on } \partial\Omega\},$$

where $V = H_o^1(\Omega) = \overline{C_k^\infty(\Omega)}$ in the Sobolev space $H^1(\Omega)$.

The equivalent formulation in variational form

$$a_o(u,v) + \varepsilon a_1(u,v) = (f,v) \text{ for } v \in V,$$

this is clear by multiplying the partial differential equations by $v \in C_k^\infty(\Omega)$ and integrating. Now we analyse the transmission condition.

Let $u_\varepsilon \in V$, then $u_\varepsilon = 0$ on $\partial\Omega$ by construction of the space V (139, p. 34) and $u_\varepsilon \in H^1(\Omega)$, as $\Omega = \bar{\Omega}_1 \cup \Omega_o$ we obtain for the trace, that $u_o = u_1$ on $\partial\Omega_1$.

Remembering Green's formula

$$-\int_{\Omega_0} u\Delta v dx = \int_{\Omega_0} \text{grad}u\cdot\text{grad}v - \int_{\partial\Omega_1} u\cdot\frac{\partial v}{\partial n} ds - \int_{\partial\Omega} u\cdot\frac{\partial v}{\partial n} ds,$$

and by multiplying the differential equation $-\Delta u_o = f_o$ by $v \in H^1(\Omega)$ and integrating, we obtain

$$\int_{\Omega_o} f_o\cdot v dx = -\int_{\Omega_o} \Delta u_o\cdot v dx = -\int_{\partial\Omega_o} \frac{\partial u_o}{\partial n} vds + a_o(u_o,v).$$

Quite analogously applied to the differential equation $-\varepsilon\Delta u_1 = f_1$, we have

$$-\int_{\Omega_1} f_1\cdot v dx = \varepsilon\int_{\Omega_1} \Delta u_1 v\, dx = -\varepsilon\int_{\Omega} \text{grad } u_1\cdot\text{grad } vdx +$$

$$+\varepsilon\int_{\partial\Omega_1} \frac{\partial u_1}{\partial n}\cdot vds = -\varepsilon a_1(u_1,v) + \varepsilon\int_{\partial\Omega_1} \frac{\partial u_1}{\partial n} vds,$$

thus

$$\int_{\Omega_1} f_1\cdot v\, dx = \varepsilon\cdot a_1(u_1,v) - \varepsilon\int_{\partial\Omega_1} \frac{\partial u_1}{\partial n}\cdot v\, ds, \text{ by adding}$$

$$\int_{\Omega_o} f_o\cdot v\, dx = a_o(u_o,v) - \int_{\partial\Omega_1} \frac{\partial u_o}{\partial n}\cdot v\, ds,$$

we obtain

$$\int_{\partial\Omega_1} (\frac{\partial u_o}{\partial n} - \frac{\partial u_1}{\partial n})\cdot v\, ds + a_o(u_o,v) + \varepsilon a_1(u_1,v) = (f,v) \text{ for } v \in V,$$

hence

$$\int_{\partial\Omega_1} (\frac{\partial u_o}{\partial n} - \frac{\partial u_1}{\partial n}) v\, ds = 0 \text{ for } v \in H^1(\Omega),$$

thus

$$\frac{\partial u_o}{\partial n} - \frac{\partial u_1}{\partial n} \text{ on } \partial D_1,$$

and the equivalence is proved.

I. 1.1 A parabolic initial-boundary-value problem in a doubly connected domain

We consider interdependent diffusion processes in a doubly connected domain, which are now time-dependent on a higher level of approximation of real diffusion processes.

The evolution process is governed by a parabolic partial differential equation in a bounded domain of $\mathbb{R}^3$.

Let the domain be

$$D = D_1 \cup D_0 \cup \partial D_0$$

and by defining

$$G_0 = D_0 x(0,T), \; G_1 = D_1 x(0,T), \partial G_1 = \partial D_1 x(0,T)$$

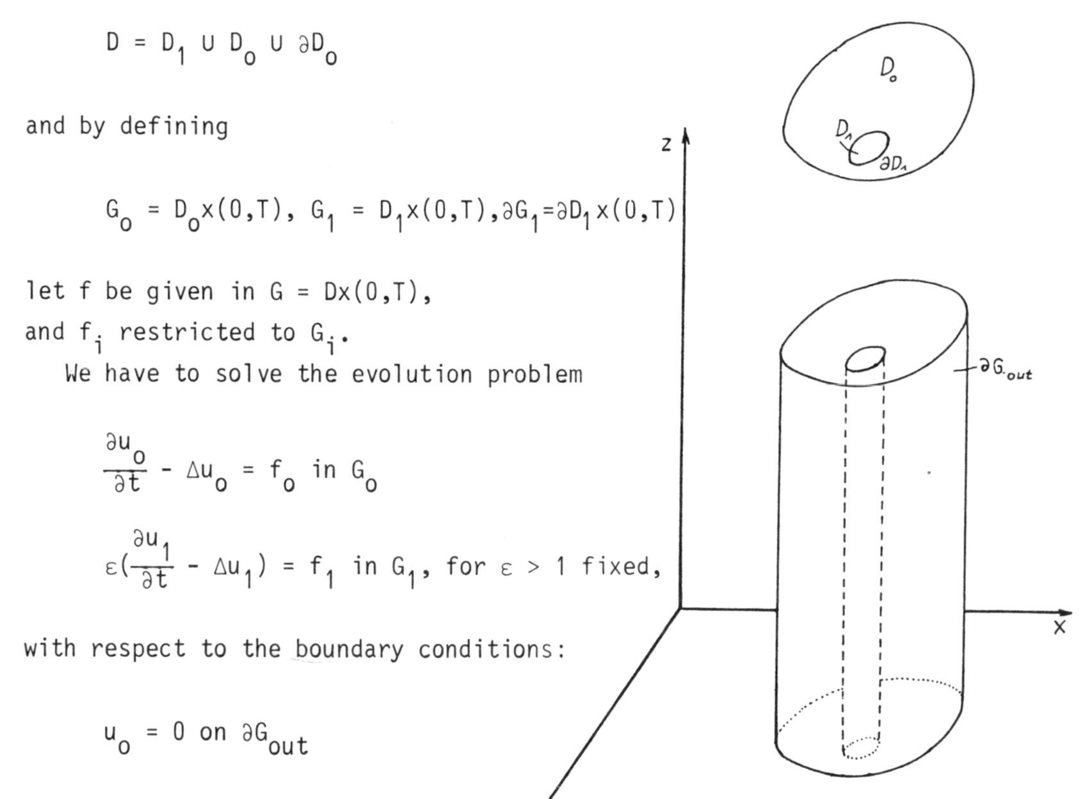

let f be given in $G = Dx(0,T)$,
and f_i restricted to G_i.

We have to solve the evolution problem

$$\frac{\partial u_0}{\partial t} - \Delta u_0 = f_0 \text{ in } G_0$$

$$\varepsilon\left(\frac{\partial u_1}{\partial t} - \Delta u_1\right) = f_1 \text{ in } G_1, \text{ for } \varepsilon > 1 \text{ fixed,}$$

with respect to the boundary conditions:

$$u_0 = 0 \text{ on } \partial G_{out}$$

the transmission condition

$$u_0 = u_1 \text{ and } \frac{\partial u_0}{\partial n} = \frac{\partial u_1}{\partial n} \text{ on } \partial G_1,$$

and the initial conditions

$$u_0 = u_{00} \text{ in } G_0 \text{ and } u_1 = u_{01} \text{ in } G_1.$$

First we show, that there exists a solution.

We can verify analogously by taking $V = H_o^1(\ )$ and $H = L^2(\ )$ and the bilinear forms

$$a_i(u,v) = \int_{G_i} \text{grad}u \cdot \text{grad}v dx, \quad b_i(u,v) = \int_{G_i} u \cdot v \, dx \text{ for } i = 0,1,$$

that the above evolution problem can be described in the following variational form

$$b_o(u',v) + \varepsilon b_1(u',v) + a_o(u,v) + \varepsilon a_1(u,v) = (f,v) \text{ for } v \in V,$$

and $u \in L^2((0,T),V)$, $U' = \frac{du}{dt} \in L^2((0,T),V')$, with $u(0) = u_o$, where $u_o = \{u_{oo}, u_{o1}\}$ are the initial conditions. To this problem can be applied the existence theorem on page 10 by taking the new scalar product of H

$$\{(u,v)\} = b_o(u,v) + \varepsilon b_1(u,v),$$

and denoting

$$a(u,v) = a_o(u,v) + \varepsilon a_1(u,v).$$

The form $v \to (f,v)$ is continuous on V and can be written as the scalar product $(f,v) = \{(\hat{f},v)\}$, that is compatible with $\{\ ,\ \}$. Now, the variational form is given by

$$\{(u',v)\} + a(u,v) = \{(\hat{f},v)\}.$$

If $f \in L^2((0,T),V')$, $u_o \in H$ and the bilinear forms $a_i(u,v)$, $b_i(u,v)$ are continuous on V, H and satisfy for $i = 0,1$

$$a_i(u,v) \geq \alpha_i p_i^2(v) \quad \text{for } \alpha_i > 0, \ v \in V,$$

where p_i is a continuous semi norm on V and

$$b_i(u,v) \geq \beta_i \ q_i^2(v) \text{ for } \beta_i > 0, \ v \in H,$$

where q_i is a continuous semi norm on H.

Noting, that $p_o + p_1$ may be an equivalent norm on V and $q_o + q_1$ an equivalent norm on H.

Then we obtain the

Theorem:

There exists a unique solution $u_\varepsilon \in L^2((0,T),V)$; $\frac{du_\varepsilon}{dt} \in L^2((0,T),V')$, of

$$b_o(u'_\varepsilon,v) + a_o(u_\varepsilon,v) + \varepsilon(b_1(u'_\varepsilon,v) + a_1(u_\varepsilon,v)) = (f,v) \text{ for } v \in V$$
$$\text{and } u_\varepsilon(0) = u_o.$$

The asymptotic expansion of the unique solution can be constructed in the following way.

Let the ansatz be $u_\varepsilon = \frac{u^{-1}}{\varepsilon} + u^o + \varepsilon \cdot u^1 + \ldots$ we show that the unique solution is approximated by

$$\|u_\varepsilon - (\frac{u^{-1}}{\varepsilon} + u^o + \varepsilon \cdot u^1 + \ldots + \varepsilon^k u^k\| \leq C.\varepsilon^{k-1}, \quad \text{where}$$

$$\|\cdot\ \cdot\| = \|\cdot\ \cdot\|_{L^\infty(0,T),H)} \text{ or } \|\cdot\ \cdot\|_{L^2(0,T),V)}.$$

Proof:

We use the ansatz $u_\varepsilon = \frac{u^{-1}}{\varepsilon} + u^o + \varepsilon u^1 + \ldots$ and $\frac{du_\varepsilon}{dt} = \frac{1}{\varepsilon}\frac{du^{-1}}{dt} + \frac{du^o}{dt} + \varepsilon \cdot \frac{du^1}{dt} + \ldots$ in the bilinear forms $a_i(u,v)$ and $b_i(u,v)$ for $i = 1,2$, then

$$\frac{1}{\varepsilon} \cdot b_o(\frac{du^{-1}}{dt},v) + b_o(\frac{du^o}{dt},v) + \varepsilon \cdot b_o(\frac{du^1}{dt},v) + \varepsilon^2 b_o(\frac{du^2}{dt},v) + \ldots$$

$$\frac{1}{\varepsilon} a_o(u^{-1},v) + a_o(u^o,v) + \varepsilon a_o(u^1,v) + \varepsilon^2 a_o(u^2,v) + \ldots$$

$$\varepsilon\{\frac{1}{\varepsilon} \cdot b_1(\frac{du^{-1}}{dt},v) + b_1(\frac{du^o}{dt},v) + \varepsilon b_1(\frac{du^1}{dt},v) + \varepsilon^2 b_1(\frac{du^2}{dt},v) + \ldots\}$$

$$\varepsilon\{\frac{1}{\varepsilon} \cdot a_1(u^{-1},v) + a_1(u^o,v) + \varepsilon \cdot a_1(u^1,v) + \varepsilon^2 a_1(u^2,v) + \ldots\},$$

we choose the ε-terms

$$\varepsilon^o \qquad b_o(\frac{du^o}{dt},v) + a_o(u^o,v) + b_1(\frac{du^{-1}}{dt},v) + a_1(u^{-1},v) = (f,v)$$

$$\varepsilon^1 \qquad b_o(\frac{du^1}{dt},v) + a_o(u^1,v) + b_1(\frac{du^o}{dt},v) + a_1(u^o,v) = 0$$

$$\varepsilon^2 \qquad b_o(\frac{du^2}{dt},v) + a_o(u^2,v) + b_1(\frac{du^1}{dt},v) + a_1(u^1,v) = 0$$

$$\varepsilon^3 \qquad b_o(\frac{du^3}{dt},v) + a_o(u^3,v) + b_1(\frac{du^2}{dt},v) + a_1(u^2,v) = 0,$$

and the general term reads

$$\varepsilon^k \qquad b_o(\frac{du^k}{dt},v) + a_o(u^k,v) + b_1(\frac{du^{k-1}}{dt},v) + a_1(u^{k-1},v) = 0.$$

The initial condition $u_\varepsilon(0) = u$ is given by the ansatz

$$u_\varepsilon(0) = \frac{1}{\varepsilon} u^{-1}(0) + u^o(0) + \varepsilon u^1(0) + \ldots + = u_o,$$

thus we obtain

$$u^{-1}(0) = 0,\ u^o(0) = 0,\ u^1(0) = u_o \text{ and } u^j(0) = 0 \text{ for } j = 2,3,\ldots .$$

Now, we introduce a subspace Y_o such that $a_o(u,v) + b_o(u,v) = 0$. Restricting the above equations to Y_o, we obtain from the first equation $u^{-1}(t) \in Y_o$.

The ε^o-term reduces to

$$b_1(\frac{du^{-1}}{dt},v) + a_1(u^{-1},v) = (f,v) \text{ for } v \in Y_o,$$

with the initial condition $u^{-1}(0) = 0$.

This means that u^{-1} is the solution of the following initial-boundary-value problem

$$\frac{\partial u_1^{-1}}{\partial t} - \Delta u_1^{-1} = f_1 \text{ in } G_1$$

$$u_1^{-1} = 0 \quad \text{on } \partial G_1$$

$$u_1^{-1}(x,0) = 0 \text{ in } G_1.$$

Let u^o be an element of V, then we consider the first equation

$$b_o(\frac{du^o}{dt},v) + a_o(u^o,v) + b_1(\frac{du^{-1}}{dt},v) + a_1(u^{-1},v) = (f_1,v) + (f_o,v)$$

$$\text{for } v \in V \qquad \text{I}$$

restricting the second equation to Y_o we obtain

$$b_1(\frac{du^o}{dt},v) + a_1(u^o,v) = 0 \text{ for } v \in Y_o. \qquad \text{II}$$

The equation above can be split. We deduce from the initial-boundary-value problem in G_1 and Green's formula

$$b_1(\frac{du^{-1}}{dt},v) + a_1(u^{-1},v) + \int_{\partial G_1} \frac{\partial u^{-1}}{\partial n} vd(\partial G_1) - \int_{G_1} f_i \cdot v\, dx,$$

and put it in the split equation

$$b_o(\frac{du^o}{dt},v) + a_o(u^o,v) + \int_{G_1} f_1\ v\, dx + \int_{\partial G_1} \frac{\partial u_1^{-1}}{\partial n} vd(\partial G_1) =$$

$$= \int_{G_1} f_1\ vdx + \int_{G_o} f_o \cdot v\, dx,$$

thus the bilinear form

$$b_o(\frac{du^o}{dt},v) + a_o(u^o,v) + \int_{\partial G_1} \frac{\partial u_1^{-1}}{\partial n} v \cdot d(\partial G_1) = \int_{G_o} f_o \cdot v\, dx$$

is equivalent to the initial-boundary-value problem

$$\frac{\partial u_o^o}{\partial t} - \Delta u_o^o = f_o \text{ in } G_o$$

$$u_o^o = 0 \text{ on } \partial G_{out}$$

$$\frac{\partial u_o^o}{\partial n} = \frac{\partial u_1^{-1}}{\partial n} \text{ on } \partial G_1$$

$$u_o^o(x,0) = u_{oo}(x) \text{ in } G_o.$$

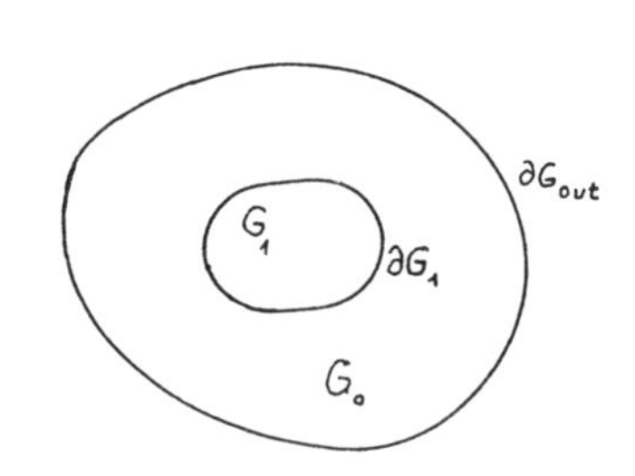

The solution u_o^o in G_o depends on $u_o^o(f_o,u_{oo},u_1^{-1})$. The bilinear form II is equivalent

$$\frac{\partial u_1^o}{\partial t} - \Delta u_1^o = 0 \text{ in } G_1$$

$$u_1^o = u_o^o \text{ on } \partial G_1$$

$$u_1^o(x,0) = u_{o1}(x) \text{ in } G_1.$$

We are taking u_o^o on the boundary ∂G_1 as boundary value of the problem in G_1. Thus, the solution u_1^o in G_1 depends on u_o^o and u_{o1}.

Let $u^1 \in V$, we consider the second equation

$$b_o(\frac{du^1}{dt},v) + a_o(u^1,v) + b_1(\frac{du^o}{dt},v) + a_1(u^o,v) = 0 \text{ for } v \in V,$$

restricting the third to Y_o, then

$$b_1(\frac{du^1}{dt},v) + a_1(u^1,v) = 0 \text{ for } v \in Y_o,$$

this is equivalent to the following initial-boundary-value problem in G_o

$$\frac{\partial u_o^1}{\partial t} - \Delta u_o^1 = 0 \text{ in } G_o$$

$$\frac{\partial u_o^1}{\partial n} = \frac{\partial u_1^o}{\partial n} \text{ on } \partial G_1$$

$$u_o^1 = 0 \quad \text{on } \partial G_{out}$$

$$U_o^1(x,0) = 0 \quad \text{in } G_o,$$

and the initial-boundary-value problem in G_1

$$\frac{\partial u_1^1}{\partial t} - \Delta u_1^1 = 0 \text{ in } G_1$$

$$u_1^1 = u_o^1 \text{ on } \partial G_1$$

$$u_1^1(x,0) = 0 \text{ in } G_1.$$

Now we consider the general term.

Let $v^j \in V$, then take the j-th equation

$$b_o(\frac{du^j}{dt},v) + a_o(u^j,v) + b_1(\frac{du^{j-1}}{dt},v) + a_1(u^{j-1},v) = 0 \text{ for } v \in V,$$

and restricting the (j+1)th-equation to Y_o, thus

$$b_1(\frac{du^j}{dt},v) + a_1(u^j,v) = 0 \text{ for } v \in Y_o.$$

The initial-boundary-value problems can be written

$$\frac{\partial u_o^j}{\partial t} - \Delta u_o^j = 0 \quad \text{in } G_o \qquad \frac{\partial u_1^j}{\partial t} - \Delta u_1^j = 0 \quad \text{in } G_1$$

$$u_o^j = 0 \quad \text{on } \partial G_{out} \qquad u_1^j = u_o^j \text{ on } \partial G_1$$

$$u_o^j(x,0) = 0 \quad \text{in } G_o \qquad u_1^j(x,0) = 0 \quad \text{in } G_1$$

$$\frac{\partial u_o^j}{\partial n} = \frac{\partial u_1^{j-1}}{\partial n} \quad \text{on } \partial G_o$$

In this way we obtain the approximated solution u_ε given by

$$u_{\varepsilon o} = u_o^o + \varepsilon u_o^1 + \dots + \varepsilon^k u_o^k \text{ and } u_{\varepsilon 1} = \frac{u_1^{-1}}{\varepsilon} + u_1^o + \varepsilon u_1^1 + \dots + \varepsilon^k u_1^k,$$

we have to show that

$$\|u_\varepsilon - (\frac{u^{-1}}{\varepsilon} + u^o + \varepsilon u^1 + \dots + \varepsilon^k u^k)\| \leqq C\cdot\varepsilon^{k+1}.$$

We define

$$\psi_\varepsilon = \frac{u^{-1}}{\varepsilon} + u^o + \varepsilon u^1 + \dots + \varepsilon^{k+1} u^{k+1},$$

then we have

$$b_o(\frac{d\psi_\varepsilon}{\partial t},v) + a_o(\psi_\varepsilon,v) + \varepsilon\{b_1(\frac{d\psi_\varepsilon}{dt},v) + a_1(\psi_\varepsilon,v)\} =$$
$$= (f,v) + \varepsilon^{k+2}\{b_1(\frac{du^{k+1}}{dt},v) + a_1(u^{k+1},v)\}.$$

By setting $\delta_\varepsilon = u_\varepsilon - \psi_\varepsilon$, we obtain

$$b_o(\frac{d\delta_\varepsilon}{dt},v) + a_o(\delta_\varepsilon,v) + \varepsilon\{b_1(\frac{d\delta_\varepsilon}{dt},v) + a_1(\delta_\varepsilon,v)\} =$$

$$= \varepsilon^{k+2}\{b_1(\frac{du^{k+1}}{dt},v) + a_1(u^{k+1},v)\} \text{ for } v \in V,$$

for $v = \delta_\varepsilon$ there follows

$$\frac{1}{2}(\frac{d}{dt}\{b_o(\delta_\varepsilon,\delta_\varepsilon) + \varepsilon b_1(\delta_\varepsilon,\delta_\varepsilon)\} + a_o(\delta_\varepsilon,\delta_\varepsilon) + \varepsilon a_1(\delta_\varepsilon,\delta_\varepsilon) =$$

$$= \varepsilon^{k+2}(\frac{d}{dt} b_1(\delta_\varepsilon,\delta_\varepsilon) + a_1(\delta_\varepsilon,\delta_\varepsilon)),$$

integrating

$$\frac{1}{2}(b_o(\delta_\varepsilon,\delta_\varepsilon) + \varepsilon b_1(\delta_\varepsilon,\delta_\varepsilon) + \int_0^t a_o(\delta_\varepsilon,\delta_\varepsilon) + \varepsilon a_1(\delta_\varepsilon,\delta_\varepsilon)dt =$$

$$= \varepsilon^{k+2}\int_0^t (\frac{d}{dt} b_1(\delta_\varepsilon,\delta_\varepsilon) + a_1(\delta_\varepsilon,\delta_\varepsilon))dt,$$

estimating the bilinear forms

$$\beta_i q_i(v) \leqq b_i(v,v),\ \alpha_i p_i(v) \leqq a_i(v,v) \text{ and } a_1(u,v) \leqq C\cdot \|u\| \cdot \|v\|,$$

we obtain

$$q_o(\delta_\varepsilon(t))^2 + \varepsilon q_1(\delta_\varepsilon(t))^2 + \int_0^t (p_o(\delta_\varepsilon(\sigma))^2 + \varepsilon p_1(\delta_\varepsilon(\sigma))^2)d\sigma \leqq$$

$$\leqq C\cdot\varepsilon^{k+2}(\int_0^t \|\delta_\varepsilon(\sigma)\|^2 d\sigma)^{\frac{1}{2}};$$

thus we have

$$q_1(\delta_\varepsilon(t))^2 + \int_0^t p_1(\delta_\varepsilon(\sigma))^2 d\sigma \leqq C\cdot\varepsilon^{k+1}(\int_0^t \|\delta_\varepsilon(\sigma)\|^2 d\sigma)^{\frac{1}{2}},$$

$$q_o(\delta_\varepsilon(t))^2 + \int_0^t p_o(\delta_\varepsilon(t))^2 d\sigma \leqq C\cdot\varepsilon^{k+1}(\int_0^t \|\delta_\varepsilon(\sigma)\|^2 d\sigma)^{\frac{1}{2}},$$

as $p_o + p_1$ is a norm equivalent to $\|\ \ \|$ on V and

$q_o + q_1$ is a norm equivalent to $|\ \ |$ on H;

we obtain finally

$$|\delta_\varepsilon(t)|^2 + \int_0^t \|\delta_\varepsilon(\sigma)\|^2 d\sigma \leqq C\cdot\varepsilon^{k+1}\left(\int_0^t \|\delta_\varepsilon(t)\|^2 d\sigma\right)^{\frac{1}{2}},$$

as $\delta_\varepsilon(t) = u_\varepsilon(t) - \psi_\varepsilon(t)$,

we have

$$|u_\varepsilon(t) - \psi_\varepsilon(t)|^2 < C\varepsilon^{k+1}\left(\int_0^t \|\delta_\varepsilon(\sigma)\|^2 d\sigma\right)^{\frac{1}{2}} \quad \text{and}$$

$$\int_0^t \|u_\varepsilon(\tau) - \psi_\varepsilon(\tau)\|^2 d\tau < C\cdot\varepsilon^{k+1}\left(\int_0^t \|\delta_\varepsilon(\sigma)\|^2 d\sigma\right)^{\frac{1}{2}}.$$

Defining $\|\ \ \|_{L^\infty((0,T),H)} = \text{ess. sup}|u(t)|$ for $t \in (0,T)$, (Yosida 367, p.34) we obtain one part of the result

$$\|u_\varepsilon - \psi_\varepsilon\|_{L^\infty(0,T),H)} < C.\varepsilon^{k+1},$$

as $\|f\|^2_{L^2((0,T),V)} = \int_0^T \|f(t)\|^2 dt$,

we obtain the other part of the result

$$\|u_\varepsilon - \psi_\varepsilon\|_{L^2((0,T),V)} < C\cdot\varepsilon^{k+1}.$$

I. 1.2 Algorithms for solving the parabolic problem

Résumé:

We have considered the evolution of diffusion processes governed by parabolic equations in two-connected domains. The domains are given by

$$G_1 = D_1 \times (0,T) \text{ and } G_o = D_o \times (0,T),$$

with the boundaries

$$\partial G_1 = \partial D_1 \times (0,T),\ \partial G_o = \partial D_o \times (0,T),$$

and ∂G_{out} = the other boundary, let $\varepsilon > 1$ be fixed. The system of parabolic equations describing the diffusion processes can be written

$$\frac{\partial u_o}{\partial t} - \Delta u_o = f_o \text{ in } G_o$$

$$\varepsilon\left(\frac{\partial u_1}{\partial t} - \Delta u_1\right) = f_1 \text{ in } G_1,$$

and u = o on the other boundary.

The interdependence of the diffusion processes is given by the transmission condition

$$u_o = u_1 \text{ and } \frac{\partial u_o}{\partial n} = \frac{\partial u_1}{\partial n} \text{ on } \partial D_1 \times (0,T),$$

and the initial conditions

$$u_o = u_{oo} \text{ in } D_o \text{ and } u_1 = u_{o1} \text{ in } D_1$$

can be interpreted as distributions u_{oo} in D_o and u_{o1} in D_1 at time t = o given in explicit form as spline-surfaces (e.g. two-dimensional, interpolating Lg-splines (139)).

The interdependence of the processes can be described by asymptotic methods.

We obtain a constructive algorithm amenable to numerical computation for the approximating solution u, where u = $u(u_o,u_1)$ and

$$u_o = u_o^o + \varepsilon u_o^1 + \varepsilon^2 u_o^2 + \dots + \varepsilon^j u_o^j$$

$$u_1 = \frac{u_1^{-1}}{\varepsilon} + u_1^o + \varepsilon u_1^1 + \dots + \varepsilon^j u_1^j.$$

u_1^{-1} is solution of the initial-boundary-value problem in G_1

$$\frac{\partial u_1^{-1}}{\partial t} - \Delta u_1^{-1} = f_1 \quad \text{in } G_1$$

$$u_1^{-1} = 0 \quad \text{on } \partial G_1$$

$$u_1^{-1}(x,0) = 0 \quad \text{in } D_1;$$

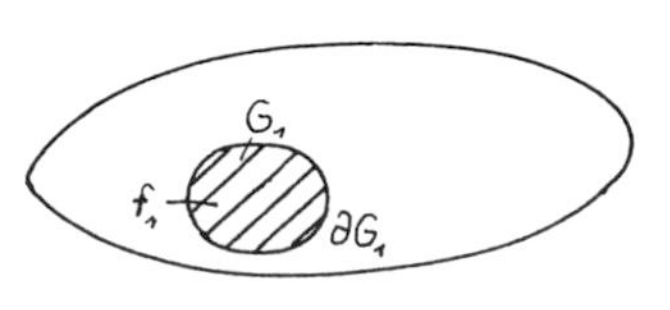

the solution $u_1^{-1} = u_1^{-1}(f_1)$, where f_1 is a diffusion distribution in D_1.
Now, taking the derivative $\frac{\partial}{\partial n}(u_1^{-1})$ of the solution u_1^{-1} on the boundary ∂G_1, we obtain the initial-boundary-value problem in the two-connected domain, G_0 by

$$\frac{\partial u_0^0}{\partial t} - \Delta u_0^0 = f_0 \quad \text{in } G_0$$

$$u_0^0 = 0 \quad \text{on } \partial G_{out}$$

$$\frac{\partial u_0^0}{\partial n} = \frac{\partial u_1^{-1}}{\partial n} \text{ on } \partial G_1$$

$$u_0^0(x,0) = u_{00}(x) \text{ in } D_0,$$

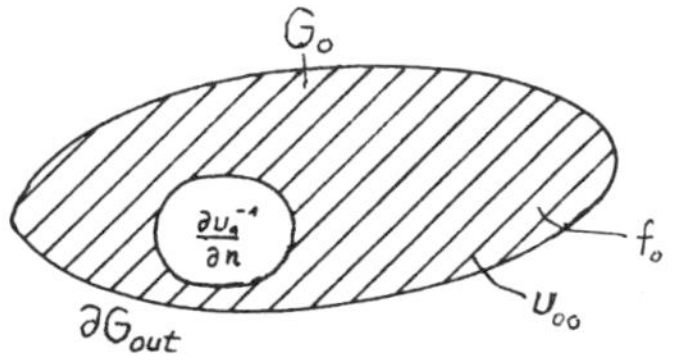

the solution $u_0^0 = u_0^0(f_0, u_{00}, u_1^{-1})$, where f_0 is the diffusion distribution in D_0,

u_{00} the initial distribution

$u_1^{-1}(f_1)$ the diffusion process in D_1,

hence the diffusion process in G_0 depends on the process in D_1 and on all diffusion distributions.

Now we take the solution u_0^0 on the boundary G_1 as boundary value for the problem in G_1, thus

$$\frac{\partial u_1^0}{\partial t} - \Delta u_1^0 = 0 \text{ in } G_1$$

$$u_1^0 = u_0^0 \text{ on } \partial G_1$$

$$u_1^0(x,0) = u_{01}(x) \text{ in } D_1$$

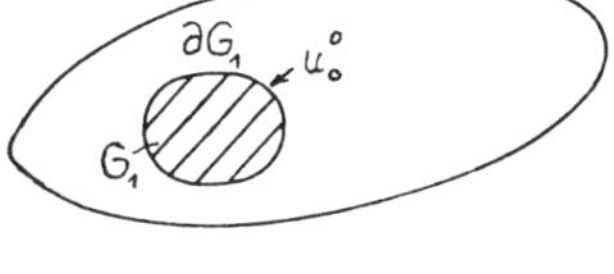

the solution depends on $u_1^0 = u_1^0(u_{01}(x), u_0^0(f_0, u_{00}, u_1^{-1}))$.

The solution u_1^0 can be given in an analytic, closed form, if u_0^0 is Hölder continuous. We take the derivative $\frac{\partial}{\partial n} u_1^0$ on the boundary ∂G_1 and solve the initial-boundary-value problem in G_0

$$\frac{\partial u_o^1}{\partial t} - \Delta u_o^1 = 0 \quad \text{in } G_o$$

$$u_o^1 = 0 \quad \text{on } \partial G_{out}$$

$$\frac{\partial u_o^1}{\partial n} = \frac{\partial u_1^o}{\partial n} \quad \text{on } \partial G_1$$

$$u_o^1(x,0) = 0 \quad \text{in } D_o;$$

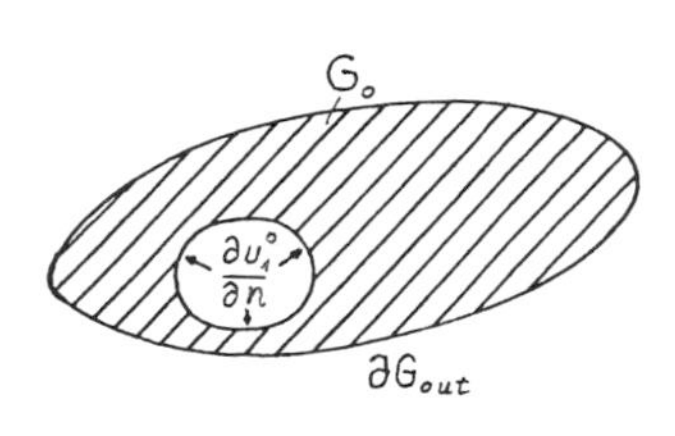

the solution u_o^1 depends on u_1^o, thus the diffusion is influenced only by the interior diffusion process u_1^o, and u_1^o depends on the initial-diffusion distributions.

I.2 Nonlinear models

The modelling methods may be applied to nonlinear parabolic partial differential equations.

The distributions f_o in D_o and f_1 in D_1 can be written in the form f(t,x,u,gradu) locally Lipschitzian in all its arguments. Let B(t,s) be a continuous function and $k \in \mathbb{R}$, such that $1 \leq k \leq 3$, assuming

$$|f(t,x,u,g)| \leq B(t,|u|)\cdot(1 + |g|^k),$$

the initial-boundary-value problem in $L^2(D)$ can be written as

$$\frac{\partial u}{\partial t} - \Delta u = f(t,x,u,\text{gradu}) \text{ in the domain D}$$

$$\frac{\partial u}{\partial n} = -h(x)u \text{ on the boundary, where } h(x) \text{ is continuous}$$

$$u(x,0) = u_o(x) \text{ for x in the domain D,}$$

has unique solution, which is also the solution in the classical sense; this means that the solution $u(t,x,u_o,h)$ is continuously differentiable in t and twice continuously differentiable in x.

I.2.1 Algorithms for the nonlinear parabolic problem in a two-connected domain

We apply this deep regularity theorem to our constructive algorithms. The nonlinear parabolic initial-boundary-value problem reads

$$\frac{\partial u_1^{-1}}{\partial t} - \Delta u_1^{-1} = f_1(x,t,u_1^{-1},\mathrm{grad}u_1^{-1}) \text{ in } G_1$$

$$u_1^{-1} = 0 \text{ on } \partial G_1$$

$$u_1^{-1}(x,0) = 0 \text{ in } D_1,$$

the solution u_1^{-1} is twice continuously differentiable in x, thus $u_1^{-1} \in C^2(D)$. Therefore we have $\frac{\partial}{\partial n} u_1^{-1}$ continuous on ∂G_1, and let u_{oo} be continuous, then there exists the nonlinear parabolic initial-boundary-value problem

$$\frac{\partial u_o^o}{\partial t} - \Delta u_o^o = f(x,t,u_o^o,\mathrm{grad}u_o^o) \text{ in } G_o$$

$$u_o^o = 0 \text{ on } \partial G_{out}$$

$$\frac{\partial u_o^o}{\partial n} = \frac{\partial u_1^{-1}}{\partial n} \text{ on } \partial G_1$$

$$u_o^o(x,0) = u_{oo}(x) \text{ in } D_o,$$

where the unique solution u_o^o is twice continuously differentiable in x. Let u_{o1} be continuous, then there exists the linear parabolic initial-boundary-value problem

$$\frac{\partial u_1^o}{\partial t} - \Delta u_1^o = 0 \text{ in } G_1$$

$$u_1^o = u_o^o \text{ on } \partial G_1$$

$$u_1^o(x,0) = u_{o1}(x) \text{ in } D_1,$$

where the solution $u_1^o(t,x,u_o^o,u_{o1}) \in C^2(D)$.

Any further term of the algorithms is now a solution of the linear parabolic partial differential equation.

The above regularity condition for the nonlinear parabolic problem needs deep mathematical results. We give a proof in the following. First we consider only the boundary value problem

$$- \Delta u = f \text{ in } D$$

$$\frac{\partial u}{\partial n} = -h(x)u \text{ on } \partial D$$

and by Green's formula, let $u \in H^1(D)$, then

$$\int_D fv(x)dx = - \int_D \Delta u(x)\cdot v(x)dx = \int_D \text{grad}u(x)\cdot\text{grad}v(x)dx - \int_{\partial D} \frac{\partial u}{\partial n}\cdot v\cdot d(\partial D)$$

$$= \int_D \text{grad}u(x)\cdot\text{grad}v(x)dx - \int_{\partial D} h(x)v(x)d(\partial D) \text{ for } v \in H^1(D),$$

thus the operator $-\Delta$ is selfadjoint, and positive definite, moreover the Laplace operator is uniformly elliptic (3,4,69,116,336). We obtain therefore a regular boundary value problem and the domain of definition is

$$D(\Delta) = \{u \in W^{2,2}(D) \mid \frac{\partial u}{\partial n} + hu = 0 \text{ on } \partial D\},$$

where the norm of $W^{k,p}(D)$ is defined by

$$\|f\|^p_{W^{k,p}(D)} = \int_D \sum_{j=1}^{k} \|(\frac{\partial}{\partial t})^j f(t)\|^p dt.$$

Now we apply the theory of semi groups.

We introduce a sectorial operator A, let $\sigma(A) > 0$ be the spectrum of A, then the fractional powers of A are defined by

$$A^{-\alpha} = \frac{1}{\Gamma(a)} \int_0^\infty t^{\alpha-1} e^{-At} dt.$$

If A is a sectorial operator in a Banach space X, define for each $\tau > 0$

$$X^\tau = D(A_I^\tau), \text{ with the graph norm } \|X\|_\alpha = \|A_I^\tau(x)\|, \quad x \in X^\tau,$$

where $A_I = A + aI$ and a is chosen so that Real $\sigma(A_I) > 0$. The space X^τ will provide the basic topology in all that follows. We have the

Theorem:

Suppose $D \in \mathbf{R}^n$ open, bounded and the boundary ∂D is a C^m hypersurface separating D from $\mathbf{R}^n \setminus \bar{D}$ (see Friedman (116), or according to Stein (327) we need only "minimally smooth", this means that ∂D is Lipschitzian). Under these assumptions let A be a sectorial operator in $X = L^p(D)$, for $1 \leq p \lneq \infty$ and $D(A) = X^1 \subset W^{m,p}(D)$ for $m \geq 1$.

Then we obtain for $0 \leq \tau \leq 1$

$$X^\tau \subset W^{k,\ell}(D), \text{ where } K - \frac{n}{\ell} < m\tau - \frac{n}{p} \quad \ell \geq p,$$

$$X^\tau \subset C^\mu(D), \quad \text{where } 0 \leq \mu \leq m\cdot\tau - \frac{n}{p} .$$

The proof can be given by the Nierenberg-Gagliardo inequality (Friedman 116)

$$||u||_{W^{k,\ell}(D)} \leq C\cdot ||u||^{v}_{W^{m,p}(D)} \cdot ||u||^{1-v}_{L^p(D)}$$

provided

$$K - \frac{n}{\ell} < v(m - \frac{n}{p}) - \frac{n(1-v)}{p} = mv - \frac{n}{p} \text{ and } \ell \geq p.$$

For $u \in D(A)$, $u \in W^{m,p}(D)$, thus $Au \in L^p(D)$, and the inequality can be rewritten

$$||u||_{W^{k,\ell}(D)} \leq C\, ||Au||^{v}_{L^p(D)} \cdot ||u||^{1-v}_{L^p(D)} .$$

As the inclusion map

$$D(A) \subset W^{m,p}(D) \to W^{k,\ell}(D)$$

extends to a continuous inclusion of

$$X^\tau \to W^{k,\ell}(D), \text{ if } \tau > v, \text{ thus } K - \frac{n}{\ell} < m\cdot v - \frac{n}{p} < m\cdot\tau - \frac{n}{p} .$$

We only consider regularity in $\mathbf{R}^3$, thus $n = 3$.

As $D(A) \subset W^{2,2}(D)$ we choose $m = 2$ and $p = 2$ and apply the theorem. The inclusion

$$X^\tau \subset W^{k,\ell}(D) \text{ if } K - \frac{3}{\ell} < 2\tau - \frac{3}{2} \text{ for } k = 1; - \frac{3}{\ell} < 2\tau - \frac{3}{2} - 1, \text{ then}$$

$$\frac{1}{\ell} > \frac{1}{2} + \frac{1}{3} - \frac{2}{3}\tau = \frac{5}{6} - \frac{4}{6}\tau \text{ and } \tau > \frac{1}{2}$$

$$X^\tau \subset C^\mu \text{ where } 0 \leqq \mu \leqq m\tau - \frac{3}{2}, \text{ for } \mu = 0, 0 \leqq 2\tau - \frac{3}{2}$$

$$\text{and } \tau > \frac{3}{4}, \text{ thus } X^\tau \subset L^\infty$$

we have the inclusion

$$X^\tau \subset W^{1,\ell}(D), \text{ if } \frac{1}{\ell} > \frac{5}{6} - \frac{4}{6}\tau \text{ and } \tau > \frac{1}{2}, \text{ take } \ell = 2k, \text{ then } \tau > \frac{3}{4}\cdot\frac{1}{k} - \frac{5}{4},$$

and finally we have the continuous inclusion

$$X^\tau \subset W^{1,2k}(D) \cap L^\infty \text{ if } \max\left(\frac{3}{4}, \frac{3}{4}\frac{1}{k} - \frac{5}{4}\right) < \tau < 1.$$

We suppose that $u \in X^\tau$, then $u \in W^{1,2k}(D)$ and $u \in L^\infty$ if $\max(\frac{3}{4}, \frac{3}{4}\frac{1}{k} - \frac{5}{4}) < \tau < 1$, by assumption f is given

$$f(t,x,u,\text{grad}u) \leqq B(t,|u(x)|)\,(1 + |\text{grad}u(x)|^k)$$

then

$$\|f(t,x,u,\text{grad}u)\|_{L^2(D)} = \int_D |f(t,x,u,\text{grad}u)|^2\,dx \leqq$$

$$\leqq (B, \|\cdot\|_\infty)\left(\left(\int_D 1dx\right)^{\frac{1}{2}} + \|u\|^k_{W^{1,2k}(D)}\right) < \infty, \text{ as } u \in L^\infty \text{ and } u \in W^{1,2}(D).$$

I.2.2 Existence of the nonlinear parabolic problem

We obtain the following estimations

$$|f(t,x,u,p_1) - f(t,x,u,p_2)| \leqq B(t,|u|)\cdot(1 + |p_1|^{k-1} + |p_2|^{k-1})|p_1-p_2|$$

$$|f(t,x,u_1,p) - f(t,x,u_2,p)| \leqq B(t,|u_1|+|u_2|)\cdot(1 + |p|^k)|u_1-u_2|.$$

Thus we can apply the following

Theorem:

Let A be a sectorial operator with $0 \leq \tau \leq 1$ and U an open subset of $\mathbf{R} \times X^\tau$, such that f is given by

$$\begin{array}{rcl} & \mathbf{R} \times X^\tau & \\ & \cap & \\ f: & U & \to L^2(D) \\ & (t,u) \mapsto f(t,u), & \end{array}$$

where f is locally Hölder continuous in t and locally Lipschitzian in u. Let $u(t_o) = u_o$ be the initial value, then for any $(t_o,u_o) \in U$ there exists a $\theta > 0$, that depends on the initial time t and the initial value $u_o \in X^\tau$, thus $\theta = \theta(t_o,u_o)$ such that

$$\frac{\partial u}{\partial t} + Au = f(t,u) \text{ for } t > t_o$$

$$u(t_o) = u_o$$

has a unique solution in $(t_o, t_o + \theta)$.

This theorem is known by Rauch (289), Henry (155), Mora (251).

The solution in abstract form is given by

$$u(t) = e^{-At}u_o + \int_0^t e^{-A(t-s)}f(s)ds,$$

thus $u(t,x,u_o)$ and we may show that

$$(t,x) \to u(t,x,u_o)$$

is continuously differentiable in t and twice continuously differentiable in x.

The parabolic equation can be written as

$$Au = f(t,u,\text{grad}u) - \frac{\partial u}{\partial t} \text{ in } D$$

$$\frac{\partial u}{\partial n} = -h(x)\cdot u \text{ on } \partial D.$$

It can be shown that

$$t \to \frac{\partial u}{\partial t} \in X^{\tau}$$

is locally Hölder continuous if $t > 0$ and also grad(u) is Hölder continuous, then $f(t,u,\text{grad } u)$ is Hölder continuous with some Hölder exponent α, denoted by $f(t,u,\text{grad } u) \in C^{\alpha}(\bar{D})$. Now we are able to apply the known Schauder theorem given in a modern statement by Gilbarg-Trudinger (135).

We consider the space $C^{\alpha}(\bar{D})$ of Hölder continuous functions on D having Hölder exponent α and the norm

$$\|u\|_{\alpha} = \sup_{\substack{x,y\in D \\ x\neq y}} \frac{|u(x) - u(y)|}{|x - y|^{\alpha}} < \infty \quad .$$

We define $C^{2+\alpha}(\bar{D})$ to be the class of C^2 functions on D whose second derivatives are in $C^{\alpha}(\bar{D})$.

Theorem:

If Λ is an uniformly elliptic differential operator in D. The boundary value problem

$$\Lambda u = f \text{ in } D$$

$$u = \varphi \text{ on } \partial D$$

has a unique solution for each $f \in C\ (\bar{D})$.

The solution is considerably smoother than the right-hand side, e.g. if $f \in C^{\alpha}(\bar{D})$ and the boundary value satisfies $\varphi \in C^{2+\alpha}(\partial D)$, then the solution $u \in C^{2+\alpha}(\bar{D})$ (135).

This theorem can be extended and applied to our problem.

I.2.3 Regularity of the nonlinear parabolic problem

Let Λ be a uniformly elliptic differential operator, the coefficients of Λ are in $C^{\alpha}(\bar{D})$, the boundary value $\varphi \in C^{0}(\partial D)$ and $f \in C^{\alpha}(\bar{D})$, then the boundary-value problem has a unique solution $u \in C^{2+\alpha}(\bar{D})$.

It remains to be shown, that grad u is Hölder continuous. Therefore we need the Sobolev imbedding theorem (39, 85, 323, 327). We suppose that D is an open set in $\mathbb{R}^n$ such that the boundary ∂D is minimally smooth in the sense of (327, p. 181). Thus we can take $D = \mathbb{R}^n$ or D is a bounded domain with C^1 boundary.

Under these conditions we have the continuous imbeddings

$$W^{m,p}(D) \subset L^{q}(D) \text{ if } \frac{1}{p} \geqq \frac{1}{q} \geqq \frac{1}{p} - \frac{m}{n} > 0$$

$$W^{m,p}(D) \subset C(D) \quad \text{if } m \cdot p > n.$$

We have that $t \to \frac{\partial u}{\partial t}$ is locally Hölder continuous, and

$$\frac{du}{dt} \in X^{\tau} \subset W^{1,2k}(D) \cap L^{\infty}.$$

The imbedding reads

$$W^{1,2k}(D) \subset L^{q}(D) \text{ if } \frac{1}{q} > \frac{1}{2k} - \frac{1}{3} = \frac{1}{6}\left(\frac{3}{k} - 2\right),$$

this inequality is satisfied, if we choose $q = \frac{6}{k}$, thus

$$W^{1,2k}(D) \subset L^{6/k}(D).$$

By construction of the abstract solution $u \in D(A)$ and by definition of $D(A)$ we have

$$\text{grad } u \in W^{1,2}(D) \cap L^{6}(D),$$

hence

$$Au = f(t,x,u,\text{grad}u) - \frac{\partial u}{\partial t} \in L^{6/k}(D),$$

applying the regularity theorem $u \in W^{2,6/k}(D)$, thus $\text{grad}(u) \in W^{1,6/k}(D)$ and by the imbedding theorem

$$W^{1,6/k}(D) \subset L^q, \text{ where } \frac{1}{q} > \frac{1}{6}(k-2).$$

If $k < 2$ we obtain that grad u is Hölder continuous.
If $k \geq 2$ we continue with the proof as above.
As grad u is Hölder continuous, we have $f(t,x,u,\text{grad } u) \in C^{\alpha}(\bar{D})$ with Hölder exponent α and the Schauder theorem can be applied. Thus, the regularity theorem is proved.

I.3 Wave phenomena in two-connected domains

I. 3.1 Model of a hyperbolic initial-boundary-value problem

We consider the evolution of processes governed by hyperbolic partial differential equations in two-connected domains. Let the domains be given by

$$G_1 = D_1 \times (0,T) \text{ and } G_o = D_o \times (0,T),$$

with the boundaries

$$\partial G_1 = \partial D_1 \times (0,T),\ \partial G_o = \partial D_o \times (0,T),\ \partial G_{out} = \text{the other boundary,}$$

and $\varepsilon > 1$ fixed.

Coupled wave phenomena can be described by

$$\frac{d^2 u_o}{dt^2} - \Delta u_o = f_o \text{ in } G_o$$

$$\varepsilon\left(\frac{d^2 u_1}{dt^2} - \Delta u_1\right) = f_1 \text{ in } G_1,$$

with the boundary condition

$$u_o = 0 \text{ on } \partial G_{out},$$

and the transmission condition

$$u_o = u_1 \text{ and } \frac{\partial u_o}{\partial n} = \frac{\partial u_1}{\partial n} \text{ on } \partial D_1 \times (0,T),$$

and the initial conditions

$$u_o(x,0) = u_{oo} \qquad u_1(x,0) = u_{o1}$$

$$\frac{\partial u_o}{\partial t}(x,0) = u_{1o} \qquad \frac{\partial u_1}{\partial t}(x,0) = u_{11},$$

where $u_{oo},\ldots,u_{11}$ are fixed initial distributions.

I.3.2 Algorithms

Introducing $V = H_o^1(D)$ and $H = L^2(D)$ and defining the bilinear forms

$$b_i(u,v) = \int_{D_i} u \cdot v dx \text{ and } a_i(u,v) = \int_{D_i} \text{grad}u \cdot \text{grad}v \, dx \text{ for } i = 0,1 \text{ and } u,v \in V,$$

the variational formulation of the initial-boundary-value problem is given by

$$b_o(u'',v) + a_o(u,v) + \varepsilon(b_1(u'',v) + a_1(u,v)) = (f,v) \text{ for } v \in V,$$

with the initial conditions

$$u(0) = u_o \in V \text{ and } \frac{du}{dt}(0) = u_1 \in V,$$

where the solution u satisfies

$$u \in L^\infty((0,T),V) \text{ and } \frac{du}{dt} \in L^\infty((0,T),H).$$

We suppose that

$$a_o(u,v) + b_o(u,v) = 0 \text{ on } Y_o.$$

We determine the asymptotic expansion of the solution u by taking the "ansätze"

$$u = \frac{u^{-1}}{\varepsilon} + u^{o} + \varepsilon u^{1} + \ldots \text{ and } \frac{d^2u}{dt^2} = \frac{1}{\varepsilon}\frac{d^2u^{-1}}{dt^2} + \frac{d^2u^{o}}{dt^2} + \varepsilon\frac{d^2u^{1}}{dt^2} + \ldots .$$

Using this in the variational formulation, and by choosing the ε-terms, we obtain (as on page 11) the ε-term:

$$\varepsilon^{o} \quad b_o(\frac{d^2u^{o}}{dt^2},v) + a_o(u^{o},v) + b_1(\frac{d^2u^{-1}}{dt^2},v) + a_1(u^{-1},v) = (f,v)$$

$$\varepsilon^{1} \quad b_o(\frac{d^2u^{1}}{dt^2},v) + a_o(u^{1},v) + b_1(\frac{d^2u^{o}}{dt^2},v) + a_1(u^{o},v) = 0$$

$$\varepsilon^{2} \quad b_o(\frac{d^2u^{2}}{dt^2},v) + a_o(u^{2},v) + b_1(\frac{d^2u^{1}}{dt^2},v) + a_1(u^{1},v) = 0$$

$$\varepsilon^{3} \quad b_o(\frac{d^2u^{3}}{dt^2},v) + a_o(u^{3},v) + b_1(\frac{d^2u^{2}}{dt^2},v) + a_1(u^{2},v) = 0;$$

$$\vdots \qquad \vdots \qquad \qquad \qquad \vdots$$

comparing with the expression on page 11 we obtain

$$u^{-1}(t) \in Y_o$$

$$b_1(\frac{d^2u^{-1}}{dt^2},v) + a_1(u^{-1},v) = (f,v) \text{ for } v \in Y_o$$

initial condition $u^{-1}(0) = 0$, $\frac{du^{-1}}{dt}(0) = 0$ then

$$u^{o}(t) \in V$$

$$b_o(\frac{d^2u^{o}}{dt^2},v) + a_o(u^{o},v) + b_1(\frac{d^2u^{-1}}{dt^2},v) + a_1(u^{-1},v) = (f,v) \quad v \in V$$

$$b_1(\frac{d^2u^{o}}{dt^2},v) + a_1(u^{o},v) = 0 \quad v \in Y_o$$

initial condition $u^{o}(0) = u_o$, $\frac{du^{o}(0)}{dt} = u_1$;

and general

$u^j(t) \in V$

$$b_o(\frac{d^2u^j}{dt^2},v) + a_o(u^j,v) + b_1(\frac{d^2u^{j-1}}{dt},v) + a_1(u^{j-1},v) = 0 \quad v \in V$$

$$b_1(\frac{d^2u^j}{dt^2},v) + a_1(u^j,v) = 0 \quad v \in Y_o$$

initial condition $u^j(0) = 0, \frac{du^j}{dt}(0) = 0$ for $j = 1,\dots$.

The solution satisfies

$$u^\tau \in L^\infty((0,T),V), \frac{du^\tau}{dt} \in L^\infty((0,T),H); \frac{d^2u^{\tau+1}}{dt^2} \in L^2((0,T)V) \frac{d^2u^{\tau+1}}{dt^2} \in L^2((0,T),H) \quad \text{for } \tau = 0,\dots,j+1,$$

finally we obtain the corresponding initial-boundary-value problems (see analogously page 12)

$$\frac{\partial^2u^{-1}}{\partial t^2} - \Delta u^{-1} = f_1 \text{ in } G_1 = D_1 \times (0,T)$$

$$u^{-1} = 0 \quad \text{on } \partial G_1$$

$$u^{-1}(x,0) = 0 \quad \text{in } D_1.$$

The solution $u^{-1} = u_1^{-1}(f_1)$ depends on f_1; taking the derivative of the solution $\frac{\partial}{\partial n} u_1^{-1}$ we have to solve the initial-boundary-value problem in the two-connected domain G_o

$$\frac{\partial^2u_o^o}{\partial t^2} - \Delta u_o^o = f_o \text{ in } G_o = D_o \times (0,T)$$

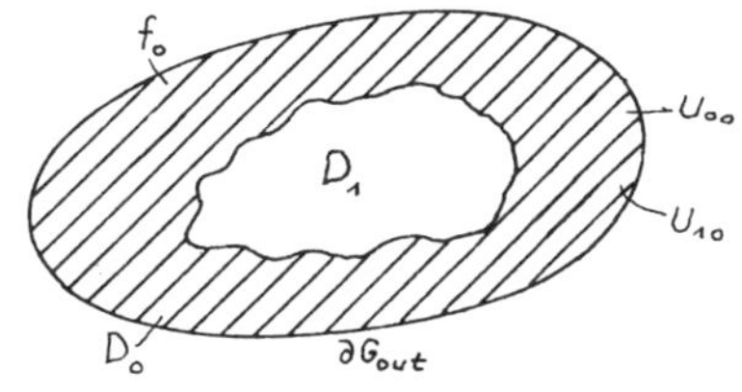

$$u_o^o = 0 \quad \text{on } \partial G_{out}$$

$$\frac{\partial u_o^o}{\partial n} = \frac{\partial u_1^{-1}}{\partial n} \text{ on } \partial G_1$$

$$u_o^o(x,0) = U_{oo}(x), \frac{du_o^o(x,0)}{dt} = u_{1o}(x)$$

The solution depends on $u_o^o = u_o^o(f_o, u_{oo}, u_{1o}, u_1^{-1})$, and can be given in a closed, analytic form. Now we restrict the solution u_o^o to the boundary ∂D_1 and solve the initial-boundary-value problem

$$\frac{\partial^2 u_1^o}{\partial t^2} - \Delta u_1^o = 0 \text{ in } G_1 = D_1 \times (0,T)$$

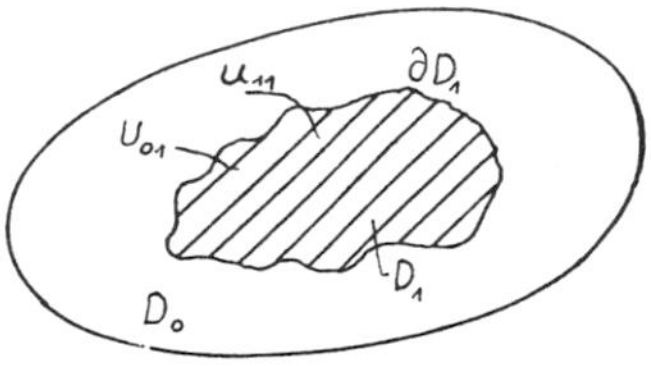

$$u_1^o = u_o^o \text{ on } \partial G_1$$

$$u_1^o(x,0) = u_{01}(x) \quad \frac{\partial u_1^o(x,0)}{\partial t} = u_{11}(x) \text{ in } D_1.$$

The solution depends on $u_1^o(u_{01}, u_{11}, u_o^o(f_o, u_{oo}, u_{1o}, u_1^{-1}(f_1)))$ and can be given in a closed, analytic form. By taking the derivative $\frac{\partial}{\partial n}(u_1^o)$ of the solution u_1^o on the boundary ∂D_1 we have to solve the initial-boundary-value problem

$$\frac{\partial^2 u_o^1}{\partial t^2} - \Delta u_o^1 = 0 \text{ in } G_o$$

$$u_o^1 = 0 \text{ on } \partial G_{out}$$

$$\frac{\partial u_o^1}{\partial n} = \frac{\partial u_1^o}{\partial n} \text{ on } \partial G_1$$

$$u_o^1(x,0) = 0 \text{ in } D_o$$

$$\frac{\partial u_o^1}{\partial t}(x,0) = 0 \text{ in } D_o.$$

I.4 Nonlinear wave phenomena

We construct a weak solution concept for nonlinear hyperbolic partial differential equations.

Given the nonlinear hyperbolic partial differential equation

$$\frac{\partial^2 y}{\partial t^2} + Ay + y \int_0^t |y(\sigma)|^2 d\sigma = f \quad \text{in } (0,T),$$

with respect to $y(0) = 0$, $y'(0) = 0$, where A is a second order elliptic differential operator.

Let Ω be an open, bounded domain in $\mathbf{R}^n$ with smooth boundary $\partial\Omega$. Let v be the control variable, $v \in L^2(\partial\Omega,(0,T))$, applied to the boundary, and $f \in L^2(\Omega\times(0,T))$.

We integrate the nonlinear equation with respect to t and take the scalar product with $\varphi \in H^1(\Omega)$,

$$(\frac{\partial^2 y}{\partial t^2},\varphi)_\Omega + (Ay,\varphi)_\Omega + \int_0^t (y\int_0^\tau |y(\sigma)|^2 d\sigma,\varphi)_\Omega d\tau =$$

$$= (\int_0^t (f(\tau)d\tau,\varphi)_\Omega + (\int_0^t v(\tau)d\tau,\varphi)_{\partial\Omega}$$

then

$$\int_\Omega (\frac{\partial^2 y}{\partial t^2},\varphi)dx + \int_\Omega Ay.\varphi\, dx + \int_0^t \int_\Omega y.\int_0^\tau |y(\sigma)|^2 d\sigma\varphi d\tau dx =$$

$$= \int_\Omega \int_0^t f\cdot\varphi\, d\tau dx + \int_{\partial\Omega}\int_0^t v(\tau)\cdot\varphi d\tau\; d(\partial\Omega) \text{ for } \varphi\in H^1(\Omega),$$

thus the exact formulation of the initial-boundary-value problem is given by

$$\frac{\partial^2 y}{\partial t^2} + Ay + y\cdot\int_0^t\int_\Omega y^2(x,\tau)dxd\tau = f \text{ in } \Omega\times(0,T)$$

$$\frac{\partial y}{\partial n} = v \text{ on } \partial\Omega\times(0,T)$$

$$y(x,0) = 0,\quad \frac{\partial y}{\partial t}(x,0) = 0 \text{ in } \Omega.$$

We want to show that there exists a unique solution of the nonlinear initial-boundary-value problem. As the nonlinear term

$$y\int_0^t |y(\tau)|^2 d\tau \text{ is given, we introduce } \tilde{y}(t) = \int_\theta^t y(\tau)d\tau$$

and obtain

$$\tilde{y}(t) - \tilde{y}(s) = -\int_t^0 y(\tau)d\tau - \int_0^s y(\tau)d\tau = -\int_t^s y(\tau)d\tau.$$

Now we define

$$\varphi(t) = \begin{cases} -\int_t^s y(\tau)d\tau & \text{for } t \leqq s \\ 0 & \text{for } t \geqq s, \end{cases}$$

then $\frac{d\varphi}{dt} = \varphi'(t) = -y(t)$ for $t \leqq s$ and $\varphi(t) = \tilde{y}(t) - \tilde{y}(s)$.

Multiplying the nonlinear partial differential equation by φ and integrating with respect to t from o to s, we have

$$\int_0^s [(y'',\varphi) + (Ay,\varphi) + \int_0^s\int_0^t |y|^2 d\sigma(y,\varphi)]dt = \int_0^s (f,\varphi)dt,$$

integrating by parts we obtain certain interesting relations; the first integral can be written as

$$\int_0^s y''\cdot\varphi dt = y'.\varphi\Big|_0^s - \int_0^s y'\cdot\varphi' dt = 0 - \int_0^s y'.ydt, \text{ as}$$

$$\int_0^s y'.ydt = y\cdot y\Big|_0^s + \int_0^s y\cdot y'dt,$$

we obtain

$$\int_0^s y''\cdot ydt = -\frac{1}{2}|y(s)|^2.$$

The second integral can be rewritten by introducing the bilinear forms

$$B : H^1(\Omega) \times H^1(\Omega) \to \mathbf{R} \text{ or } \mathbb{C}$$

$$(u,v) \longmapsto B(u,v) = (Au,v),$$

where A is the differential operator and $A \in L(H^1(\Omega), (H^1(\Omega))')$, for v fixed, $B(u,v)$ is a linear functional on $H^1(\Omega)$, and

$$B(u,v) = B(v,u) \text{ for } u,v \in H^1(\Omega),$$

$B(u,v)$ satisfies the inequality

$$B(v,v) \geqq \alpha(v,v) = \alpha \cdot \|v\|^2 \text{ for } \alpha = \text{constant} > 0,\ v \in H^1(\Omega),$$

we obtain

$$\int_0^s (Ay,\varphi)dt = \int_0^s (A\varphi',\varphi)dt = \int_0^s B(\varphi',\varphi)dt = \frac{1}{2}\cdot\int_0^s \frac{d}{dt} B(\varphi,\varphi)dt =$$

$$= \frac{1}{2}(B(\varphi(s),\varphi(s)) - B(\varphi(0),\varphi(0))$$

$$= -\frac{1}{2} B(\tilde{y}(s),\tilde{y}(s)) \text{ as } \varphi(0) = -\int_0^s y(\tau)d\tau = \tilde{y}(s)$$

We have $\frac{1}{2}\cdot\frac{d}{dt}(\varphi,\varphi) = \frac{1}{2}\cdot\frac{d}{dt}|\varphi|^2 = |\varphi|$, therefore the third integral can be written

$$-\int_0^s\int_0^t |y|^2 d\tau(y,\varphi)dt = \frac{1}{2}\int_0^s (\int_0^t |y|^2 d\tau)\frac{d}{dt}|\varphi|^2 dt =$$

$$= \frac{1}{2}|\varphi(t)|^2 \int_0^t |y|^2 d\tau\Big|_{t=0}^{t=s} - \frac{1}{2}\int_0^s |\varphi(t)|^2|y(t)|^2 dt$$

$$= -\frac{1}{2}\int_0^s |y(t)|^2 |\varphi(t)|^2 dt,$$

and finally we obtain

$$-\frac{1}{2}|y(s)|^2 - \frac{1}{2} B(\tilde{y}(s),\tilde{y}(s)) - \frac{1}{2}\int_0^s |y(t)|^2|\varphi(t)|^2 dt =$$

$$= \int_0^s (f,\varphi)dt = \int_0^s f(\tilde{y}(t)-\tilde{y}(s))dt.$$

Using the nonnegativity of the third expression on the left-hand side, and the estimation

$$B(\tilde{y}(s),\tilde{y}(s)) \geqq \alpha\cdot \|\tilde{y}(s)\| \text{ for } \alpha > 0,$$

we obtain

$$|y(s)|^2 + \alpha\|\tilde{y}(s)\|^2 \leqq C\int_0^s f\cdot\tilde{y}(t)dt - \int_0^s f\cdot\tilde{y}(s)dt \leqq$$

$$\leqq C.\int_0^s \|f\|\cdot\|\tilde{y}(t)\| dt + \int_0^s \|f(t)\| dt \|\tilde{y}(s)\|,$$

applying the known Gronwall's inequality, then

$$|y(s)|^2 + \|\tilde{y}(s)\|^2 \leq C \int_0^s \|f(t)\|^2 dt.$$

By assumption $f, f', f'' \in L^2((0,T),(H^1(\Omega))')$, thus we can prove the

Lemma:

There exists a solution of the nonlinear partial differential equation, such that

$$y, y' \in L^\infty((0,T),H^1(\Omega)) \text{ and } y'' \in L^\infty((0,T),H).$$

Proof:

Differentiating

$$\frac{\partial y}{\partial t} + Ay + y \int_0^t |y(\tau)|^2 d\tau = f,$$

with respect to t, we obtain by introducing $w = y'$

$$w'' + Aw + w \int_0^t |y(\tau)|^2 d\tau = f' - y \cdot |y(t)|^2 \text{ with } w(0) = 0 = w'(0).$$

By taking the scalar product with w'

$$(w'',w') + (Aw,w') + (w \cdot \int_0^t |y(\tau)|^2 d\tau, w') = (f',w') - (y \cdot |y(\tau)|^2, w')$$

$$\frac{1}{2}\frac{d}{dt}(w',w') + B(w,w)) + \frac{1}{2}\int_0^t |y(\tau)|^2 \frac{d}{dt}|w(t)|^2 d\tau = (f',w') - (y \cdot |y(\tau)|^2, w')$$

by integration

$$\frac{1}{2}\int_0^t \frac{d}{dt}(|W'(t)|^2 + B(w,w))dt + \frac{1}{2}\int_0^t \int_0^t |y(\tau)|^2 d\tau \frac{d}{dt}|w(t)|^2 dt =$$

$$= \int_0^t (f',w')d\tau - \int_0^t (y \cdot |y(\tau)|^2, w')d\tau,$$

integration by parts yields

$$\int_0^t \int_0^t |y(\tau)|^2 d\tau \cdot \frac{d}{dt}|w(t)|^2 dt = \int_0^{t(\tau)} \int_0^t |y(\tau)|^2 \frac{d}{dt}|w(t)|^2 d\tau dt =$$

$$= \int_0^{t(\tau)} |y(\tau)|^2 d|w(t)|^2\Big|_0^t - \int_0^t |y(t)|^2 |w(t)|^2 \, dt \quad \text{as } w(0) = 0,$$

$$= \int_0^t |w(t)|^2 |y(\tau)|^2 d\tau - \int_0^t |y(t)|^2 |w(t)|^2 dt = |w(t)|^2 \int_0^t |y(\tau)|^2 d\tau -$$

$$- \int_0^t |y(\tau)|^2 |w(\tau)|^2 d\tau,$$

$$2 \int_0^t f'(\tau)\cdot w'(\tau) d\tau = 2f'(\tau)\cdot w(\tau)\Big|_0^t - 2 \int_0^t f''(\tau)\cdot w(\tau) d\tau =$$

$$2(f'(t)\cdot w(t) - 2 \int_0^t (f'',w) d\tau$$

and finally

$$|w'(t)|^2 + B(w,w) + |w(t)|^2 \int_0^t |y(\tau)|^2 d\tau = \int_0^t |y(\tau)|^2 |w(\tau)|^2 d\tau +$$

$$2(f'(t),w(t)) - 2 \int_0^t f''(\tau)\cdot w(\tau) d\tau - 2 \int_0^t |y(\tau)|^2 y(\tau)\cdot w'(\tau) d\tau;$$

it can be estimated

$$\left|\int_0^t f'(t)\cdot w(t) dt\right| \leq \|f'\|_{L^2((0,T),(H^1(\Omega))')} \cdot \|w\|_{L^\infty(0,T),(H'(\Omega))')}$$

and finally we obtain by applying Gronwall's inequality

$$|w'(t)|^2 + \|w(t)\|^2 \leq \chi(\|f\| , \|f'\| , \|f''\|_{L^2(0,T),(H^1(\Omega))')})$$

where χ is some known function.

This estimation can be applied to the Galerkin approximation of the nonlinear partial differential equation.

Let S_m be a finite dimensional subspace of $H^1(\Omega)$, such that for $s \in H^1(\Omega)$, there exists $s_m \in S_m$, such that $s_m \to s$. We define z_m as the solution of the finite dimensional problem

$$\left(\frac{d^2 z_m}{dt^2} + Az_m + z_m \cdot \int_0^t (|z_m(\tau)|^2 d\tau, \phi) = (f,\phi) \text{ for } \phi \in S_m,\ z_m(t) \in S_m\right.$$

with $z_m(0) = 0$ and $\frac{dz_m}{dt}(0) = 0$.

By the above estimates, we obtain

$$\left|\frac{dz_m}{dt}\right|^2 + \|z_m\|^2 \leq \chi(\|f\|, \|f'\|, \|f''\|)$$

is bounded in $L^\infty((0,T),(H^1(\Omega))')$, hence $z_m(t)$ and $\frac{dz_m(t)}{dt}$ remain in a bounded set of $L^\infty((0,T),H^1(\Omega))$ $\frac{d^2 z_m(t)}{dt^2}$ remain in a bounded set of $L^\infty((0,T),K)$, where K is a space, such that the mapping $H^1(\Omega) \to K$ is compact. By known facts from functional analysis, we can extract a subsequence of z_m, denoted again by z_m, such that

$$z_m \to y \quad \text{weak star in } L^\infty((0,T);H^1(\Omega)),$$

$$\frac{dz_m}{dt} \to \frac{dy}{dt} \text{ weak star in } L^\infty((0,T);H^1(\Omega)),$$

$$\frac{d^2 z_m}{dt^2} \to \frac{d^2 y}{dt^2} \text{ weak star in } L^\infty((0,T);K).$$

Since $H^1(\Omega) \to K$ is compact, it follows that $z_m \to y$ in $L^2((0,T),K)$ strongly; this means

$$\int_0^t |z_m(\tau)|^2 d\tau \to \int_0^t |y(\tau)|^2 d\tau$$

uniformly in t.

Now, for $\phi \in S_m$ fixed, we pass to the limit in S_m, the finite dimensional subspace of $H^1(\Omega)$, and obtain

$$\left(\left(\frac{d^2 z_m}{dt^2} + Az_m + z_m \cdot \int_0^t |z_m(\tau)|^2 d\tau\right), \phi\right) = (f,\phi)$$

$$\downarrow \qquad \downarrow \qquad \downarrow$$

$$\left(\left(\frac{d^2 y}{dt^2} - Ay - y \cdot \int_0^t |y(\tau)|^2 d\tau\right), \phi\right) = (f,\phi)$$

hence, there exists the solution of the nonlinear partial differential equation, such that

$$y, y' \in L^\infty((0,T);H^1(\Omega)) \text{ and } y'' \in L^\infty((0,T),K), \text{ if } f,f',f'' \in L^2((0,T);(H^1(\Omega))').$$

We consider the nonlinear equation

$$\frac{\partial^2 y}{\partial t^2} + Ay + y \int_0^t \int_\Omega y(x,\tau)^2 dx d\tau = f \text{ in } \Omega \times (0,T),$$

with boundary and initial conditions

$$\frac{\partial y}{\partial n} = v \text{ on } \partial\Omega \times (0,T), \; y(x,0) = 0 = \frac{\partial y(x,0)}{\partial t} \text{ in } \Omega.$$

I.4.1 Regularity of the nonlinear hyperbolic problem and the unique solution $y(v) \in L^2((0,T) \times \Omega)$ and $\int_0^t y(\tau)d\tau \in L^\infty((0,T),H^1(\Omega))$. We obtain certain interesting results, if f, f', f" $\in L^2((0,T);(H^1(\Omega))')$.

Theorem:

Let f and $\tilde{f}$ be in $L^2((0,T),(H^1(\Omega))')$ and remain there in a bounded set, if y, $\tilde{y}$ are the corresponding solutions of the nonlinear hyperbolic partial differential equation, then

$$\|y-\tilde{y}\|_{L^\infty((0,T),K)} + \|y_n-\tilde{y}_n\|_{L^\infty((0,T),H^1(\Omega))} \leq$$

$$\leq \text{ constant } \|\tilde{f}-f\|_{L^2((0,T),(H^1(\Omega))')},$$

where

$$y_n(t) = \int_0^t y(\tau)d\tau$$

is a part of the nonlinear term of the equation.

Proof:

By assumption

$$\frac{\partial^2 y}{\partial t^2} + Ay + y \int_0^t |y(\tau)|^2 d\tau = f$$

$$\frac{d^2 \tilde{y}}{dt^2} + A\tilde{y} + \tilde{y} \int_0^t |\tilde{y}(\tau)|^2 d\tau = \tilde{f},$$

defining $\eta = y - \tilde{y}$, we obtain

$$\frac{\partial^2 \eta}{\partial t^2} + A\eta + (y-\tilde{y}) \int_0^t |y|^2 d\tau - \tilde{y} \int_0^t (|y|^2 - |\tilde{y}|^2) d \quad = \tilde{f} - f$$

where $\eta(0) = 0 = \eta'(0)$.

According to the proof on page 33 we introduce

$$\varphi(t) = \begin{cases} -\int_t^s \eta(\sigma) d\sigma & \text{for } t \leqq s \\ 0 & \text{for } t \geqq s, \end{cases}$$

and $\eta_1(t) = \int_0^t \eta(\sigma) d\sigma$, then $\varphi(t) = \eta_1(t) - \eta_1(s)$ for $t \leqq s$ and $\varphi'(t) = -\eta(t)$.

Multiplying the equation above by $\varphi(t)$,

$$\eta''.\varphi(t) + A\eta.\varphi(t) + \eta.\varphi(t). \int_0^t |y|^2 d\sigma + \tilde{y}\cdot\varphi(t). \int_0^t (|y|^2 - |\tilde{y}|^2) d\sigma = (\tilde{f} - f)\varphi(t),$$

and integrating with respect to t from 0 to s,

$$\int_0^s \eta''\cdot\varphi(t)dt + \int_0^s A\eta\cdot\varphi(t)dt + \int_0^s \eta\varphi(t)\left(\int_0^t |y|^2 d\sigma\right)dt +$$

$$+ \int_0^s \tilde{y}\varphi(t)\left(\int_0^t (|y|^2 - |\tilde{y}|^2) d\sigma\right)dt - \int_0^s (\tilde{f} - f)\varphi(t)dt = 0$$

integration by parts yields

$$\int_0^s \eta''\varphi(t)dt = \eta'(t)\cdot\varphi(t)\Big|_0^s - \int_0^s \eta'(t)\varphi'(t)dt = \eta'(s)\cdot\varphi(s) +$$

$$+ \int_0^s \eta'(t)\eta(t)dt, \text{ as } \eta'(0) = 0 \text{ and } \varphi'(t) = -\eta(t),$$

$\frac{1}{2}|\eta(t)|^2 = \eta(t)\cdot\eta'$ and $\varphi(s) = \eta_1(s)-\eta_1(s) = 0$, then

$$\int_0^s \eta''\varphi(t)dt = + \int_0^s \eta'(t)\varphi(t)dt = \eta\cdot\eta \Big|_0^s - \int_0^s \eta'\cdot\eta\, dt, \quad \text{thus}$$

$$\int_0^s \eta'(t)\cdot\eta(t)dt = \frac{1}{2}|\eta(s)|^2.$$

As $\eta_1(s) = \int_0^s \eta(\tau)d\tau$, $A\eta_1(s) = \int_0^s A\eta(\tau)d\tau$ and $\varphi(\tau) = \eta_1(\tau)-\eta_1(s)$, then

$$\int_0^s A\eta(\tau)\cdot\varphi(\tau)d\tau = (A\eta_1(s)\cdot\varphi(\tau)) = (A\eta_1(s)(\eta_1(\tau)-\eta_1(s))) =$$

$$= (A\eta_1(s)\eta_1(\tau))-(A\eta_1(s)\eta_1(s)) = B(\eta_1(s)\cdot\eta_1(\tau))-B(\eta_1(s)\cdot\eta_1(s)).$$

The third integral can be written, if $\phi'(t) = -\eta(t)$; $\frac{1}{2}\frac{d}{dt}|\varphi|^2 = \varphi\cdot\frac{\partial\varphi}{\partial t} = -\varphi\cdot\eta$,

$$\int_0^s \eta\varphi(t)\left(\int_0^t |y|^2 d\tau\right)dt = - \int_0^s \left(\int_0^t |y|^2 d\tau\cdot\frac{1}{2}\frac{d}{dt}|\varphi|^2 dt\right) =$$

$$- \int_0^t |y|^2 d\tau\cdot\frac{1}{2}(\varphi(t)^2)\Big|_{t=0}^{t=s} - \frac{1}{2}\int_0^s |y(t)|^2\cdot|\varphi(t)|^2 dt, \text{ as } \varphi(s) = 0$$

$$= - \frac{1}{2}\int_0^s |y(t)|^2\cdot|\varphi(t)|^2 dt;$$

the last integral

$$\int_0^s \tilde{y}\cdot\varphi(t)\left(\int_0^t (|y|^2 - |\tilde{y}|^2)d\sigma\right)dt$$

can be rewritten as follows:

as $\quad \eta = y-\tilde{y}$, $\eta^2 = y^2-2y\tilde{y} + \tilde{y}^2$ and $|y|^2 - |\tilde{y}|^2 = (y-\tilde{y})(y+\tilde{y}) = \eta(y+\tilde{y})$,

hence

$$\int_0^t (|y|^2 - |\tilde{y}|^2)d\sigma = \int_0^t \eta(\sigma)(y + \tilde{y})d\sigma$$

and estimating

$$\left|\int_0^t (|y|^2 - |\tilde{y}|^2)d\sigma\right| \leq \left(\int |y|d\tau + \int |y|d\tau\right)\cdot\int_0^t |\eta(\sigma)d\sigma$$

$$\leq C(\ \|y\|_{L^\infty((0,T),K)} + \|\tilde{y}\|_{L^\infty((0,T),K)})\cdot\int_0^t |\eta(\sigma)|\,d\sigma.$$

Now the integral can be estimated by

$$\int_0^s \left(\int_0^t \eta(\sigma)(y+\tilde{y})d\sigma\right)\tilde{y}(\eta_1(t)-\eta_1(s))dt \leq$$

$$\leq C\cdot(\ \|y\|_{L^\infty((0,T),K)} + \|\tilde{y}\|_{L^\infty((0,T),K)})\cdot\int_0^s \eta_1(t)-\eta_1(s)\ \cdot\left(\int_0^t |\eta(\sigma)|d\sigma\right)dt,$$

this can be estimated by the Gronwall inequality

$$\leq C\cdot\int_0^s |\eta_1(t)|^2 dt + \int_0^s |\eta(\sigma)|^2 d\sigma;$$

we have

$$\int_0^s \eta''\varphi(t)dt + \int_0^s A\eta\cdot\varphi dt + \int_0^s \eta\varphi(t)\left(\int_0^t |y|^2 d\sigma\right)dt +$$

$$+ \int_0^s \tilde{y}\cdot\varphi(t)\left(\int_0^t (|y|^2-|\tilde{y}|^2)d\sigma\right)dt - \int_0^s (\tilde{f}-f)\varphi(t)dt = 0,$$

thus

$$\tfrac{1}{2}|\eta(s)|^2 + B(\eta_1(s),\eta_1(\tau))-B(\eta_1(s),\eta_1(s)) + \frac{1}{2}\int_0^s |y(t)|^2\cdot|\varphi(t)|^2 dt \leq$$

$$\leq\ -\int_0^s (f-\tilde{f})(\eta_1(t) - \eta_1(s)dt + C\cdot\left(\int_0^s |\eta_1(t)|^2 dt + \int_0^s |\eta(\sigma)|^2 d\sigma\right),$$

then

$$|\eta(s)|^2 + \|\eta_1(s)\|^2 \leq C\cdot\int_0^s \|\tilde{f}-f\|\cdot\|\eta_1(t)-\eta_1(s)\|\,dt +$$

$$+ C\cdot\left(\int_0^s |\eta_1(t)|^2 dt + \int_0^s |\eta(\sigma)|^2 d\sigma,\right.$$

as

$$\eta(t) = y(t) - \tilde{y}(t) \text{ and } \eta_1(t) = \int_0^t \eta(\sigma)d\sigma =$$

$$= \int_0^t y(\tau)d\tau - \int_0^t \tilde{y}(\tau)d\tau \underset{\text{def.}}{=} y_n(t) - \tilde{y}_n(t),$$

we obtain finally

$$|y(s) - \tilde{y}(s)|^2 + \|y_n(s)-\tilde{y}_n(s)\|^2 \leq \text{constant} \int_0^s \|\tilde{f}-f\| \, dt.$$

For all $f, \tilde{f} \in L^2((0,T);(H^1(\Omega))')$ we have

$$\|y-\tilde{y}\|_{L^\infty((0,T),K)} + \|y_n-\tilde{y}_n\|_{L^\infty((0,T);H^1(\Omega))} \leq$$

$$\leq \text{constant} \|\tilde{f}-f\|_{L^2((0,T);(H^1(\Omega))')}.$$

I.5 Modelling methods in three-connected domains

The following deals with the description of diffusion processes in 3-connected and multiconnected regions (on a first level of approximation).

Let Ω be open in $\mathbb{R}^n$ of the form

$$\Omega = \Omega_0 \cup \Omega_1 \cup \partial\Omega_1 \cup \Omega_2 \cup \partial\Omega_2,$$

where the boundary of the 3-connected domain Ω_0 is given by

$$\partial\Omega_0 = \partial\Omega_1 \cup \Omega_2 \cup \Gamma.$$

According to our general theory we have $V = H_0^1(\Omega)$ and the bilinear forms

$$a_i(u,v) = \int_{\Omega_i} \text{grad}u \cdot \text{grad}v \, dx.$$

The boundary-value problem, for $\varepsilon > 1$ fixed, is given by

$$-\Delta u_{\varepsilon o} = f_o \text{ in } \Omega_o$$

$$-\varepsilon\Delta u_{\varepsilon 1} = f_1 \text{ in } \Omega_1$$

$$-\varepsilon\Delta u_{\varepsilon 2} = f_2 \text{ in } \Omega_2$$

and on the outer boundary Γ $u_{\varepsilon o} = 0$.

On the boundary of the regions Ω_1, Ω_2 we consider transmission conditions, which describe processes going on in the multiconnected domain Ω_o and influenced by processes in Ω_1 and Ω_2,

$$u_{\varepsilon o} = u_{\varepsilon 1}, \quad \frac{\partial u_{\varepsilon o}}{\partial n} = \varepsilon \cdot \frac{\partial u_{\varepsilon 1}}{\partial n} \text{ on } \partial\Omega_1$$

and

$$u_{\varepsilon o} = u_{\varepsilon 2}, \quad \frac{\partial u_{\varepsilon o}}{\partial n} = \varepsilon^2 \cdot \frac{\partial u_{\varepsilon 2}}{\partial n} \text{ on } \partial\Omega_2,$$

the equivalent formulation in bilinear forms is given by

$$a_o(u_\varepsilon, v) + \varepsilon a_1(u_\varepsilon, v) + \varepsilon^2 a_2(u_\varepsilon, v) = (f,v) \text{ for } v \in V.$$

I.5.1 Asymptotic expansion

Now we construct the asymptotic representation of the unique solution. We are looking for an expansion in the form (ansatz)

$$u_\varepsilon = \frac{u^{-2}}{\varepsilon^2} + \frac{u^{-1}}{\varepsilon} + u^o + \varepsilon u^1.$$

Using this in the bilinear form

$$a_o(u_\varepsilon, v) = \frac{1}{\varepsilon^2} a_o(u^{-2}, v) + \frac{1}{\varepsilon} a_o(u^{-1}, v) + a_o(u^o, v) + \varepsilon a_o(u^1, v) + \ldots$$

$$\varepsilon a_1(u_\varepsilon, v) = \frac{1}{\varepsilon} a_1(u^{-2}, v) + a_1(u^{-1}, v) + \varepsilon a_1(u^o, v) + \varepsilon^2 a_1(u^1, v) + \ldots$$

$$\varepsilon^2 a_2(u_\varepsilon, v) = + a_2(u^{-2}, v) + \varepsilon a_2(u^{-1}, v) + \varepsilon^2 a_2(u^o, v) + \varepsilon^3 a_2(u^1, v) + \ldots$$

Identifying various powers of ε, we obtain

$$\frac{1}{\varepsilon} \quad a_o(u^{-1},v) + a_1(u^{-2},v) = 0$$

$$\varepsilon^o \quad a_o(u^o,v) + a_1(u^{-1},v) + a_2(u^{-2},v) = (f,v)$$

$$\varepsilon \quad a_o(u^1,v) + a_1(u^o,v) + a_2(u^{-1},v) = 0$$

$$\varepsilon^2 \quad a_o(u^2,v) + a_1(u^1,v) + a_2(u^o,v) = 0$$

$$\varepsilon^3 \quad a_o(u^3,v) + a_1(u^2,v) + a_2(u^1,v) = 0$$

this equation-system is considered with respect to the subspaces

$$Y_o = \{v \in H_o^1(\Omega)/v = 0 \text{ in } \Omega_o\},$$

$$Y_1 = \{v \in H_o^1(\Omega)/v = 0 \text{ in } \Omega_o \cup \Omega_1\}.$$

Let v be in Y_1, then $v = 0$ in Ω_o and Ω_1, thus, so we have only

$$a_2(u^{-2},v) = (f,v),$$

and the equivalent partial differential equation

$$-\Delta u_2^{-2} = f_2 \text{ in } \Omega_2$$

$$u_2^{-2} = 0 \quad \text{on } \partial\Omega_2,$$

thus u^{-2} is known.

Now calculate u^{-1}.

In the first equation we set $a_o(u^{-1},v) = 0$, then $u^{-1} \in Y_o$. In the second equation we restrict the bilinear form to Y_o and obtain

$$a_1(u^{-1},v) + a_2(u^{-2},v) = (f,v) \text{ for } v \in Y_o,$$

and in the third we restrict to Y_1, thus

$$a_2(u^{-2},v) = 0 \text{ for } v \in Y_1,$$

Determine u^0.

We take the second equation

$$a_0(u^0,v) + a_1(u^{-1},v) + a_2(u^{-2},v) = (f,v) \text{ for } v \in V.$$

In the second equation we choose $v \in Y_0$, thus $a_0(\cdot,v) = 0$ and

$$a_1(u^{-1},v) + a_2(u^{-2},v) = (f,v).$$

In the fourth we restrict the bilinear form to Y_1, hence

$$a_2(u^{-1},v) = 0 \text{ for } v \in Y_1.$$

To calculate u^{-1} we have the system

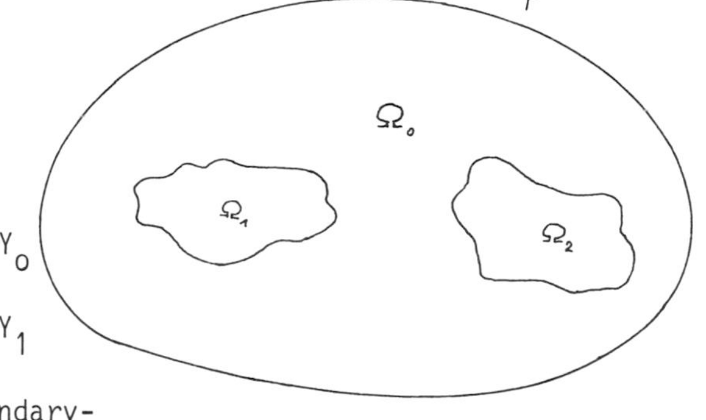

$$u^{-1} \in Y_0$$

$$a_1(u^{-1},v)+a_2(u^{-2},v) = f(v) \quad v \in Y_0$$

$$a_2(u^{-1},v) = 0 \qquad v \in Y_1$$

where u^{-2} is the solution of the boundary-value problem given above

$$-\Delta u^{-2} = f_2 \text{ in } \Omega_2$$

$$u^{-2} = 0 \quad \text{on } \partial\Omega_2.$$

The bilinear form $a_2(u^{-1},v) = 0$ for $v \in Y_1$ is equivalent to

$$-\Delta u_2^{-1} = 0 \text{ in } \Omega_2$$

$$u_2^{-1} = u_0^1 \text{ on } \partial\Omega_2.$$

By construction $u_0 = 0$ on Γ, and $-\Delta u_0 = f_0$ in Ω_0, but $u_0 = 0$ on the boundaries $\partial\Omega_1$, $\partial\Omega_2$; thus $u_0^1 = 0$, as the solution is unique in Ω_2, we have

$u_2^{-1} = 0.$

The boundary-value problem

$$-\Delta u^{-2} = f_2 \text{ in } \Omega_2$$
$$u^{-2} = 0 \text{ on } \partial\Omega_2$$

is equivalent to

$$a_2(u^{-2},v) = \int_{\Omega_2} f_2 \cdot v dx.$$

We determined the relation

$$a_1(u^{-1},v) + a_2(u^{-2},v) = (f_1,v) + (f_2,v),$$

that

$$a_1(u^{-1},v) = \int_{\Omega_1} f_1 \cdot v \; dx$$

and the equivalent formulation

$$-\Delta u_1^{-1} = f_1 \text{ in } \Omega_1$$
$$u_1^{-1} = 0 \text{ on } \partial\Omega_1,$$

the solution u_1^{-1} depends on f_1 and the total solution is given by

$$u^{-1} = \{u_o^{-1}, u_1^{-1}, u_2^{-1}\} = \{0, u_1^{-1}, 0\}.$$

Calculating u^o we may consider the equation system

$$a_o(u^o,v) + a_1(u^{-1},v) + a_2(u^{-2},v) = (f,v) \text{ for } v \in V$$
$$a_1(u^o,v) + a_2(u^{-1},v) = 0 \qquad \text{for } v \in Y_o$$
$$a_2(u^o,v) = 0 \qquad \text{for } v \in Y_1.$$

Above we have shown that

$$-\Delta u_2^{-2} = f_2 \text{ in } \Omega_2$$

$$u_2^{-2} = 0 \text{ on } \partial\Omega_2$$

and by Green's formula

$$-\int_{\partial\Omega_2} \frac{\partial u_2^{-2}}{\partial n} v\, ds + a_2(u_2^{-2}, v) = \int_{\Omega_2} f_2 \cdot v\, dx \text{ for } v \in V$$

and

$$-\Delta u_1^{-1} = f_1 \text{ in } \Omega_1$$

$$u_1^{-1} = 0 \quad \text{on } \partial\Omega_1,$$

rewritten by Green's formula,

$$-\int_{\partial\Omega_1} \frac{\partial u_1^{-1}}{\partial n} v\, ds + a_1(u^1, v) = \int_{\Omega_1} f_1 \cdot v\, dx \text{ for } v \in V.$$

Using these expressions in the first relation

$$a_o(u^o, v) + a_1(u^{-1}, v) + a_2(u^{-2}, v) = (f, v),$$

we obtain

$$a_o(u^o, v) + \int_{\Omega_1} f_1 \cdot v dx + \int_{\partial\Omega_1} \frac{\partial u_2^{-1}}{\partial n} v\, ds = \int_{\Omega_2} f_2 \cdot v\, dx +$$

$$+ \int_{\partial\Omega_2} \frac{\partial u_2^{-2}}{\partial n} v\, ds = \quad (f_1 + f_2 + f_o) v\, dx,$$

hence

$$a_o(u^o, v) = \int_{\Omega_o} f_o \cdot v\, dx - \int_{\partial\Omega_1} \frac{\partial u_1^{-1}}{\partial n} \cdot v\, ds - \int_{\partial\Omega_2} \frac{\partial u_2^{-2}}{\partial n} v\, ds,$$

and equivalent

$$-\Delta u_o^o = f_o \text{ in } \Omega_o$$

$$u_o^o = 0 \text{ on } \Gamma$$

$$\frac{\partial u_o^o}{\partial n} = \frac{\partial u_1^{-1}}{\partial n} \text{ on } \partial\Omega_1$$

$$\frac{\partial u_o^o}{\partial n} = \frac{\partial u_2^{-1}}{\partial n} \text{ on } \partial\Omega_2.$$

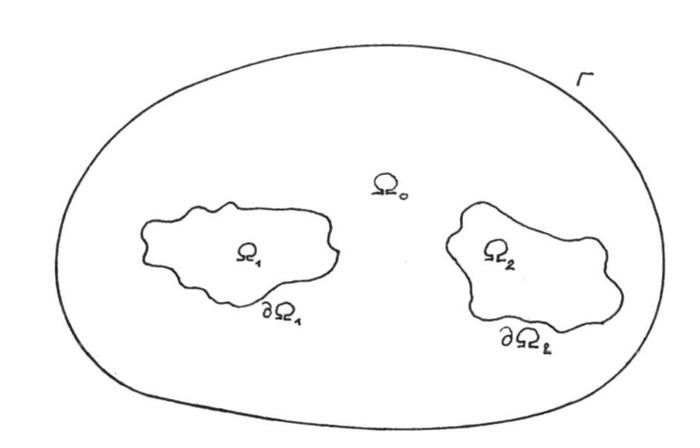

Now we consider the equations

$$a_1(u^o,v) + a_2(u^{-1},v) = 0 \text{ for } v \in Y_o$$

$$a_2(u^o,v) = 0 \quad \text{for } v \in Y_1$$

We have shown above, that

$$-\Delta u_2^{-1} = 0 \text{ in } \Omega_2$$

$$u_2^{-1} = 0 \text{ on } \partial\Omega_2,$$

then $u_2^1 = 0$, thus $a_2(u^{-1},v) = 0$ for $v \in Y_o$.

The bilinear form

$$a_1(u^o,v) = 0 \quad v \in Y_o$$

is equivalent to

$$-\Delta u_1^o = 0 \text{ in } \Omega_1$$

$$u_1^o = u_o^o \text{ on } \partial\Omega_1,$$

and the bilinear form

$$a_2(u^o,v) = 0 \text{ for } v \in Y_1$$

is equivalent to

$$-\Delta u_2^o = 0 \text{ in } \Omega_2$$

$$u_2^o = u_o^o \text{ on } \partial\Omega_2.$$

The solution u^o is representable as

$$u^o = \{u_o^o, u_1^o, u_2^o\}.$$

Calculating u^1 we consider the equation system

$$a_o(u^1,v) + a_1(u^o,v) + a_2(u^{-1},v) = 0 \quad v \in V$$

$$a_1(u^1,v) + a_2(u^o,v) = 0 \quad v \in Y_o$$

$$a_2(u^1,v) = 0 \quad v \in Y_1.$$

If $-\Delta u_1^o = 0$ in Ω_1, applying Green's formula, then

$$-\int_{\partial\Omega_1} \frac{\partial u_1^o}{\partial n} v \, ds + a_1(u^o,v) = 0,$$

hence

$$a_1(u^o,v) = \int_{\partial\Omega_1} \frac{\partial u_1^o}{\partial n} v \, ds.$$

Above it is shown that

$$-\Delta u_2^{-1} = 0 \text{ in } \Omega_2$$

$$u_2^{-1} = 0 \text{ on } \partial\Omega_2$$

can be written by Green's formula

$$-\int_{\partial\Omega_2} \frac{\partial u_2^{-1}}{\partial n} v \, ds + a_2(u^{-1},v) = 0,$$

where

$$a_2(u^{-1},v) = 0 \text{ and } - \int_{\partial\Omega_2} \frac{\partial u_2^{-1}}{\partial n} v \, ds = 0.$$

Using these expressions the bilinear form

$$a_o(u^1,v) + a_1(u^o,v) + a_2(u^{-1},v) = 0$$

can be written as

$$a_o(u^1,v) + \int_{\partial\Omega_1} \frac{\partial u_1^o}{\partial n} v \, ds + \int_{\partial\Omega_2} \frac{\partial u_2^{-1}}{\partial n} v \, ds = 0,$$

and we obtain the equivalent boundary-value problem

$$-\Delta u_o^1 = 0 \text{ in } \Omega_o$$

$$u_o^1 = 0 \text{ on } \Gamma$$

$$\frac{\partial u_o^1}{\partial n} = \frac{\partial u_1^o}{\partial n} \text{ on } \partial\Omega_1$$

$$\frac{\partial u_o^1}{\partial n} = 0 \quad \text{on } \partial\Omega_2.$$

Now we analyse the other equations

$$a_1(u^1,v) + a_2(u^o,v) = 0 \quad v \in Y_o$$

$$a_2(u^1,v) = 0 \quad v \in Y_1,$$

and we have the bilinear form

$$a_1(u^1,v) = 0 \quad v \in Y_o$$

and the equivalent boundary-value problem

$$- \Delta u_1^1 = 0 \text{ in } \Omega_1$$

$$u_1^1 = u_o^1 \text{ in } \partial\Omega_1.$$

I.5.2 Algorithms of the elliptic problem

The bilinear form

$$a_2(u^1,v) = 0 \quad v \in Y_1$$

is equivalent to

$$-\Delta u_2^1 = 0 \text{ in } \Omega_2$$
$$u_2^1 = u_o^1 \text{ on } \partial\Omega_2$$

Finally the solution is given by

$$u^1 = \{u_o^1, u_1^1, u_2^1\}.$$

The asymptotic representation of the solution in the 3-connected domain is given by

$$u_{\varepsilon 0} = u_o^o + \varepsilon u_o^1 \text{ in } \Omega_o$$
$$u_{\varepsilon 1} = \frac{u_1^{-1}}{\varepsilon} + u_1^o + \varepsilon u_1^1 \text{ in } \Omega_1$$
$$u_{\varepsilon 2} = \frac{u_2^{-1}}{\varepsilon} + u_2^o + \varepsilon u_2^1 \text{ in } \Omega_2.$$

Résumé:

In the following there is given only one simple model describing the evolution of diffusion processes on a first level of approximation.

The boundary-value problems in the domains Ω_1 and Ω_2 have unique solutions:

$$-\Delta u_1^{-1} = f_1 \text{ in } \Omega_1$$
$$u_1^{-1} = 0 \quad \text{on } \partial\Omega_1,$$

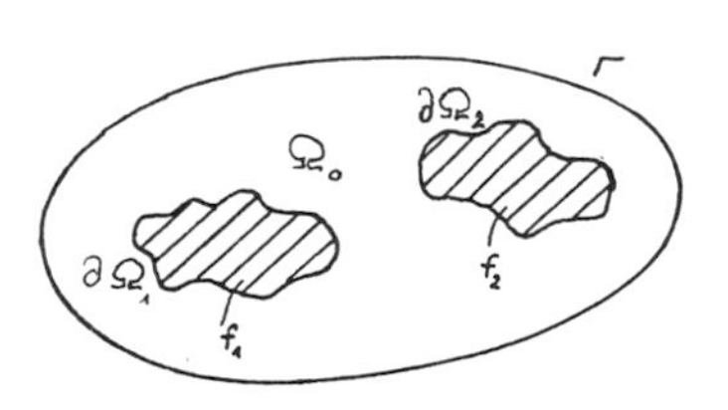

the solution u_1^{-1} depends on f_1, and

$$-\Delta u_2^{-2} = f_2 \text{ in } \Omega_2$$
$$u_2^{-2} = 0 \text{ on } \partial\Omega_2,$$

the solution u_2^{-2} depends on f_2.

We determine the solution in the 3-connected domain

$$-\Delta u_o^o = f_o \text{ in } \Omega_o$$

$$u_o^o = 0 \text{ on } \Gamma$$

$$\frac{\partial u_o^o}{\partial n} = \frac{\partial u_1^{-1}}{\partial n} \text{ on } \partial\Omega_1$$

$$\frac{\partial u_o^o}{\partial n} = \frac{\partial u_2^{-2}}{\partial n} \text{ on } \partial\Omega_2$$

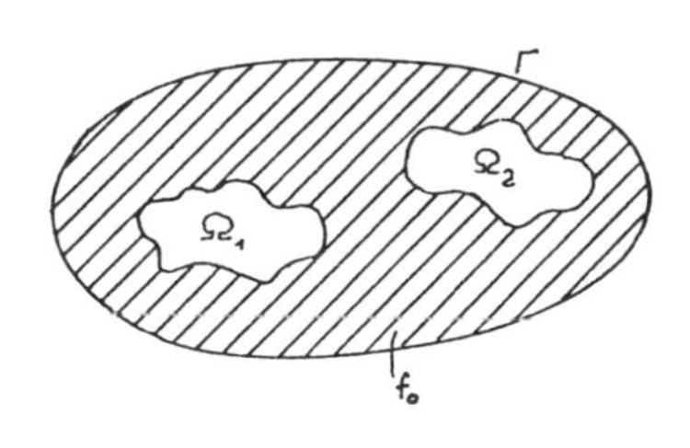

the solution u_o^o depends on the solutions u_1^{-1}, u_2^{-2}.

The solution in the 3-connected domain now influences the potential distributions in Ω_1 and Ω_2 (e.g. the vortex distributions in Ω_1 and Ω_2 are influenced by the fluid process described by u_o^o):

$$-\Delta u_1^o = 0 \text{ in } \Omega_1$$

$$u_1^o = u_o^o \text{ on } \partial\Omega_1,$$

the solution u_1^o depends on u_o^o;

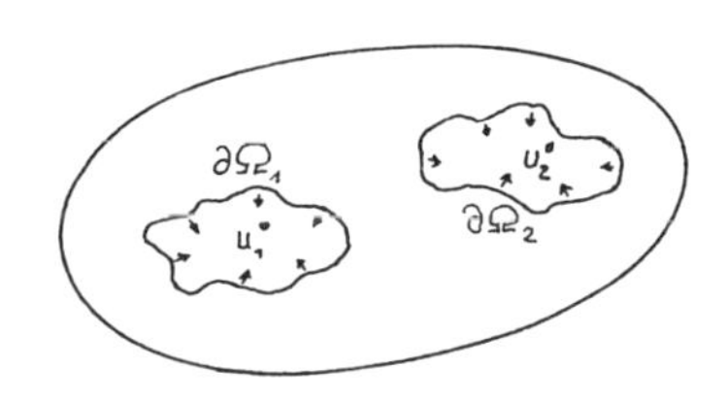

$$-\Delta u_2^o = 0 \text{ in } \Omega_2$$

$$u_2^o = u_o^o \text{ on } \partial\Omega_2$$

the solution u_2^o depends on u_o^o.

The solution in the 3-connected domain Ω_o is influenced by the potential distribution in Ω_1 (e.g. the vortex distribution in the domain Ω_1 is influencing the fluid or diffusion process in Ω_o).

$$-\ \Delta u_o^1 = 0 \text{ in } \Omega_o$$

$$u_o^1 = 0 \text{ on } \Gamma$$

$$\frac{\partial u_o^1}{\partial n} = \frac{\partial u_1^o}{\partial n} \text{ on } \partial\Omega_1$$

$$\frac{\partial u_o^1}{\partial n} = 0 \quad \text{on } \partial\Omega_2$$

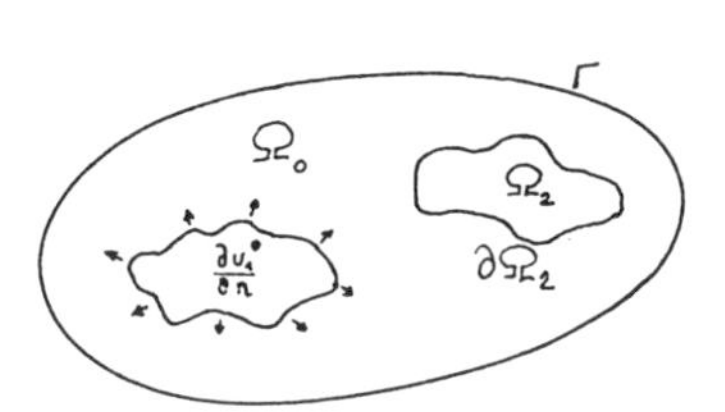

the solution u_o^1 depends on u_1^o.
The solution u_o^1 in the 3-connected domain influences again the potential distributions in Ω_1 and Ω_2:

$$-\Delta u_1^1 = 0 \text{ in } \Omega_1$$

$$u_1^1 = u_o^1 \text{ on } \partial\Omega_1,$$

the solution u_1^1 depends on u_o^1, and

$$-\Delta u_2^1 = 0 \text{ in } \Omega_2$$

$$u_2^1 = u_o^1 \text{ on } \partial\Omega_1,$$

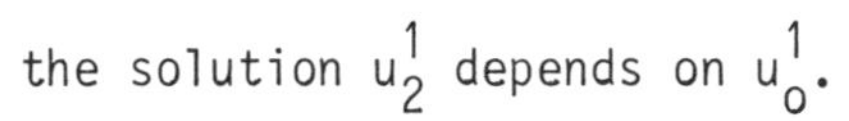

the solution u_2^1 depends on u_o^1.

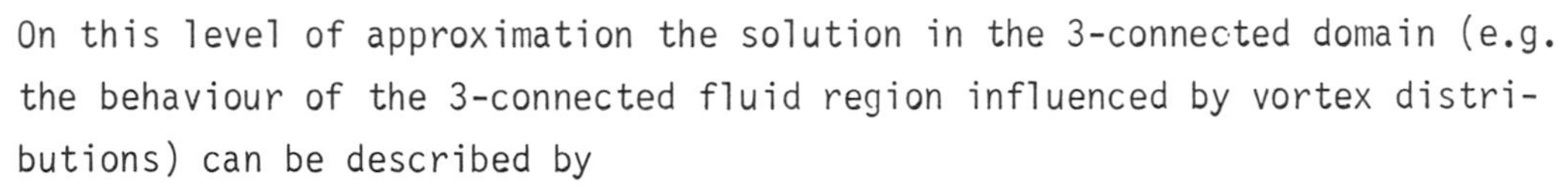

On this level of approximation the solution in the 3-connected domain (e.g. the behaviour of the 3-connected fluid region influenced by vortex distributions) can be described by

$$-\Delta u_o^2 = 0 \text{ in } \Omega_o$$

$$u_o^2 = 0 \text{ on } \Gamma$$

$$\frac{\partial u_o^2}{\partial n} = \frac{\partial u_1^1}{\partial n} \text{ on } \partial\Omega_1$$

$$\frac{\partial u_o^2}{\partial n} = \frac{\partial u_2^1}{\partial n} \text{ on } \partial\Omega_2.$$

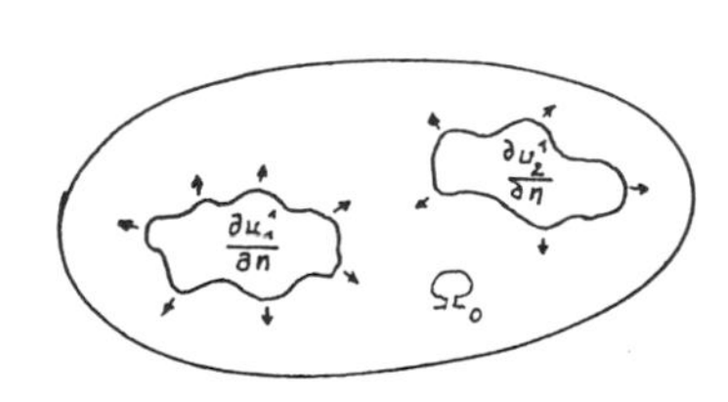

The asymptotic solution can be written as

$$u_{\varepsilon o} = u_o^o + \varepsilon u_o^1 + \varepsilon^2 u_o^2 + \dots \text{ in } \Omega_o,$$

$$u_{\varepsilon 1} = \frac{u_1^{-1}}{\varepsilon} + u_1^o + \varepsilon u_1^1 + \dots \text{ in } \Omega_1,$$

$$u_{\varepsilon 2} = \frac{u_2^{-1}}{\varepsilon} + u_2^o + \varepsilon u_2^1 + \dots \text{ in } \Omega_2.$$

The asymptotic methods can be applied to describe possible evolutions of diffusion processes on a first level of approximation.

The method can be generalized to diffusion processes in multi-connected regions.

I.6 Model-building in multiconnected domains

We consider only one possible modelling-process in a multiconnected domain of $\mathbb{R}^n$. (Applications to real diffusion processes in $\mathbb{R}^3$ will be treated in a later work.)

Let D_i for $i = 0,\dots,q$ be open, disjoint and simple-connected in $\mathbb{R}^n$. Define the form

$$D = \bar{D}_1 \cup \bar{D}_2 \cup \dots \cup \bar{D}_q \cup D_1,$$

with the boundaries $\partial D_j = B_j$ and $\partial D_1 = B_1 \cup B_2 \cup \dots \cup B_q \cup \Gamma$, where Γ = outer boundary, such that $D_o = D_1 \cup D_2 \cup \dots \cup D_j$, $D = D_o \cup D_1$, then D_1 is a q+1-connected domain.

Introducing the bilinear form

$$a_i(u,v) = \int_{D_i} \text{grad}u \cdot \text{grad}v \, dx \text{ for } i = 0,1;$$

and the Hilbert space $H_o^1(D)$.

We consider the boundary value problem

$$-\Delta u_{\varepsilon o} = f_o \text{ in } D_o \quad \text{and} \quad -\varepsilon \Delta u_{\varepsilon 1} = f_1 \text{ in } D_1, \ \varepsilon \text{ fixed},$$

with respect to the transmission condition

$$u_{\varepsilon o} = u_{\varepsilon 1} \text{ and } \frac{\partial u_{\varepsilon o}}{\partial n} = \frac{\partial u_{\varepsilon 1}}{\partial n} \text{ on } B_j \text{ for } j = 0,1,\dots,q$$

and the outer boundary condition $u_{\varepsilon 1} = 0$ on Γ.

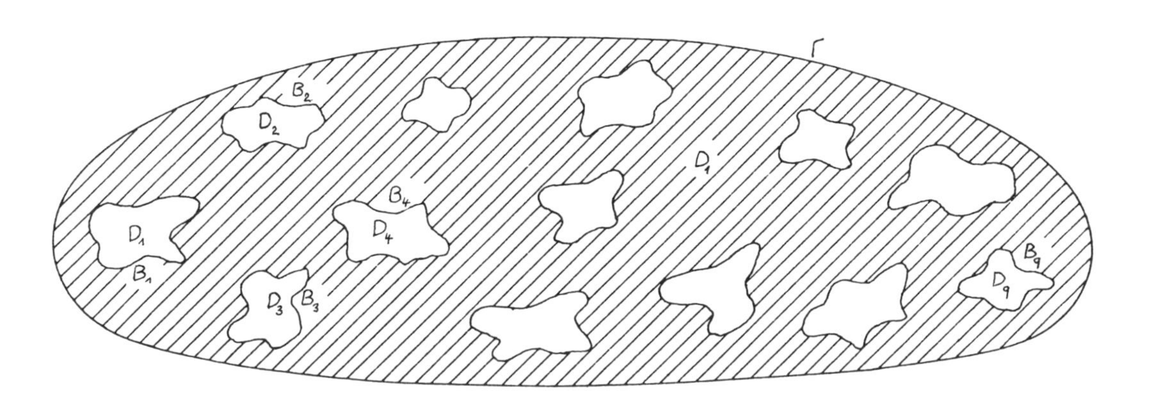

I.6.1 Asymptotic expansion and algorithms

The equivalent bilinear form is given by

$$a_o(u_\varepsilon,v) + a_1(u_\varepsilon,v) = (f,v) \text{ for } v \in V.$$

Let the bilinear form be $a_o(u,v) = 0$, then grad $v = 0$, hence $v = \text{constant} = b_i$ in D_i, thus we introduce the space

$$Y_o = \{v | v = b_i \text{ in } D_i \text{ for } i = 0,1,\dots,q\}.$$

The unique solution can be expressed as an asymptotic expansion. We choose the ansatz

$$u_\varepsilon = \frac{u^{-1}}{\varepsilon} + u^o + \varepsilon u^1 + \varepsilon^2 u^2 + \dots,$$

use it in the bilinear form

$$a_o(u_\varepsilon,v) = \frac{1}{\varepsilon} a_o(u^{-1},v) + a_o(u^o,v) + \varepsilon a_o(u^1,v) + \varepsilon^2 a_o(u^2,v) + \dots$$

$$\varepsilon a_1(u_\varepsilon,v) = \qquad a_1(u^{-1},v) + \varepsilon a_1(u^o,v) + \varepsilon^2 a_1(u^1,v) + \dots,$$

and identify powers of ε:

ε-term

$$\frac{1}{\varepsilon} \quad a_o(u^{-1},v) = 0 \text{ for } v \in V$$

$$\varepsilon^o \quad a_1(u^{-1},v) + a_o(u^o,v) = (f,v)$$

$$\varepsilon \quad a_1(u^o,v) + a_o(u^1,v) = 0$$

$$\varepsilon^2 \quad a_1(u^1,v) + a_o(u^2,v) = 0$$

$$\vdots$$

$$\varepsilon^j \quad a_1(u^{j-1},v) + a_o(u^j,v) = 0 \text{ for } j = 1,2,\dots \text{ and } v \in V.$$

The bilinear form $a_o(u^{-1},v) = 0$ implies $u^{-1} \in Y_o$. For $v \in Y_o$ we obtain from the second equation $a_1(u^{-1},v) = (f,v)$. As we can show that

$$a_1(v,v) \geq \alpha \|v\|^2 \text{ for } \alpha > 0 \text{ constant,}$$

the bilinear form has a unique solution.

Let $v \in C_k^\infty(D_1)$, the unique solution of the bilinear form

$$a_1(u^{-1},v) = (f,v)$$

is solution of the partial differential equation

$$-\Delta u^{-1} = f_1 \text{ in } D_1.$$

Applying Green's formula we obtain

$$-\sum_{j=1}^{q} \int_{B_j} \frac{\partial u_1^{-1}}{\partial n} v \, d(B_j) + a_1(u^{-1},v) = \int_{D_1} f_1 \, v \, dx$$

as

$$a_1(u^{-1},v) = (f_1,v) + (f_o,v) = \int_{D_1} f_1 \cdot v \, dx + \int_{D_o} f_o \cdot v \, dx$$

it follows

$$\sum_{j=1}^{q} \int_{B_j} \frac{\partial u_1^{-1}}{\partial n} \cdot v \, d(B_j) = \int_{D_o} f_o \cdot v \, dx \text{ for } v \in Y_o,$$

now we take

$$v = \begin{cases} 1 & \text{for } i = j \\ 0 & i \neq j \end{cases} \quad \text{in the domains } D_j,$$

hence

$$\int_{B_j} \frac{\partial u_1^{-1}}{\partial n} d(B_j) = \int_{D_j} f \, dx \text{ for } j = 1,\ldots,q.$$

Finally we obtain the boundary value problem

$$- \Delta u_1^{-1} = f_1 \text{ in } D_1$$

$$u_1^{-1} = b_i \text{ on } B_i \text{ for } i = 1,\ldots,q$$

$$u_1^{-1} = 0 \quad \text{on } \Gamma$$

$$\int_{B_i} \frac{\partial u_1^{-1}}{\partial n} d(B_i) = \int_{D_i} f \, dx$$

We determine the solution u^o: by considering the second equation and restricting the first to Y_o

$$a_o(u^o,v) + a_1(u^{-1},v) = (f,v) \quad \text{for } v \in V$$

$$a_1(u^o,v) = 0 \quad \cdot \qquad v \in Y_o.$$

Above we analysed the partial differential equation $-\Delta u_1^{-1} = f_1$ in D_1, remember that the boundary is written by $B \cup \Gamma$, where $B = B_1 \cup B_2 \cup \ldots \cup B_q$ and now apply Green's formula

$$- \int_B \frac{\partial u_1^{-1}}{\partial n} v \, ds + a_1(u^{-1},v) = (f_1,v).$$

By using

$$a_o(u^o,v) + a_1(u^{-1},v) = (f_1,v) + (f_o,v),$$

we obtain

$$a_o(u^o,v) = \int_{D_o} f \cdot v \, dx - \int_B \frac{\partial u_1^{-1}}{\partial n} v \, ds \quad \text{for } v \in V,$$

and the equivalent boundary value problem

$$-\Delta u_i^o = f_i \text{ in } D_i$$

$$\frac{\partial u_i^o}{\partial n} = \frac{\partial u_1^{-1}}{\partial n} \text{ on } B_i \text{ for } i = 1,\ldots,q.$$

This is the known "von Neumann" problem. The solution exists, if the condition

$$\int_{B_j} \frac{\partial u_1^{-1}}{\partial n} ds = \int_{D_j} f \, dx$$

is satisfied. The solution is unique up to a constant. Thus we have a representation of the solution u^o

$$u^o = \hat{u}^o + \mathring{y}, \text{ where } \mathring{y} \in Y_o,$$

this means that y^o is constant in D_i.

We are able to write the second equation in the form

$$a_1(\hat{u}^o,v) + a_1(\mathring{y},v) = a_1(u^o,v) = 0.$$

Let v be an element of $C_k^\infty(D_1)$ and $a_1(u^o,v) = 0$, thus u^o is a solution of $-\Delta u_1^o = 0$ in D_1 and by assumption $u_1^o = 0$ on Γ, now we apply Green's formula

$$-\int_B \frac{\partial u_1^o}{\partial n} vd(B) + a_1(u^o,v) = 0,$$

then

$$\int_B \frac{\partial u_1^o}{\partial n} vd(B) = 0 \text{ for } v \in Y_o,$$

and finally we obtain the boundary value problem

$$-\Delta u_1^o = 0 \text{ in } D_1$$

$$u_1^o = 0 \text{ on } \Gamma$$

$$u_1^o = u_i^o \text{ on } B_i$$

$$\int_{B_i} \frac{\partial u_1^o}{\partial n} d(B_i) = 0.$$

We take the third equation

$$a_1(u^o,v) + a_o(u^1,v) = 0$$

and express the solution of the bilinear form $a_o(u^1,v)$ by the solution of the problem $-\Delta u_1^o = 0$ in D_1, applying Green's formula

$$-\int_{B\cup\Gamma} \frac{\partial u_1^o}{\partial n} v\, ds + a_1(u^o,v) = 0$$

hence

$$a_o(u^1,v) = -\int_{B\cup\Gamma} \frac{\partial u_1^o}{\partial n} v\, ds,$$

put it into the equation above, we obtain

$$a_o(u^1,v) - \int_{B\cup\Gamma} \frac{\partial u_1^o}{\partial n} v\, ds,$$

and the corresponding boundary value problem follows

$$-\Delta u_i^1 = 0 \text{ in } D_i$$

$$\frac{\partial u_i^1}{\partial n} = \frac{\partial u_1^o}{\partial n} \text{ on } B_i.$$

Now take the fourth equation

$$a_1(u^1,v) + a_o(u^2,v) = 0 \text{ for } v \in V.$$

As $v \in Y_o$, it is $a_o(u^2,v) = 0$ and also $a_1(u^1,v) = 0$ in D_1, this means that u^1 is a solution of $-\Delta u_1^1 = 0$ in D_1. By assumption wehave $u_1^1 = 0$ on Γ and by construction $u_1^1 = u_i^1$ in B_i.

Applying Green's formula to the bilinear form, we have

$$-\int_{B\cup\Gamma} \frac{\partial u^1}{\partial n} v \, ds + a_1(u^1,v) = 0.$$

As $a_1(u^1,v) = 0$, we obtain

$$\int_{B_i} \frac{\partial u_1^1}{\partial n} ds = 0 \text{ for } i = 1,\dots,q,$$

and finally we obtain the boundary value problem in the q+1 connected domain D_1

$$-\Delta u_1^1 = 0 \text{ in } D_1$$

$$u_1^1 = u_i^1 \text{ on } B_i \text{ for } i = 1,\dots,q$$

$$u_1^1 = 0 \quad \text{on } \Gamma$$

$$\int_{B_i} \frac{\partial u_1^1}{\partial n} ds = 0 \text{ for all } i.$$

Closed, analytic solutions of multiconnected boundary value problems in D_1 can be constructed by methods given in (Grusa (141)).

I.6.2 Optimal control in a multiconnected domain

We consider a simple model of certain optimal control problems in a multiconnected domain.

Let $\{D_k\}_k$ for $k = 0,1,\ldots,n$ be a family of simple connected bounded domains in $\mathbb{R}^n$ with smooth boundaries ∂D_k for $k = 0,1,\ldots,n$. Assuming $\bar{D}_k \subset D_o$ for all k, then

$$D = D_o \setminus \bigcup_{k=1}^{n} \bar{D}_k$$

is an n-connected domain with boundary $\partial D = \partial D_o \bigcup_{k=1}^{n} \partial D_k$.

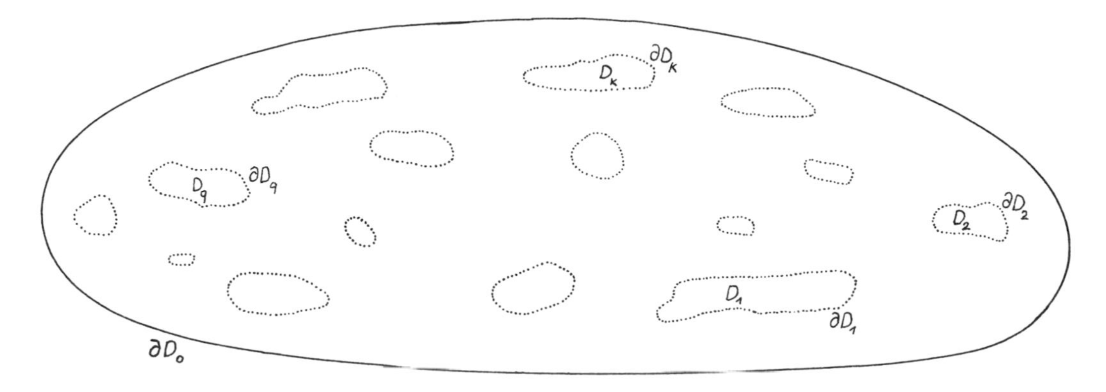

We are able to consider optimal control problems in n-connected domains and their explicit solutions. We restrict ourselves to comparatively simple examples in this part.

However our methods of model-building are quite general and hence more general problems can be solved by the reader using our modelling-methods. The following model can be seen under the heading of the minimization of positive definite or semi-definite quadratic forms defined on a closed, convex subset of a Hilbert space.

The optimal control problem in which the control variable is C, and C is unique solution of

$$F(\tilde{c}) \leq F(c) \text{ for all } c \in \mathbb{R}^n \text{ and } \tilde{c} \in \mathbb{R}^n,$$

where the cost functional

$$F(c) = \frac{1}{2}\int_D |\Delta\phi|^2 dx - \int_D f\cdot\phi dx - \sum_{k=1}^{n} \mu_k C_k,$$

and the state variable satisfies the following state equation

$$\Delta\Delta\phi = f \quad \text{in the n-connected domain } D$$

$$\phi = g_{1o} \text{ on the outer boundary } \partial D_o$$

$$\phi = g_{1k} + C_k \text{ on the boundaries } \partial D_k \text{ for } k = 1,\dots,n$$

$$\frac{\partial\phi}{\partial n} = g_2 \quad \text{on the boundary } \partial D.$$

This can be expressed in an equivalent form.
Let $\tilde{C}$ be the unique solution of the optimal control problem

$$F(\tilde{c}) = \min_{c \in \mathbb{R}^n} F(c),$$

where the cost functional is given by

$$F(c) = \frac{1}{2}\int_D |\Delta\phi|^2 dx - \int_D f\cdot\phi dx - \sum_{k=1}^{n} \mu_k c_k$$

and ϕ depends on C via the state equation (the biharmonic boundary value problem) in the equivalent form

$$\min_{\psi \in N_g} S(\psi),$$

with the closed, convex subset N_g of the Hilbert space H^2,

$$N_g = \{\psi \in H^2(D) \,|\, \psi|_{\partial D_o} = g_{1o}, \psi|_{\partial D_k} = g_{1k} + \text{const.} \quad \text{for } k = 1,\dots,n,$$
$$\text{and } \frac{\partial\psi}{\partial n} = g_2\}$$

and the cost functional

$$S(\psi) = \frac{1}{2}\int_D |\Delta\psi|^2 dx - \int_D f\cdot\psi dx - \sum_{k=1}^{n} \mu_k c_k,$$

with

$$\psi|_{\partial D_k} = g_{1k} + c_k \text{ for } k = 1,\dots,n.$$

Therefore the unique solution $\tilde{\psi}$ of the minimization problem

$$S(\tilde{\psi}) = \min_{\psi \in N_g} S(\psi)$$

is given and satisfies

$$\tilde{\psi}|_{\partial D_k} - g_{ik} = \tilde{C}_k.$$

The equivalence is clear, hence it remains to be shown, that if ψ satisfies the biharmonic problem, then

$$F(\tilde{c}) \leqq F(c) \text{ for } c \in \mathbf{R}^n.$$

Let $c \in \mathbf{R}^n$ and ψ be the solutions of the biharmonic problem, then $\psi \in N_g$ by construction, and $S(\tilde{\psi}) = \min_{\psi \in N_g} S(\psi)$, where $\psi \in N_g$. The cost functional can be written as

$$F(c) = S(\phi),$$

and by construction

$$S(\phi) \geqq \min_{\psi \in N_g} S(\psi) = S(\tilde{\psi}) \text{ for } c \in \mathbf{R}^n.$$

We know (page 6), that

$$C_k^\infty(D) \subset N_o = \{\phi \in H^2(D) | \frac{\partial \phi}{\partial n}|_{\partial D} = 0; \ \phi|_{\partial D_o} = 0; \ \phi|_{\partial D_k} = 0 \text{ for } k = 1,\dots,n\},$$

hence the biharmonic problem reduces to

$$\int_D \Delta\tilde{\psi} \, \Delta\phi \, dx = \int_D f \cdot \phi \, dx \text{ for all } \phi \in C_k^\infty(D),$$

this implies (by integration by parts)

$$\Delta\Delta\tilde{\psi} = f \quad \text{in} \quad D.$$

By construction $\tilde{\psi} \in N_g$, thus $\frac{\partial \tilde{\psi}}{\partial n}|_{\partial D} = g_2$ and $\tilde{\psi}|_{\partial D_o} = g_{1o}$, and we have $\tilde{\psi}|_{\partial D_k} = g_{1k} + \tilde{c}_k$ for $k = 1,\dots,n$, applying S we obtain

$$S(\tilde{\psi}) = F(\tilde{c}) \geqq \min_{c \in \mathbb{R}^n} F(c)$$

and finally

$$F(\tilde{c}) \leqq F(c) \text{ for all } c \in \mathbb{R}^n, \text{ where } \tilde{c} \in \mathbb{R}^n.$$

Lemma:

The uniq ue solution of the minimization problem is also a solution of the linear system

$$\frac{\partial F}{\partial c_k}(\tilde{c}) = 0 \text{ for } k = 1,\dots,.$$

The matrix of this linear system is symmetric, positive definite and can be computed as follows

$$\frac{\partial F}{\partial c_k}(c) = -\int_{\partial D_k} \frac{\partial \Delta\psi}{\partial n} d(D_k) - \mu_k \text{ for } k = 1,\dots,n.$$

Proof:

Let C and $\delta C \in \mathbb{R}^n$, then by definition

$$F(c) = \frac{1}{2}\int_D |\Delta\psi|^2 dx - \int_D f\cdot\psi dx - \sum_{k=1}^{n} \mu_k c_k$$

$$F(\delta c) = \frac{1}{2}\int_D |\Delta\delta\psi|^2 dx - \int_D f\cdot\delta\psi dx - \sum_{k=1}^{n} \mu_k\, \delta c_k$$

$$F(c+\delta c) = \frac{1}{2}\int_D |\Delta(\psi+\delta\psi)|^2 dx - \int_D f(\psi+\delta\psi) dx - \sum_{k=1}^{n} \mu_k (c_k+\delta c_k)$$

then

$$\frac{F(c+t\delta c)-F(c)}{t} = \frac{1}{t}\{\frac{1}{2}\int_D t\cdot 2\Delta\psi\cdot\Delta\delta\psi dx - t\int_D f\cdot\delta\psi dx - t\sum_{k=1}^{n} \mu_k \delta C_k +$$

$$+ \frac{1}{2}\int_D t^2 |\Delta\delta\psi|^2 dx\}.$$

Let $\delta\psi$ be the unique solution in $H^2(D)$ of the boundary value problem

$$\Delta\Delta(\delta\psi) = 0 \text{ in } D$$

$$\delta\psi|_{\partial D_o} = 0, \ \frac{\partial\delta\psi}{\partial n}\Big|_{\partial D} = 0$$

$$\text{but } \delta\psi|_{\partial D_k} = \delta C_k \text{ for } k = 1,\ldots,n,$$

we finally obtain the expression

$$\lim_{\substack{t\to 0\\ t\neq 0}} \frac{F(c+t\delta C)-F(C)}{t} = \int_D \Delta\psi\Delta\delta\psi dx - \int_D f(\delta\psi)dx - \sum_{k=1}^{n} \mu_k \ \delta C_k$$

denoted by

$$\operatorname{grad} F(C)\cdot\delta C$$

Applying Green's formula, we obtain

$$\int_D \Delta\psi\Delta\delta\psi \ dx = \int_D \Delta\Delta\psi\cdot\delta\psi \ dx + \int_{\partial D} \Delta\psi \ \frac{\partial\delta\psi}{\partial n} \ d(\partial D) - \int_{\partial D} \frac{\partial\Delta\psi}{\partial n} \ \delta\psi \ d(\partial D).$$

We have $\frac{\partial\delta\psi}{\partial n}\Big|_{\partial D} = 0$ and $\delta\psi = \delta C_k$ on ∂D_k for $k = 1,\ldots,n$, then the boundary integral can be written as

$$0 - \int_{\partial D} \frac{\partial\Delta\psi}{\partial n} \cdot\delta C_k \ d(\partial D).$$

Let ψ be the solution of the biharmonic equation

$$\Delta\Delta\psi = f \text{ in } D$$

$$\psi|_{\partial D_o} = g_{1o}, \ \psi|_{\partial D_k} = g_{1k} + C_k, \ \frac{\partial\psi}{\partial n}\Big|_{\partial D} = g_2,$$

by defining $\omega = -\Delta\psi$, we obtain

$$-\int_{\partial D} \frac{\partial\Delta\psi}{\partial n} \ \delta C_k \ d(\partial D) = \int_{\partial D} \frac{\partial\omega}{\partial n}\cdot\delta C_k d(\partial D),$$

and

$$\int_D \Delta\psi\Delta\delta\psi\, dx = \int_D f\cdot\delta\psi\, dx + \int_{\partial D} \frac{\partial\omega}{\partial n}\,\delta C_k\, d(\partial D),$$

using this in grad $F(c)\cdot\delta C$ we finally have

$$\text{grad } F(c)\cdot\delta C = \lim_{\substack{t\to 0\\ t\neq 0}} \frac{F(c+t.\delta c)-F(c)}{t} = \int_{\partial D} \frac{\partial\omega}{\partial n}\,\delta C_k d(\partial D) - \sum_{k=1}^{n} \mu_k\delta C_k =$$

$$= \sum_{k=1}^{n} \left(\int_{\partial D_k} \frac{\partial\omega}{\partial n}\, d(\partial D_k) - \mu_k\right)\delta C_k, \text{ for } \delta C_k \in \mathbf{R}^n.$$

Introducing a function Ξ defined on the boundary ∂D for $k = 0,1,\ldots,n$

$$\Xi_k\Big|_{\partial D_i} = \delta_{ki} = \begin{cases} 1 & \text{for } k = i \\ 0 & \text{for } k \neq i \end{cases}$$

then

$$\text{grad } F\cdot\delta C = \sum_{k+1}^{n} \left(\int_{\partial D_k} \frac{\partial\omega}{\partial n}\,\Xi_k\, d(\partial D_k) - \mu_k\right)\delta C_k.$$

Applying Green's formula

$$\int_D \text{grad}\omega\cdot\text{grad}\Xi_k dx = -\int_D \Delta\omega\ .\ \Xi_k dx + \int_{\partial D} \frac{\partial\omega}{\partial n}\,\Xi_k\, d(\partial D), \text{ as } \Delta\omega = -f,$$

$$= \int_D f.\ \Xi_k\, dx + \int_{\partial D} \frac{\partial\omega}{\partial n}\,\Xi_k\, d(\partial D),$$

hence

$$\int_{\partial D} \frac{\partial\omega}{\partial n}\,\Xi_k\, d(\partial D) = \int_D \text{grad}\omega\cdot\text{grad}\ \Xi_k dx - \int_D f.\Xi_k\, dx,$$

and the expression of grad $F\cdot\delta C$ reads

$$\text{grad}F\cdot\delta C = \sum_{k=1}^{n} \int_{\partial D_k} \frac{\partial\omega}{\partial n}\cdot\Xi_k\delta C_k\, d(\partial D_k) - \mu_k\delta C_k =$$

$$\sum_{k=1}^{n} \left(\int_D \text{grad}\omega\cdot\text{grad}\ \Xi_k\, dx - \int_D f\cdot\Xi_k dx - \mu_k\right)\delta C_k.$$

This expression may be rewritten as follows:

Let $\omega = \omega_h$ be a solution of the boundary value problem

$$\Delta\Delta\omega_h = 0 \text{ in } D$$

$$\frac{\partial\omega_h}{\partial n}\Big|_{\partial D} = 0, \quad \omega_h\Big|_{\partial D_k} = \delta_{h,k},$$

and $\omega = \omega_b$ be a solution of the boundary value problem

$$\Delta\Delta\omega_b = f \text{ in } D$$

$$\frac{\partial\omega_b}{\partial n}\Big|_{\partial D} = 0, \; \omega_b\Big|_{\partial D_k} = g_{hk}, \text{ where } C_k = 0.$$

Defining

$$A_k = -\sum_k \int_{\partial D_k} \frac{\partial\omega_b}{\partial n} \cdot \Xi_k d(\partial D_k) = \sum_k \int_D \text{grad } \omega_b \cdot \text{grad}\Xi_k dx - \sum_k \int_{\partial D_k} f.\Xi_k d(\partial D_k),$$

where $c_k = 0$, and

$$E_k = \sum_{k=1}^{n} \int_{\partial D_k} \frac{\partial\omega_h}{\partial n} \cdot \Xi_k d(\partial D_k) = \sum_k \int_D \text{grad}\omega_h \cdot \text{grad}\Xi_k dx \delta C_k = \sum_{k=1}^{n} E_{h,k}\delta C_k = E \cdot C$$

where the matrix

$$E_{h,k} = \int_D \text{grad}\omega_h \cdot \text{grad}\Xi_k d(dx) \quad \text{for } k,h = 1,\ldots,n$$

is symmetric, positive definite, and finally we have

$$\text{grad } F \cdot \delta C = E_k - A_k - \mu_k = E \cdot C - (A + \mu)$$

I.6.3 Result:

The optimal control problem

$$F(\tilde{c}) = \min_{c \in \mathbb{R}^n} F(c)$$

has the unique solution $\tilde{C}$, which is also a solution of the linear system

$$\frac{\partial F}{\partial c_k}(\tilde{C}) = 0.$$

We have shown that

$$\frac{\partial F}{\partial C_k}(C) = E \cdot C - (A + \mu),$$

hence $E \cdot \tilde{C} = A + \mu$ and finally $\tilde{C} = E^{-1}(A + \mu)$, where the matrix E is given by

$$E_{h,k} = \int_{\partial D_k} \frac{\partial \omega_h}{\partial n} d(\partial D_k) = \int_D \text{grad}\omega_h \cdot \text{grad } \Xi_k dx \text{ for } h,k = 1,\dots,n$$

and ω_n can be expressed by $\omega_h = -\Delta\psi_h$, where ψ_h is a solution of the boundary value problem

$$\Delta\Delta\psi_h = 0 \text{ in } D$$

$$\psi_h|_{\partial D_k} = \delta_{h,k} \text{ on } \partial D_k \text{ for } K = 0,1,\dots,n; \frac{\partial \psi_h}{\partial n}|_{\partial D} = 0.$$

Thus the matrix $E_{h,k}$ depends only on the domain D and is independent of the values f, g_{ik}. The expression

$$A_k = -\int_{\partial D_k} \frac{\partial \omega_b}{\partial n} d(\partial D_k) = -\int_D \text{grad}\omega_b \cdot \text{grad } \Xi_k - \int_{\partial D} f \cdot \Xi d(\partial D) \text{ for } k = 1,\dots,n$$

and ω_b given by $\omega_b = -\Delta\psi_b$, where ψ_b is the solution of the boundary value problem

$$\Delta\Delta\psi_b = f \text{ in } D$$

$$\psi_b|_{\partial D_k} = g_{1k} \text{ for } k = 1,\dots,n, \frac{\partial \psi_b}{\partial n}|_{\partial D} = g_2.$$

There follows, that the optimal solution of the control problem is given by

$$\tilde{C} = E^{-1}(A + \mu),$$

thus we are able to compute $\tilde{C}_k$ for $k = 1,\dots,n$ explicitly.

The solution $\tilde{\psi}$ of the biharmonic problem $S(\tilde{\psi}) = \min_{\psi \in N_g} S(\psi)$ is given by

$\Delta\Delta\psi = f$ in the n-connected domain D

$\psi = g_{1o}$ on the outer boundary ∂D_o

$\psi = g_{1k} + \tilde{C}_k$ on the boundaries ∂D_k, where $\tilde{C}_k$ is the optimal control

$\frac{\partial \psi}{\partial n} = g_2$ on the boundary D.

$\tilde{C}_k$ are the values of the optimal control computed above, and applied to the boundary values on ∂D_k for $k = 1,\ldots,n$.

$\tilde{C}$ depends:

on the multiconnected region via E,

on the fixed cost on the boundaries ∂D_k given by μ_k.

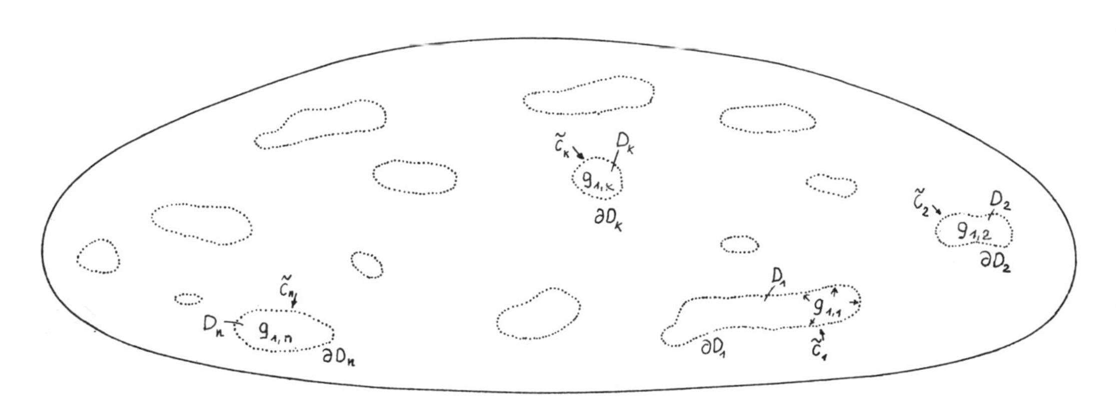

I.7 Control applied to points in $\mathbb{R}^3$

1.7.1 Existence problems

Let D be an open set in $\mathbb{R}^3$ with smooth boundary ∂D. Let there be a finite set of internal points in D, denoted by $a_1, a_2, \ldots, a_m$. We consider controls concentrated on the internal points in D. Therefore we denoted by $\delta(x-a_i)$ the Dirac's measure at the point a_i for $i = 1,\ldots,m$, and $v_i \in L^2(0,T)$ are the control parameters at a_i and f is given in $L^2(D)$. Thus, the expression of the control is given by

$$\sum_{i=1}^{m} v_i(t)\delta(x-a_i).$$

For any $v_i \in L^2(0,T)$ the control v is denoted by $v = (v_1,\ldots,v_m) \in (L^2(0,T))^m$.

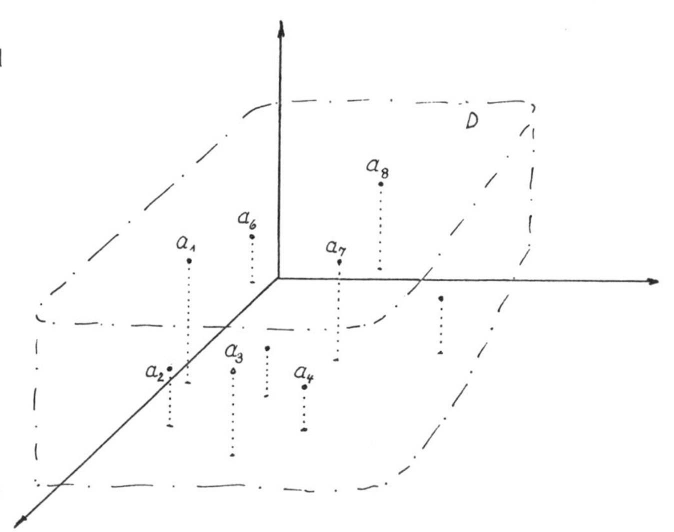

The state of the parabolic system (e.g. describing general diffusion processes which are now controlled) is given by

$$\frac{\partial y}{\partial t} - \Delta y = f + \sum_{i=1}^{m} v_i(t)(\delta(x-a_i) \text{ in } Dx(0,T)$$

$$y = 0 \quad \text{on the boundary } \partial Dx(0,T)$$

$$y(x,0) = y_0(x) \text{ the initial condition in } D.$$

We have to show that there exists a unique solution y, depending on the control v, $y(t,x,y_0,v)$.

Introducing $g \in L^2(Dx(0,T))$, then there exists a solution φ of the initial-boundary-value problem

$$-\frac{\partial \varphi}{\partial t}\Delta\varphi = g \quad \text{in } Dx(0,T)$$

$$\varphi = 0 \quad \text{on } \partial Dx(0,T)$$

$$\varphi(x,T) = 0 \quad \text{in } D.$$

Multiplying the state equation by φ and integrating over $Dx(0,T)$, we have

$$\int_D \int_0^T \frac{\partial y}{\partial t}\varphi\, dtdx - \int_D \int_0^T \Delta y \cdot \varphi\, dtdx = \int_D \int_0^T f \cdot \varphi\, dtdx + \\ + \sum_{k=1}^{m} \int_D \int_0^T v_i(t) \cdot \delta(x-a_i)\varphi(x,t)\, dtdx.$$

Integration by parts

$$\int_0^T \frac{\partial y}{\partial t} \cdot \varphi\, dt = y(x,t)\cdot\varphi(x,t)\Big|_{t=0}^{t=T} - \int_0^T y\, \frac{\partial \varphi}{\partial t}\, dt$$

$$= y(x,T)\varphi(x,T) - y(x,0)\cdot\varphi(x,0) - \int_0^T y\, \frac{\partial \varphi}{\partial t} dt, \text{ as } \varphi(x,T) = 0 \text{ and } y(x,0) = y_0,$$

we obtain by integrating with respect to D

$$\int_D \int_0^T \frac{\partial y}{\partial t} \cdot \varphi\, dtdx = -\int_D y_0(x)\varphi(x,0)dx - \int_D \int_0^T y\, \frac{\partial \varphi}{\partial t} \cdot dtdx.$$

As $y = 0$ and $\varphi = 0$ on the boundary $\partial Dx(0,T)$, we obtain

$$-\int_{Dx(0,T)} \Delta y \cdot \varphi\, dtdx = -\int_{Dx(0,T)} y \cdot \Delta\varphi\, dtdx = \int_{Dx(0,T)} y \cdot \frac{\partial \varphi}{\partial t} dxdt + \int_{Dx(0,T)} y \cdot g\, dxdt.$$

We have

$$\int_D \varphi(x,t)\delta(x-a_i)dx = \varphi(a_i,t),$$

and then

$$\sum_{i=1}^{m} \int_D \int_0^T v_i(t)\cdot\varphi(x,t)\delta(x-a_i)dxdt = \sum_{i=1}^{m} \int_0^T v_i(t)\varphi(a_i,t)dt.$$

Using all this in the first expression, we obtain

$$\int_{Dx(0,T)} g\cdot y(x,t)dxdt = \int_D y_0(x)\varphi(x,0)dx + \int_{Dx(0,T)} f\cdot\varphi dxdt + \sum_{i=1}^{m}\int_0^T v_i(t)\varphi(a_i,t)dt$$

Denoting the right-hand side by $L(\varphi)$, we are able to show that the functional $L(\varphi)$ is continuous on $L^2(D)$, thus

$$\int_{Dx(0,T)} y\cdot g \; dxdt = L(\varphi)$$

has a unique solution.

Applying the representation theorem of Riesz ((157), p. 234f.), there exists a unique solution $y \in L^2(D)$, which is also a solution of the state equation.

Lemma:

The functional $L(\varphi)$ is continuous on $L^2(D)$.

The solution φ of the reduced problem $\frac{\partial\varphi}{\partial t} - \Delta\varphi = g$ in $Dx(0,T)$ satisfies

$$\varphi \in L^2((0,T);H^2(D)); \; \frac{\partial\varphi}{\partial t} \in L^2(Dx(0,T)), \text{ then } \varphi(\cdot,0) \in L^2(D) \quad \text{and}$$

$$\left\|\frac{\partial\varphi}{\partial t}\right\|_{L^2(Dx(0,T))} + \|\varphi\|_{L^2((0,T);H^2(D))} \leqq C\,\|g\|_{L^2(Dx(0,T))}. \qquad (216)$$

Since the dimension of the space $n \leqq 3$, we have, by the imbedding theorem of Sobolev,

$$H^2(D) \subset C^0(\bar{D}) \text{ and thus } \varphi(a_i,t) \in L^2(0,T)$$

and

$$\int_0^T |\varphi(a_i,t)|^2 dt \leqq C\,\|g\|_{L^2(Dx(0,T))}.$$

Finally the linear functional $L(\varphi)$ can be estimated by

$$|L(\varphi)| \leq c(\|\varphi\|_{L^2(Dx(0,T))} + \|\varphi(\cdot,0)\|_{L^2(D)} + \sum_{i=1}^{m} \|\varphi(a_i,t)\|_{L^2(0,T)} \leq$$

$$\leq c \cdot \|g\|_{L^2(Dx(0,T))}$$

and is continuous on $L^2(Dx(0,T))$.

The explicit solution $y = y(x,t,v,a_i)$ depends on the distribution of the points a_i in D and the corresponding control parameter.

Now, we can introduce a cost function

$$J(v,a) = \int_{Dx(0,T)} |y(x,t,v,a)-z_a(x,t)|^2 dxdt + \sum_{i=1}^{m} M_i \int_0^T |v_i(t)|^2 dt$$

and the corresponding optimal control problem is given by

$$\inf_v J(v,a),$$

for $v \in U_{cc}$ = a closed, convex set in $(L^2((0,T)))^m$.

We prove that there exist $\bar{v} \in U_{cc}$, $\bar{a} \subset B^m$ (see page 77), where the a_i are rearranged in a, such that

$$J(\bar{v},\bar{a}) = \inf_{\substack{v \in U_{cc} \\ a \in B^m}} J(v,a).$$

We have to show that

$$J(\bar{v},\bar{a}) \leq J(v,a) \text{ for all } v \in U_{cc},\ a \in B^m.$$

By construction of the cost function, we get

$$J(v,a) \geq \inf_i M_i \sum_{i=1}^{m} \int_0^T |v_i(t)|^2 dt,$$

Let $(v_i)_\mu$ and $(a_i)_\mu$ be minimizing sequences, then $(v_i)_\mu$ remains in a bounded set of U_{cc} by the above estimate.

We can extract a subsequence, also denoted by $(v_i)_\mu$, such that

$(v_i)_\mu \to \bar{v}$ in $(L^2(0,T))^m$ weakly and $u \in U_{cc}$

$(a_i)_\mu \to \bar{a}$ in B^m.

In the sense of distributions we have

$$(v_i)_\mu \ (\delta(x-(a_i)_\mu)) \to \bar{v}(\delta-a).$$

Let $(y)_\mu$ be the solution of the state equation with respect to the control variable $(v_i)_\mu$ and the points $(a_i)_\mu$, then the functional relation on page 70 reads

$$\int_{Dx(0,T)} g\cdot y_\mu(x,t)dtdx = \int_{Dx(0,T)} f\cdot\varphi dxdt + \int_D y_o(x)\cdot\varphi(x,0)dx + \sum_{i=1}^{m}\int_0^T (v_i)_\mu(t)\cdot\varphi((a_i)_\mu,t)dt.$$

By taking the limits

$$y_\mu \to y \quad \text{in } L^2(Dx(0,T)) \text{ weakly}$$

$$(v_i)_\mu \to v \quad \text{in } (L^2(0,T))^m \quad \text{weakly}$$

$$(a_i)_\mu \to a \quad \text{in } B^m = \mathop{X}_{i=1}^{m} B_i$$

we obtain

$$\int_{Dx(0,T)} g\cdot ydtdx = \int_{Dx(0,T)} \varphi \ f\cdot \ dxdt + \int_D y_o(x)\varphi(x,0)dx + \sum_{i=1}^{m}\int_0^T \bar{v}_i(t) \ (\bar{a}_i,t)dt,$$

thus, there exists the solution $y = y(\bar{v}_i,\bar{a}_i) = y(\bar{v},\bar{a})$, and finally we have

$$\liminf_\mu J((v_i)_\mu, (a_i)_\mu) \geqq J(\bar{v},\bar{a}).$$

Thus, the control problem is solved. Now we generalize the problem.

I.7.2 Control applied to balls in $\mathbb{R}^3$: Existence problem

We suppose that the points a_i satisfy $a_i \in B_i$, where B_i are disjoint, bounded and closed sets in D. We choose

B_i = a ball with centre a_i and radius r, with finite volume.

We set $a = (a_1,a_2,\ldots,a_m) \in B^m = \mathop{X}_{i=1}^{m} B_i$.

Let V be a Hilbert space with norm $\| \ \|$ and H be a Hilbert space with norm $| \ |$, such that the inclusion

$V \subset H$ is continuous and dense.

Then

$$V \subset H \cong H' \subset V'$$

by identifying H with its dual H'; thus $V \subset V'$ is continuous and dense. We assume, that there exists $\lambda \in \mathbb{R}$ and $\alpha > 0$, such that the bilinear form satisfies

$$a(\varphi,\varphi) + \lambda|\varphi|^2 \geqq \alpha \,\|\varphi\|^2 \text{ for } \varphi \in V.$$

Let $\mathcal{B}_i$ be a family of operators $\mathcal{B}_i \in L(V,V')$ such that

$$|(\mathcal{B}_i\varphi,\varphi)| \leqq C\cdot|\varphi| \cdot \|\varphi\| \cdot$$

In applications we choose

$$\mathcal{B}_i\varphi = \frac{1}{\mathrm{vol}(B_i)} \int_{B_i} \varphi(x)dx.$$

We consider the initial boundary-value problem

$$\frac{\partial y}{\partial t} - \Delta y = f + \sum_{i=1}^{m} v_i(t)\cdot \mathcal{B}_i(y) \text{ in } Dx(0,T)$$

$$y = 0 \quad \text{on } \partial Dx(0,T)$$

$$y(x,0) = y_0(x) \text{ in } D,$$

written in variational form

$$(\tfrac{\partial y}{\partial t},\varphi) + a(y,\varphi) - \sum_{i=1}^{m} v_i(t)(\mathcal{B}_i y,\varphi) = (f,\varphi) \text{ for } \varphi \in V.$$

By known methods (see Lions, Magenes (216)) we obtain a

Theorem:

If $f \in L^2((0,T),V')$, $v_i \in L^2(0,T)$, $y_0 \in H$, then the solution y satisfies

$$y \in L^2((0,T),V); \ \frac{\partial y}{\partial t} \in L^2((0,T),V').$$

Now, we define the corresponding cost function

$$J(v) = \int_0^T |y(t,v)-z_d(t)|^2 dt + \sum_{i=1}^{m} M_i \int_0^T |v_i(t)|^2 dt,$$

where the constant $M_i > 0$, and $z_d(t)$ are the time-dependent fixed costs,

$z_d \in L^2((0,T),H)$. Let U_{cc} be a closed, convex subset of $(L^2(0,T))^m$, we have the

Theorem:

The optimal control

$$J(u) = \inf_{v \in U_{cc}} J(v)$$

has unique solution $u \in U_{cc}$.

By taking $\varphi = y$ the variational form reads

$$(\frac{\partial y}{\partial t}, y) + a(y,y) - \sum_i v_i(t)(B_i y \cdot y) = (f,y).$$

According to our assumptions we have

$$-|(B_i y \cdot y)| \geqq -c \cdot |y| \cdot \|y\|$$

and

$$a(y,y) \geqq \alpha \|y\|^2 - \lambda |y|^2 \quad |(f,y)| \leqq \|f\|^2_{V'} \cdot \|y\|^2,$$

applied to the variational form

$$\frac{1}{2}\frac{d}{dt} |y(t)|^2 + \alpha \|y(t)\|^2 - \lambda |y(t)|^2 - c \cdot \sum_i |v_i(t)| \cdot |y| \cdot \|y\| \leqq \ldots \leqq$$
$$|(f,y)| \leqq \|f\|^2_{V'} \cdot \|y\|^2,$$

hence

$$\frac{1}{2}\frac{d}{dt} |y(t)|^2 + \alpha \cdot \|y(t)\|^2 \leqq \|f(t)\|^2_{V'} \cdot \|y(t)\|^2 + \lambda |y(t)|^2 +$$
$$+ c \cdot \sum_i |v_i(t)| \cdot |y(t)| \cdot \|y(t)\|.$$

As

$$\frac{1}{2}\int_0^t \frac{d}{dt} |y(t)|^2 dt = |y(t)|^2 - |y(0)|^2 = |y(t)|^2 - |y_0(x)|^2,$$

with the initial value $y_0(x)$ and $\sum_i v_i(t) \leqq c(1 + v^2(t))$. We find

$$|y(t)|^2 + \alpha \int \|y(t)\|^2 dt \leqq |y_0(x)|^2 + \int_0^t \|f(t)\|^2 \cdot \|y(t)\|^2 dt +$$
$$+ \int_0^t (\lambda + C(1 + v^2(\sigma)))|y(\sigma)|^2 d\sigma,$$

thus

$$|y(t)|^2 < |y(0)|^2 + \int_0^t \{\|f(t)|_{V'}^2 + (\lambda+c(1+v^2(\sigma))\}\cdot|y(t)|^2 dt;$$

and applying Gronwall's inequality, which holds for any function $E(t) \geqq 0$ and constant $C > 0$,

$$E(h) \leqq E(0) + c\cdot\int_0^h E(t)dt$$

implies

$$E(h) \leqq \exp(c\cdot h)\cdot E(0).$$

We obtain finally

$$|y(t)|^2 \leqq |y(0)|^2\cdot\exp\left(\int_0^T (\|f\|_{V'}^2 + \lambda + c(1+v^2(\sigma))d\sigma\right),$$

defining the right-hand side by Ξ we obtain

$$\int \|y(t)\|^2 dt < \frac{1}{\alpha}\cdot\Xi,$$

this means, that the solution y remains in a bounded set of $L^2((0,T),V)$. For considering the control problem, we introduce

$$Q(0,T) = \{\varphi|\varphi \in L^2((0,T),V), \frac{\partial\varphi}{\partial t} \in L^2((0,T),V')\}$$

with the norm

$$\left(\int_0^T (\|\varphi(t)\|^2 + \|\frac{\partial\varphi}{\partial t}\|_{V'}^2)dt\right)^{1/2}.$$

Defining the operator $\mathcal{B}_i$

$$\mathcal{B}_i : Q(0,T) \to L^2((0,T),V')$$

$$\varphi \to (\mathcal{B}_i\varphi)(t) = \mathcal{B}_i(\varphi(t)),$$

and let $\mathcal{B}_i$ be compact, the problem is solved in the following way.

Let v_μ be a minimizing sequence of the control variable, bounded in $(L^2(0,T))^m$, then there exists a unique solution $y(v_\mu)$ of the state equation, denoted by $y(v_\mu) = y_\mu$.

Applying our considerations, we have

$$\int \|y_\mu(t)\|^2 dt < \frac{1}{\alpha}\cdot\Xi,$$

thus y_μ remains in a bounded set of $L^2((0,T),V)$.

Using the variational formulation, we obtain

$$\frac{\partial y_\mu}{\partial t} \text{ remains in a bounded set of } L^2((0,T),V').$$

By definition of the set Q, the solution

$$y_\mu \text{ remains in a bounded set of } Q(0,T).$$

Thus, we are able to extract a subsequence, again denoted by y_μ, such that

$$y_\mu \to y \quad \text{in } Q(0,T) \text{ weakly},$$

$$v_\mu \to u \quad \text{in } (L^2(0,T))^m \text{ weakly}.$$

As the $\mathcal{B}_i$ are compact, we obtain by a known theorem, that

$$\mathcal{B}_i y_\mu \to \mathcal{B}_i y \text{ in } L^2((0,T),V') \text{ strongly},$$

$$(v_i)_\mu \mathcal{B}_i y_\mu \to u_i \mathcal{B}_i y \text{ in } L^1((0,T),V') \text{ weakly}.$$

By taking the limit in

$$(\frac{\partial y_\mu}{\partial t}, y_\mu) + a(y_\mu, y_\mu) + \sum_i (v_i)_\mu \cdot (\mathcal{B}_i y_\mu, y_\mu) = (f, y_\mu)$$

we have

$$(\frac{\partial y}{\partial t}, y) + a(y,y) + \sum_i u_i(t) \cdot (\mathcal{B}_i y, y) = (f,y),$$

thus, the unique solution depends on u, hence $y = y(u)$, and we obtain

$$\liminf_\mu J(v_\mu) \geqq J(u),$$

One possible application of the theorem.

We define the operators $\mathcal{B}_i$ by

$$\mathcal{B}_i \varphi = \frac{1}{\mathrm{Vol.}(B_i)} \int_{B_i(a_i)} \varphi(x) dx,$$

where $B(a_i)$ = the ball with centre a_i and radius r, then $\mathcal{B}_i$ satisfies the inequality with constant $\frac{1}{\mathrm{Vol} \cdot (B_i)}$. We assume, that the distance (B_i, boundary of the region D) $\geqq$ r.

I.7.3 Optimal rearrangement processes within the balls in $\mathbb{R}^3$

Balls in D — Time - dependent control parameter

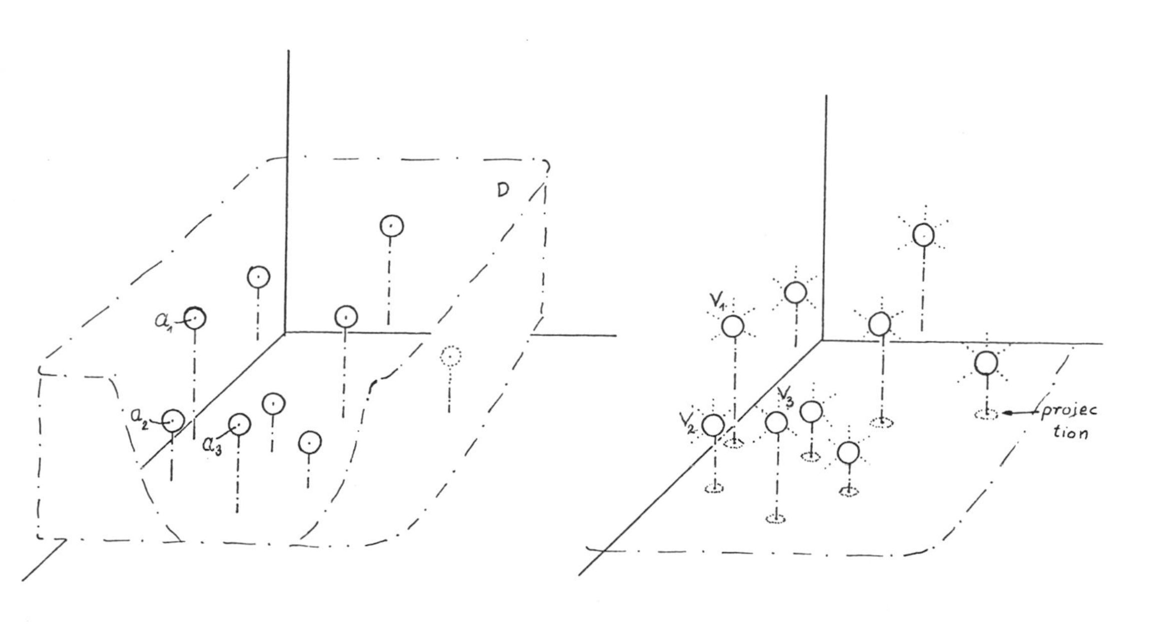

We set

$$a = (a_1, a_2, \dots, a_m) \in B^m = \mathop{X}_{i=1}^{m} B_i, \qquad \begin{matrix} (v_1)_\mu \to u_1 \\ (v_2)_\mu \to u_2 \\ \vdots \qquad \vdots \\ (v_m)_\mu \to u_m \end{matrix} \quad \text{and } u = (u_1, u_2, \dots, u_m) \in (L^2(0,T))^m.$$

The rearrangement of the points within the balls:

$$\begin{matrix} (a_1)_\mu \to \bar{a}_1 \\ (a_2)_\mu \to \bar{a}_2 \\ \vdots \qquad \vdots \\ (a_m)_\mu \to \bar{a}_m \end{matrix} \quad \text{and } \bar{a} = (\bar{a}_1, \bar{a}_2, \dots, \bar{a}_m) \in B^m = \mathop{X}_{i=1}^{m} B_i \text{ is the optimal position,}$$

Let $v = (v_1, v_2, \dots, v_m)$ be the given control parameter,

$a = (a_1, a_2, \dots, a_m)$ be the position of the balls in space.

The state equation reads

$$\frac{\partial y}{\partial t} - \Delta y = f + \sum_{i=1}^{m} v_i(t) \frac{1}{Vol \cdot B_i} \int_{B_i(a_i)} y(x)dx \text{ in } Dx(0,T)$$

$$y = 0 \text{ on } \partial Dx(0,T)$$

$$y(x,0) = 0 \text{ in } D,$$

the solution $y = y(x,t,v,a)$ is unique and determines the cost function

$$J(v,a) = \int_{Dx(0,T)} |y(x,t,v,a)-z_d(x,t)|^2 dxdt + \sum_{i=1}^{m} M_i \int_0^T |v_i(t)|^2 dt.$$

We are able to show that there exists $u \in (L^2(0,T))^m$ and $\bar{a} \in B^m$, such that

$$J(u,\bar{a}) \leqq J(v,a) \text{ for all } v \in (L^2(0,T)^m \text{ and } a \in B^m = \mathop{X}_{i=1}^{m} B_i,$$

in words, there exists an influence control u and a rearrangement of the points $(a_1,a_2,...,a_m)$ within the balls making the cost $J(v,a)$ minimal. Or we can say, that there exist two parameters (u,a), such that

u = the influence control,

a = the optimal position of the geographical points.

The rearrangement of the points $(a_1,a_2,...,a_m)$ to $(\bar{a}_1,\bar{a}_2,...,\bar{a}_m)$ will in general depend on the $z_d(x,t)$, seen as certain fixed desired costs.

Thus the optimal position a of the points $(a_1,a_2,...,a_m)$ will depend on the desired cost $z_d(,xt)$.

The rearrangement phenomena can also be extended by similar methods to systems described by nonlinear partial differential equations, with applications in economics, physics, biology,...

Proof:

We suppose that v_μ and a_μ be minimizing sequences. The cost function satisfies

$$J(v,a) \geqq \inf \sum_i M_i \int_0^T |v_i(t)|^2 dt,$$

by construction, hence v_μ remains in a bounded set of $(L^2(0,T))^m$. The operators $\mathcal{B}_i$ are chosen

$$\mathcal{B}_i y = \frac{1}{Vol \cdot B_i} \int_{B_i(a_i)} y(x)dx$$

and satisfy the estimate with the constant

$\frac{1}{Vol \cdot B_i}$.

The solution of the state equation (page 69) with respect to the minimizing sequences v_μ, a_μ given by $y = y(x,t,v_\mu,a_\mu)$ and according to the proof of the theorem on page 74 we obtain the solution (page 76)

$$y_\mu \to y \text{ in } Q(0,T) \text{ weakly,}$$

$$v_\mu \to u \text{ in } (L^2(0,T))^m \text{ weakly.}$$

Let $(a)_\mu \in B^m$ and $(a)_\mu \to \bar{a}$, where $\bar{a} \in B^m$, we have $y_\mu \to y$ in $Q(0,T)$ weakly, by using a compactness result of Lions ((219), chap. 1, §5.2) we obtain

$$y_\mu \to y \text{ in } L^2(Dx(0,T)) \text{ strongly}$$

and finally

$$\mathcal{B}_i y_\mu = \frac{1}{Vol \cdot B_i} \int_{B_i(a_i)} y_\mu(x),t)dx$$

converges in $L^2(Dx(0,T))$ strongly.

According to the proof on page 76 we only have to show that

$$(v_i)_\mu \mathcal{B}_i y_\mu \to u_i \mathcal{B}_i y \text{ in } L^1((0,T),V') \text{ weakly.}$$

We have to consider

$$(v_i)_\mu(t) \int_{B_i((a_i)_\mu)} y_\mu(x,t)dx - u_i(t) \int_{B_i(\bar{a}_i)} y(x,t)dx,$$

since $(v)_\mu \to u$ in $L^2(0,T)$ weakly, it remains to prove, dropping the index i,

$$\left|\int_{B(a_\mu)} y_\mu(x,t)dx - \int_{B(a)} y(x,t)dx\right| = \left|\int_{B(a_\mu)} y_\mu dx - \int_{B(a_\mu)} y\, dx + \right.$$

$$\left. + \int_{B(a_\mu)} y\, dx - \int_{B(a)} y\, dx\right| \leq$$

$$\leq \int_{B(a_\mu)} |y_\mu - y| dx + \int_{B(a_\mu)-B(a)} |y| \cdot 1\, dx$$

$$\leq \int_D |y_\mu - y|\, dx + \int_{B(a_\mu)-B(a)} |y| \cdot 1\, dx$$

As $y_\mu \to y$ in $L^2(Dx(0,T))$ strongly, we have

$$\int_D |y_\mu - y|^2 dx \to 0,$$

since $(a)_\mu \to \bar{a}$ in B^m, we have

$$\int_{B(a_\mu)-B(\bar{a})} y dx \to 0.$$

Applying the Cauchy-Schwarz inequality, then

$$\left|\int_{B(a_\mu)-B(\bar{a})} y \cdot 1 d\right| \leq \left(\int |y|^2 dx\right)^{\frac{1}{2}} \cdot \left(\int 1^2 dx\right)^{\frac{1}{2}} < c\left(\int_D |y|^2 dx\right)^{\frac{1}{2}} \text{ measure}$$

$$(B(a_\mu)-B(\bar{a}))^{\frac{1}{2}} = c \cdot \left(\int_D |y|^2\right)^{\frac{1}{2}} \cdot |a_\mu - \bar{a}|^{\frac{1}{2}}.$$

Thus we have proved that there exist $u \in (L^2(0,T))^m$ and $\bar{a} \in B^m$, such that

$$J(u,\bar{a}) \leqq J(v,a) \text{ for } v \in (L^2(0,T))^m \text{ and } a \in B^m.$$

I.7.4 Domain variation

Let D be an open, bounded set in $\mathbb{R}^n$ with subdomains D_μ, such that $D_\mu \subset D$, then there exists a fixed part Γ of the boundary ∂D, such that the boundary ∂D_μ of the subdomains is decomposed in

$$\partial D_\mu = \Gamma \cup F_\mu,$$

where F_μ is a free boundary.

The family of subdomains D_μ are satisfying if $\mu_n \to \bar{\mu}$ the

$$\text{distance } (D_{\mu_n}, D_{\bar{\mu}}) \to 0.$$

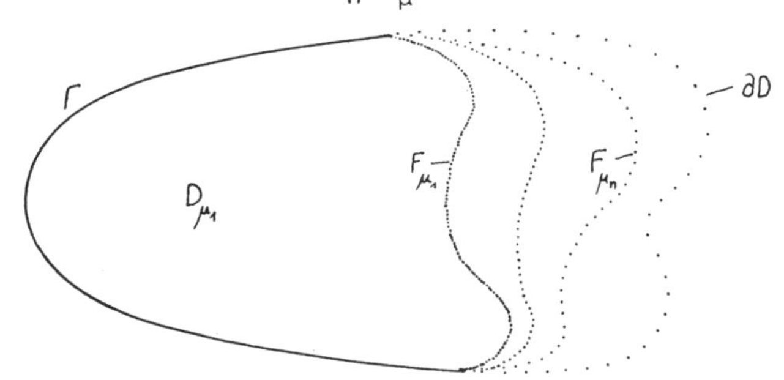

The state of the process governed by a parabolic differential equation in a subdomain D_μ is given by

$$\frac{\partial y}{\partial t} - \Delta y = f \quad \text{in } D_\mu x(0,T) \text{ restricting } f \text{ to } D_\mu$$

$$\frac{\partial y}{\partial n} = v \quad \text{on the fixed boundary } \Gamma x(0,T)$$

$$y = 0 \quad \text{on the free boundary } F_\mu x(0,T)$$

$$y(x,0) = y_0(x) \text{ in } D_\mu, \text{ restricting } y_0(x) \text{ to } D_\mu.$$

Thus, the control v is applied to the fixed boundary Γ. Since $f \in L^2(D_\mu x(0,T))$ and $v \in L^2(\Gamma x(0,T))$, $y_0 \in L^2(D)$, there exists a unique solution, depending on

$$y = y(x,t,v,\mu,f,y_0).$$

We are now interested in the solution as a function of the control parameter v, and the parameter μ describing the domain variation with respect to the free boundary, hence $y = y(v,\mu)$.

The initial-boundary-value problem can be rewritten in variational form

$$(\frac{\partial \varphi}{\partial t},\varphi) + a(\varphi,\varphi)_{D_\mu} = (f,\varphi)_{D_\mu} + \int_\Gamma v\cdot\varphi \, d\Gamma \text{ for } \varphi \in V_\mu,$$

where

$$V_\mu = \{u|u \in H^1(D_\mu); u = 0 \text{ on } F_\mu\}$$

and the solution satisfies

$$y \in L^2((0,T),V_\mu); \frac{\partial y}{\partial t} \in L^2((0,T),V_\mu') \text{ and } y(x,0) = y_0(x) \in L^2(D).$$

The cost function may be given by

$$J(v,\mu) = \int_{\Gamma x(0,T)} |y(x,t,v,\mu)-z_d(t)|^2 d\Gamma dt + M\int_{\Gamma\times(0,T)} |v|^2 d\Gamma dt, \text{ constant } M > 0.$$

We assume, that U_{cc} be a closed, convex subset of $L^2(\Gamma\times(0,T))$ and consider

$$\inf_{v\in U_{cc}} J(v,\mu) \text{ and } \mu \in [0,1].$$

We are able to prove, that there exist $u \in U_{cc}$ and $\mu \in [0,1]$ such that

$$J(u,\bar{\mu}) = \inf_{v\in u_{cc}} J(v,\mu) \text{ and } \mu \in [0,1],$$

$\bar{\mu}$ means, there exists an optimal domain $D_{\bar{\mu}}$ where the cost function is minimal.

There exists one possible application to diffusion processes and free boundary value problems.

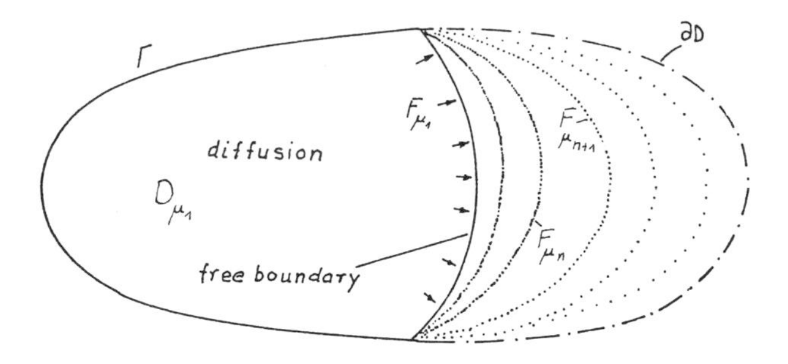

The domain variation can be interpreted in economics as expanding trade areas.

II Global control phenomena

II.1 Minimization problems and variational inequalities

Let $\mathcal{H}$ be a real Hilbert space with norm $\|\ \|$. We shall assume a continuous, symmetric, bilinear form B on $\mathcal{H}$ such that $u,v \to B(u,v)$ for $u,v \in \mathcal{H}$ and a continuous linear form

$$v \to O(v) \text{ on } \mathcal{H} \text{ and a closed, convex set } U_{cc} \text{ in } \mathcal{H}.$$

Introducing the quadratic functional

$$J(v) = B(u,v) - 2\cdot O(v), \text{ we have the}$$

Theorem 1

Let $B(u,v)$ be a continuous, symmetric bilinear form on $\mathcal{H}$, which satisfies

$$B(v,v) \geqq c\cdot \|v\|^2 \text{ for all } v \in \mathcal{H} \text{ and } c = \text{constant} > 0.$$

Then there exists a unique element u in U_{cc}, such that

$$J(u) = \inf_{v\in U_{cc}} J(v) .$$

Remark:

In possible applications (e.g. the optimal control theory) of the theory $\mathcal{H}$ will be the space of controls and U_{cc} the set of admissible controls where $J(v)$ is the known cost function (economic function).

Proof:

We take $\{v_n\} \in U_{cc}$ as a minimizing sequence, such that

$$J(v_n) \underset{n\to\infty}{\to} \inf_{v\in U_{cc}} J(v).$$

As O is a continuous linear form on $\mathcal{H}$, we have $\|O(v)\| \leq c_1\cdot \|v\|$ and by assumption

$$J(v) = B(v,v) - 2O(v) \geqq c. \|v\|^2 - c_1 \|v\|$$

thus we obtain

$$\inf J(v) \underset{n\to\infty}{\longleftarrow} J(v_n) \geqq c. \|v_n\|^2 - c_1 \|v_n\|, \text{ hence } \|v_n\| \leqq \text{constant.}$$

It is known in functional analysis, that if H is a reflexive Banach space and v_n is a sequence in H, such that v_n is bounded, then there exists a subsequence v_μ and an element w in H for which

$$(f,v_\mu) \xrightarrow[\mu\to\infty]{} (f,w) \text{ for } f \in H' = \text{the dual space of } H;$$

this means

$$v_\mu \to w \text{ in H weakly.}$$

As Hilbert spaces are reflexive, we may apply this.

Since U_{cc} is a closed convex set in $\mathfrak{H}$, U_{cc} is weakly closed, this implies that $w \in U_{cc}$.

The function $v \to B(v,v)$ is lower semi-continuous in the weak topology of $\mathfrak{H}$ and $v \to O(v)$ is continuous in the weak topology, thus

$$J(v) = B(v,v) - 2\cdot O(v)$$

is weakly lower semi-continuous, written as

$$\liminf_{v_\mu \to w} J(v_\mu) \geqq J(w).$$

So we obtain by the minimizing sequence

$$J(w) \leqq \inf_{v\in U_{cc}} J(v) \text{ and } w \in U_{cc},$$

then we must have $J(w) = \inf_{v\in U_{cc}} J(v)$, and taking u = w, the theorem is proved.

It can be shown that, if $v \to J(v)$ is a convex function from $U_{cc} \to \mathbf{R}$, such that

$$J(v) \to +\infty, \text{ as } \|v\| \to +\infty \text{ for } v \in U_{cc}$$

and $v \to J(v)$ is strongly lower semi-continuous, then there exists $u \in U_{cc}$ such that

$$J(u) = \inf_v J(v),$$

we have uniqueness if $v \to J(v)$ is strictly convex.

We now obtain a theorem characterizing the solution of the minimization problem.

As $J(u) = \inf\limits_{v} J(v)$, we have for any $v \in U_{cc}$ and $\mathfrak{D} \in (0,1)$

$$J(u) \leqq J((1-\mathfrak{D})u + \mathfrak{D}u) = J(u + \mathfrak{D}(v-u), \text{ thus}$$

$$\frac{1}{\mathfrak{D}} \cdot (J(u+\mathfrak{D}(v-u)) - J(u)) \geqq 0, \quad \text{if the limit exists, we get}$$

$$\lim_{\mathfrak{D}\to 0} \frac{1}{\mathfrak{D}} (J(u+\mathfrak{D}(v-u))-J(u)) \geqq 0, \text{ this is denoted by}$$

$$(J'(u)\cdot(v-u)) \geqq 0 \text{ for } v \in U_{cc}.$$

As $J(v) = B(v,v) - 2O(v)$, then

$$J(v+\mathfrak{D}(v-u))=B(u+\mathfrak{D}(v-u),u)-B(u+\mathfrak{D}(v-u),\mathfrak{D}(v-u))-2\cdot O(u)-2\mathfrak{D}O(v-u) \text{ by linearity,}$$

$$=B(u,u)-2\cdot O(u)+2\mathfrak{D}\cdot B(u,v-u)+\mathfrak{D}^2B(v-u,v-u)-2\mathfrak{D}O(v-u) \text{ by symmetry,}$$

hence

$$\frac{1}{\mathfrak{D}}(J(u+\mathfrak{D}(v-u))-J(u)) = 2B(u,v-u) + \mathfrak{D}B(v-u,v-u) - 2O(v-u),$$

taking the limit $\mathfrak{D} \to 0$, then

$$0 \leqq (J'(u)\cdot(v-u)) = 2(B(u,v-u) - O(v-u)).$$

If the assumptions of the above theorem remain valid, we obtain the

Theorem 2

The unique solution $u \in U_{cc}$ of $J(u) = \inf\limits_{v \in U_{cc}} J(v)$ is characterized by

$$B(u,v-u) \geqq O(v-u) \text{ for } v \in U_{cc}.$$

For a direct solution of certain variational inequalities we have the known theorem due to G. Stampacchia (325) and Lions-Stampacchia (214).

Theorem 3

Let $B(u,v)$ be a continuous, not necessarily symmetric, bilinear form satisfying

$$B(w,w) \geqq c \; \|w\| \text{ for } w \in U_{cc} \text{ and } c > 0,$$

then there exists a unique $u \in U_{cc}$ which satisfies

$$B(u,v-u) \geqq \mathcal{O}(v-u) \text{ for } v \in U_{cc}.$$

In the special case $U_{cc} = \mathcal{H}$ we have the

Theorem 4

Let $B(u,v)$ be a continuous, not necessarily symmetric, bilinear form satisfying $B(w,w) \geqq c \cdot \|w\|$ for $w \in \mathcal{H}$ and $c > 0$, then there exists a unique $u \in \mathcal{H}$ such that $B(u,v) = \mathcal{O}(v)$ for all $v \in \mathcal{H}$.

This is known as the Lax-Milgram Theorem c.f. Lax-Milgram (210), I.M. Visik (346, 347).

The framework of variational inequalities can be applied to the following control problems.

Let there be a general system. The word "system" will be applied to any physical or economical or social phenomenon whose evolution changes with time.

The state of the system, denoted by y, is to be controlled by a given parameter v, thus y(v).

The controlled state of the system is a solution of an equation

$$Ay(v) = \text{given function of } v,$$

where A is an operator specifying the general system. A can be a differential operator which specifies, for example, reaction-diffusion processes in multi-connected regions.

The observations z(v) of the process, which are a function of the state y(v), are known exactly.

Defining a numerical function $z \to \psi(z) \geqq 0$ on the space of observation, then the cost function is given by

$$J(v) = \psi(z(v)).$$

We are looking for unique solution of

$$J(u) = \inf_{v \in U_{cc}} J(v),$$

and obtain algorithms amenable to numerical computation for the approximation of the control $u \in U_{cc}$, which determines the inf J(v). This control is termed an optimal control.

II.1.1 Characterization of a parabolic control process

For example, let us suppose that the state is given by the parabolic equation

$$\frac{\partial y(v)}{\partial t} - \Delta y(v) = v(t)\cdot g(x) \quad \text{in } D \times (0,T)$$

$$y = 0 \quad \text{on the boundary } D \times (0,T)$$

$$y(x,0) = 0 \quad \text{the initial condition in } D,$$

where $v(t)$ is the time-dependent control parameter.

The solution $y(t,x,v)$ satisfies

$$y \in L^2((0,T);H_0^1(D)), \ \frac{\partial y}{\partial t} \in L^2(0,T);(H^1(D))') \text{ (see page 9).}$$

The map $[0,T] \to L^2(D)$

$t \to y(t,v)$ is continuous, then $y(T,v) \in L^2(D)$ and therefore the map $L^2([0,T]) \longrightarrow L^2(D)$

$v \to y(t,v)$ is continuous, thus we can define the cost function for $z_d \in L^2(D)$ by

$$J(v) = |y(T,v) - z_d^2|^2 + N.\int_0^T v^2 dt.$$

We consider the problem

$$J(u) = \inf_{v \in U_{cc}} J(v).$$

Then there exists a unique solution $u \in U_{cc}$.

As the cost function $J(v)$ is quadratic, thus strictly convex, and

$$J(v) > N\cdot\int_0^T v^2 \, dt,$$

so we have, if $\|v\|_{L^2([0,T])} \longrightarrow +\infty$ then $J(v) \to +\infty$, and by the remark of Theorem 1 the existence is clear.

The unique solution is characterized by

$$u \in U_{cc} \ (J'(u),(v-u)) \geq 0 \text{ for } v \in U_{cc}$$

$$(J'(u).(v-u)) = \frac{d}{d\mu}(J(u+\mu\cdot(v-u))|_{\mu=0}$$

Using the cost function, we have

$$\frac{d}{d\mu}J(u+\mu(v-u))\Big|_{\mu=0} = \frac{d}{d\mu}\left(|y(T,u+\mu(v-u))-z_d|^2 + N\cdot\int_0^T (u+\mu(v-u))^2dt\right)$$

$$= 2(y(T,u+\mu(v-u)-z_d|_{\mu=0}\cdot y(T,v-u) + N\cdot 2\int_0^T u(v-u)dt$$

$$= 2(y(T,u)-z_d,y(T,v)-y(T,u)) + 2N\int_0^T u(v-u)dt$$

and finally

$$\frac{1}{2}(J'(u),v-u)-N\cdot\int_0^T u(v-u)dt = (y(T)-z_d,y(T,v)-y(T)),$$

the right-hand side can be expressed as follows:

Introducing the adjoint state p by

$$-\frac{\partial p}{\partial t} - \Delta p = 0 \text{ in } D \times (0,T)$$

$$p(T) = y(T)-z_d \text{ in } D$$

$$p = 0 \text{ on } \partial D \times (0,T);$$

and multiplying the adjoint equation by $y(v)-y$, we obtain by applying Green's formula

$$-\int_D \frac{\partial p}{\partial t}(y(v)-y)dx = \int_D \Delta p(y(v)-y)dx = \int_D \text{div gradp}(y(0)-y)dx =$$

$$\int_D \text{gradp}\cdot\text{grad}(y(v)-y)dx = \int_D p\cdot\Delta(y(v)-y)dx,$$

integration by parts gives

$$\int_0^T \frac{\partial p}{\partial t}(y(v)-y)dt = p\cdot(y(v)-y)\Big|_0^T - \int_0^T p.\,\frac{\partial}{\partial t}(y(v)-y)dt$$

$$= p(T)\cdot(y(T,v)-y(T))-0-\int_0^T p\left(\frac{\partial y(v)}{\partial t} - \frac{\partial y}{\partial t}\right)dt, \text{ as } p(T)=y(T)-z_d \text{ in } D$$

$$= (y(T)-z_d,y(T,v)-y(T)) - \int_0^T p\left(\frac{\partial y(v)}{\partial t} - \frac{\partial y}{\partial t}\right)dt,$$

as $-\int_D \frac{\partial p}{\partial t}(y(v)-y)dx = \int_D p\cdot\Delta(y(v)-y)dx$, then

$$(y(T)-z_d, y(T,v)-y(T)) = \int_0^T \int_D (p(\frac{\partial y(v)}{\partial t} - \frac{\partial y}{\partial t}) - p(\Delta y(v)-\Delta y))dxdt$$

$$= \int_0^T \int_D p(\frac{\partial y(v)}{\partial t} - \Delta y(v) - (\frac{\partial y}{\partial t} - \Delta y))dxdt$$

$$= \int_0^T \int_D p(v(t)\cdot g(x) - u(t)\cdot g(x))dxdt =$$

$$\int_0^T \int_D p\cdot g(v-u)dxdt,$$

by applying the state equation and the optimality system written as

$$\frac{\partial y}{\partial t} - \Delta y = u(t)\cdot g(x) \text{ in } Dx(0,T) \qquad -\frac{\partial p}{\partial t} - \Delta p = 0 \text{ in } Dx(0,T)$$

$$y = 0 \text{ on } \partial Dx(0,T) \qquad p = 0 \text{ on } \partial Dx(0,T)$$

$$y(0) = 0 \text{ in } D, \qquad p(T) = y(T)-z_d \text{ in } D,$$

and finally we obtain

$$0 \leqq \frac{1}{2}(J'(u),v-u) - N.\int_0^T u(v-u)dt = (y(T)-z_d, y(T,v)-y(v)) = \int_0^T p\cdot g(v-u)dt.$$

The optimal control u is characterized by the optimal system and the inequality

$$\int_0^T (p.g + Nu)\cdot(v-u)dt \geqq 0.$$

In the unconstrained case, $U_{cc} = L^2(0,T)$, we have $p\cdot g + Nu = 0$ and the optimal control is given by

$$u = -\frac{1}{N}\cdot p\cdot g.$$

II.2 Control phenomena

II.2.1 Modelling methods by Dirac's distribution

In the model, we take $g(x)$ to be concentrated in the neighbourhood of a point a, we write

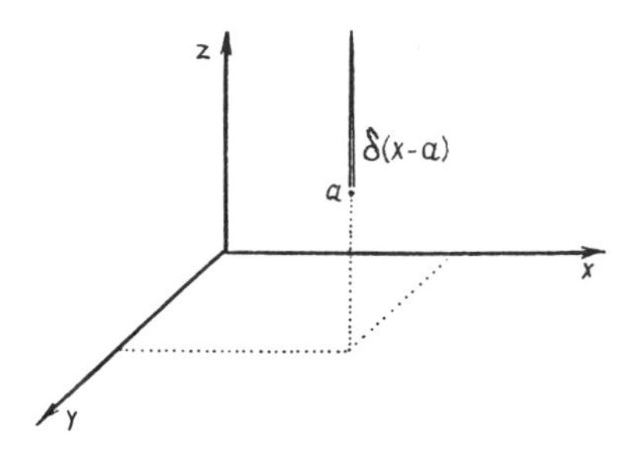

$g(x) = \delta(x-a)$ the Dirac mass at the point a.

$g(x)$ can be seen as a density function concentrated at the point a. But the Dirac mass at the point a makes the right-hand side of the parabolic partial differential equation singular.

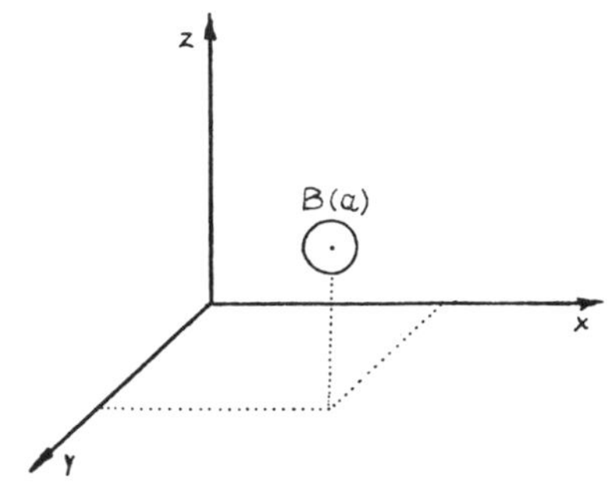

The singularity can be modelled in the following way. By excision of a small domain (e.g. a ball of radius $\mu \geqq 0$) in the neighbourhood of the point a. We apply to the boundary surface of the ball the control variable $v_\mu(t)$.

Now, for $\mu > 0$ and $\mu \to 0$ the ball B(a) converges to the singular point a.

The state of a process governed by a parabolic partial differential equation can be controlled at a point $a \in D$ by the variable $v(t)$ as follows

$$\frac{\partial y}{\partial t} - \Delta y = v(t) \cdot \delta(x-a) \text{ in } Dx(0,T),$$

where the solution y is subject to the condition

$$y = 0 \text{ on the boundary } \partial D \times (0,T)$$

and to the initial condition

$$y(x,0) = 0 \text{ in } D.$$

The solution depends on

$$y = y(x,t,v)$$

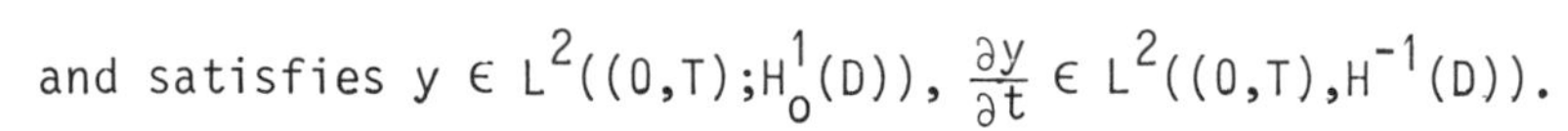

and satisfies $y \in L^2((0,T);H_0^1(D))$, $\frac{\partial y}{\partial t} \in L^2((0,T),H^{-1}(D))$.

II.2.2 Approximation of the pointwise control

This problem can be approximated in the following way.

Let $B_\mu(a)$ be a ball with centre a and radius μ and let $\partial B_\mu(a)$ be the surface of the ball (spheroid).

We introduce the two-connected domain $D \setminus B_\mu(a)$ denoted by D_μ.

The approximating solution y_μ, where $\lim_{\mu \to 0} y_\mu = y$, is a solution of the initial boundary-value problem for $\mu > 0$ small enough:

$$\frac{\partial y_\mu}{\partial t} - \Delta y_\mu = 0 \text{ in } D_\mu \times (0,T)$$

$$y_\mu = N(t) \text{ (unknown) on } \partial B_\mu(a)$$

$$\int_{\partial B_\mu(a)} \frac{\partial y_\mu}{\partial n} \, ds = v_\mu(t) \text{ for a.e. } t \in (0,T)$$

$$y_\mu = 0 \text{ on } \partial D \times (0,T)$$

$$y_\mu(x,0) = 0 \text{ in } D_\mu.$$

Diffusion process controlled at a finite number of internal points

We obtain the following generalization.

The state of a diffusion process can be controlled at the points $a_1, a_2, \ldots, a_m \in D$ by the control variables $v_1, v_2, \ldots, v_m$ given by

$$\frac{\partial y}{\partial t} - \Delta y = \sum_{i=1}^{m} v_i(t)\delta(x-a_i) \text{ in } D \times (0,T)$$

$$y = 0 \text{ on the outer boundary } \partial D$$

$$y(x,0) = 0 \text{ in } D.$$

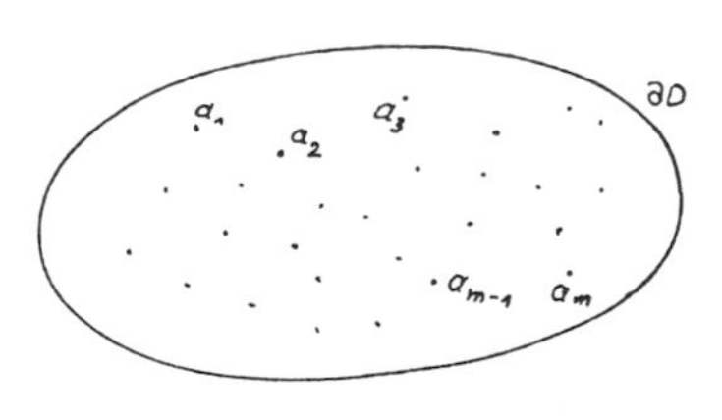

This problem can be approximated as follows.

Let $B_\mu(a_i)$ be a ball with centre a_i and radius μ, and let $\partial B_\mu(a_i)$ be the corresponding surface of the ball (spheroid).

Now we introduce the m+1-connected domain $D \setminus \bigcup_{i=1}^{m} B_\mu(a_i)$, where the $B_\mu(a_i)$ are disjoint, denoted by

$$D_\mu^m = D_\mu \setminus \bigcup_{i=1}^{m} B_\mu(a_i).$$

For $\mu > 0$ small enough, the approximating solution y_μ, where $\lim_{\mu\to 0} y_\mu = y$, is a solution of the initial-boundary-value problem

$$\frac{\partial y_\mu}{\partial t} - \Delta y_\mu = 0 \text{ in } D_\mu^m \times (0,T)$$

$$y_\mu = N_\mu^i(t) \text{ unknown function on } \partial B_\mu(a_i) \text{ for } i = 1,\dots,m$$

$$\int_{\partial B_\mu(a_i)} \frac{\partial y_\mu}{\partial n}\, ds = (v_i)_\mu \text{ for a.e. } t \in (0,T),\ i = 1,\dots,m \text{ as control variables}$$

$$y_\mu = 0 \text{ on the outer boundary } \partial D \times (0,T)$$

$$y_\mu(x,0) = 0 \text{ the initial condition in } D_\mu^m.$$

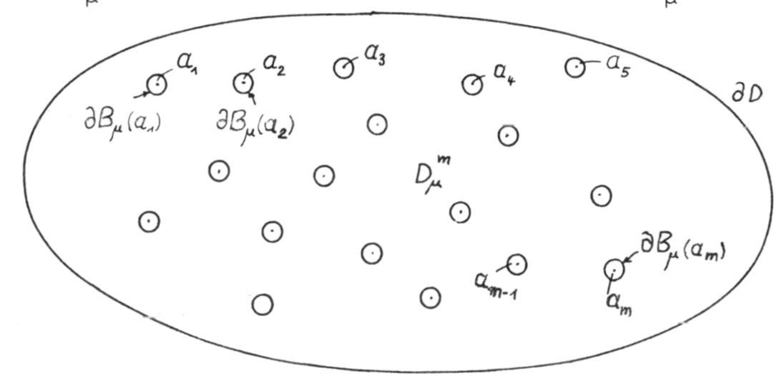

The solution of this problem in the m+1-connected domain D_μ^m can be given in closed, analytic form (by methods constructed in Grusa (141)) if the boundary values and the initial values are Hölder continuous.

The model given above can be formulated for D a bounded, open set in $\mathbb{R}^n$ with smooth boundary, if A is a second order linear, elliptic differential operator with variable coefficients $b_{ij}(x) \in W^{1,\infty}(D)$ for i, j = 1,...,n, where

$$A\varphi = \frac{\partial}{\partial x_i}\left(b_{ij}(x)\frac{\partial}{\partial x_j}\varphi\right).$$

The approximated problem can be written

$$\frac{\partial y_\mu}{\partial t} + Ay = 0 \text{ in } D^m_\mu \times (0,T)$$

$$y_\mu = N^i_\mu(t) \text{ on } \partial B_\mu(a_i) \text{ for } i = 1,\dots,m$$

$$\int_{\partial B_\mu(a_i)} \frac{\partial y_\mu}{\partial n} ds = (v_i)_\mu \quad \text{a.e. } t \in (0,T) \text{ for } i = 1,\dots,m$$

$$y_\mu = 0 \text{ on } \partial D \times (0,T)$$

$$y_\mu(x,0) = 0 \text{ in } D^m_\mu.$$

Now, $\frac{\partial y_\mu}{\partial n}$ is the conormal derivative given by

$$\frac{\partial y}{\partial n} = a_{ij}\, n_i \frac{\partial y}{\partial x_j},$$

where $n = (n_1, n_2, \dots, n_n)$ is the unit normal on the surface of the ball $B_\mu(a_i)$ directed towards the interior of B_μ.

II.3 Control concentrated on one internal point

In the next step of constructing dynamic models, we formulate the optimal control problem only for one point $a \in D$.

The state of the general diffusion process controlled at the point $a \in D$ is a solution of

$$\frac{\partial y}{\partial t} + Ay = v(t)\cdot\delta(x-a) \text{ in } D\times(0,T) \qquad \text{I.}$$

$$y = 0 \text{ on } \partial D \times (0,T)$$

$$y(x,0) = 0 \text{ in } D,$$

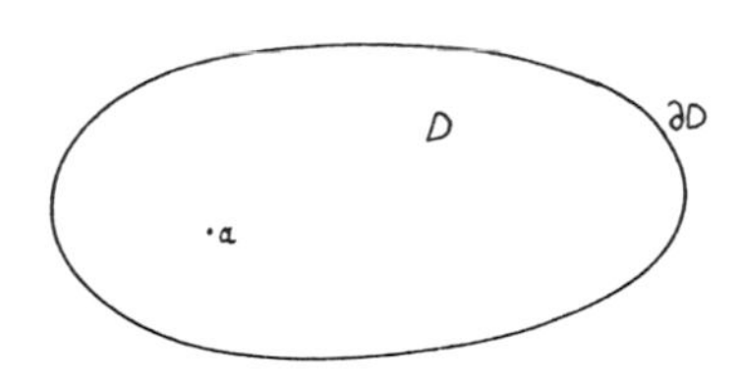

this problem has a unique solution $y(x,t,v)$ for $v \in L^2(0,T)$.

We introduce a new function space (which can be characterized (221, 222)

$$U = \{v \mid v(t) \in L^2(0,T),\ y(T,v) \in L^2(D)\},$$

with the graph norm

$$\|v\|_u = (\|v\|_{L^2(0,T)} + \|y(T,v)\|_{L^2(D)})^{\frac{1}{2}}.$$

We define the cost function

$$J(v) = \int_D |y(T,v)-z_d|^2 dx + M\cdot\int_0^T v^2 dt \text{ for } v \in U.$$

and z_d is a fixed element in $L^2(D)$.

If $M > 0$, then there exists a unique element $u(t)$ of the optimal control problem

$$J(u) = \inf_{v\in U} J(v)$$

and the optimal control u is characterized by the equation system and the corresponding adjoint problem

$$\frac{\partial y}{\partial t} + Ay = u(t)\cdot\delta(x-a) \text{ in } Dx(0,T) \qquad -\frac{\partial p}{\partial t} + A^*(p) = 0 \text{ in } Dx(0,T)$$

$$y = 0 \text{ on } \partial Dx(0,T) \qquad p = 0 \text{ on } \partial Dx(0,T)$$

$$y(x,0) = 0 \text{ in } D, \qquad p(x,T) = y(x,T)-z_d,$$

A* is the adjoint operator of A, and we have (see page 89)

$$\int_0^T (p\cdot(\delta(x-a)+M\cdot u)\cdot(v-u) = \int_0^T (p(a,t)+Mu)\cdot(v-u)dt \geqq 0$$

In the unconstrained case we have

$$u = -\frac{1}{M}\cdot p(a,t),$$

thus the state equation follows to

$$\frac{\partial y}{\partial t} + Ay = -\frac{1}{M}\cdot p(a,t)\cdot\delta(x-a) \text{ in } Dx(0,T)$$

$$y = 0 \text{ on } \partial Dx(0,T)$$

$$y(x,0) = 0 \text{ in } D.$$

The approximated form of this general optimal diffusion process reads as follows.

The initial-boundary-value problem I can be rewritten

$$\frac{\partial y_\mu}{\partial t} + Ay_\mu = 0 \text{ in } D_\mu \times (0,T)$$

$$y_\mu = N_\mu(t) \text{ on } \partial B_\mu(a)$$

$$\int_{\partial B(a)} \frac{\partial y_\mu}{\partial n} \cdot d(\partial B_\mu) = v_\mu(t)$$

$$y_\mu = 0 \text{ on } \partial D \times (0,T)$$

$$y_\mu(x,0) = 0 \text{ in } D.$$

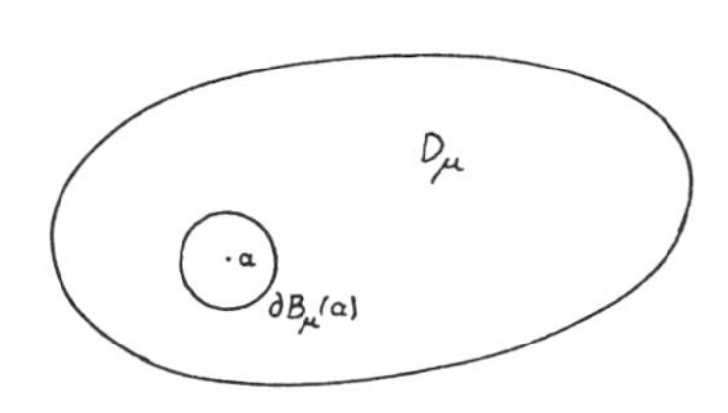

There exists a unique solution $y_\mu(t,v_\mu)$, if $v_\mu(t) \in L^2(0,T)$. Introducing the subspace

$$K(D_\mu) = \{w | w \in H^1(D_\mu), w = 0 \text{ on } \partial D x(0,T), w = \begin{matrix}\text{constant which} \\ \text{depends on } w\end{matrix} \text{ on } \partial B_\mu(a)\},$$

denoting by $K'(D_\mu)$ the dual of $K(D_\mu)$, the solution y_μ satisfies

$$y_\mu \in L^2(0,T), K(D_\mu)) \text{ and } \frac{\partial y_\mu}{\partial t} \in L^2(0,T), K'(D_\mu)),$$

hence there exists a continuous operator

$$L^2(0,T) \to L^2((0,T), K(D_\mu)) \times L^2((0,T), K'(D_\mu))$$

$$v_\mu \longrightarrow (y_\mu, \frac{\partial y_\mu}{\partial t}),$$

thus we have $y_\mu(T,v_\mu) \in L^2(D_\mu)$ and therefore the operator

$$\Lambda_\mu : L^2(0,T) \to L^2(D_\mu)$$

$$v_\mu \longrightarrow y_\mu(T,v_\mu).$$

II.3.1 Characterization of the optimal control

For $v \in L^2(0,T)$ the cost function

$$J_\mu(v) = \int_{D_\mu} |y_\mu(T,v) - z_d|^2 dx + M \int_0^T v^2(t)dt$$

is uniquely defined by

$$J_\mu(v) = \int_{D_\mu} |\Lambda_\mu(v) - z_d|^2 dx + M \int_v^T v^2(t)dt$$

and the optimal control problem

$$J_\mu(u_\mu) = \inf_{v\in L^2(0,T)} J_\mu(v)$$

has a unique solution u_μ.

Now, the optimal solution can be characterized by the following

Theorem:

Let the state equation with the corresponding adjoint equation be given

$$\frac{\partial y_\mu}{\partial t} + Ay_\mu = 0 \text{ in } D_\mu x(0,T) \qquad -\frac{\partial p_\mu}{\partial t} + A^*p_\mu = 0 \text{ in } D_\mu \text{ x } (0,T)$$

$$y_\mu = N_\mu(t) \text{ on } \partial B_\mu \qquad p_\mu = N_\mu(t)$$

$$\int_{\partial B_\mu} -\frac{\partial y_\mu}{\partial n} d(\partial B_\mu) = v_\mu(t) \qquad \int_{\partial B_\mu} -\frac{\partial p_\mu}{\partial n} d(\partial B_\mu) = 0$$

$$y_\mu = 0 \text{ on } \partial D \text{ x } (0,T) \qquad p_\mu = 0 \text{ on } \partial D \text{ x } (0,T)$$

$$y_\mu(x,0) = 0 \text{ in } D_\mu \qquad p_\mu(x,T) = y_\mu(x,T)-z_d \text{ in } D_\mu,$$

then the inequality

$$\int_0^T (N_\mu(t) + M\cdot u_\mu(t))\cdot(v-u_\mu)dt \geq 0$$

is satisfied.

Proof:

For the sake of clarity, we take

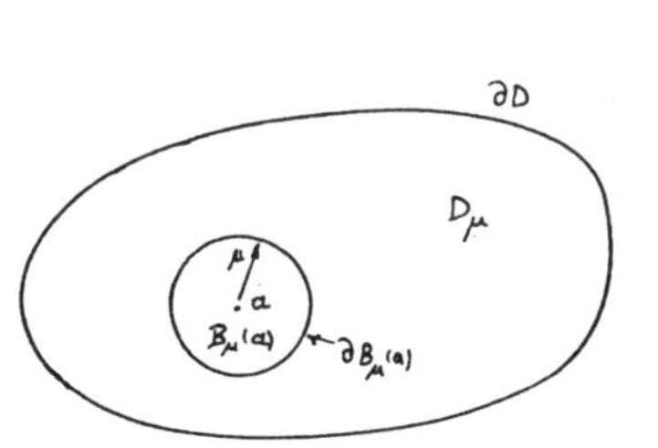

$$y_\mu \to y,\ D_\mu \to D,\ \partial D \to \Gamma\ ,\ \partial B_\mu(a_i) \to \partial B$$

multiplying the adjoint equation by $y(v)-y$ and integrating, yields

$$\int_0^T \int_D \frac{\partial p}{\partial t} (y(v)-y)dxdt = \int_0^T \int_D A^*p(y(v)-y)dxdt$$

The differential operator is defined by

$$A = -\frac{\partial}{\partial x_i}(a_{ij}(x) \frac{\partial}{\partial x_j}),$$

and applying Green's formula, we obtain

$$\int_D \frac{\partial}{\partial x_i}(a_{ij}(x)\frac{\partial p}{\partial x_j})(y(v)-y)dx = \int_{\partial D} n_i a_{ij}(x)\frac{\partial p}{\partial x_j}(y(v)-y)ds -$$

$$- \int_{\partial D} p\cdot n_i a_{ij}(x)\frac{\partial}{\partial x_i}(y(v)-y)ds + \int_D p\cdot\frac{\partial}{\partial x_j}(a_{ij}(x)\cdot\frac{\partial}{\partial x_i}(y(v)-y)dx$$

hence

$$\int_D A^*p(y(v)-y)dx = \int_{\partial D} \frac{\partial p}{\partial n}(y(v)-y)ds - \int_{\partial D} p\cdot\frac{\partial}{\partial n}(y(v)-y)ds + \int_D p\cdot A(y(v)-y)dx.$$

We apply integration by parts

$$\int_0^T \frac{\partial p(u,t)}{\partial t} (y(v)\cdot y)dt = p(u,t)\cdot(y(v,t)-y(t))\Big|_{t=0}^{t=T} -$$

$$- \int_0^T p(u,t)\cdot\frac{\partial}{\partial t}(y(v)-y)dt, \text{ as } y(x,0) = 0 \text{ and } p(x,T) = (y(x,T)-Z_d) \text{ in } D_\mu$$

$$= (y(u,T)-z_d)\cdot(y(v,T)-y(T)) - \int_0^T p(u,t)\cdot\frac{\partial}{\partial t}(y(v)-y)dt.$$

As

$$\int_0^T \int_D \frac{\partial p}{\partial t}(y(v)-y)dtdx = \int_0^T \int_D A^*p(y(v)-y)dtdx, \text{ we obtain by integration}$$

$$= \int_0^T (\int_{\partial D} \frac{\partial p}{\partial n}(y(v)-y)ds - \int_{\partial D} p\cdot\frac{\partial}{\partial n}(y(v)-y)ds)dt$$

$$+ \int_0^T \int_D p\cdot A(y(v)-y)dx,$$

$$\int_0^T \int_D p\cdot((Ay(v) + \frac{\partial y(v)}{\partial t})-(Ay + \frac{\partial y}{\partial t}))dxdt = (y(T)-z_d)(y(T,v)-y(T)) +$$

$$\int_0^T \int_{\partial D} \frac{\partial p}{\partial n}\cdot(y(v)-y)dsdt - \int_0^T \int_{\partial D} p.\frac{\partial}{\partial n}(y(v)-y)dsdt.$$

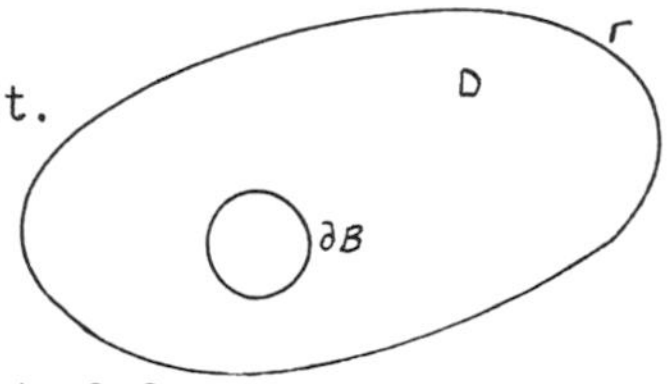

The boundary integrals are derived as follows:

$$\int_{\partial D} \frac{\partial p}{\partial n}(y(v)-y)ds = \int_\Gamma \frac{\partial p}{\partial n}(y(v)-y)ds + \int_{\partial B}\frac{\partial p}{\partial n}(y(v)-y)ds=0+0,$$

as $y = 0$, $y(v) = 0$ on Γ and $y(v)=N(t)$ on ∂B, $\int_{\partial B} \frac{\partial p}{\partial n} d(\partial B)=0$,

$$\int_{\partial D} p\cdot(\frac{\partial}{\partial n}(y(v)- \frac{\partial y}{\partial n})ds = \int_{\Gamma} p\cdot(\frac{\partial}{\partial n} y(v)- \frac{\partial y}{\partial n})ds +$$

$$+ \int_{\partial B} p\cdot(\frac{\partial}{\partial n} y(v) - \frac{\partial y}{\partial n})ds = N(t)\cdot(v-u),$$

as $p = 0$ on Γ, the first integral vanishes; in the second integral

$$p = N(t) \text{ on } \partial B \text{ and} \int_{\partial B} \frac{\partial y(v)}{\partial n} ds = v, \int_{\partial B} \frac{\partial y}{\partial n} ds = u.$$

As we are considering homogeneous problems

$$Ay(v) + \frac{\partial y(v)}{\partial t} = 0 \text{ and } \frac{\partial y}{\partial t} + Ay = 0.$$

The above relation reduces to

$$0 = (y(T,u)-z_d)(y(T,v)-y(T)) - \int_0^T N(t)\cdot(v-u)dt,$$

and the defined cost function satisfied

$$\frac{1}{2}(J'(u).v-u) = (y(T,u)-z_d)\cdot(y(T,v)-y(T) + M \int_0^T u\cdot(v-u)dt$$

$$= \int_0^T N(t)\cdot(v-u)dt + M\cdot\int_0^T u(v-u)dt = \int_0^T (N(t) + Mu)\cdot(v-u)dt,$$

hence, the solution u of the optimal control problem is characterized by the state equation and the corresponding adjoint problem and the integral inequality

$$\int_0^T (N(t) + Mu).(v-u)dt \geqq 0,$$

in the unconstrained case $N(t) + Mu = 0$, hence

$$u = - \frac{1}{M} N(t).$$

II.3.2 Control concentrated on a finite number of internal points

The state of a diffusion process may be controlled at the points $a_1,a_2,\dots,a_m \in D$ by the control variables $v_1,v_2,\dots,v_m$ given by

$$\frac{\partial y}{\partial t} + Ay = \sum_{i=1}^{m} v_i(t)\cdot\delta(x-a_i) \text{ in } D \times (0,T)$$

$$y = 0 \text{ on } \partial D \times (0,T)$$

$$y(x,0) = 0 \text{ in } D.$$

Introducing the cost function

$$J(v) = \int_D |y(x,t,v)-z_d|^2 dxdt + \sum_{i=1}^{m} M_i \int_0^T v_i^2 dt \text{ for } M_i > 0,\ z_d \in L^2(D).$$

Let v be $v = (v_1,v_2,\ldots,v_m) \in (L^2(0,T))^m$, then there exists a unique optimal control $u = (u_1,u_2,\ldots,u_m) \in (L^2(0,T))^m$ satisfying

$$J(u) = \inf_{v\in(L^2(0,T))^m} J(v),$$

and $u = u(u_1,u_2,\ldots,u_m)$ is characterized by

$$\frac{\partial y}{\partial t} + Ay = \sum_{i=1}^{m} u_i(t)\cdot\delta(x-a_i) \qquad -\frac{\partial p}{\partial t} + A^*p = 0$$

$$y = 0 \text{ on } \partial D \times (0,T) \qquad p = 0 \text{ on } \partial D \times (0,T)$$

$$y(x,0) = 0 \text{ in } D, \qquad p(x,T) = y(x,T)-z_d \text{ in } D,$$

with the integral inequality

$$\sum_{i=1}^{m} \int_0^T (p(a_i,t) + M_i u_i(t))\cdot(v_i(t)-u_i(t))dt \geq 0 \text{ for } v \in (L^2(0,T))^m.$$

Proof:

On page 97 we derived the relation

$$\int_D p(\frac{\partial y(v)}{\partial t} + Ay(v)) - (Ay + \frac{\partial y}{\partial t})dxdt = (y(T)-z_d)\cdot(y(T,v)-y(T)) + $$
$$+ \int_0^T \int_{\partial D} \frac{\partial p}{\partial n}(y(v)-y)dsdt - \int_0^T \int_{\partial D} p\cdot\frac{\partial}{\partial n}(y(v)-y)dsdt.$$

Integrating the state equation

$$\frac{\partial y}{\partial t} + Ay = \sum_i v_i(t)\cdot\delta(x-a_i), \int_0^T \int_D p(\frac{\partial y}{\partial t} + Ay)dxdt =$$

$$= \sum_{i=1}^m \int_0^T \int_{\partial D} p(x,t)\cdot\delta(x-a_i)dx\cdot v_i(t)dt$$

$$= \sum_{i=1}^m \int_0^T p(a_i,t)\cdot v_i(t)\cdot v_i(t)dt,$$

and analogously

$$\int_0^T \int_0 p(Ay + \frac{\partial y}{\partial t})dxdt = \sum_i \int_0^T p(a_i,t)u_i(t)dt,$$

the boundary integrals vanish by construction. By the definition of the cost function we have

$$\frac{1}{2}(I'(u),v-u) = (y(T)-z_d)\cdot(y(T,v)-y(T)) + \sum_{i=1}^m M_i \int_0^T u_i(v_i-u_i)dt,$$

using this in the relation, we obtain finally

$$\sum_{i=1}^m \int_0^T p(a_i,t)\cdot(v_i(t)-u_i(t))dt = -\sum_{i=1}^m M_i \int_0^T u_i(v_i-u_i)dt + \frac{1}{2}(I'(u), v-u) \geq 0$$

hence there follows the characterizing integral. Without constraints we obtain the optimal control

$$u_i(t) = -\frac{1}{M_i} p(a_i,t) \text{ for } i = 1,\ldots,m,$$

with the characterizing equations

$$\frac{\partial y}{\partial t} + Ay = -\sum_{i=1}^m \frac{1}{M_i} p(a_i,t)\delta(x-a_i) \text{ in } Dx(0,T); \quad -\frac{\partial p}{\partial t} + A^*p = 0 \text{ in } Dx(0,T)$$

$$y = 0 \text{ on } \partial Dx(0,T) \qquad\qquad p = 0 \text{ on } \partial Dx(0,T)$$

$$y(x,0) = 0 \text{ in } D \qquad\qquad p(x,T) = y(x,T)-z_d \text{ in } D.$$

II.3.3 Controlled diffusion process

The specialized equation system for A = Laplace operator reads as

$$\frac{\partial y}{\partial t} - \Delta y + \sum_{i=1}^{m} \frac{1}{M_i} p(a_i,t)\delta(x-a_i) = f \text{ in } Dx(0,T); \quad -\frac{\partial p}{\partial t} - \Delta p - y = -z_d \text{ in } Dx(0,T)$$

$$y = 0 \text{ on } \partial Dx(0,T) \qquad\qquad p = 0 \text{ on } \partial Dx(0,T)$$

$$y(x,0) = 0 \text{ in } D \qquad\qquad p(x,T) = 0 \text{ in } D.$$

Solving the above system, we take the ansatz

$$p(t) = Q(t)y(t) + R(t)$$

and compute Q and R directly by applying the time derivative to the ansatz, we have

$$\frac{\partial p}{\partial t} = \frac{\partial Q}{\partial t}\cdot y + Q\cdot\frac{\partial y}{\partial t} - \frac{\partial R}{\partial t} \text{ using } -\frac{\partial p}{\partial t} - \Delta p - y = -z_d \text{ and } \Delta p - \Delta(Q\cdot y+R) = \Delta(Q\cdot y) + \Delta R,$$

we obtain

$$-\frac{\partial Q}{\partial t}\cdot y - Q\cdot\frac{\partial y}{\partial t} - \frac{\partial R}{\partial t} - \Delta(Qy+R) - y = -z_d.$$

Now we replace

$$\frac{\partial y}{\partial t} \text{ by } \frac{\partial y}{\partial t} - \Delta y + \sum_i \frac{1}{M_i} p(a_i,t)\delta(x-a_i) = f,$$

then we obtain

$$-\frac{\partial Q}{\partial t}y - Q(\Delta y - \sum_{i=1}^{m} \frac{1}{M_i}\cdot p(a_i,t)\cdot\delta(x-a_i) - Qf - \frac{\partial R}{\partial t} - \Delta(Qy+R) - y = -z_d. \qquad \boxed{A}$$

Introducing unique kernels $Q(x,\xi,t)$ by known methods of L. Schwartz (312), such that

$$Q(t)h = \int_D Q(x,\xi,t)\cdot h(\xi)d\xi,$$

where Q satisfies a nonlinear partial differential equation. The "ansatz" can be written as

$$p(x,t) = Q(x,t)\cdot y(..,t) + R(x,t),$$

we specialize

$$p(a_i,t) = Q(a_i,t)y(..,t) + R(a_i,t)$$

and obtain, as

$$Q(t)\, y(..,t) = \int_D Q(x,\xi,t)y(\xi,t)d\xi,$$

the expression

$$p(a_i,t) = \int_D Q(a_i,\xi,t)y(\xi,t)d\xi + R(a_i,t).$$

Specializing the kernel-definition, we have

$$Q(t)\delta(x-a) = \int_D Q(x,\xi,t)\cdot\delta(\xi-a_i)d\xi = Q(x,a_i,t)$$

by definition of the Dirac distribution.

The expression

$$Q(\Delta y - \sum_i \frac{1}{M_i}\, p(a_i,t)\delta(x-a_i))$$

reads as follows:

if $p(a_i,t)\delta(x-a_i) = \int_D Q(a_i,\xi,t)\cdot y(\xi,t)\delta(x-a_i)d\xi +$

$R(a_i,t)\cdot\delta(x-a_i)$, and $Q(a_i,\xi,t)\delta(x-a_i) = Q(x,a_i,t)$,

we obtain

$$\sum_{i=1}^{m} \frac{1}{M_i}\, p(a_i,t)\cdot\delta(x-a_i) = \sum_i \frac{1}{M_i} \int_D Q(a_i,\xi,t)\cdot\delta(x-a_i)y(\xi,t)dt -$$

$$\sum_i \frac{1}{M_i}\cdot R(a_i,t)\cdot\delta(x-a_i);$$

applying Q

$$Q(\sum_{i=1}^{m} \frac{1}{M_i}\, p(a_i,t)\delta(x-a_i)) = \sum_i \frac{1}{M_i} \int_D Q(a_i,\xi,t)y(\xi,t)Q\cdot\delta(x-a_i)d\xi -$$

$$\sum_i \frac{1}{M_i}\cdot R(a_i,t)Q\cdot\delta(x-a_i), \text{ as } Q\cdot\delta(x-a_i) = Q(x,a_i,t),$$

$$= \sum_i \frac{1}{M_i}\cdot Q(x,a_i,t) \int_D Q(a_i,\xi,t)y(\xi,t)d\xi - \sum_i \frac{1}{M_i}\cdot R(a_i,t)\cdot Q(x,a_i,t).$$

Using this in $\boxed{A}$, we obtain finally

$$- \frac{\partial Q}{\partial t}\cdot y - Q\cdot\Delta y - \Delta(Q\cdot y) + \sum_i \frac{1}{M_i} Q(x,a_i,t) \int_D Q(a_i,\xi,t)y(\xi,t)d\xi -$$

$$- \frac{\partial R}{\partial t} - \Delta R + \sum_i \frac{1}{M_i} R(a_i,t)Q(x,a_i,t) = Q\cdot f - z_d, \quad \text{thus}$$

$$\frac{\partial Q}{\partial t} + (\Delta_x + \Delta_\xi)Q + \sum_{i=1}^m \frac{1}{M_i} Q(x,a_i,t)\cdot Q(a_i,\xi,t) = \delta(x-\xi)$$

$$\frac{\partial R}{\partial t} - \Delta R + \sum_{i=1}^m \frac{1}{M_i} Q(x,a_i,t)\cdot R(a_i,t) = Qf - z_d.$$

We have

$$p(x,T) = \int_D Q(x,\xi,T)y(\xi,T)d\xi + R(x,T),$$

if $p(x,T) = 0$, then there follow the initial conditions

$$Q(x,\xi,T) = 0 \text{ on } D \times D \text{ and } R(x,T) = 0 \text{ on } D.$$

Let x be an element of the boundary ∂D, but $\xi \in D$, then

$$p(x,t) = \int_D Q(x,\xi,t)\cdot y(\xi,t)d\xi + R(x,t),$$

from $p(x,t) = 0$ we obtain the boundary conditions

$$Q(x,\xi,t) = 0 \text{ for } x \in \partial D,\ \xi \in D \text{ and } R(x,t) = 0 \text{ for } x \in \partial D.$$

II. 3.4 Solution

We have derived finally:

Let Q be a solution of

$$- \frac{\partial Q}{\partial t} + (\Delta_x + \Delta_\xi)Q + \sum_{i=1}^m \frac{1}{M_i} Q(x,a_i,t)\cdot Q(a_i,\xi,t) = \delta(x-\xi) \text{ in } D \times (0,T),$$

with boundary-initial conditions

$$Q(x,\xi,t) = 0 \text{ for } x \in \partial D,\ \xi \in D,\ Q(x,\xi,T) = 0 \text{ on } D \times D,$$

and let R be a solution of the linear parabolic problem

$$- \frac{\partial R}{\partial t} - \Delta R + \sum_{i=1}^m \frac{1}{M_i} Q(x,a_i,t)R(a_i,t) = Qf - z_d \text{ in } D \times (0,T),$$

with boundary-initial conditions

$$R(x,t) = 0 \text{ for } x \in \partial D \text{ and } R(x,T) = 0 \text{ in } D.$$

Taking these solutions, we obtain

$$p(a_i,t) = \int_D Q(a_i,\xi,t)y(\xi,t)d\xi + R(a_i,t)$$

and the optimal control is given by

$$u_i(a_i,t) = -\frac{1}{M_i} p(a_i,t).$$

Thus we have minimized the cost functional with respect to the control parameters $v_1,v_2,\ldots,v_m$ acting at the chosen points $a_1,a_2,\ldots,a_m$. The minimum is given by

$$u = (u_1,u_2,\ldots,u_m),$$

where each u_i can be expressed in the closed form given above.

Methods for solving nonlinear partial differential equations of Riccati type, are introduced G. Da Prato (73, 74, 75) and R. Temam (330).

II. 4 Optimal control concentrated on a finite number of internal points

II.4.1 Approximation problem

On the right-hand side of the state equation Dirac's δ-functions appear as singularities.

The singularity at the point a_i for $i = 1,\ldots,m$ can be approximated as follows.

For $\mu > 0$ small enough, we take around the point a_i balls $B_\mu(a_i)$ with radius μ, then

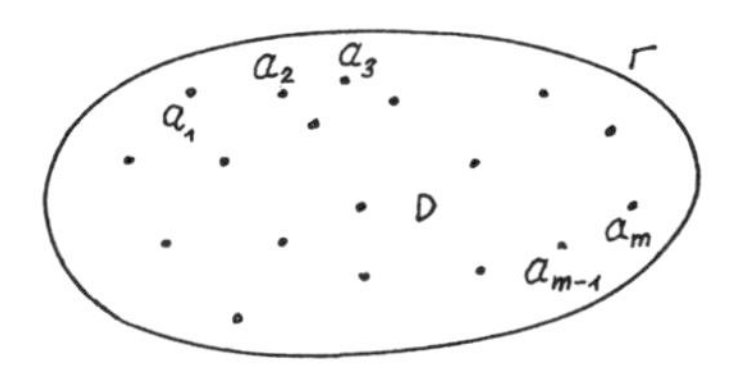

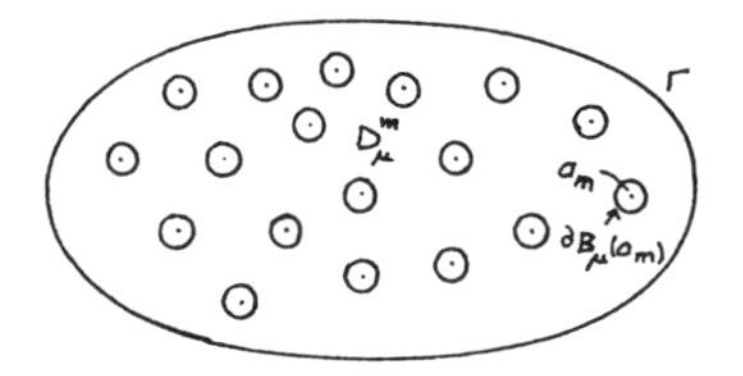

there follows an m+1 connected domain $D \setminus \bigcup_{i=1}^{m} B_\mu(a_i) = D_\mu^m$ by construction. Taking the radius μ of the balls $B_\mu(a_i)$ and tending $\mu \to 0$, then the balls $B_\mu(a_i)$ reduce to the point a_i. The approximation problem of the singular state equation can be written as

$$\frac{\partial y_\mu}{\partial t} + Ay_\mu = 0 \text{ on } D_\mu^m \times (0,T)$$

$$y_\mu = N_\mu^i(t) \text{ (unknown function) on } \partial B_\mu(a_i) \text{ for } i = 1,\dots,m$$

$$\int_{\partial B_\mu(a_i)} \frac{\partial y_\mu}{\partial n} ds = (v_i)_\mu \text{ a.e. } t \in (0,T) \text{ for } i = 1,\dots,m$$

$$y_\mu = 0 \text{ on the outer boundary } \Gamma \times (0,T)$$

$$y_\mu(x,0) = 0 \text{ in } D_\mu^m.$$

If $(v_i)_\mu \in L^2(0,T)$ for $i = 1,\dots,m$, there exists a unique solution $y_\mu(t,(v_i)_\mu)$. Introducing the subspace

$$K(D_\mu^m) = \{w \in H^1(D_\mu^m) | w = 0 \text{ on } \Gamma \times (0,T),\ \omega = \text{constant which depends on } w$$

$$\text{on } \partial B_\mu(a_i) \text{ for } i = 1,\dots,m\}.$$

Denoting $K'(D_\mu^m)$ the dual of $K(D_\mu^m)$, the solution y_μ satisfies

$$y_\mu \in L^2((0,T),K(D_\mu^m)),\ \frac{\partial y_\mu}{\partial t} \in L^2((0,T),K'(D_\mu^m)),$$

hence we obtain a continuous operator

$$L^2(0,T) \to L^2((0,T),K(D_\mu^m)) \times L^2((0,T),K'(D_\mu^m))$$

$$(v_i)_\mu \to (y_\mu, \frac{\partial y_\mu}{\partial t}), \text{ thus we have } y_\mu(T,(v_i)_\mu \in L^2(D_\mu^m),$$

for any $(v_i)_\mu \in L^2(0,T)$ we define a cost function

$$J_\mu(v) = \int_{D_\mu^m} |y_\mu(T,v)-z_d|^2 dt + \sum_{i=1}^m M_i \int_0^T v_i^2(t)dt.$$

The optimal control problem

$$J_\mu(u_\mu) = \inf_{v_\mu \in (L2(0,T))^m} J(v)$$

has a unique solution $u_\mu = ((u_1)_\mu, (u_2)_\mu,\dots,(u_m)_\mu)$.

The optimal solution is characterized by the state equation and the adjoint equation

$$\frac{\partial y_\mu}{\partial t} + Ay_\mu = 0 \text{ in } D_\mu^m x(0,T) \qquad -\frac{\partial p_\mu}{\partial t} + A^*(p_\mu) = 0 \text{ in } D_\mu^m \times (0,T)$$

$$y_\mu = N_\mu^i(t) \text{ on } \partial B_\mu(a_i) \text{ for } i = 1,\dots,m \qquad p_\mu = N_\mu^i(t) \text{ on } \partial B_\mu(a_i) \text{ for } i=1,\dots,m$$

$$\int_{\partial B_\mu(a_i)} \frac{\partial y_\mu}{\partial n} d(\partial B_\mu) = (v_i)_\mu(t) \text{ for } i = 1,\dots,m \qquad \int_{\partial B_\mu(a_i)} -\frac{\partial p_\mu}{\partial n} d(\partial B_\mu) = 0 \text{ for } i=1,\dots,m$$

$$y_\mu = 0 \text{ on } \Gamma \times (0,T) \qquad p_\mu = 0 \text{ on } \Gamma \times (0,T)$$

$$y_\mu(x,0) = 0 \text{ in } D_\mu^m, \qquad p_\mu(x,T) = y_\mu(x,T) - z_d \text{ in } D_\mu^m,$$

with the integral inequality

$$\sum_{i=1}^m \int_0^T (N_i(t)+M_i u_i(t))\cdot(v_i(t)-u_i(t))dt \geq 0 \text{ for } v_i \in L^2(0,T)$$

Proof:

For the sake of clarity, we take:

$$y_\mu \to y,\ \partial D \to \Gamma,\ \partial B_\mu(a_i) \to \partial B_i,$$

$$p_\mu \to p,\ (v_i)_\mu \to v_i.$$

We multiply the adjoint equation by $y(v)-y$, integrate and apply Green's formula.

$$\int_{D_\mu^m} \frac{\partial p}{\partial t}(y(v)-y)dx = \int_{D_\mu^m} A^*p(y(v)-y)dx$$

$$= \int_{\partial D_\mu^m} \frac{\partial p}{\partial n}(y(v)-y)ds - \int_{\partial D_\mu^m} p\cdot\frac{\partial}{\partial n}(y(v)-y)ds + \int_{D_\mu^m} p\cdot A(y(v)-y)dx$$

$$= \int_\Gamma \frac{\partial p}{\partial n}(y(v)-y)ds + \sum_{i=1}^m \int_{\partial B_i} \frac{\partial p}{\partial n}(y(v)-y)ds - \int_\Gamma p\cdot\frac{\partial}{\partial n}(y(v)-y)ds -$$

$$- \sum_{i=1}^m \int_{\partial B_i} p\cdot\frac{\partial}{\partial n}(y(v)-y)ds + \int_{D_\mu^m} p\cdot A(y(v)-y)dx,$$

as $y = 0$, $p = 0$ on the boundary $\Gamma \times (0,T)$, the corresponding integrals vanish. As $\int_{\partial B_i} \frac{\partial p}{\partial n} . d(\partial B_i) = 0$ for $i = 1,\dots,m$ and $y = N_i(t)$ on ∂B_i, then the boundary integrals are vanishing. But

$$\int_{\partial B_i} \frac{\partial y(v)}{\partial n} ds = v_i, \int_{\partial B_i} \frac{\partial y}{\partial n} ds = u_i \text{ for } i=1,\dots,m, \text{ and } p = N_i(t) \text{ on } \partial B_i \text{ for } i = 1,\dots,m,$$

we obtain

$$\sum_{i=1}^{m} \int_{\partial B_i} p \cdot \frac{\partial}{\partial n}(y(v)-y)ds = \sum_{i=1}^{m} N_i(t) \cdot (v_i(t)-u_i(t)),$$

hence

$$\int_0^T \int_{D_\mu^m} \frac{\partial p}{\partial t}(y(v)-y)dxdt = - \sum_{i=1}^{m} \int_0^T N_i(t)(v_i(t)-u_i(t))dt + + \int_0^T \int_{D_\mu^m} p \cdot A(y(v)-y)dxdt.$$

On the other hand, integration by parts yields

$$\int_0^T \frac{\partial p(x,t)}{\partial t}(y(v)-y)dt = (y(T)-z_d) \cdot (y(T,v)-y(T)) - \int_0^T p . \frac{\partial}{\partial t}(y(v)-y)dt$$

$$\int_0^T \int_{D_\mu^m} \frac{\partial p}{\partial t}(y(v)-y)dtdx = \int_{D_\mu^m} (y(T)-z_d)(y(v,T)-y(T))dx - - \int_{D_\mu^m} \int_0^T p \frac{\partial}{\partial t}(y(v)-y)dtdx$$

$$= - \sum_{i=1}^{m} \int_0^T N_i(t)(v_i(t)-u_i(t))dt + \int_{D_\mu^m} \int_0^T p \cdot A(y(v)-y)dtdx,$$

$$\int_0^T \int_{D_\mu^m} ((\frac{\partial y(v)}{\partial t} + Ay(v)) - (\frac{\partial y}{\partial t} - Ay))dxdt = \int_{D_\mu^m} (y(T)-z_d) \cdot (y(T,v)-y(T))dx - - \sum_{i=1}^{m} \int_0^T N_i(t) \cdot (v_i(t)-u_i(t))dt.$$

By definition of the cost function, we obtain

$$\frac{1}{2}(J'(u).(v-u)) = \int_{D_\mu^m} (y(T)-z_d)(y(T,v)-y(T))dx + \sum_{i=1}^{m} M_i \int_0^T u_i(t)(v_i(t)-u_i(t))dt,$$

considering the homogeneous problem $\frac{\partial y(v)}{\partial t} + Ay(v) = 0$ $\frac{\partial y}{\partial t} + Ay = 0$, we have

$$\int_{D^m_\mu} (y(T)-z_d)(y(T,v)-y(T))dx = \sum_{i=1}^{m} \int_0^T N_i(t)\cdot(v_i(t)-u_i(t))dt$$

hence

$$\frac{1}{2}(J'(u), v-u) = \sum_{i=1}^{m} \int_0^T (N_i(t)+M_i u_i(t))\cdot(v_i(t)-u_i(t))dt.$$

If $u = (u_1, u_2, \ldots, u_m)$ is the optimal control, then we obtain the integral relation

$$\sum_{i=1}^{m} \int_0^T (N_i(t)+M_i u_i(t))\cdot(v_i(t)-u_i(t))dt \geqq 0 \text{ for } v_i \in L^2(0,T).$$

Without constraints the optimal control is given by $N_i(t)+M_i u_i(t) = 0$, hence

$$u_i(t) = -\frac{1}{M_i} N_i(t) \text{ for } i = 1,\ldots,m,$$

and the characterizing equations can be written

$$\frac{\partial y}{\partial t} + Ay = 0 \text{ in } D^m_\mu \times (0,T) \qquad -\frac{\partial p}{\partial t} + A^*p = 0 \text{ in } D^m_\mu \times (0,T)$$

$$y = N_i(t) \text{ on } \partial B_i \text{ for } i = 1,\ldots,m \qquad p = N_i(t) \text{ on } \partial B_i \text{ for } i = 1,\ldots,m$$

$$\int_{\partial B_i} \frac{\partial y}{\partial n} d(\partial B_i) = u_i(t) = -\frac{1}{M_i} N_i(t) \qquad \int_{\partial B_i} \frac{\partial p}{\partial n} d(\partial B_i) = 0 \text{ i.e. } t \in (0,T),\ i = 1,\ldots,m$$

$$y = 0 \text{ on } \Gamma \times (0,T) \qquad p = 0 \text{ on } \Gamma \times (0,T)$$

$$y(x,0) = 0 \text{ in } D^m_\mu, \qquad p(x,T) = y(x,T)-z_d \text{ in } D^m_\mu,$$

thus the state equation reduces to

$$\frac{\partial y}{\partial t} + Ay = 0 \text{ in } D_\mu^m \times (0,T)$$

$$y = p_i \text{ on } \partial B_i \text{ for } i = 1,\dots,m$$

$$\int_{\partial B_i} \frac{\partial y}{\partial n}\, d(\partial B_i) = -\frac{1}{M_i}\, p_i(t)$$

$$y = 0 \text{ on } \Gamma \times (0,T)$$

$$y(x,0) = 0 \text{ in } D_\mu^m.$$

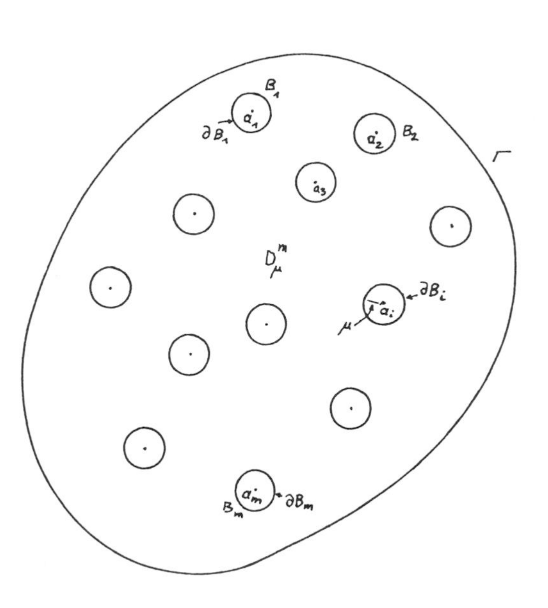

II. 4.2 Approximation process

There is an interesting approximation theorem for parabolic initial-boundary-value problems in the three dimensional space. We only formulate the two connected case.

Let the cost function be defined by

$$J_\mu(v) = \int_{D_\mu} |y_\mu(T,v)-z_d|^2 dx + M \int_0^T v^2 dt.$$

Let u_μ be the unique solution of

$$J_\mu(u_\mu) = \inf_{v \in L^2(0,T)} J_\mu(v),$$

then we are able to derive the

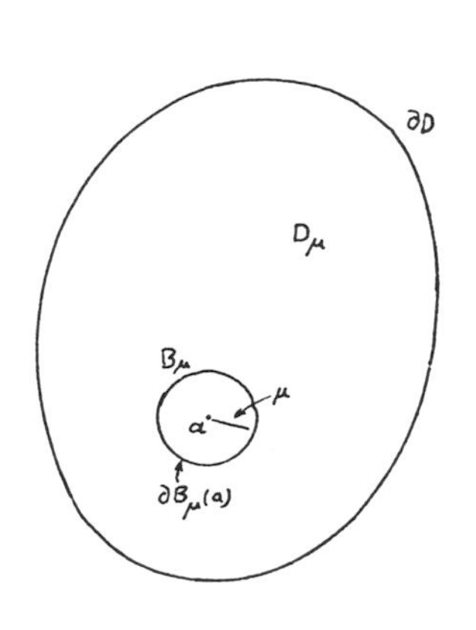

Theorem:

$$J_\mu(u_\mu) \xrightarrow[\mu \to 0]{} J(u), \text{ where } u_\mu \xrightarrow[\mu \to 0]{} u \text{ in } L^2(0,T) \text{ strongly}$$

Proof:

The specialized cost function for $v = 0$ can be written

$$J_\mu(0) = \int_{D_\mu} |y_\mu(T,0)-z_d|^2 dx = \int_{D_\mu} |z_d|^2 dx,$$

as the solution $y(t,v)$ of the initial-boundary-value (zero) problem $y(t,0) = 0$ for $v = 0$, then

$$0 \leqq J_\mu(u_\mu) = \inf_v J_\mu(v) \leqq J_\mu(0) = \int_{D_\mu} |z_d|^2 dx \leqq \int_D |z_d|^2 dx = c_0 = \text{constant}.$$

Taking the prolongation $\tilde{y}$, then

$$\int_{D_\mu} |y_\mu(T,u_\mu)-z_d|^2 dx = \int_D |\tilde{y}_\mu(T,u_\mu)-z_d|^2 dx$$

and $0 \leqq J_\mu(u_\mu) \leqq c_0$ is bounded, thus there exists a subsequence, denoted again by u_μ, such that

$$u_\mu(t) \xrightarrow[\mu\to 0]{} m(t) \text{ in } L^2(0,T) \text{ weakly.}$$

Applying a known theorem (216) we obtain that the solution

$$\tilde{y}_\mu(T,u_\mu) \xrightarrow[\mu\to 0]{} y(T,m) \text{ in } L^2(D) \text{ weakly and } m \in U$$ (see page 93)

By construction $J(u) = \inf_v J(v)$,

as $m \in U$ we have $J(u) \leqq J(m)$, then we obtain by applying the weak convergence

$$J_\mu(u_\mu) = \int_{D_\mu} |\tilde{y}(T,u_\mu)-z_d|^2 dx + M \int_0^T u_\mu^2 dt$$

$$\downarrow \mu\to 0 \qquad\qquad \downarrow \mu\to 0$$

$$\int_D |y(T,m)-z_d|^2 dx \qquad M \int_0^T m^2(t)dt$$

and $J(m)$ can be written as

$$J(m) = \int_D |y(T,m)-z_d|^2 dx + M \int_0^T m^2 dt,$$

thus we obtain

$$\liminf_{\mu\to 0} J_\mu(u_\mu) \geqq J(m) \geqq J(u)$$

by the inferior semi-continuity for the weak topology.

It remains to be shown that $\lim_{\mu\to 0} J_\mu(u_\mu) \leqq J(u)$, by methods given in (72, 221, 222, 223, 292).

II. 5 Optimal cost

We consider in some detail the optimal cost as a function of internal points of the domain. This modelling method is important for various applications.

The optimal control u is characterized by

$$(y(T)-z_d, y(T,v)-y(T)) + M\int_0^T u(v-u)dt \geqq 0 \text{ for } v \in U_{cc}, u \in U_{cc} \text{ (page 89)}$$

in the case of pointwise control (at x = a) the optimal control is given by

$$\int_0^T p(a,t) + Mu(t)\cdot(v-u)dt \geqq 0, \text{ for } v \in U_{cc}, u \in U_{cc} \text{ (page 94).}$$

In the unconstrained case $U_{cc} = U$, the integral inequality reduces to

$$p(a,t) + Mu(t) = 0$$

hence

$$u(t) = -\frac{1}{M} p(a,t) \text{ (page 94).}$$

Now we determine the optimal cost as a function of the point a.

The cost function is defined by

$$J(v) = |y(T,v)-z_d|^2 + M\int_0^T v^2 dt$$

and the solution of

$$J(u) = \inf_{v \in U_{cc}} J(v)$$

can be written as

$$u = -\frac{1}{M}\cdot p(a,t).$$

Now the optimal cost function is given by

$$J_{opt}(a) = \int_D |y(T)-z_d|^2 + M\int_0^T u^2 dt.$$

As $p(T) = y(T)-z_d$ in D and $u(t) = -\frac{1}{M} p(a,t)$,

we obtain

$$J_{opt}(a) = \int_D p^2(x,T)dx + \frac{1}{M}\int_0^T p^2(a,t)dt$$

and the optimal system can be written

$$\frac{\partial y}{\partial t} - \Delta y + \frac{1}{M} \cdot p \cdot \delta(x-a) = 0 \text{ in } D \times (0,T) \qquad -\frac{\partial p}{\partial t} - \Delta p = 0 \text{ in } D \times (0,T)$$

$$y = 0 \text{ on } \partial D \times (0,T) \qquad p = 0 \text{ on } \partial D \times (0,T)$$

$$y(x,0) = 0 \text{ in } D, \qquad p(T) = y(T) - z_d \text{ in } D.$$

The solutions are given by

$$y = y(x,t,a) \text{ and } p = p(x,t,a).$$

The optimal cost as a function of the point a can be written as

$$J_{opt}(a) = \int_D p^2(x,T,a)dx + \frac{1}{M}\int_0^t p^2(a,t,a)dt.$$

We determine the formal derivative of $J_{opt}(a)$,

$$\frac{\partial J_{opt}(a)}{\partial a} = 2 \cdot \int_D p \cdot (\frac{\partial p}{\partial x} \cdot \frac{\partial x}{\partial a} + \frac{\partial p}{\partial T} \cdot \frac{\partial T}{\partial a} + \frac{p}{a} \cdot 1)dx + \frac{1}{M}\int_0^T \frac{\partial}{\partial x_i} p^2(a,t,a)\Big|_{x=a} +$$

$$\frac{2}{M}\int_0^T p(a,t) \cdot \dot{p}(a,t)dt$$

$$= 2(p(T).\dot{p}(T)) + \frac{1}{M}\int_0^T \frac{\partial}{\partial x_i} p^2(a,t,a)dt.$$

As $y(T) - z_d = p(T)$, then $y'(T) = p'(T)$, the optimal state reads

$$\frac{\partial y}{\partial t} - \Delta y = -\frac{1}{M} \cdot p \cdot \delta(x-a),$$

differentiating with respect to a

$$\frac{\partial}{\partial a}(\frac{\partial y}{\partial t}) - \Delta \dot{y} = -\frac{1}{M}\frac{\partial}{\partial a}(p \cdot \delta(x-a)) = -\frac{1}{M}(\dot{p} \cdot \delta(x-a) - p \cdot \frac{\partial}{\partial a_i}(\delta(x-a)) = 0,$$

$$\text{as } \frac{\partial}{\partial a_i}(\delta(x-a)) = -\frac{\partial}{\partial x_i}(\delta(x-a))$$

$$\frac{\partial \dot{y}}{\partial t} - \Delta \dot{y} = -\frac{1}{M} - (\dot{p} \cdot \delta(x-a) - \frac{\partial}{\partial x_i}(\delta(x-a) \cdot p).$$

Now we take the scalar product with p

$$\int_{D\times(0,T)} \left(\frac{\partial \dot{y}}{\partial t} - \Delta\dot{y}\right) p\,dx\,dt = -\frac{1}{M}\int_{D\times(0,T)} p\cdot\dot{p}\,\delta(x-a)\,dx\,dt +$$

$$\int_{D\times(0,T)} p\cdot p\cdot\frac{\partial}{\partial x}(\delta(x-a))\,dx\,dt.$$

If $a \in D$ and $\delta(x-a)$ be the Dirac distribution concentrated at a, then

$$\int_D \varphi(x)\cdot\delta(x-a)\,dx = \varphi(a) \text{ for } \varphi \in C^0(D),$$

then

$$\int_{D\times(0,T)} p\cdot\dot{p}\,\delta(x-a)\,dx\,dt = \int_0^T p(a,t)\cdot\dot{p}(a,t)\,dt,$$

and by using the Green's formula

$$\int_{D\times(0,T)} p^2 \frac{\partial}{\partial x_i}(\delta(x-a))\,dx\,dt = \int_0^T p^2.\delta(x-a)\Big|_{\partial D} dt -$$

$$- \int_{D\times(0,T)} \left(\frac{\partial}{\partial x_i}\cdot p^2\right)\cdot\delta(x-a)\,dx\,dt = -\int_0^T \left(\frac{\partial}{\partial x_i} p^2\right)\Big|_{x=a} dt, \text{ as } p = 0 \text{ on } \partial D\times(0,T).$$

The integral satisfies

$$\int_{D\times(0,T)} \left(\frac{\partial}{\partial t} - \Delta\right)(\dot{y})p\,dx\,dt = \dot{y}p\Big|_0^T - \int_{D\times(0,T)} \dot{y}\left(\frac{\partial}{\partial t} - \Delta\right)p\,dx\cdot dt$$

$$= \dot{y}(T)\cdot p(T) = \dot{p}(T)\cdot p(T),$$

we have $\left(\frac{\partial}{\partial t} - \Delta\right)p = 0$ by the adjoint state and as $p(0) = 0$ on the boundary and $y(T) - z_d = p(T)$ in D, hence $\dot{y}(T) = \dot{p}(T)$. Finally we obtain

$$(\dot{p}(T)\cdot p(T)) = -\frac{1}{M}\cdot\int_0^T p(a,t).\dot{p}(a,t)\,dt - \frac{1}{M}\int_0^T \left(\frac{\partial}{\partial x_i} p^2\right)\Big|_{x=a} dt,$$

using this in $\dfrac{\partial J_{opt}(a)}{\partial a}$ we have

$$\frac{\partial}{\partial a} J_{opt}(a) = -\frac{1}{M}\int_0^T \left(\frac{\partial}{\partial x_i} p^2\right)\Big|_{x=a} dt = -\frac{2}{M}\int_0^T p(a,t)\frac{\partial}{\partial x_i}(p(x,t)\,dt.$$

This formal expression makes sense, one can show that $a \to J_{opt}(a)$ is a $C^1(D)$ function, if $z_d \in H_0^1(D)$. It is possible to prove that J_{opt} is C^1 up to the

the boundary (319).

A partial answer is given in the following theorem describing redistribution processes.

II.5.1 Redistribution processes

<u>Theorem</u>:

Let z_d be in $L^2(D)$, the $J_{opt} \in C^0(D)$ and $J_{opt} = |z_d|^2$ on the boundary.

Let b_k be a sequence in D, such that $b_k \to a$ in D, and v_o be the control parameter, $v_o \in \mathcal{D}(0,T)$ the space of infinitely differentiable functions in D (see page 2).

We have the state equation and the adjoint equation

$$\frac{\partial y}{\partial t} - \Delta y = v_o(t)\cdot\delta(x-b_k) \text{ in } Dx(0,T) \qquad -\frac{\partial p}{\partial t} - \Delta p = 0 \text{ in } D \times (0,T)$$

$$y = 0 \text{ on } \partial D \times (0,T) \qquad p = 0 \text{ on } \partial D \times (0,T)$$

$$y(x,0) = 0 \text{ in } D, \qquad p(T) = y(T)-z_d \text{ in } D.$$

The cost function is given by

$$J(v) = |y(T,v)-z_d|^2 + M\int_0^T v^2 dt,$$

and the optimal cost function

$$J_{opt}(a) = \int_D p^2(x,T,a)dx + \frac{1}{m}\int_0^T p^2(a,t,a)dt.$$

The solution of the state equation depends on

$$y = y(t,v_o,b_k)$$

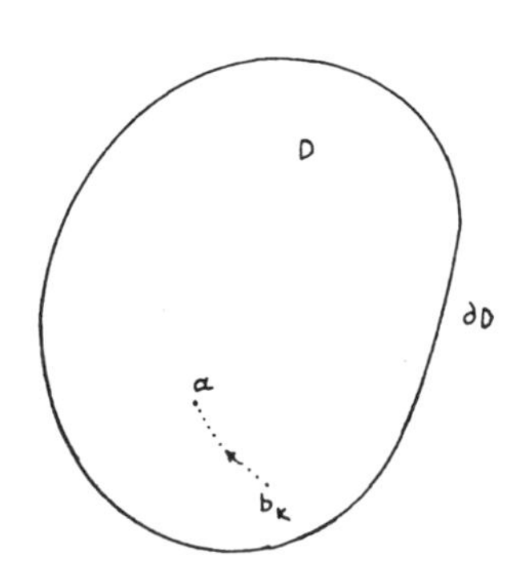

and the corresponding cost function

$$J(v_o,b_k) = |y(T,v_o.b_k)-z_d|^2 + M\int_0^T v_o^2 dt,$$

then we obtain by construction

$$J_{opt}(b_k) \leq J(v_o,b_k).$$

As $v_0 \in \mathcal{D}(0,T)$ the solution $y(T,v_0,b_k) \xrightarrow{\text{in } L^2(D)} y(T,v_0,a)$ if $b_k \xrightarrow[k\to\infty]{} a$, thus $J(v_0,b_k) \to J(v_0,a)$ with $J_{opt}(b_k) \leq J(v_0,b_k)$, we obtain

$$\lim_k J_{opt}(b_k) \leq J(v_0,a) \text{ for } v_0 \in \mathcal{D}(0,T),$$

as $\mathcal{D}$ is dense in U, we have

$$\lim_k J_{opt}(b_k) \leq J_{opt}(a).$$

Now we use the duality theory of Rockafellar (292).

We have the unique solution $y(T,v)$, thus there exists an operator

$$\Lambda: L^2(0,T) \to L^2(D)$$
$$v \to y(T,v),$$

where $\Lambda v = y(T,v)$ and Λ is an unbounded operator, we define (see page 93)

$$D(\Lambda) = \{v \in L^2(0,T) \mid \Lambda v \in L^2(D)\} = U.$$

The cost function can be written

$$J(v) = |\Lambda v - z_d|^2 + M\cdot \int_0^T v^2 dt,$$

defining $\mathcal{E}(v) = M \int_0^T v^2 dt$ for $v \in L^2(0,T)$ and $\mathcal{B}(k) = |k-z_d|^2$ for $k \in L^2(D)$,

$$J(v) = \mathcal{B}(\Lambda v) + \mathcal{E}(v) \text{ for } v \in D(\Lambda).$$

The conjugate functionals $\mathcal{E}^*$ and $\mathcal{B}^*$ of $\mathcal{E}$ and $\mathcal{B}$ are given by

$$\mathcal{E}^*(v) = \frac{1}{4M} \int_0^T v^2 dt \text{ for } v \in L^2(0,T) \text{ and } \mathcal{B}^* = \frac{1}{4} |k|^2 + (k,z_d) \text{ for } k \in L^2(D).$$

An application of Rockafellars duality theorem yields (see I. Ekeland and R. Teman (97)):

$$\inf_{v \in D(\Lambda)} \{\mathcal{E}(v) + \mathcal{B}(\Lambda v)\} = - \inf_{k \in D(\Lambda^*)} \{\mathcal{E}^*(\Lambda^* k) + \mathcal{B}^*(-k)\}.$$

The construction of the operator Λ^*.

We have the state equation on page 114. Let k be given in $L^2(D)$, we define $z(k)$ as the solution of

$$\frac{\partial z}{\partial t} - \Delta z = 0 \text{ in } D \times (0,T)$$

$$z = 0 \text{ on } \partial D \times (0,T)$$

$$z(T) = k \text{ in } D,$$

multiplying by $y(v)$ and applying the Green's formula, we have

$$-\int \frac{\partial z}{\partial t} \cdot y(v)dxdt = \int \Delta z \cdot y(v)dxdt = \int z \cdot \Delta y(v)\, dxdt,$$

now integration by parts yields

$$\int_0^T \frac{\partial z}{\partial t} \cdot y(v)dt = z(T) \cdot y(T,v) - z(0) \cdot y(0) - \int_0^T z(t) . \frac{\partial y(v)}{\partial t}\, dt,$$

as $z(T) = k$ in D and $y(0) = 0$ in D, we obtain

$$-\int z \cdot \Delta y(v)dxdt = k \cdot y(T,v) - \int_0^T z(t) \cdot \frac{\partial y(v)}{\partial t}\, dt,$$

thus for $k \in \mathcal{D}(D)$ and $v \in \mathcal{D}(0,T)$ we obtain

$$(k,y(T,v)) = \int_0^T \int_D z(x,t)(\frac{\partial y(v)}{\partial t} - \Delta y(v))dxdt =$$

$$\int_0^T \int_D z(x,t) \cdot v(t) \delta(x-a)dxdt = \int_0^T z(a,t)v(t)dt.$$

As $\mathcal{D}(0,T)$ is dense in $U = D(\Lambda)$, we have

$$(k,y(T,v)) = \int_0^T z(a,t)v(t)dt \text{ for } v \in U,$$

and $z(a,t) \in U'$, then the integral $\int_0^T z(a,t) \cdot v(t)dt$ denotes the duality between U' and U.

Finally we obtain, that $\Lambda^*k = z(a,t)$, as we have defined $\Lambda v = y(T,v)$ and by applying it to

$$(k,y(T,v)) = \int_0^T z(a,t)v(t)dt,$$

we obtain the result

$$(K, y(T,v)) = (K, \Lambda v) = (\Lambda^* K, v) = \int_0^T z(a,t)v(t)dt.$$

The operator Λ^* satisfies

$$\Lambda^* \in L(L^2(D), U')$$

and we have the subspace

$$D(\Lambda^*) = \{k \in L^2(D) \mid \Lambda^* k = z(a,t) \in L^2(0,T)\}.$$

The duality theorem reads

$$\inf_{v \in D(\Lambda)} \{\mathbb{E}(v) + \mathbb{B}(\Lambda v)\} = - \inf_{k \in D(\Lambda^*)} \{\mathbb{E}^*(\Lambda^* k) + \mathbb{B}^*(-k)\} =$$

$$\inf_{k \in D(\Lambda^*)} \{\frac{1}{4M} \int_0^T z^2(a,t,k)dt + \frac{1}{4} |k|^2 - (k, z_d)\}.$$

Let z be a solution of the initial-boundary-value problem and by setting

$$\mathbb{A}(k,a) = \frac{1}{4M} \int_0^T z^2(a,t,k)dt + \frac{1}{4} |k|^2 - (k, z_d),$$

Rockafellar's duality theorem can be written

$$J_{opt}(a) = - \inf_{k \in D(\Lambda^*)} \mathbb{A}(k,a) \underset{def.}{=} -Q(a).$$

We have to prove that $J_{opt} \in C(\bar{D})$, therefore it remains to be shown, that J_{opt} is lower semi-continuous, this is characterized by the inequality

$$\liminf_{b_k \to a} J_{opt}(b_k) \geqq J_{opt}(a),$$

and this is equivalent to $\overline{\lim}\, Q(b_k) \leqq Q(a)$.

We take $k = m$, where $m \in \mathcal{D}(D)$ then

$$Q(b_k) \leqq \mathbb{A}(m, b_k) \text{ by construction.}$$

For $b_k \to a$ in D, we obtain $\mathbb{A}(m,b_k) \to \mathbb{A}(m,a)$, as the solution $z(a,t,k)$ is continuous, thus we can take the limit as $m \in \mathcal{D}(D)$. Therefore

$$Q(b_k) \leqq \mathbb{A}(m,b_k) \leqq \mathbb{A}(m,a) \text{ for } m \in \mathcal{D}(D)$$

and

$$\limsup_{k\to\infty} Q(b_k) \leq \mathbb{A}(m,a) \text{ for } m \in \mathcal{D}(D),$$

as $\mathcal{D}(D)$ is dense in $D(\Lambda^*)$.

II.5.2 Optimal cost on the boundary of the domain

Now we determine the optimal cost on the boundary of a domain. The state equation reads for $v_o \in \mathcal{D}(0,T)$

$$\frac{\partial y}{\partial t} - \Delta y = v_o(t)\delta(x-b_k) \text{ in } D \times (0,T)$$

$$y = 0 \text{ on } \partial D \times (0,T)$$

$$y(0) = 0 \text{ in } D.$$

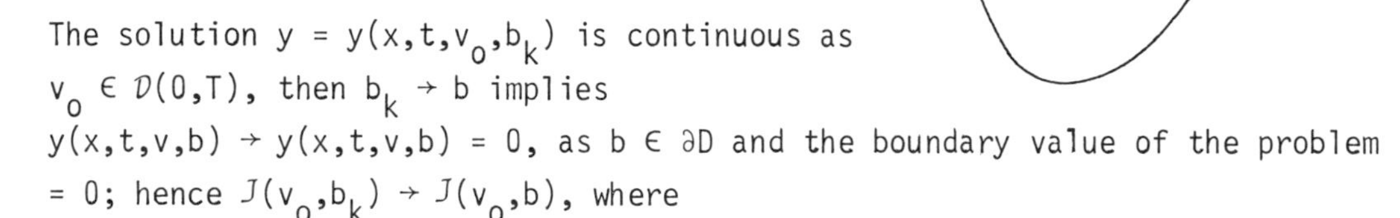

The solution $y = y(x,t,v_o,b_k)$ is continuous as $v_o \in \mathcal{D}(0,T)$, then $b_k \to b$ implies $y(x,t,v,b) \to y(x,t,v,b) = 0$, as $b \in \partial D$ and the boundary value of the problem $= 0$; hence $J(v_o,b_k) \to J(v_o,b)$, where

$$J(v_o,b_k) = |y(T,v_o,b_k)-z_d|^2 + M\int_0^T v_o^2 \, dt.$$

By construction $J_{opt}(b_k) \leq J(v_o,b_k)$, as $y(x,t,v_o,b_k) \to 0$ for $b_k \to b$, we obtain

$$\limsup_k J_{opt}(b_k) \leq \limsup_k J(v_o,b_k) \text{ for } v_o \in \mathcal{D}(0,T), \text{ thus}$$

$$\limsup_k J_{opt}(b_k) \leq |z_d|^2.$$

II. 5.3 Evolution of the minimal cost approaching the boundary

On the other hand, we consider the solution of

$$\frac{\partial z}{\partial t} - \Delta z = 0 \text{ in } D \times (0,T)$$

$$z = 0 \text{ on } \partial D \times (0,T)$$

$$z(T) = m_o \text{ in } D.$$

Let m_o be in $\mathcal{D}(D)$, then the solution $z(b_k,m_o)$ is continuous, and for $b_k \to b \in \partial D$, $z(b_k,t,m_o) \to$ the boundary value zero in $L^2(0,T)$. We have

$$\mathbb{A}(m_o,b_k) = \frac{1}{4M}\int_0^T z^2(b_k,t,m_o)dt + \frac{1}{4}|m_o|^2 - (m_o,z_d) \text{ for } m_o \in \mathcal{D}(D).$$

By construction

$$Q(b_k) \leqq \mathbb{A}(m_o,b_k), \text{ for } b_k \to b \in \partial D,$$

then we obtain

$$\limsup_k Q(b_k) \leqq \frac{1}{4}|m_o|^2 - (m_o,z_d) \text{ for } m_o \in \mathcal{D}(D),$$

thus

$$\limsup_k Q(b_k) \leqq -|z_d|^2, \text{ this is equivalent to } \limsup_k (-Q(b_k) \geqq |z_d|^2,$$

as $J_{opt}(b_k) = -Q(b_k)$ we obtain $\limsup_k J_{opt}(b_k) \geqq |z_d|^2$, and finally there follows

$$\lim_{b_k\to b} J_{opt}(b_k) = |z_d|^2.$$

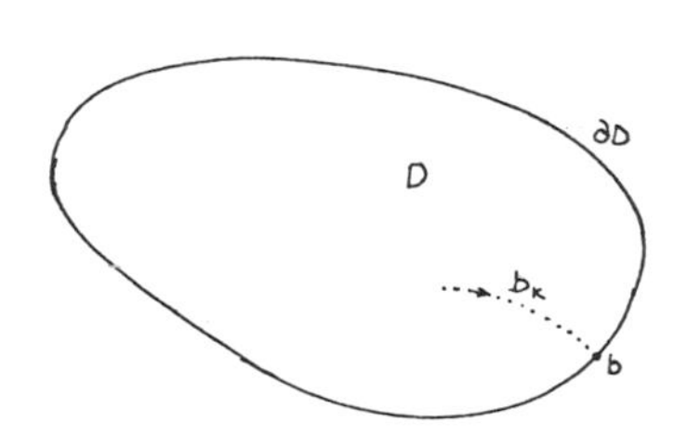

$J_{opt}(b_k)$ is the optimal cost at the points b_k.
The result shows the evolution of the optimal cost (= the minimal cost) by approaching the boundary.
The minimal cost J_{opt} is fixed at the boundary.
One possible interpretation in economics. Approaching the boundary ∂D of the domain D, the minimal cost J_{opt} is fixed, thus the minimal cost is reduced to the fixed cost $|z_d|^2$.

II. 6 Some aspects of game theory in multiconnected domains

II. 6.1 Variational inequalities

We are giv en a game with two players, who are acting on two control parameters v_1 and v_2.

Let the control parameters or functions be elements of real Hilbert spaces $\mathcal{H}_i$ for i = 1,2 with norm $\| \quad \|_{\mathcal{H}_i}$. A third real Hilbert space $\mathbb{E}$ with norm $\| \quad \|_{\mathbb{E}}$ is given. We assume that there are linear operators O_i i = 1,2 from Hilbert space $\mathcal{H}_i$ to $\mathbb{E}$, denoted by $O_i \in L(\mathcal{H}_i,\mathbb{E})$.

Now we introduce a cost function which depends on the control parameters v_1,v_2

$$J(v_1,v_2) = \|O(v_1) + O(v_2)-z_d\|^2_{\mathbb{E}} + C_1\|v_1\|_{\mathcal{H}_1} - C_2\|v_2\|_{\mathcal{H}_2},$$

where z_{d_i} are the fixed cost in $\mathfrak{E}$ and C_1, C_2 are positive constants.

Let U^i_{cc} be a closed, convex, non empty subset of $\mathfrak{H}_i$. We consider the game

$$\inf_{v_1} \sup_{v_2} J(v_1,v_2) = \sup_{v_2} \inf_{v_1} J(v_1,v_2) = J(u_1,u_2),$$

with control parameters $v_i \in U^i_{cc}$.

We have to find the unique solution (u_1,u_2) of the game, where $u_i \in U^i_{cc}$, for $i = 1,2$.

First we are showing the existence of a solution.

Theorem: The problem of finding the element $(u_1,u_2) = u$ is equivalent to the variational inequalities

$$\left(\frac{\partial}{\partial u_1} J(u), v_1-u_1\right) \geqq 0 \text{ for } v_1 \in U^1_{cc}$$

$$\left(\frac{\partial}{\partial u_2} J(u), v_2-u_2\right) \leqq 0 \text{ for } v_2 \in U^2_{cc}$$

where $u = (u_1,u_2) \in U^1_{cc} \times U^2_{cc}$.

Proof: We know (see page 85), that the unique solution of

$$J(u) = \inf_{v \in U_{cc}} \geqq J(v),$$

where $u \in U_{cc}$ is characterized by

$$\frac{1}{2}\,(J'\,(u), v-u) \geqq 0 \text{ for } v \in U_{cc},$$

where we defined

$$(J'(u),v) = \frac{d}{d\mu}\,(J(u+\mu v))\Big|_{\mu=0} .$$

Consider the game, then

$$\inf_{v_1 \in U_{cc}} J(v_1,\cdot) = J(u_1,\cdot)$$

is characterized by

$$\frac{1}{2}\,(J'(u_1), v_1-u_1) \geqq 0 \text{ for } v_1 \in U_{cc}.$$

and

$$\sup_{v_2 \in U_{cc}} \ldots J(\cdot,v_2) = J(\cdot,u_2)$$

is characterized by

$$\frac{1}{2}(J'(u_2),v_2-u_2) \leqq 0 \text{ for } v_2 \in U_{cc}.$$

These variational inequalities can be written in the following equivalent form. The derivative of the cost functional

$$J(v_1,v_2) = \|O_1(v_1) + O_2(v_2)-z_d\|_{\mathbb{E}}^2 + C_1\|v_1\|_{H_1} - C_2\|v_2\|_{H_2}$$

can be evaluated

$$\frac{1}{2}(J'(u_1),v_1-u_1) = \frac{d}{d\mu}J(u_1+\mu(v_1-u_1),u_2)\Big|_{\mu=0},$$

hence

$$\frac{d}{d\mu}J(u_1+\mu(v_1-u_1)u_2)\Big|_{\mu=0} = \frac{d}{d\mu}(\|O_1((u_1)+\mu(v_1-u_1)) + O_2(u_2)-z_d\|_{\mathbb{E}}^2$$

$$+ C_1\|u_1+\mu(v_1-u_1)\|_{H_1} - C_2\|v_2\|_{H_2})\Big|_{\mu=0}$$

$$= (O_1(u_1) + O_2(u_2)-z_d,\ O_1(v_1-u_1))_{\mathbb{E}} + C_1(u_1,v_1-u_1)_{H_1}$$

and

$$\frac{d}{d\mu}(J(u_2+\mu(v_2-u_2),u_1))\Big|_{\mu=0} = (O_2(u_2)+O_1(u_1)-z_d,O_2(v_2-u_2))_{\mathbb{E}}-C_2\cdot(u_2,v_2-u_2)_{H_2};$$

the variational inequality follows

$$(O_2(u_2) + O_1(u_1)-z_d,\ O_2(v_2-u_2))_{\mathbb{E}} - C_2(u_2,v_2-u_2)_{H_2} \leqq 0$$

and equivalent

$$-(O_2(u_2) + O_1(u_1)-z_d,\ O_2(v_2-u_2))_{\mathbb{E}} + C_2(u_2,v_2-u_2)) \geqq 0,$$

this characterizes

$$\frac{1}{2}(J'(u_2),v_2-u_2) \leqq 0 \text{ for } v_2 \in U_{cc}.$$

By the first evaluation

$$\frac{1}{2}(J'(u_1),v_1-u_1) \geqq 0 \text{ for } v_1 \in U_{cc}$$

is characterized by

$$(O_1(u_1)+O_2(u_2)-z_d,\ O_1(v_1-u_1))_{\mathbb{E}} + C_1(u_1,v_1-u_1)_{H_1} \geqq 0.$$

This can be written as

$$(O_1(u_1)+O_2(u_2),O_1(v_1-u_1))_{\mathbb{E}}-(z_d,O_1(v_1-u_1))_{\mathbb{E}}+C_1(u_1,v_1-u_1)_{\mathbb{H}_1} \geqq 0,$$

$$-(O_1(u_1)+O_2(u_2),O_2(v_2-u_2))_{\mathbb{E}}+(z_d,O_2(v_2-u_2))_{\mathbb{E}}+C_2(u_2,v_2-u_2)_{\mathbb{H}_2} \geqq 0$$

and equivalent

$$(O_1(u_1)+O_2(u_2),\ O_1(v_1-u_1))_{\mathbb{E}} + C_1(u_1,v_1-u_1)_{\mathbb{H}_1} \geqq (z_d,O_1(v_1-u_1),$$

$$-(O_1(u_1)+O_2(u_2),O_2(v_2-u_2))_{\mathbb{E}} + C_2(u_2,v_2-u_2)_{\mathbb{H}_2} \geqq -(z_d,O_2(v_2-u_2));$$

adding these inequalities, we obtain the variational inequality

$$B(u,v-u) \geqq F(\cdot,v-u),$$

where the bilinear forms are

$$B(u,v-u) = (O_1(u_1)+O_2(u_2),O_1(v_1-u_1)-O_2(v_2-u_2))_{\mathbb{E}}+C_1(u_1,v_1-u_1)_{\mathbb{H}_1} + C_2(u_2,v_2-u_2)_{\mathbb{H}_2}$$

and

$$F(\cdot,v-u) = (z_d,O_1(v_1-u_1)-O_2(v_2-u_2)).$$

Now, we apply the general theorem 3 (see page 85) to

$$B(u,v-u) \geqq F(\cdot,v-u) \text{ for } v \in U^1_{cc} \times U^2_{cc}.$$

As the linear form $v \rightarrow F(\cdot,v)$ is continuous on $\mathbb{H}_1 \times \mathbb{H}_2$, the variational inequality admits a unique solution, if the bilinear-form B satisfies

$$B(v,v) \geqq \alpha(\|v_1\|^2_{\mathbb{H}_1} + \|v_2\|^2_{\mathbb{H}_2}) \text{ for } \alpha > 0 \text{ and } v_i \in \mathbb{H}_i.$$

According to the definition of B

$$\begin{aligned} B(v,v) &= C_1(v_1,v_1)_{\mathbb{H}_1} + C_2(v_2,v_2)_{\mathbb{H}_2} + (O_1(v_1)+O_2(v_2),O_1(v_1)-O_2(v_2))_{\mathbb{E}} \\ &= C_1 \|v_1\|^2_{\mathbb{H}_1} + C_2 \|v_2\|^2_{\mathbb{H}_2} + \|O_1(v_1)\|^2_{\mathbb{E}} - \|O_2(v_2)\|^2_{\mathbb{E}}. \end{aligned}$$

The operators $O_i \in L(\mathbb{H}_i,\mathbb{E})$ are bounded and satisfy $\|O_2(v_2)\|^2_{\mathbb{E}} \leqq 1_2 \|v_2\|_{\mathbb{H}_2}$, or equivalent $- \|O_2(v_2)\|^2_{\mathbb{E}} \geqq - 1_2 \|v_2\|^2_{\mathbb{H}_2}$.

For $C_2 > 1_2$, we have $(C_2-1_2) > 0$, then there exists $\alpha > 0$ and we are able to estimate

$$B(v,v) \geqq C_1 \|v_1\|^2_{\mathbb{H}_1} +(C_2-1_2) \|v_2\|^2_{\mathbb{H}_2}+ \|O_1(v_1)\|^2_{\mathbb{E}} >\alpha(\|v_1\|^2_{\mathbb{H}_1}+ \|v_2\|^2_{\mathbb{H}_2}).$$

If $C_2 > 1_2$, e.g. C_2 large enough, there exists $\alpha > 0$, such that the problem

$$\inf_{v_1} \sup_{v_2} J(v_1,v_2) = \sup_{v_2} \inf_{v_1} J(v_1,v_2) = J(u_1,u_2)$$

admits a unique solution $u = (u_1,u_2)$ characterized by the variational inequality

$$(0_1(u_1) + 0_2(u_2)-z_d)_{\mathcal{E}} + C_1(u_1,v_1-u_1)_{\mathcal{H}_1} \geq 0 \text{ for } v_1 \in U^1_{cc}$$
$$-(0_1(u_1) + 0_2(u_2)-z_d)_{\mathcal{E}} + C_2(u_2,v_2-u_2)_{\mathcal{H}_2} \geq 0 \text{ for } v_2 \in U^2_{cc}.$$

II. 6.2 Diffusion processes governed by differential games

Let the diffusion process be described by the parabolic system

$$\frac{\partial y(x,t)}{\partial t} - \Delta y(x,t) = v_1(x,t)\cdot g(x,t) \text{ in } D \times (0,T)$$

$$\frac{\partial y(x,t)}{\partial n} = v_2(x,t) \text{ on the boundary } \partial D \times (0,T)$$

$$y(x,0) = 0 \quad \text{initial condition in } D.$$

Let g, v_1 be in $L^2(D \times (0,T))$ and $v_2 \in L^2(D \times (0,T))$, the system admits a unique solution $y = y(x,t,v_1,v_2)$, such that

$$y(x,T,v_1,v_2) \in L^2(D \times (0,T)),$$

where v_1 and v_2 are the strategy parameters of the game.

The strategy v_1 is applied to D and v_2 to the boundary ∂D. In possible applications g can be chosen e.g. as Dirac's distribution, $\delta(x-a)$, and v_1 only depending on t, hence $v_1(t)$.

The diffusion process, given by the solution $y(T,g,v_1,v_2)$ is used in the cost functional

$$J(v) = J(v_1,v_2) = |y(T,v_1,v_2)-z_d|^2 + C_1 \int_{Dx(0,T)} v_1^2 dxdt$$
$$- C_2 \int_{\partial Dx(0,T)} v_2^2 \, d(\partial D)dt.$$

The solution $y(T,v_1,v_2)$ can be written in operator form:
We assume that $v_1 = 0$, then the system $\frac{\partial y}{\partial t} = \Delta y$ may be written in variational form

$$(\tfrac{\partial y}{\partial t},y) = (\Delta y,y)_D.$$

By Green's formula we have

$$\int_D \Delta y \cdot y dx = \int_{\partial D} \frac{\partial y}{\partial n} \cdot y ds - \int_D \text{grad} y \cdot \text{grad} y \, dx = \int_{\partial D} v_2 \cdot y \, ds - \int_D \text{grad} y \cdot \text{grad} y dx,$$

hence

$$(\frac{dy}{dt}, y) + \int_D \text{grad} y \cdot \text{grad} y \, dx = \int_{\partial D} v_2 \cdot y ds.$$

This can be written

$$\frac{1}{2} \frac{d}{dt} |y(t)|^2 + |\text{grad} y(T, g, v_1, v_2)|^2 = (v_2, y(T, \ldots))_{\partial D},$$

as

$$\int_0^t \frac{d}{dt} |y(\tau)|^2 d\tau = |y(t)|^2 - |y(0)|^2 = |y(t)|^2, \; y(x,0) = y(0) = 0.$$

We obtain by integrating and taking $t = T$

$$|y(T)|^2 + 2 \int_0^T |\text{grad } y(T, \ldots)|^2 \, d\tau = 2 \int_0^T \int_{\partial D} v_2(\tau).y(\tau, \ldots) d\tau dx,$$

if $y(\tau) \leqq \max_{0<\tau\leqq T} y(\tau) = \text{constant} = l_2$, we have finally

$$|y(T)|^2 < 2 \cdot l_2 \int_0^T |v_2(\tau)|^2_{\partial D} dt, \text{ hence } |y(T)|^2 \leqq l_2 \, \|v_2\|^2_{\mathcal{H}_2}.$$

Thus we can apply the above characterization of the differential game. The corresponding adjoint equation of the parabolic system is given by

$$-\frac{\partial p}{\partial t} - \Delta p = 0 \text{ in } D \times (0,T)$$

$$\frac{\partial p}{\partial n} = 0 \text{ on } \partial D \times (0,T), \text{ the boundary condition}$$

$$p(T) = y(T) - z_d, \text{ the initial condition;}$$

multiplying by $y(v)-y$ and applying Green's formula, we obtain

$$\int_D \Delta p(y(v)-y) dx = \int_{\partial D} \frac{\partial p}{\partial n} (y(v)-y) ds - \int_{\partial D} p \cdot \frac{\partial}{\partial n} (y(v)-y) ds +$$

$$+ \int_D p \cdot \Delta(y(v)-y) ds, \text{ as } \frac{\partial p}{\partial n} = 0 \text{ on } \partial D,$$

$$= \int_{\partial D} p(\frac{\partial y(v)}{\partial n} - \frac{\partial y}{\partial n}) ds + \int_D p(\Delta y(v) - \Delta y) dx,$$

as $\frac{\partial y(v)}{\partial n} = v_2$ on the boundary and let $\frac{\partial y}{\partial n} = u_2$ be the optimal solution, then

$$= -\int_{\partial D} p(v_2-u_2)ds + \int_D p(\Delta y(v)-\Delta y)dx.$$

Integrating by parts gives

$$\int_0^T \frac{\partial p}{\partial t}(y(v)-y)dt = p\cdot(y(t,v)-y(t))\Big|_{t=0}^{t=T} - \int_0^T p\frac{\partial}{\partial t}(y(v)-y)dt$$

$$= p(T)\cdot(y(T,v)-y(T))-p(0)(y(0,v)-y(0)) - \ldots$$

the parabolic system and the optimal parabolic system can be written as

$$\frac{\partial y(v)}{\partial t} - \Delta y(v) = v_1\cdot g \qquad \frac{\partial y}{\partial t} - \Delta y = u_1\circ g$$

$$\frac{\partial y(v)}{\partial n} = v_2 \qquad \frac{\partial y}{\partial n} = u_2$$

$$y(0,v) = 0, \qquad y(0) = 0, \text{ and } P(T) = y(T)-z_d,$$

so we obtain

$$= (y(T)-z_d)(y(T,v)-y(T) - \int_0^T p\frac{\partial}{\partial t}(y(v)-y)dt, \text{ then}$$

$$\int_{Dx(0,T)} \frac{\partial p}{\partial t}(y(v)-y)dtdx = ((y(T)-z_d)(y(T,v)-y(T)))_D -$$

$$- \int_{Dx(0,T)} p\frac{\partial}{\partial t}(y(v)-y)dtdx.$$

Now we are combining the results on page 124 and 125

$$\int_0^T\int_D \Delta p(y(v)-y)dxdt = \int_0^T\int_D p\cdot\Delta(y(v)-y)dxdt - \int_0^T\int_{\partial D} p(v_2-u_2)dsdt$$

$$= \int_D\int_0^T \frac{\partial p}{\partial t}(y(v)-y)dxdt$$

$$= - ((y(T)-z_d)\cdot(y(T,v)-y(T))_D + \int_0^T\int_D p\frac{\partial}{\partial t}(y(v)-y)dtdx.$$

Thus we have

$$-\int_0^T\int_D p\cdot\Delta(y(v)-y)dxdt + \int_0^T\int_{\partial D} p(v_2-u_2)dsdt +$$

$$\int_0^T\int_D p\cdot(\frac{\partial y(v)}{\partial t} - \frac{\partial y}{\partial t})dtdx = ((y(T)-z_d)(y(T,v)-y(T))_D,$$

hence

$$\int_0^T \int_D p\left(\left(\frac{\partial y(v)}{\partial t} - \Delta y(v)\right) - \left(\frac{\partial y}{\partial t} - \Delta y\right)\right)dxdt + \int_0^T \int_{\partial D} p(v_2-u_2)dsdt = (\qquad)_D.$$

As the parabolic and optimal parabolic systems are satisfied, we have

$$\int_0^T \int_D p\cdot g(v_1-u_1)dxdt + \int_0^T \int_{\partial D}^T p(v_2-u_2)dsdt = ((y(T)-z_d)(y(T,v)-y(T))_D.$$

The right-hand side can be expressed as follows:

We have shown on page 111 that the cost functional

$$J(v) = |y(T,v)-z_d|^2 + M\cdot\int_0^T v^2 dt$$

satisfies

$$\frac{1}{2}(J'(u),v-u) = (y(T)-z_d,y(T,v)-y(T)) + M\int_0^T u(v-u)dt.$$

It can be shown in the same way, that the cost functional of the differential game satisfies

$$\frac{1}{2}(J'(u_1),v_1-u_1) = (\qquad\qquad) + C_1\int_{Dx(0,T)} u_1(v_1-u_1)dtdx$$

and analogously

$$\frac{1}{2}(J'(u_2),v_2-u_2) = (\qquad\qquad) - C_2\int_{\partial Dx(0,T)} u_2(v_2-u_2)dtdx.$$

Using this, we can write

$$\int_0^T \int_D p\cdot g(v_1-u_1)dxdt = (\qquad\qquad)_D = \frac{1}{2}(J'(u_1),v_1-u_1)-C_1\int_{Dx(0,T)} u_1(v_1-u_1)dtdx$$

Thus

$$\int_0^T \int_D (p\cdot g+C_1u_1)\cdot(v_1-u_1)dxdt = \frac{1}{2}(J'(u_1),v_1-u_1) \geqq 0$$

by the variational inequality and for $v_1 \in U^1_{cc}$ (page 120). On the other hand

$$\int_0^T \int_{\partial D} p(v_2-u_2)dsdt = (\qquad)_D = \frac{1}{2}(J'(u_2),v_2-u_2)+C_2\int_{\partial Dx(0,T)} u_2(v_2-u_2)dsdt,$$

then

$$\int_0^T \int_{\partial D} (-p+C_2u_2)\cdot(v_2-u_2)dsdt = -\frac{1}{2}(J'(u_2),v_2-u_2) \geqq 0,$$

as $(J'(u_2), v_2-u_2) \leqq 0$ for $v_2 \in U^2_{cc}$ (page 121).

II. 6.3 Result

The diffusion process with the control variables v_1 and v_2

$$\frac{\partial y}{\partial t} - \Delta y = v_1 \cdot g \text{ in } D \times (0,T)$$

$$\frac{\partial y}{\partial n} = v_2 \quad \text{on } \partial D \times (0,T)$$

$$y(x,0) = 0 \quad \text{in } D,$$

has a unique solution $y(T, v_1, v_2)$, by introducing the cost functional

$$J(v_1,v_2) = |y(T,v_1,v_2)-z_d|^2 + C_1 \int_{Dx(0,T)} v_1^2 dxdt - C_2 \int_{\partial Dx(0,T)} v_2^2 d(\partial D)dt,$$

the optimal strategy $u = (u_1,u_2)$ of the differential game (for C_2 large enough)

$$J(u_1,u_2) = \inf_{v_1} \sup_{v_2} J(v_1,v_2) = \sup_{v_2} \inf_{v_1} J(v_1,v_2)$$

is given by the parabolic state equation and the corresponding adjoint equation

$$\frac{\partial y}{\partial t} - \Delta y = u_1 g \text{ in } Dx(0,T) \qquad -\frac{\partial p}{\partial t} - \Delta p = 0 \text{ in } D \times (0,T)$$

$$\frac{\partial y}{\partial n} = u_2 \quad \text{on } \partial Dx(0,T) \qquad \frac{\partial p}{\partial n} = 0 \text{ on } \partial D \times (0,T)$$

$$y(x,0) = 0 \quad \text{in } D, \qquad p(T) = y(T)-z_d \text{ in } D,$$

with the integral inequalities

$$\int_0^T \int_D (pg+C_1u_1)\cdot(v_1-u_1)dxdt \geqq 0 \text{ for } v_1 \in U^1_{cc},$$

$$\int_0^T \int_{\partial D} (-p+C_2u_2)(v_2-u_2)d(\partial D)dt \geqq 0 \text{ for } v_2 \in U^2_{cc}.$$

Remark 1:

In the unconstrained case, where $U^1_{cc} = L^2(Dx(0,T))$, $U^2_{cc} = L^2(\partial Dx(0,T))$, we obtain

$$p\cdot g + C_1u_1 = 0, \text{ then } u_1 = -\frac{1}{C_1}\cdot pg,$$

$$-p+C_2u_2 = 0 \text{ then } u_2 = \frac{1}{C_2}\cdot p,$$

the optimal strategy $u = (u_1,u_2)$

Remark 2:

Let g be g = 1, then the equation system reduces to

$$\frac{\partial y}{\partial t} - \Delta y + \frac{1}{c_1}p = 0 \text{ in } D \times (0,T) \qquad -\frac{\partial p}{\partial t} - \Delta p = 0 \text{ in } D \times (0,T)$$

$$\frac{\partial y}{\partial n} = \frac{1}{C_2}\cdot p \text{ on } Dx(0,T) \qquad \frac{\partial p}{\partial n} = 0 \text{ on } \partial D \times (0,T)$$

$$y(0) = 0 \text{ in } D, \qquad p(T) = y(T)-z_d \text{ in } D.$$

Remark 3:

For $g = \delta(x-a)$ the equation system reduces to

$$\frac{\partial y}{\partial t} - \Delta y = u_1\delta(x-a) \text{ in } D \times (0,T) \qquad -\frac{\partial p}{\partial t} - \Delta p = 0 \text{ in } D \times (0,T)$$

$$\frac{\partial y}{\partial n} = u_2 \text{ on } \partial D \times (0,T) \qquad \frac{\partial p}{\partial n} = 0 \text{ on } \partial D \times (0,T)$$

$$y(0) = 0, \text{ in } D, \qquad p(T) = y(T)-z_d \text{ in } D,$$

with the integral inequalities

$$\int_0^T (p(a,t)+C_1u_1)\cdot(v_1-u_1)dt \geqq 0 \text{ for } v_1 \in U_{cc}^1,$$

$$\int_{\partial D} (-p(a,t)+C_2u_2)\cdot(v_2-u_2)d(\partial D) \geqq 0 \text{ for } v_2 \in U_{cc}^2.$$

II. 6.4 Innovation processes governed by differential games

The innovation diffusion may be described by

$$Ay = v_1 \text{ in } D$$

$$\frac{\partial y}{\partial n} = v_2 \text{ on } \partial D,$$

and has a unique solution $y(v_1,v_2)$, where v_1 and v_2 are the control variables.

We introduction the cost functional

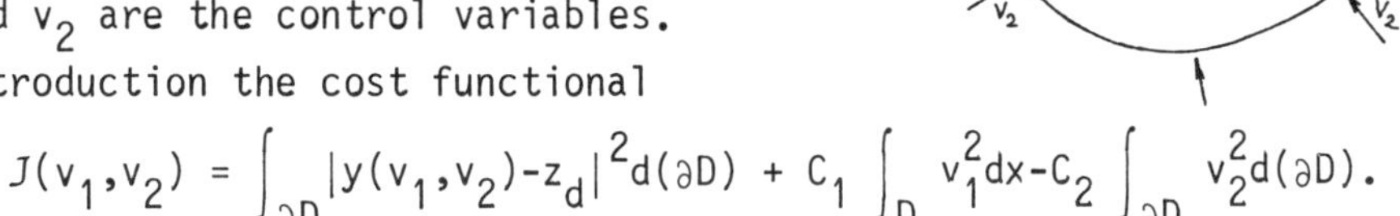

$$J(v_1,v_2) = \int_{\partial D} |y(v_1,v_2)-z_d|^2 d(\partial D) + C_1 \int_D v_1^2 dx - C_2 \int_{\partial D} v_2^2 d(\partial D).$$

The optimal strategy $u = (u_1,u_2)$ of the differential game (for C_2 large enough)

$$J(u_1,u_2) = \inf_{v_1} \sup_{v_2} J(v_1,v_2) = \sup_{v_2} \inf_{v_1} J(v_1,v_2)$$

is characterized by the solution of

$$Ay = u_1 \text{ in } D \qquad A^*p = 0 \text{ in } D$$

$$\frac{\partial y}{\partial n} = u_2 \text{ on } \partial D, \qquad \frac{\partial p}{\partial n} = y - z_d \text{ on } \partial D,$$

with the variational inequalities

$$\int_D (p + C_1 u_1)\cdot(v_1 - u_1)dx \geqq 0 \text{ for } v_1 \in U_{cc}^1,$$

$$\int_{\partial D} (-p + C_2 u_2)\cdot(v_2 - u_2)d(\partial D) \geqq 0 \text{ for } v_2 \in U_{cc}^2.$$

In the case without constraints

$$p + C_1 u_1 = 0 \qquad u_1 = -\frac{1}{C_1}\cdot p$$

and the optimal strategy

$$-p + C_2 u_2 = 0 \qquad u_2 = \frac{1}{C_2}\cdot p,$$

thus, the equation system reduces to

$$Ay + \frac{1}{C_1}\cdot p = 0 \text{ in } D \qquad A^*p = 0 \text{ in } D$$

$$\frac{\partial y}{\partial n} = \frac{1}{C_2} p \text{ on } \partial D, \qquad \frac{\partial p}{\partial n} = y - z_d \text{ on } \partial D.$$

II. 6.5 Asymptotic expansion of the optimal strategy

In the case $C_1 = \varepsilon$, ε small enough, we obtain a very interesting asymptotic expansion of the optimal strategy of the differential game. The equations are

$$Ay + \frac{1}{\varepsilon} p = 0 \text{ in } D \qquad A^*p = 0 \text{ in } D$$

$$\frac{\partial y}{\partial n} = \frac{1}{C_2}\cdot p \text{ on } \partial D, \qquad \frac{\partial p}{\partial n} = y - z_d \text{ on } \partial D.$$

Using the ansatz of the asymptotic expansion

$$y = y^o + \varepsilon y^1 + \ldots \quad \text{and} \quad p = \varepsilon p^1 + \varepsilon^2 p^2 + \ldots$$

we obtain

$$Ay^o + \varepsilon A y^1 + \frac{1}{\varepsilon}\cdot\varepsilon p^1 + \varepsilon p^2 = 0 \qquad \varepsilon A^* p^1 + \varepsilon^2 A^* p^2 = 0$$

$$\frac{\partial y^o}{\partial n} + \frac{\partial y^1}{\partial n} = \frac{1}{C_2}\varepsilon p^1 + \frac{1}{C_2}\varepsilon^2\cdot p^2 \qquad \varepsilon\cdot\frac{\partial}{\partial n}p^1 + \varepsilon^2\cdot\frac{\partial}{\partial n}p^2 = y^o + \varepsilon y^1 - z_d$$

Identifying corresponding ε-terms

1) $Ay^o + p^1 = 0$ in D

$\frac{\partial y^o}{\partial n} = 0$ on ∂D,

2) $Ay^1 + p^2 = 0$ in D

$\frac{\partial y^1}{\partial n} = \frac{1}{c_1}\cdot p^1$ on ∂D,

3) $A^*p^1 = 0$ in D

$y^o - z_d = 0$ on ∂D,

4) $A^*p^2 = 0$ in D

$\frac{\partial p}{\partial n} = y^1$ on ∂D.

Applying the adjoint operator A* to 1) and observing 3), then

$$A^*Ay^o = 0 \text{ in } D$$

$$\frac{\partial y^o}{\partial n} = 0 \text{ on } \partial D$$

$$y^o = z_d \text{ on } \partial D.$$

Applying the adjoint operator A* to 2) and observing 4), there follows

$$A^*Ay^1 = 0 \text{ in } D$$

$$\frac{\partial y^1}{\partial n} = \frac{1}{c_2}\cdot p^1 \text{ on } \partial D$$

$$y^1 = \frac{\partial p^1}{\partial n} \text{ on } \partial D.$$

The optimal strategy is given by

$$u_1 = \frac{1}{\varepsilon} p \approx -\frac{1}{\varepsilon}(\varepsilon p^1 + \varepsilon^2 p^2 + \dots) = -p^1 - \varepsilon p^2 - \dots$$

and

$$u_2 = \frac{1}{c_2} p \approx \frac{1}{c_2}(\varepsilon p^1 + \varepsilon^2 p^2 - \dots) = \frac{\varepsilon}{c_2} p^1 + \frac{\varepsilon^2}{c_2} p^2 + \dots$$

We are taking the 'ansätze' of the optimal strategy

$$u_1 = u_1^o + \varepsilon u_1^1 + \varepsilon^2 u_1^2,$$

$$u_2 = u_2^o + \varepsilon u_2^1 + \varepsilon^2 u_2^2,$$

and by comparison

$$u_1^o = -p^1, \quad u_1^1 = -p^2$$

$$u_2^o = 0, \quad u_2^1 = \frac{1}{c_2}\cdot p^1, \quad u_2^2 = \frac{1}{c_2}\cdot p^2.$$

The part $u_1^o = -p^1$ of the optimal strategy can be determined in the following way:

There exists a unique solution of

$$A^*Ay^o = 0 \quad \text{in } D$$

$$\frac{\partial y^o}{\partial n} = 0 \text{ on } \partial D$$

$$y^o = z_d \text{ in } D,$$

which depends on z_d, thus $y(z_d)$.
As $p^1 = -A(y^o)$ in D, p^1 is known and u_1^o, u_2^o are given explicitly

$$u_1^o = -p^1 \text{ and } u_2^1 = \frac{1}{c_2}\cdot p^1.$$

As p^1 is known, we solve

$$A^*Ay^1 = 0 \text{ in } D$$

$$\frac{\partial y^1}{\partial n} = \frac{1}{c_2}\cdot p^1 \quad \text{on } \partial D$$

$$y^1 = \frac{\partial p^1}{\partial n} \quad \text{on } \partial D,$$

and the solution depends on p^1 and C_2, thus $y^1(p^1, C_2)$.
As $p^2 = -A(y^1)$ in D, p^2 is known and we obtain

$$u_1^1 = -p^2 \text{ and } u_2^2 = \frac{1}{c_2}\cdot p^2$$

hence, the optimal strategy of the differential game is given explicitly by

$$u_1 = u_1^o + \varepsilon u_1^1$$

$$u_2 = u_2^o + \varepsilon u_2^1 + \varepsilon^2 u_2^2, \text{ where } u_2^o = 0.$$

It is obvious that we can give many varieties of problems of the preceding nature.

II. 6.6 Wave phenomena governed by differential games

Let there be two distinct, disjoint open sets D_1 and D_2 in $\mathbb{R}^n$, with smooth boundaries ∂D_1 and ∂D_2

We are given diffusion distributions $g_1 \in L^2(D_1)$ and $g_2 \in L^2(D_2)$ and control variables v_1, v_2 depending only on the time.

The wave phenomena in the domains D_1 and D_2 are described by the following initial-boundary-value problems

$$\frac{\partial^2 y_1(v_1)}{\partial t^2} - \Delta y_1(v_1) = g_1(x)\cdot v_1(t) \text{ in } D_1 \times (0,T) \qquad \frac{\partial^2 y_2(v_2)}{\partial t^2} - \Delta y_2(v_2) = g_2(x)\cdot v_2(t) \text{ in } D_2 \times (0,T$$

$$\frac{\partial y_1(v_1)}{\partial n} = 0 \text{ on } \partial D_1 \times (0,T) \qquad \frac{\partial y_2(v_2)}{\partial n} = 0 \text{ on } \partial D_2 \times (0,T)$$

$$\frac{\partial y_1(v_1)}{\partial t} = 0 \text{ in } D_1 \qquad \frac{\partial y_2(v_2)}{\partial t} = 0 \text{ in } D_2$$

$$y_1(v_1)(x,0) = 0 \text{ in } D_1 \qquad y_2(v_2)(x,0) = 0 \text{ in } D_2$$

For $g_1 \in L^2(D_1)$, $g_2 \in L^2(D_2)$ and $v_1, v_2 \in L^2(0,T)$ there exist unique solutions $y_1(T,x,v_1,g_1)$ and $y_2(T,x,v_2,g_2)$, such that the cost functional is defined by

$$J(v_1,v_2) = |O_1(v_1) - O_2(v_2) - z_d|^2 + C_1 \int_0^T v_1^2 dt - C_2 \int_0^T v_2^2 dt,$$

where $O_1(v_1) = \int_{D_1} y_1(T,x,v_1,g_1)dx$ and $O_2(v_2) = \int_{D_2} y_2(T,x,v_2,g_2)dx$.

The optimal strategy $u = (u_1,u_2)$ of the differential game (for C_2 large enough)

$$J(u_1,u_2) = \inf_{v_1} \sup_{v_2} J(v_1,v_2) = \sup_{v_2} \inf_{v_1} J(v_1,v_2)$$

is characterized as follows:

Introducing the corresponding adjoint equations

$$\frac{\partial^2 p_1}{\partial t^2} - \Delta p_1 = 0 \text{ in } D_1 \times (0,T) \qquad \frac{\partial^2 p_2}{\partial t^2} - \Delta p_2 = 0 \text{ in } D_2 \times (0,T)$$

$$\frac{\partial p_1}{\partial n} = 0 \text{ on } \partial D_1 \times (0,T) \qquad \frac{\partial p_2}{\partial n} = 0 \text{ on } \partial D_2 \times (0,T)$$

$$\frac{\partial p_1}{\partial n}(x,T) = (O_1(v_1) - O_2(v_2) - z_d \text{ in } D_1 \qquad \frac{\partial p_2}{\partial n}(x,T) = (O_1(v_1) - O_2(v_2)) - z_d \text{ in } D_2$$

$$p_1(x,T) = 0 \text{ in } D_1, \qquad p_2(x,T) = 0 \text{ in } D_2,$$

the explicit solution of these equations can be given by

$$p_1(t) = p_2(t) = (T-t)(O_1(v_1)-O_2(v_2)-z_d) = (T-t)\Big(\int_{D_1} y_1(T,x,v_1,g_1)dx -$$

$$- \int_{D_2} y_2(T,x,v_2,g_2)dx - z_d\Big),$$

and the variational inequalities according to page 127 can be written as

$$\int_0^T \int_D (p_1 \cdot g_1 + C_1 u_1) \cdot (v_1 - u_1) dxdt = \int_0^T (p_1(t) \int_D (g_1(x) +$$

$$+ C_1 u_1) dx (v_1(t) - u_1(t)) dt \geqq 0 \text{ for } v_1 \in U^1_{cc},$$

$$\int_0^T \int_D (-p_2 \cdot g_2 + C_1 u_2) \cdot (v_2 - u_2) dxdt = \int_0^T (-p_2(t) \int_D (g_2(x) +$$

$$+ C_2 u_2) dx (v_2(t) - u_2(t)) dt \geqq 0 \text{ for } v_2 \in U^2_{cc}.$$

II. 6.7 Optimal strategy

Now, the optimal system reads

$$\frac{\partial^2 y_1}{\partial t^2} - \Delta y_1 = g_1(x) \cdot u_1(t) \text{ in } Dx(0,T) \qquad \frac{\partial^2 y_2}{\partial t^2} - \Delta y_2 = g_2(x) \cdot u_2(t) \text{ in } D \times (0,T)$$

$$\frac{\partial y_1}{\partial n} = 0 \text{ on } \partial D \times (0,T) \qquad \frac{\partial y_2}{\partial n} = 0 \text{ on } \partial D \times (0,T)$$

$$\frac{\partial y_1}{\partial n}(x,0) = 0 \text{ in } D \qquad \frac{\partial y_2}{\partial t}(x,0) = 0 \text{ in } D$$

$$y_1(x,0) = 0 \text{ in } D. \qquad y_2(x,0) = 0 \text{ in } D.$$

In the case without constraints we obtain the very interesting optimal strategy of the differential game

$$u_1 = -\frac{1}{c_1} \cdot p_1(t) \int_D g_1(x)dx \quad \text{and} \quad u_2 = \frac{1}{c_2} \cdot p_2(t) \int_D g_2(x)dx,$$

where $p_1(t)$ and $p_2(t)$ are given explicitly.

Proof of the considerations:

Applying the theorem on page 124f, we only have to prove the following:

$$\int_0^T \frac{\partial^2 p}{\partial t^2}(y(v)-y)dt = \frac{\partial p}{\partial t}(x,t)\cdot(y(v)-y)\Big|_{t=0}^{t=T} - \int_0^T \frac{\partial p}{\partial t}\cdot\frac{\partial}{\partial t}(y(v)-y)dt$$

$$= \frac{\partial p}{\partial t}(x,T)\cdot(y(T,v)-y(T)) - \frac{\partial p}{\partial x}(x,0)(y(v)(x,0)-y(x,0)) - \int_0^T \frac{\partial p}{\partial t}\cdot\frac{\partial}{\partial t}(y(v)-y)$$

$$= \frac{\partial p}{\partial t}(x,T)\cdot(y(T,v)-y(T) - \int_0^T \frac{\partial p}{\partial t}\cdot\frac{\partial}{\partial t}(y(v)-y)dt, \text{ as } y(v)(x,0)=0 \text{ and}$$

$$y(x,0) = 0,$$

the integral can be expressed by

$$\int_0^T \frac{\partial p}{\partial t}\cdot\frac{\partial}{\partial t}(y(v)-y)dt = p(x,t)\left(\frac{\partial y(v)}{\partial t}(x,t) - \frac{\partial y}{\partial t}(x,t)\right)\Big|_{t=0}^{t=T} -$$

$$- \int_0^T p\cdot\frac{\partial^2}{\partial t^2}(y(v)-y)dt = - \int_0^T p.\frac{\partial^2}{\partial t^2}(y(v)-y)dt, \text{ as } p(x,T) = 0$$

and the initial conditions $\frac{\partial y(v)}{\partial t}(x,0) = 0$, $\frac{\partial y}{\partial t}(x,0) = 0$ are satisfied, hence

$$\int_0^T \frac{\partial^2 p}{\partial t^2}(y(v)-y)dt = \frac{\partial p}{\partial t}(x,T)\cdot(y(T,v)-y(T)) + \int_0^T p\cdot\frac{\partial^2}{\partial t^2}(y(v)-y)dt.$$

In this connection we remember some very interesting examples (Grusa (141)) concerning differential game problems and von Stakelberg control phenomena in multiconnected regions.

II.7 Diffusion processes in multiconnected domains governed by von Stackelberg strategies

In order not to overburden this work, we have restricted ourselves to comparatively simple models. However, the modelling methods we have introduced are quite general and more general problems can be solved using the same modelling methods.

Let D_o be a bounded, open set in $\mathbb{R}^2$ with smooth boundary ∂D_o. Let D_1, D_2 and D_3 be open, bounded, disjoint subsets of D_o with smooth boundaries ∂D_1, ∂D_2 and ∂D_3, such that there exists a 4-connected domain, given by

$$D = D_o \setminus \bigcup_{k=1}^{3} D_k.$$

The diffusion process in the multiconnected domain is influenced by the diffusion in D_3 and by the boundary values v_1 on ∂D_1 and v_2 on ∂D_2.

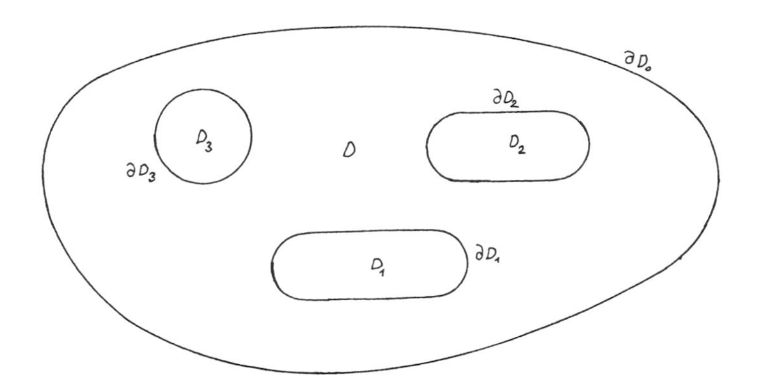

V_1 and v_2 are control variables, where v_1 can be seen as the strategy parameter of a leader, located in D_1, dominating the follower, and v_2 as the strategy parameter of the follower, located in D_2.

The leader and the follower are minimizing their corresponding cost functions. The optimal strategy of the leader and the follower are given in explicit form. The diffusion process may be described e.g. by second order elliptic differential operators A_o in the domain D_3 and A_1 in the 4-connected domain D. There are diffusion distributions f_o in D_3 and f_1 in D, where

$$f_o \in L^2(D_3), \ f_1 \in L^2(D)$$

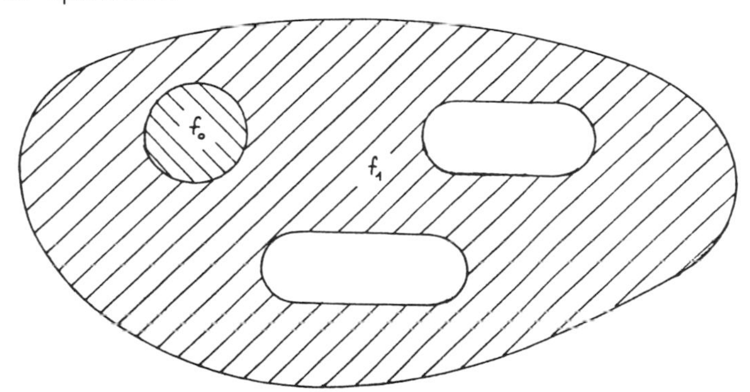

Let the diffusion distributions f_o and f_1 be known at certain points, then we can construct the corresponding interpolating surface splines (e.g. as two-dimensional, inter-polating Lg-splines (139)). These diffusion distributions (f_o, f_1) constructed by two-dimensional splines, are used on the right-hand side of the diffusion equation:

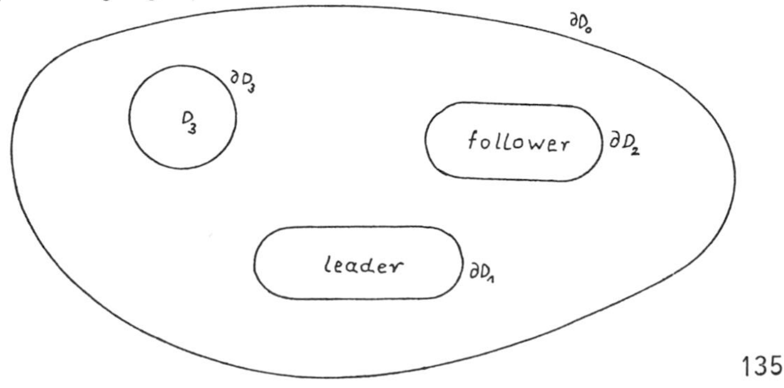

$$A_1y_1 = f_1 \text{ in } D$$
$$\frac{\partial y_1}{\partial n} = \frac{\partial y^o}{\partial n} \text{ on } \partial D_3$$
$$y_1 = y_o \text{ on } \partial D_3 \qquad A_oy_o = f_o \text{ in } D_3$$
$$\frac{\partial y_1}{\partial n} = v_1 \text{ on } \partial D_1 \qquad \frac{\partial y_o}{\partial n} = \frac{\partial y_1}{\partial n} \text{ on } \partial D_3$$
$$-\frac{\partial y_1}{\partial n} = v_2 \text{ on } \partial D_2 \qquad y_o = y_1 \text{ on } \partial D_3.$$
$$\frac{\partial y_1}{\partial n} = 0 \text{ on } \partial D_o,$$

For $f_o \in L^2(D_3)$, $f_1 \in L^2(D)$, and $v_i \in L^2(\partial D_i)$ and $i = 1,2$, there exists a unique solution $y_1 = y_1(f_1,v_1,v_2)$.

Now, let the strategy v_1 of the leader be given, the follower (v_2) wants to minimize the cost functional

$$J_2(v_1,v_2) = |y_1(v_1,v_2)-z_{d2}^2|^2_{\partial D_2} + C_2|v_2|^2_{\partial D_2}, \text{ where } z_{d_2} \in L^2(\partial D_2),$$

on the boundary of his domain D_2, hence

$$J_2(u_2) = \inf_{v_2 \in L^2(\partial D_2)} J_2(v_1,v_2).$$

The strategy v_2 of the follower may be chosen. The leader (v_1) wants to minimize the cost functional

$$J_1(v_1,v_2) = |y_1(v_1,v_2)-z^2_{d_1}|^2_{\partial D_1} + C_1|v_2|^2_{\partial D_1}, \text{ where } z_{d_1} \in L^2(\partial D_1),$$

on the boundary of his domain D_1, thus

$$J_1(u_1) = \inf_{v_1 \in L^2(\partial D_1)} J_1(v_1,v_2)$$

According to the theory the optimal strategy $u = (u_1,u_2)$ can be characterized by

$$A_1 y_1 = f_1 \text{ in } D \qquad \text{with the transmission condition}$$

$$y_1 = y_o \text{ on } \partial D_3 \qquad \text{of the domain } D_3$$

$$\frac{\partial y_1}{\partial n} = \frac{\partial y_o}{\partial n} \text{ on } \partial D_3 \qquad A_o y_o = f_o \text{ in } D_3$$

$$\frac{\partial y_1}{\partial n} = u_1 \text{ on } \partial D_1 \qquad y_o = y_1 \text{ on } \partial D_3$$

$$\frac{\partial y_1}{\partial n} = u_2 \text{ on } \partial D_2 \qquad \frac{\partial y_o}{\partial n} = \frac{\partial y_1}{\partial n} \text{ on } \partial D_3$$

$$\frac{\partial y_1}{\partial n} = 0 \text{ on } \partial D_o$$

and the equivalent variational formulation (see page 42f)

$$a_o(y_1,\varphi) + \varepsilon a_1(y_1,\varphi) = (u_1,\varphi)_{\partial D_1} + (u_2,\varphi)_{\partial D_2} + (f,\varphi) \text{ for } \varphi \in H^1(D).$$

The first adjoint problem can be written as

$$A_1^* p_1 = 0 \text{ in } D \qquad \text{with the transmission condition}$$

$$p_1 = p_o \text{ on } \partial D_3 \qquad \text{of the domain } D_3$$

$$\frac{\partial p_1}{\partial n} = \frac{\partial p_o}{\partial n} \text{ on } \partial D_3 \qquad A_o^* p_o = 0 \text{ in } D_3$$

$$\frac{\partial p_o}{\partial n} = y_1 - z_{d_2}^2 \text{ on } \partial D_2 \qquad \frac{\partial p_o}{\partial n} = \frac{\partial p_1}{\partial n} \text{ on } \partial D_3,$$

$$\frac{\partial p_1}{\partial n} = 0 \text{ on } \partial D_o,$$

with the variational inequality

$$\int_{\partial D_2} (p_1 + C_2 u_2)(v_2 - u_2)\, d(\partial D_2) \geqq 0 \text{ for } v_2 \in L^2(\partial D_2).$$

The equivalent variational formulation is given by

$$a_o^*(p_1,\psi) + \varepsilon \cdot a_1^*(p_1,\psi) = (y_1 - z_{d_2}^2, \psi) \text{ for } \psi \in H^1(D_o).$$

The second adjoint problem can be written as

$$A_1^* q_1 = 0 \text{ in } D \qquad \text{with the transmission condition}$$

$$q_1 = q_o \text{ on } \partial D_3 \qquad \text{of the domain } D_3$$

$$\frac{\partial q_1}{\partial n} = \frac{\partial q_o}{\partial n} \text{ on } \partial D_3 \qquad A_o^* q_o = 0 \text{ in } D_3$$

$$\frac{\partial q_1}{\partial n} = 0 \text{ on } \partial D_2 \qquad q_o = q_1 \text{ on } \partial D_3$$

$$\frac{\partial q_1}{\partial n} = y_1 - z_{d_1}^2 \text{ on } \partial D_1 \qquad \frac{\partial q_o}{\partial n} = \frac{\partial q_1}{\partial n} \text{ on } \partial D_3,$$

$$\frac{\partial q_1}{\partial n} = 0 \text{ on } \partial D_o,$$

with the variational inequality

$$\int_{\partial D_1} (q_1 + C_1 u_1)\cdot(v_1 - u_1)d(\partial B_1) \geqq 0 \text{ for } v_1 \in L^2(\partial D_1).$$

The equivalent variational formulation is given by

$$a_o^*(q_1,\theta) + \varepsilon a_1^*(q_1,\theta) = (y_1 - z_{d_1}^2,\theta) \text{ for } \theta \in H^1(D_o).$$

In the case without constraints we obtain the optimal strategy by

$$p_1 + C_2 u_2 = 0 \quad \text{hence} \quad u_2 = -\frac{1}{C_2}\cdot p_1$$

$$q_1 + C_2 u_1 = 0 \quad \text{hence} \quad u_1 = -\frac{1}{C_1}\cdot q_1.$$

Using this in the variational formulation, we finally obtain

$$a_o(y,\varphi) + \varepsilon a_1(y,\varphi) + \frac{1}{C_1}(q,\varphi) + \frac{1}{C_2}(p,\varphi) = (f,\varphi) \text{ for } y \in H^1(D_o),\ \varphi \in H^1(D_o)$$

$$a_o^*(p,\psi) + \varepsilon a_1^*(p,\psi) = (y - z_{d_2}^2,\psi) \text{ for } p \in H^1(D_o),\ \psi \in H^1(D_o)$$

$$a_o^*(q,\theta) + \varepsilon a_1^*(q,\theta) = (y - z_{d_1}^2,\theta) \text{ for } q \in H^1(D_o),\ \theta \in H^1(D_o).$$

We try to give an asymptotic expansion of the above problem by taking the "ansätze"

$$y = \frac{1}{\varepsilon}\cdot y^{-1} + y^o + \varepsilon y^1, \quad p = \frac{1}{\varepsilon}\cdot p^{-1} + p_o + \varepsilon p_1, \text{ and } q = \frac{1}{\varepsilon}\cdot q^{-1} + q^o + \varepsilon q^1$$

in the variational form.

II. 7.1 Asymptotic expansion of the bilinear form

$$a_o(y,\varphi) = \frac{1}{\varepsilon}\cdot a_o(y^{-1},\varphi) + a_o(y^o,\varphi) + \varepsilon a_o(y^1,\varphi)$$

$$\varepsilon a_1(y,\varphi) = a_1(y^{-1},\varphi) + \varepsilon a_1(y^o,\varphi) + \varepsilon^2 a_1(y^1,\varphi)$$

$$\frac{1}{C_1}(q,\varphi) = \frac{1}{C_1}\frac{1}{\varepsilon}(q^{-1},\varphi)_{\partial D_1} + \frac{1}{C_1}(q^o,\varphi)_{\partial D_1} + \varepsilon\frac{1}{C_1}\cdot(q^1,\varphi)_{\partial D_1}$$

$$\frac{1}{C_2}\cdot(p,\varphi) = \frac{1}{C_2}\frac{1}{\varepsilon}(p^{-1},\varphi)_{\partial D_2} + \frac{1}{C_2}(p^o,\varphi)_{\partial D_2} + \varepsilon\frac{1}{C_2}(p^1,\varphi)_{\partial D_2},$$

$$a_o^*(p,\psi) = \frac{1}{\varepsilon}\cdot a_o^*(p^{-1},\psi) + a_o^*(p^o,\psi) + \varepsilon\cdot a_o^*(p^1,\psi)$$

$$\varepsilon a_1^*(p,\psi) = a_1^*(p^1,\psi) + \varepsilon a_1^*(p^o,\psi) +$$

$$-(y,\psi)_{\partial D_2} = -\frac{1}{\varepsilon}(y^{-1},\psi)_{\partial D_2} - (y^o,\psi)_{\partial D_2} - \varepsilon(y^1,\psi)_{\partial D_2},$$

$$a_o^*(q,\theta) = \frac{1}{\varepsilon}a_o^*(q^{-1},\theta) + a_o^*(q_o,\theta) + \varepsilon a_o^*(q^1,\theta)$$

$$\varepsilon a_1^*(q,\theta) = a_1^*(q^{-1},\theta) + \varepsilon a_1^*(q^o,\theta)$$

$$-(y,\theta)_{\partial D_1} = -\frac{1}{\varepsilon}\cdot(y^{-1},\theta)_{\partial D_1} - (y^o,\theta)_{\partial D_1} - \varepsilon(y^1,\theta)_{\partial D_1}.$$

Now we are gathering the ε-terms and taking into account the variational formulation;

the $\frac{1}{\varepsilon}$-terms:

$$a_o(y^{-1},\varphi) + \frac{1}{C_1}(q^{-1},\varphi)_{\partial D_1} + \frac{1}{C_2}(p^{-1},\varphi)_{\partial D_2} = 0$$

$$a_o^*(p^{-1},\psi) - (y^{-1},\psi)_{\partial D_2} = 0$$

$$a_o^*(q^{-1},\theta) - (y^{-1},\theta)_{\partial D_1} = 0;$$

the ε^o-terms:

$$a_o(y^o,\varphi) + a_1(y^{-1},\varphi) + \frac{1}{C_1}(q^o,\varphi)_{\partial D_1} + \frac{1}{C_2}(p^o,\varphi)_{\partial D_2} = (f,\varphi)$$

$$a_o^*(p^o,\psi) + a_1^*(p^1,\psi) - (y^o,\psi)_{\partial D_2} = (-z_{d_2}^2,\psi)$$

$$a_o^*(q^o,\theta) + a_1^*(q^{-1},\theta) - (y^o,\theta)_{\partial D_1} = (-z_{d_1}^2,\theta);$$

the ε-terms:

$$a_o(y^1,\varphi) + a_1(y^o,\varphi) + \frac{1}{C_1}(q^1,\varphi)_{\partial D_1} + \frac{1}{C_2}\cdot(p^1,\varphi)_{\partial D_2} = 0$$

$$a_o^*(p^1,\psi) + a_1^*(p^o,\psi) = (y^1,\psi)_{\partial D_2}$$

$$a_o^*(q^1,\Theta) + a_1^*(q^o,\Theta) = (y^1,\Theta)_{\partial D_1}.$$

Let Ψ be arbitrary on ∂D_2 and a_o^* be given

$$a_o^*(p^{-1},\psi) = (y^{-1},\psi)_{\partial D_2},$$

and $\psi = 0$ in the domain D_3, then $a_o^*(p^{-1},\psi) = 0$, hence $(y^{-1},\psi)_{\partial D_2} = 0$, so $y^{-1} = 0$ on ∂D_2.

Let Θ be arbitrary on ∂D_1 and

$$a_o^*(q^{-1},\Theta) = (y^{-1},\Theta)_{\partial D_1}$$

and $\Theta = 0$ in the domain D_3, then $a_o^*(q^{-1},\Theta) = 0$, hence $y^{-1} = 0$ on ∂D_1. Let φ be arbirary on ∂D_1 and ∂D_2, then we have

$$a_o(y^{-1},\varphi) = -\frac{1}{C_1}(q^{-1},\varphi)_{\partial D_1} - \frac{1}{C_2}(p^{-1},\varphi)_{\partial D_2}$$

and $\varphi = 0$ in the domain D_3, then $a_o(y^{-1},\varphi) = 0$ and we obtain

$$\frac{1}{C_1}(q^{-1},\varphi)_{\partial D_1} = -\frac{1}{C_1}(p^{-1},\varphi)_{\partial D_2}, \text{ as } C_1 \neq 0,\ C_2 \neq 0,$$

we have

$$q^{-1} = 0 \text{ on } \partial D_1 \text{ and } p^{-1} = 0 \text{ on } \partial D_2;$$

and quite analogously

$$q_1^{-1} = 0 \text{ and } p_1^{-1} = 0 \text{ on } \partial D_2,$$

$$q_1^{-1} = 0 \text{ and } y_1^{-1} = 0 \text{ on } \partial D_1.$$

We have shown, that $y^{-1} = 0$ on ∂D_1 and ∂D_2, then $y_o^{-1} = 0$ in D_3, and analogously $p_o^{-1} = q_o^{-1} = 0$ in D_3.

The first equation reads

$$a_o(y^o,\varphi) + a_1(y^{-1},\varphi) + \frac{1}{C_1}(q^o,\varphi)_{\partial D_1} + \frac{1}{C_2}(p^o,\varphi)_{\partial D_2} = (f,\varphi),$$

by taking $\varphi = 0$ in D_3, then $a_o(y^o,\varphi) = 0$ and

$$a_1(y^{-1},\varphi) + \frac{1}{C_1}(q^o,\varphi)_{\partial D_1} + \frac{1}{C_2}(p^o,\varphi)_{\partial D_2} = (f,\varphi),$$

the equivalent formulation as a boundary-value problem is:

$$A_1 y_1^{-1} = f \text{ in } D$$

$$\frac{\partial y_1^{-1}}{\partial n} = -\frac{1}{C_1}\, q_1^o \text{ on } \partial D_1$$

$$\frac{\partial y_1^{-1}}{\partial n} = -\frac{1}{C_2}\, p_1^o \text{ on } \partial D_2$$

$$\frac{\partial y_1^{-1}}{\partial n} = 0 \qquad \text{on } \partial D_o;$$

on ∂D_3 there is $y_o^{-1} = 0$ (see above), since $y_1^{-1} = y_o^{-1}$ on ∂D_3, then there follows, that $y_1^{-1} = 0$ on ∂D_3.

Above we have shown, that $y_1^{-1} = 0$ on ∂D_2 and $y^{-1} = 0$ on ∂D_1, thus we obtain the boundary-value problem

$$A_1(y_1^{-1}) = f \text{ in } D$$

$$y_1^{-1} = 0 \text{ on } \partial D_1$$

$$y_1^{-1} = 0 \text{ on } \partial D_2$$

$$y_1^{-1} = 0 \text{ on } \partial D_3$$

$$y_1^{-1} = 0 \text{ on } \partial D_o,$$

which has a unique solution.

Let φ be $\varphi \neq 0$ on ∂D_3, then the first equation gives

$$A_o y_o^o = f_o \text{ in } D_3$$

$$\frac{\partial y_o^o}{\partial n} = \frac{\partial y_1^{-1}}{\partial n} \text{ on } \partial D_3,$$

where y_1^{-1} is the above solution.

The second equation reads

$$a_o^*(p^o,\psi) + a_1^*(p^{-1},\psi) = (y^o - z_{d_2},\psi)_{\partial D_2},$$

taking $\psi = 0$ in the domain D_3, there follows

$$a_1^*(p^{-1},\psi) = (y^o - z_{d_2},\psi)_{\partial D_2},$$

the equivalent formulation as a boundary-value problem is:

$$A_1^* p_1^{-1} = 0 \text{ in } D$$

$$\frac{\partial p_1^{-1}}{-\partial n} = 0 \text{ on } \partial D_1$$

$$\frac{\partial p_1^{-1}}{\partial n} = 0 \text{ on } \partial D_o$$

$$\frac{\partial p_1^{-1}}{\partial n} = y_1^o - z_{d_2} \text{ on } \partial D_2,$$

as $p_o^{-1} = 0$ and the transmission condition $p_1^{-1} = p_o^{-1}$ on ∂D_3 are satisfied, then $p_1^{-1} = 0$ on ∂D_3.

We have shown above, that $p_1^{-1} = 0$ on ∂D_2, therefore we obtain the boundary-value problem

$$A_1^* p_1^{-1} = 0 \text{ in } D$$

$$\frac{\partial p_1^{-1}}{\partial n} = 0 \text{ on } \partial D_1$$

$$p_1^{-1} = 0 \text{ on } \partial D_2$$

$$p_1^{-1} = 0 \text{ on } \partial D_3$$

$$\frac{\partial p_1^{-1}}{-\partial n} = 0 \text{ on } \partial D_o,$$

it follows that the solution $p_1^{-1} = 0$ in D and the condition on the boundary ∂D_2 gives $y_1^o = z_{d_2}$ on ∂D_2.

The third equation reads

$$a_o^*(q^o,\Theta) + a_1^*(q^{-1},\Theta) = (y^o - z_{d_1}^2,\Theta)_{\partial D_1},$$

taking $\Theta = 0$ in the domain D_3, then

$$a_1^*(q^{-1},\Theta) = (y^o - z_{d_1}^2,\Theta)_{\partial D_1},$$

in the equivalent form

$$A_1^* q_1^{-1} = 0 \text{ in } D$$

$$\frac{\partial q_1^{-1}}{\partial n} = y_1^o - z_{d_1}^2 \text{ on } \partial D_1$$

$$\frac{\partial q_1^{-1}}{\partial n} = 0 \text{ on } \partial D_2$$

$$\frac{\partial q_1^{-1}}{\partial n} = 0 \text{ on } \partial D_o$$

$$q_1^{-1} = 0 \text{ on } \partial D_3,$$

as $q_o^{-1} = 0$ on ∂D_3 and the transmission condition $q_o^{-1} = q_1^{-1}$ on ∂D_3 is satisfied, then $q_1^{-1} = 0$ on ∂D_3.

We know from page 140, that $q_1^{-1} = 0$ on the boundaries ∂D_1 and ∂D_2. The new rewritten boundary-value problem

$$A_1^* q_1^{-1} = 0 \text{ in } D$$

$$q_1^{-1} = 0 \text{ on } \partial D_1$$

$$q_1^{-1} = 0 \text{ on } \partial D_2$$

$$q_1^{-1} = 0 \text{ on } \partial D_3$$

$$-\frac{\partial q_1^{-1}}{\partial n} = 0 \text{ on } \partial D_o$$

has a unique solution $q_1^{-1} = 0$ in D, thus we obtain $y_1^o = z_{d_1}^2$ on ∂D_1.

The first equation (of the ε-term) reads

$$a_o(y^1,\varphi) + a_1(y^o,\varphi) + \frac{1}{C_1}(q^1,\varphi)_{\partial D_1} + \frac{1}{C_2}(p^1,\varphi)_{\partial D_2} = 0$$

by taking $\varphi = 0$ in the domain D_3, we obtain

$$a_1(y^o,\varphi) + \frac{1}{C_1}(q^1,\varphi)_{\partial D_1} + \frac{1}{C_2}(p^1,\varphi)_{\partial D_2} = 0$$

calling to mind that $y_1^o = z_{d_2}$ on ∂D_2, $y_1^o = z_{d_1}$ on ∂D_1, and the transmission condition $y_1^o = y_o^o$ on ∂D_3, we finally obtain:

The boundary-value problem

$$A_1 y_1^o = 0 \text{ in } D$$

$$y_1^o = z_{d_2} \text{ on } \partial D_2$$

$$y_1^o = z_{d_1} \text{ on } \partial D_1$$

$$y_1^o = y_o^o \text{ on } \partial D_3$$

$$\frac{\partial y_1^o}{\partial n} = 0 \text{ on } \partial D_o$$

has a unique solution $y_1^o(z_{d_1}, z_{d_2}, y_o^o)$.

We derived on page 138 the optimal strategy $u = (u_1, u_2)$, and on page 137 that the normal derivatives are given by

$$\frac{\partial y_1^o}{\partial n} = u_1 \text{ on } \partial D_1, \text{ and } \frac{\partial y_1^o}{\partial n} = u_2 \text{ on } \partial D_2.$$

II. 7.2 Résumé:

The diffusion process in the 4-connected domain D may be governed by the strategy parameters v_1 of the leader and v_2 of the follower, such that the leader is minimizing his cost function $J_1(v_1, v_2)$,

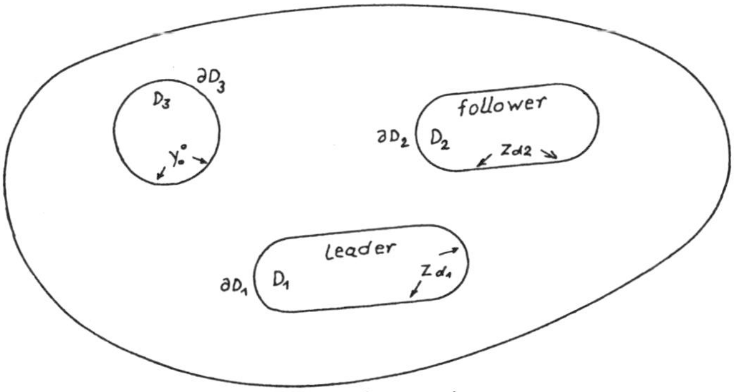

$$J_1(u_1) = \inf_{v_1} J_1(v_1, v_2),$$

and the follower is minimizing his cost function $J_2(v_1, v_2)$

$$J_2(u_2) = \inf_{v_2} J_2(v_1, v_2).$$

The optimal strategies $u_1(u_2)$ of the leader (follower) are characterized by the solution of the boundary-value problem (see above).

It can be explained that $z_{d_1}(z_{d_2})$ are certain fixed costs of the leader (follower) and y_o^o a diffusion process generated in D_3 influencing the diffusion in D.

Let the functions z_{d_1}, z_{d_2}, and y_o^o be Hölder continuous, then there exists (Grusa (141)) the closed, analytic solution $y(y_o^o, z_{d_1}, z_{d_2})$ of the boundary-value problem.

For applications in economics see our models in ((141), p. 112f). The optimal strategy u_1 of the leader is given by

$$\frac{\partial y(y_o^o, z_{d_1}, z_{d_2})}{\partial n} \quad \text{on his boundary } D_1.$$

The best choice for the follower is now to take

$$\frac{\partial y(y_o^o, z_{d_1}, z_{d_2})}{\partial n} \quad \text{on his boundary } D_2.$$

II. 7.3 Some remarks on the movement of pedestrians along the street and advertising

Here only one of the possible applications in economics is given: a few remarks on the dynamic process of advertising (291, 164).

Let us imagine a large city like New York with shopping streets, departmental stores, big shopping centres and smaller ones:

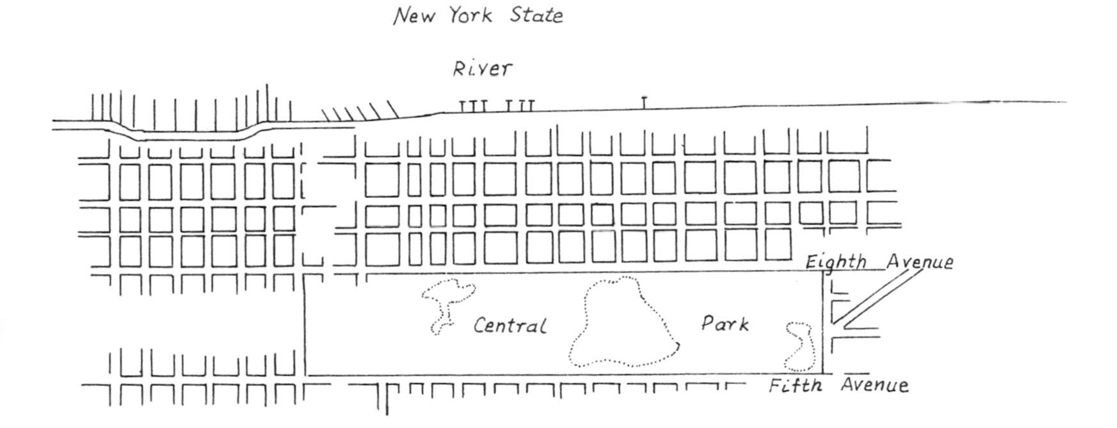

Let us consider the following model-building process taken from the field of advertising without going into too much depth.

We are interested only in one shopping street and the influence of advertising on the movement of the pedestrians along the street:

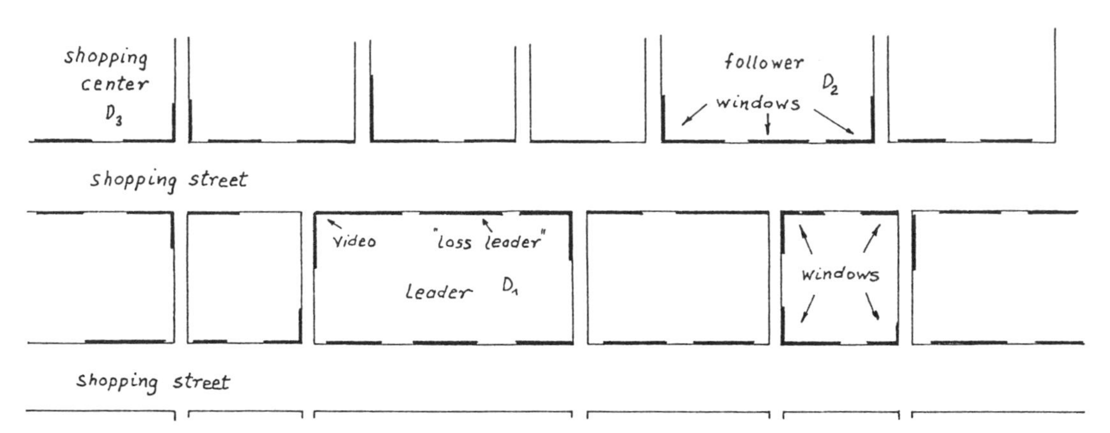

The latest summer fashion is now on view in the big departmental stores, these stores can be seen as fashion leaders.

There is a fashion leader located in D_1 and a department store located in D_2 acting as follower. The advertising cost of the leader and the follower are denoted by z_{d_1} and z_{d_2}, where the cost functions are given by $J_1(v_1,v_2)$ and $J_2(v_1,v_2)$ and v_1,v_2 can be seen as advertising strategies (almost identical to the window display).

The solution y_0^0 describes diffusion phenomena in D_3 depending on f_0, the initial diffusion distribution in D_3. The movements of the pedestrians along the street are modelled by the diffusion processes described above. The diffusion phenomena can be seen as customers influenced by the departmental stores who then in turn influence the movement of the pedestrians along the street.

The controlled diffusion phenomena can be applied to the influenced pedestrians. The dynamic process of influence is modelled e.g. by von Stackelberg strategies, denoted by v_1 and v_2.

Solving the diffusion process written as a boundary-value problem:

$$A_1 y_1^o = 0 \text{ in } D^4$$
$$y_1^o = z_{d_1} \text{ on } \partial D_1$$
$$y_1^o = z_{d_2} \text{ on } \partial D_2$$
$$y_1^o = y_o^o \text{ on } \partial D_3$$
$$\frac{\partial y_1^o}{\partial n} = 0 \text{ on } \partial D_o,$$

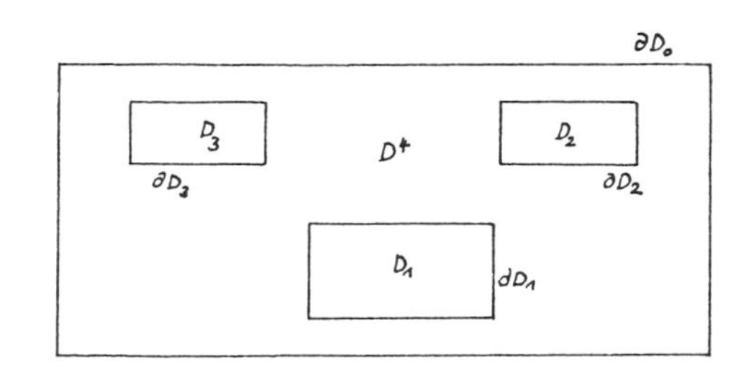

we obtain a unique solution $y_1^o = y_1^o(z_{d_1}, z_{d_2}, y_o^o)$.

The optimal advertising strategies of the leader (u_1) and the follower (u_2) are given by

$$u_1 = \frac{\partial}{\partial n} y_1^o(z_{d_1}, z_{d_2}, y_o^o) \text{ on the boundary } \partial D_1 = \text{window front of the leader}$$
$$u_2 = \frac{\partial}{\partial n} y_1^o(z_{d_1}, z_{d_2}, y_o^o) \text{ on the boundary } \partial D_2 = \text{window front of the follower.}$$

A more realistic model is given, by introducing domains D_4, D_5, D_6, D_7, D_8 where the other shops are located.

These shops do not influence the dynamic process of advertising, this is described by the boundary conditions

$$\frac{\partial y_1^o}{\partial n} = 0 \text{ on } \partial D_4,\ \partial D_5,\ \partial D_6,\ \partial D_7,\ \partial D_8.$$

The shops are only leading the movement of the pedestrians along the shopfront. The optimal advertising strategies (u_1, u_2) may be characterized by the boundary-value problem:

$$A_1 y_1 = 0 \text{ in } D^9$$
$$y_1 = z_{d_1} \text{ on } \partial D_1$$
$$y_1 = z_{d_2} \text{ on } \partial D_2$$
$$y_1 = y_o^o \text{ on } \partial D_3$$
$$\frac{\partial y_1}{\partial n} = 0 \text{ on } \partial D_4, \partial D_5, \partial D_6, \partial D_7, \partial D_8$$
$$\frac{\partial y_1}{\partial n} = 0 \text{ on } \partial D_o$$

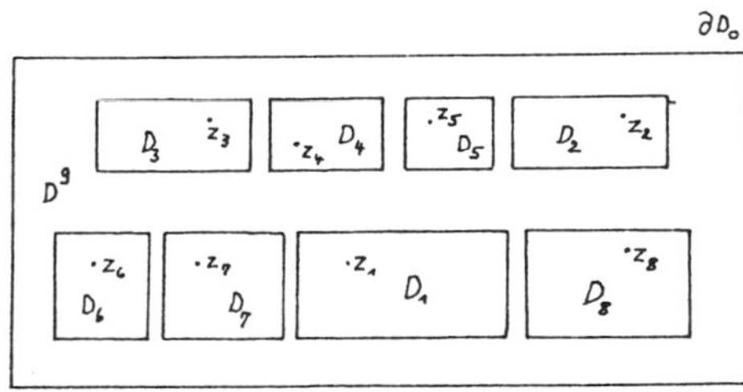

II. 7.4 Solution of a controlled diffusion process in a multiconnected domain

Let the diffusion phenomena of the population movement be governed by the partial differential equation

$$A_1 y = \left(\frac{\partial^2}{\partial x^2} + \frac{\partial^2}{\partial y^2}\right)y - C\cdot y = 0 \text{ in } D^9,$$

where the coefficient C = constant, and C is interpretable as

$$C = \text{constant}\cdot P\cdot Q,$$

where P and Q are given economic indices, the solution reads:

$$y(x,y) = \sum_{k=0}^{8} \psi(x,y,x_k,y_k)\int_{\partial D_k} \mu_k(t)ds + \text{Real}\left(\int \frac{t}{t-z}\frac{\partial}{\partial t_1}(J_0^*(2\cdot\sqrt{C(z-t_1)\bar{z}}))dt_1 \; \mu_0(t)ds\right) +$$

$$\text{Real} \sum_{k=0}^{8} \int_{\partial D_k} \left(O_k(z,t) - \int_0^z O_k(z,t)\frac{\partial}{\partial t_1}(J_0^*(2\cdot\sqrt{C(t_1-z)(\bar{z}_k-\bar{z})})dt_1\right)\mu_k(t)ds,$$

where $z_k = x_k + iy_k$, $\bar{z}_k = x_k - iy_k$ are fixed points within the domains D_k, and $\psi(x,y,x_k,y_k)$ a fundamental solution, J_0^* the Bessel function of order zero. The density functions $\mu_k(t)$ are constructed by integral equations of the boundary conditions (see (141)) and are depending on the boundary values e.g. the fixed cost z_{d_1}, z_{d_2}. The diffusion phenomena y_0^0 in the domain D_3 are describing the influenced customers. The kernels $O_k(z,t)$ are explicitly constructed by integral equation methods given in (141), p. 307ff.)

The influenced population movement depends on

i) the density functions $\mu_k(t)$ for $k = 0,1,\ldots,8$,

the advertising cost z_{d_1}, z_{d_2},

the influence y_0^0 of the centre in D_3,

ii) the constant C, where P and Q are economic indices,

iii) the arrangement of the domains D_k.

Thus the closed, analytic solution y_1 depends on $y_1(z_{d_1},z_{d_2},y_0^0,D_k)$. The optimal strategy of the leader is given by

$$u_1 = \frac{\partial}{\partial n} y_1(z_{d_1},z_{d_2},y_0^0,D_k) \text{ on the boundary } \partial D_1 = \text{window front of the leader.}$$

The best choice for the follower is now to take

$u_2 = \frac{\partial}{\partial n} y_1(z_{d_1}, z_{d_2}, y_o^o, D_k)$ on the boundary ∂D_2 = window front of the follower.

II. 8 Optimal processes in multiconnected regions

This part deals with methods describing processes in a simple connected domain and in two- and three-connected domains.

Summary:

The statement of the problem. Let Ω be open, bounded in $\mathbb{R}^n$ with regular boundary $\partial\Omega$. Let

$$A = \sum_{i,j=1}^{n} \frac{\partial}{\partial x_i} (a_{ij} \frac{\partial}{\partial x_j}(\, ,)) + a_o,$$

be an elliptic differential operator, where the coefficients a_{ij}, a_o are regular in $\bar{\Omega}$ and

$$\sum_{i,j=1}^{n} a_{ij}(x) \, \xi_i \xi_j > \alpha \cdot \sum_{i=1}^{n} \xi_i^2,$$

with α = constant > 0, $a_o(x) \geq \alpha$.

Considering a process in Ω governed by the partial differential equation

$$A(y(v)) = f \text{ in } \Omega, \; f \in L^2(\Omega),$$

$$\frac{\partial y(v)}{\partial n} = v \text{ on the boundary } \partial\Omega,$$

where $v \in L^2(\partial\Omega)$ is the control variable.

The unique solution, denoted by $y_\varepsilon(v)$ is the state of the process.

Let $z_d \in L^2(\partial\Omega)$ be fixed, and $\varepsilon > 0$ a constant, then we introduce the cost function

$$J_\varepsilon(v) = \int_{\partial\Omega} |y_\varepsilon(v) - z_d|^2 d(\partial\Omega) + \varepsilon \int_{\partial\Omega} |v|^2 d(\partial\Omega).$$

Let U_{cc} = a closed, convex, non empty subset of $L^2(\partial\Omega)$.

The optimal control phenomena of the process can be written as

$$\inf_{v \in U_{cc}} J_\varepsilon(v),$$

and the unique solution $u_\varepsilon \in U_{cc}$ is given by

$$J_\varepsilon(u_\varepsilon) = \inf_{v \in U_{cc}} J_\varepsilon(v).$$

II.8.1 Optimal control-processes in two- and three-connected domains

We obtain the following optimality system for a simple connected domain. The optimal control u_ε is characterized by

$$\int_{\partial\Omega}(y(u_\varepsilon)-z_d)(y(v)-y(u_\varepsilon)d(\partial\Omega) + \varepsilon\int_{\partial\Omega}u_\varepsilon(v-u_\varepsilon)d(\partial\Omega) \geq 0 \text{ for } v,u_\varepsilon \in U_{cc},$$

by introducing $y(0)$ we obtain the system

$$\int_{\partial\Omega}(y(u_\varepsilon)-y(0))\cdot(y(v)-y(u_\varepsilon))d(\partial\Omega) + \varepsilon\int_{\partial\Omega}u_\varepsilon(v-u_\varepsilon)d(\partial\Omega) \geq$$

$$\geq \int_{\partial\Omega}((z_d-y(0))\cdot(y(v) - y(u_\varepsilon))d(\partial\Omega)) \quad \text{for } v,u_\varepsilon \in U_{cc}.$$

This method is applicable to multiconnected regions.

We treat in the following only one method for describing complex optimal processes. We construct the asymptotic solution, which can be interpreted as the evolution of the complex optimal process. Let Ω be an open, bounded set in $\mathbf{R}^n$ with smooth boundary

$$\partial\Omega = \partial B_0 \cup \Gamma,$$

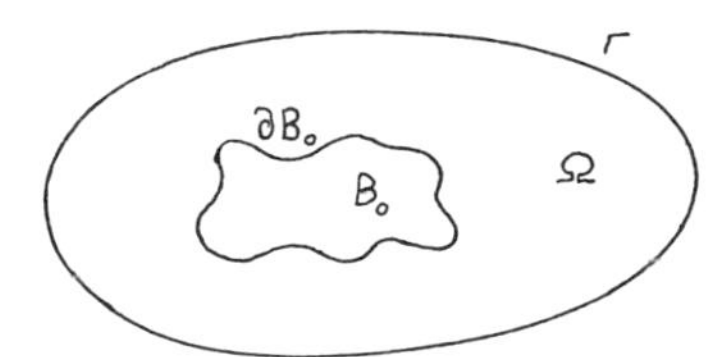

and g fixed on the boundary Γ,
and $z_{B_0} \in L^2(\partial B_0)$.
The corresponding cost function is given by

$$J_\varepsilon(v) = \int_{\partial B_0}|y(v)-z_{B_0}|^2 d(\partial B_0) + \varepsilon\int_{\partial\Omega}v^2\, d(\partial\Omega)-2\int_\Gamma g\cdot y(v)d(\Gamma),$$

and the optimal control problem

$$J_\varepsilon(u_\varepsilon) = \inf_{v\in U_{cc}} J_\varepsilon(v)$$

is characterized by

$$\int_{\partial B_0}(y(u_\varepsilon)-y(0))\cdot(y(v)-y(0))d(\partial B_0) + \varepsilon\int_{\partial\Omega}u_\varepsilon\cdot v\cdot d(\partial\Omega) \geq$$

$$\int_{\partial B_0}(z_{B_0}-y(0))\cdot(y(v)-y(0))d(\partial B_0) + \int_\Gamma g\cdot(y(v)-y(0))d(\Gamma) \text{ for } v,u_\varepsilon \in U_{cc}$$

The optimal process in the three-connected domain.

Let Ω be an open, bounded set in $\mathbf{R}^n$ with smooth boundary

$$\partial\Omega = \partial B_0 \cup \partial B_1 \cup \Gamma,$$

and given functions (fixed) on the boundaries: $z_{B_o} \in L^2(\partial B_o)$,

$z_{B_1} \in L^2(\partial B_1)$, $g_1 \in L^2(\partial B_1)$, $g_2 \in L^2(\Gamma)$.

The corresponding cost function is given by

$$J_\varepsilon(v) = \int_{\partial B_o} |y(v)-z_{B_o}|^2 d(\partial B_o) + \varepsilon \int_{\partial B_1} |y(v)-z_{B_1}|^2 d(\partial B_1) + \varepsilon^2 \int_{\partial\Omega} v^2 d(\partial\Omega) -$$

$$- 2\int_{\partial B_1} g_1 y(v) d(\partial B_1) - 2\int_\Gamma g_2\, y(v) d(\Gamma),$$

and the optimal control problem

$$J_\varepsilon(u_\varepsilon) = \inf_{v \in U_{cc}} J_\varepsilon(v)$$

is characterized by

$$\int_{\partial B_o} (y(u_\varepsilon)-y(0))\cdot(y(v)-y(0)) d(\partial B_o) + \varepsilon \int_{\partial B_1} u_\varepsilon \cdot v d(\partial B_1) + \varepsilon^2 \int_{\partial\Omega} u_\varepsilon (v-u_\varepsilon) d(\partial\Omega)$$

$$\geq \int_{\partial B_1} (z_{B_1} - y(0))\cdot(y(v)-y(0)) d(\partial B_1) + \int_{\partial B_o} (z_{B_o} - y(0))\cdot(y(v)-y(u_\varepsilon)) d(\partial B_o) +$$

$$+ \int_{\partial B_1} g_1 (y(v)-y(0)) d(\partial B_1) + \int_{\partial B_2} g_2 (y(v)-y(0)) d(\partial B_2).$$

We are able to construct in the unconstrained case, $U_{cc} = L^2(\partial\Omega)$, explicit expressions of the optimal control u_ε.

We obtain asymptotic expansions of the u_ε and their error estimates.

II. 8.2 Optimal process in a three-connected domain

We consider the optimal process in the three-connected domain with respect to the cost function

$$J_\varepsilon(v) = \int_{\partial B_o} |y(v)-z_{B_o}|^2 d(\partial B_o) + \varepsilon \int_{\partial B_1} |y(v)-z_{B_1}|^2 d(\partial B_1) +$$

$$+ \varepsilon^2 \int_{\partial\Omega} v^2 d(\partial\Omega) - 2\int_{\partial B_1} g_1 y(v) d(\partial B_1) - 2\int_\Gamma g_2 y(v) d\Gamma.$$

It can be shown, that there exists a unique solution $u \in U_{cc}$, such that

$$J_\varepsilon(u_\varepsilon) = \inf_{v\in U_{cc}} J_\varepsilon(v)$$

is satisfied. Defining the derivative $J'(u)$ as

$$(J'(u),v) = \frac{d}{d\lambda} J(u + \lambda v)\Big|_{\lambda=0}$$

the optimal control $u_\varepsilon = u$ can be characterized by

$$u \in U_{cc}, \text{ and } \frac{1}{2}\,(J'(u),v-u) \geqq 0 \text{ for all } v \in U_{cc},$$

applying this, we get

$$\frac{d}{d\lambda}(J(u+\lambda(v-u))) = \frac{d}{d\lambda}\Big(\int_{\partial B_o} |y(u+\lambda(v-u))-z_{B_o}|^2 d(\partial B_o) +$$

$$\varepsilon \int_{\partial B_1} |y(u+\lambda(v-u)|^2 d(\partial B_1) + \varepsilon^2 \int_{\partial\Omega} (u+\lambda(v-u)|^2 d(\partial\Omega)$$

$$- 2 \int_{\partial B_1} g_1 y(u+\lambda(v-u))d(\partial B_1) - 2 \int_\Gamma g_2\cdot y(u+\lambda(v-u))d(\Gamma),$$

$$\frac{d}{d\lambda}(J(u+\lambda(v-u)))\Big|_{\lambda=0} = 2 \int_{\partial B_o} (y(u)-z_{B_o})\cdot(y(v-u)d(\partial B_o)$$

$$+ \varepsilon\cdot 2\int_{\partial B_1} (y(u)-Z_{B_1})\cdot(y(v-u)d(\partial B_1)$$

$$+ 2\,\varepsilon^2 \int_{\partial\Omega} u(v-u)d(\partial\Omega) - 2 \int_{\partial B_1} g_1 y(v-u)d(\partial B_1) - 2 \int_\Gamma g_2\cdot y(v-u)d\Gamma.$$

The optimal control process is characterized by

$$0 \leqq \int_{\partial B_o} (y(u_\varepsilon)-z_{B_o})\cdot(y(v)-y(u_\varepsilon))d(\partial B_o) + \int_{\partial B_1} \varepsilon\cdot(y(u_\varepsilon)-z_{B_1})(y(v)-y(u_\varepsilon))d(\partial B_1)$$

$$+ 2\varepsilon^2 \int_{\partial\Omega} u_\varepsilon(v-u_\varepsilon)d(\partial\Omega) - 2 \int_{\partial B_1} g_1(y(v)-y(u_\varepsilon))d(\partial B_1) - 2\int_\Gamma g_2(y(v)-y(u_\varepsilon))d\Gamma.$$

This linear functional can be used to introduce a bilinear form in relation to the adjoint state equation

$$a_o^*(p_\varepsilon,\phi) + \varepsilon a_1^*(p_\varepsilon,\phi) + \varepsilon^2 a_2(p_\varepsilon,\phi) = (y(u_\varepsilon)-z_{B_1},\phi)_{\partial B_1} +$$

$$\varepsilon(y(u_\varepsilon)-z_{B_o},\phi)_{\partial B_o} + (-g_1,\phi)_{\partial B_1} + (-g_2,\phi)_\Gamma$$

taking $\phi = y(v)-y(u_\varepsilon) = y-y_\varepsilon$, we obtain

$$(p_\varepsilon, v-u_\varepsilon)_{\partial\Omega} = a_o^*(p_\varepsilon, y-y_\varepsilon) + \varepsilon a_1^*(p_\varepsilon, y-y_\varepsilon) + \varepsilon^2(a_2(p_\varepsilon, y-y_\varepsilon) =$$

$$= (y(u_\varepsilon)-z_{B_1}, y-y_\varepsilon)\partial B_1 + \varepsilon(y(u_\varepsilon)-z_{B_o}, y-y_\varepsilon)_{\partial B_o} - (g_1, y-y_\varepsilon)_{\partial B_1} - (g_2, y-y_\varepsilon)_\Gamma .$$

Using this in the above inequality, we obtain

$$\int_{\partial\Omega} p_\varepsilon + \varepsilon^2 u_\varepsilon (v-u_\varepsilon) d(\partial\Omega) \geqq 0$$

and in the unconstrained case

$$p_\varepsilon + \varepsilon^2 u_\varepsilon = 0.$$

Thus, the optimal control process u_ε is given by

$$u_\varepsilon = -\frac{1}{\varepsilon} \cdot p_\varepsilon,$$

where p_ε is the solution of the adjoint state equation given explicitly

$$A^* p_\varepsilon = 0 \text{ in } \Omega$$

$$\frac{\partial p_\varepsilon}{\partial n^*} = y_\varepsilon - z_{B_o} \quad \text{on } \partial B_o$$

$$\frac{\partial p_\varepsilon}{\partial n^*} = y_\varepsilon - z_{B_1} \quad \text{on } \partial B_1$$

$$\frac{\partial p_\varepsilon}{\partial n^*} = -g_1 \quad \text{on } \partial B_1$$

$$\frac{\partial p_\varepsilon}{\partial n^*} = -g_2 \quad \text{on } \Gamma.$$

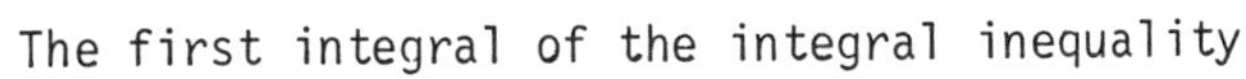

The first integral of the integral inequality

$$\int_{\partial B_o} (y(u_\varepsilon)-z_{B_o}) \cdot (y(v)-y(u_\varepsilon)) d(\partial B_o)$$

can be written as

$$\int_{\partial B_o} (y(u_\varepsilon)-y(0) + y(0)-z_{B_o}) \cdot (y(v)-y(0) + y(0)-y(u_\varepsilon)) d(\partial B_o),$$

abbreviating $\varphi_\varepsilon = y(u_\varepsilon)-y(0)$, $\varphi = y(v)-y(0)$, we obtain finally

$$\int_{\partial B_o} (\varphi_\varepsilon \cdot \varphi + (-1)\varphi_\varepsilon^2 + y(0)-z_{B_o})\cdot\varphi + (y(0)-z_{B_o})\varphi_\varepsilon)d(\partial B_o).$$

Thus, the whole integral inequality is given in abbreviated form:

$$0 \leqq \int_{\partial B_o} \varphi_\varepsilon \cdot \varphi - \int_{\partial B_o} \varphi_\varepsilon^2 + \int_{\partial B_o} (y(0)-z_{B_o})\varphi + \int_{\partial B_o} (y(0)-z_{B_o})\varphi_\varepsilon +$$

$$\varepsilon^2 \int_{\partial\Omega} u_\varepsilon(v-u_\varepsilon)d\partial\Omega + \varepsilon(\int_{\partial B_1} \varphi_\varepsilon \cdot \varphi - \int_{\partial B_1} \varphi_\varepsilon^2 + \int_{\partial B_1} (y(0)-z_{B_1})\varphi +$$

$$+\int_{\partial B_1} (y(0)-z_{B_1})\varphi_\varepsilon) - \int_{\partial B_1} g_1\varphi - \int_{\partial B_1} \varphi_1 \cdot \varphi_\varepsilon - \int_\Gamma g_2 \cdot \varphi - \int_\Gamma g_2\varphi_\varepsilon ,$$

defining the bilinear forms:

$$a_o(\varphi_\varepsilon,\varphi) = \int_{\partial B_o} \varphi_\varepsilon \cdot \varphi \, d(\partial B_o), \quad a_1(\varphi_\varepsilon,\varphi) = \int_{\partial B_1} \varphi_\varepsilon \cdot \varphi \, d(\partial B_1) \quad \text{and}$$

$a_2(\varphi_\varepsilon,\varphi) = \int_{\partial\Omega} \Lambda\varphi_\varepsilon \cdot \Lambda(\varphi-\varphi_\varepsilon)d(\partial\Omega)$, where $\Lambda\varphi_\varepsilon = u_\varepsilon$ and $\Lambda\varphi = v$ (see page 158) thus, we obtain

$$0 \leqq a_o(\varphi_\varepsilon,\varphi) - a_o(\varphi_\varepsilon,\varphi_\varepsilon) + \int_{\partial B_o} (y(0)-z_{B_o})\varphi_\varepsilon + \int_{\partial B_o} (y(0)-z_{B_o})\varphi$$

$$+ \varepsilon a_1(\varphi_\varepsilon,\varphi) - \varepsilon a_1(\varphi_\varepsilon,\varphi_\varepsilon) + \int_{\partial B_1} (y(0)-z_{B_1})\varphi_\varepsilon + \int_{\partial B_1} (y(0)-z_{B_1})\varphi +$$

$$+ \varepsilon^2 a_2(\varphi_\varepsilon,\varphi - \varphi_\varepsilon) - \int_{\partial B_1} g_1 \varphi_\varepsilon - \int_\Gamma g_2 \varphi_\varepsilon - \int_{\partial B_1} g_1\varphi_\varepsilon - \int_\Gamma g_2\varphi.$$

II. 8.3 Solution of the integral inequality by asymptotic expansion

This can be written as

$$a_o(\varphi_\varepsilon,\varphi) + \varepsilon a_1(\varphi_\varepsilon,\varphi) + \varepsilon^2 a_2(\varphi_\varepsilon,\varphi-\varphi_\varepsilon) \geqq a_o(\varphi_\varepsilon,\varphi_\varepsilon) + \varepsilon a_1(\varphi_\varepsilon,\varphi_\varepsilon) +$$

$$+ \int_{\partial B_o} (z_{B_o}-y(0))\varphi_\varepsilon + \int_{\partial B_1} (z_{B_1}-y(0))\varphi_\varepsilon +$$

$$\int_{\partial B_1} g_1 \varphi_\varepsilon + \int_\Gamma g_2 \varphi_\varepsilon + \varepsilon^2 a_2(\varphi_\varepsilon,\varphi_\varepsilon) +$$

$$+ \int_{\partial B_o} (z_{B_o} - y(0))\varphi + \int_{\partial B_1} (z_{B_1} - y(0))\varphi + \int_{\partial B_1} g_1\varphi + \int_\Gamma g_2\varphi$$

$$\geqq \int_{\partial B_o} (z_{B_o} - y(0))\varphi + \int_{\partial B_1} (z_{B_1} - y(0))\varphi + \int_{\partial B_1} g_1\varphi + \int_\Gamma g_2\varphi$$

finally we have the integral inèquality

$$a_o(\varphi_\varepsilon, \varphi) + \varepsilon a_1(\varphi_\varepsilon, \varphi) + \varepsilon^2 a_2(\varphi_\varepsilon, \varphi) \geqq \int_{\partial B_o} (z_{B_o} - y(0))\varphi d(\partial B_o) +$$

$$+ \int_{\partial B_1} (z_{B_1} - y(0))\varphi d(\partial B_1) + \int_{\partial B_1} g_1 \cdot \varphi d(\partial B_1) + \int_\Gamma g_2\varphi \, d\Gamma,$$

if

$$a_2(\varphi, \varphi) \geqq \alpha_1 \, ||\varphi||_{H^1(\Gamma)},$$

there exists a unique solution by the theorem of (Stampacchia, Lions (214, 215)). We obtain the approximate solution by the asymptotic development

$$\varphi_\varepsilon = \frac{1}{\varepsilon^2} \varphi^{-2} + \frac{1}{\varepsilon} \cdot \varphi^{-1} + \varphi^o + \dots \quad .$$

Using this ansatz the integral inequality reads

$$a_o(\frac{1}{\varepsilon^2} \varphi^{-2} + \frac{1}{\varepsilon}\varphi^{-1} + \varphi^o, \varphi) + \varepsilon \cdot a_1(\frac{1}{\varepsilon^2} \cdot \varphi^{-2} + \frac{1}{\varepsilon} \varphi^{-1} + \varphi^o, \varphi) + \varepsilon^2 a_2(\frac{1}{\varepsilon^2}\varphi^{-2} + \frac{1}{\varepsilon} \varphi^{-1} + \varphi^o, \varphi) \geqq$$

$$\geqq \int_{\partial B_1} (z_{B_1} - y(0))\varphi d(\partial B_1) + \int_{\partial B_o} (z_{B_o} - y(0))\varphi d(\partial B_o) + \int_{\partial B_1} g_1 \cdot \varphi d(\partial B_1) + \int_\Gamma g_2 \varphi d(\Gamma),$$

introducing the subspaces

$$Y_o = \{\varphi | \varphi \in H^1, \ \varphi = 0 \text{ on } \partial B_o\} \quad \text{and} \quad Y_1 = \{\varphi | \varphi \in H^1, \ \varphi = 0 \text{ on } \partial B_o, \ \varphi = 0 \text{ on } \partial B_1\},$$

we obtain the following ε-terms:

ε^o-term

$$a_o(\varphi^o, \varphi) + a_1(\varphi^{-1}, \varphi) + a_2(\varphi^{-2}, \varphi) = \int_{\partial B_o} (z_{B_o} - y(0))\varphi d(\partial B_o) + \int_{\partial B_1} g_1 \varphi d(\partial B_1) + \int_\Gamma g_2 \varphi \, d(\Gamma),$$

the ε^1-term

$$a_1(\varphi^o, \varphi) + a_2(\varphi^{-1}, \varphi) = \int_{\partial B_1} (z_{B_1} - y(0))\varphi d(\partial B_1),$$

the $\frac{1}{\varepsilon}$ -term

$$a_o(\varphi^{-1},\varphi) + a_1(\varphi^{-2},\varphi) = 0,$$

the $\frac{1}{\varepsilon^2}$-term

$$a_o(\varphi^{-2},\varphi) = 0.$$

The ε^o-term can be written

$$\int_{\partial B_o} \varphi^o\cdot\varphi d(\partial B_o) + \int_{\partial B_1} \varphi^{-1}\cdot\varphi d(\partial B_1) + \int_{\partial B_o \cup \partial B_1 \cup \Gamma} \Lambda\varphi^{-2}\Lambda\varphi d(\partial\Omega) =$$

$$\int_{\partial B_o} (z_{B_o}-y(0))\varphi d(\partial B_o) + \int_{\partial B_1} g_1\cdot\varphi d(\partial B_1) + \int_\Gamma g_2\cdot\varphi d(\Gamma),$$

hence

$$\int_{\partial B_o} (\varphi^o+\Lambda^*\Lambda\varphi^{-2}-(z_{B_o}-y(0)))\varphi d(\partial B_o) = \int_{\partial B_1} (-\varphi^{-1}-\Lambda^*\Lambda\varphi^{-2}-g_1)\varphi d(\partial B_1) + \\ + \int_\Gamma (g_2-\Lambda^*\Lambda\varphi^{-2})\varphi d(\Gamma).$$

We have

$$\varphi^o = z_{B_o} - y(0) - \Lambda^*\Lambda\varphi^{-2} \text{ on } \partial B_o$$

$$\varphi^{-1} = g_1-\Lambda^*\Lambda\varphi^{-2} \qquad \text{on } \partial B_1$$

$$g_2 = \Lambda^*\Lambda\varphi^{-2} \qquad \text{on } \Gamma.$$

The ε^1-term can be written as

$$\int_{\partial B_1} \varphi^o\cdot\varphi d(\partial B_1) + \int_{\partial B_o \cup \partial B_1 \cup \Gamma} \Lambda\varphi^{-1}\Lambda\varphi d(\partial\Omega) = \int_{\partial B_1} (z_{B_1}-y(0))\varphi d(\partial B_1),$$

then

$$\int_{\partial B_1} [\varphi^o+\Lambda^*\Lambda\varphi^{-1}-(z_{B_1}-y(0)]\varphi d(\partial B_1) + \int_{\partial B_o} \Lambda^*\Lambda\varphi^{-1}\varphi d(\partial B_o) + \int_\Gamma \Lambda^*\Lambda\varphi^{-1}\varphi d(\Gamma),$$

thus, on the boundaries we have

$$\varphi_1^o = z_{B_1} - y(0) - \Lambda^*\Lambda\varphi^{-1} \text{ on } \partial B_1$$

$$\varphi^{-1} = 0 \text{ on } \partial B_o$$

$$\Lambda^*\Lambda\varphi^{-1} = 0 \text{ on } \Gamma.$$

The $\frac{1}{\varepsilon}$ - term gives

$$a_o(\varphi^{-1},\varphi) + a_1(\varphi^{-2},\varphi) = 0,$$

hence

$$\int_{\partial B_o} \varphi^{-1} \varphi d(\partial B_o) = - \int_{\partial B_1} \varphi^{-2} \varphi d(\partial B_1),$$

it follows that

$$\varphi_o^{-1} = 0 \text{ and } \varphi_1^{-2} = 0;$$

and the $\frac{1}{\varepsilon^2}$ - term gives $a_o(\varphi^{-2},\varphi) = 0$, thus $\int_{\partial B_o} \varphi^{-2}\varphi d(\partial B_o) = 0$, and $\varphi_o^{-2} = 0$.

Finally the ansatz of the asymptotic expansion

$$\varphi_\varepsilon = \frac{\varphi^{-2}}{\varepsilon^2} + \frac{\varphi^{-1}}{\varepsilon} + \varphi^o + \ldots$$

is given explicitly:

$$\varphi_o^{-2} = 0 \text{ on } \partial B_o,\ \varphi_1^{-2} = 0 \text{ on } \partial B_1 \text{ and } \Lambda^*\Lambda\varphi^{-2} = g_2 \text{ on } \Gamma$$

$$\varphi_o^{-1} = 0 \text{ on } \partial B_o,\ \varphi_1^{-1} = g_1-\Lambda^*\Lambda\varphi^{-2} \text{ on } \partial B_1 \text{ and } \Lambda^*\Lambda\varphi^{-1} = 0 \text{ on } \Gamma,$$

$$\varphi_o^o = z_{B_o}-y(0)-\Lambda^*\Lambda\varphi^{-2} \text{ on } \partial B_o,\ \varphi_1^o = z_{B_1}-y(0)-\Lambda^*\Lambda\varphi^{-1} \text{ on } \partial B_1 \text{ and } \Lambda^*\Lambda\varphi^o=0 \text{ on } \Gamma.$$

The asymptotic expansion can be estimated as follows

$$\|\varphi_\varepsilon - (\frac{1}{\varepsilon^2}\varphi^{-2} + \frac{1}{\varepsilon}\varphi^{-1} + \varphi^o - \ldots + \varepsilon^j\varphi^j)\|_{H^{-2j-4}} < C\cdot\varepsilon^{j+1}.$$

Proof:

The state equation is given by

$$A(y(v)) = f \text{ in } \Omega$$

$$\frac{\partial y(v)}{\partial n} = v \text{ on } \partial\Omega.$$

Let $f \in L^2(\Omega)$ and $v \in L^2(\partial\Omega)$, there exists a unique solution $y(v)$ depending on the control parameter v.

Now, we introduce the problem $A\phi = 0$ in Ω and $\phi = \varphi$ on the boundary, where φ **is fixed.**

Defining the boundary operator Λ by $\Lambda\varphi = \frac{\partial\phi}{\partial n}$.

To construct Λ, we have to solve the von Neumann boundary value problem

$$A\psi = 0 \text{ in } \Omega$$

$$\frac{\partial\psi}{\partial n} = \xi \text{ on } \partial\Omega.$$

By definition $\Lambda\psi = \frac{\partial\psi}{\partial n} = \xi$; as the state equation is a von Neumann boundary value problem, we obtain by comparing

$$\Lambda\varphi = \frac{\partial y(v)}{\partial n} = v, \text{ hence } \Lambda\varphi = v \text{ or } \Lambda^{-1}v = \varphi$$

and analogously

$$\Lambda\varphi_\varepsilon = u_\varepsilon \quad \text{hence } \Lambda^{-1}u_\varepsilon = \varphi_\varepsilon.$$

We have $\Lambda\varphi = \frac{\partial y(v)}{\partial n} = v$, applying Λ^{-1}, then $\varphi = \Lambda^{-1}\Lambda\varphi = \Lambda^{-1}\frac{\partial y(v)}{\partial n} = \Lambda^{-1}v$, and $\Lambda^{-1}\frac{\partial y(v)}{\partial n}$ is interpretable as $y(v)-y(0)|_{\partial\Omega} = \Lambda^{-1}\frac{\partial y(v)}{\partial n}$, thus we obtain $y(v)-y(0)|_{\partial\Omega} = \Lambda^{-1}v = \varphi$ and analogously $y(u_\varepsilon)-y(0)|_{\partial\Omega} = \varphi_\varepsilon$, and finally

$$\varphi - \varphi_\varepsilon = y(v)-y(0)-y(u_\varepsilon) + y(0) = y(v)-y(u_\varepsilon).$$

One can show, that the operator

$\Lambda\Lambda^*$ is an isomorphism from $H^2(\Gamma)$ to $H^{s-2}(\partial\Omega)$.

The operator $\Lambda^*\Lambda$ is defined as follows.

Let $A\phi = 0$ in Ω, such that $\phi = 0$ on ∂B_0 and $\phi = \varphi$ on Γ, the adjoint problem is given by

$$A^*\psi = 0 \text{ in } \Omega$$

$$\psi = \frac{\partial\phi}{\partial n} \text{ on } \partial\Omega.$$

According to the definition of Λ we have $\Lambda\varphi = \frac{\partial\phi}{\partial n}\Big|_\Gamma = \psi$ by the adjoint problem. Applying Λ^*, but now on the boundary Γ, we obtain

$$\Lambda^*\Lambda\varphi = \Lambda^*\left(\frac{\partial\phi}{\partial n}\right) = \Lambda^*\psi = \frac{\partial\psi}{\partial n^*}, \text{ thus } \Lambda^*\Lambda \in L(H^s(\Gamma);H^{s-2}(\Gamma));$$

we have finally

$$(\Lambda^*\Lambda,\varphi,\varphi)_\Gamma \underset{\text{def.}}{=} \int_\Gamma (\Lambda^*\Lambda\varphi)\varphi d(\Gamma) \underset{\text{def.}}{=} \int_\Gamma \frac{\partial\psi}{\partial n^*}\cdot\varphi d(\Gamma) \underset{\text{on}\,\Gamma}{=} \int_\Gamma \frac{\partial\psi}{\partial n^*} \phi d(\Gamma) =$$

$$\int_\Gamma \psi\cdot \frac{\partial\phi}{\partial n^*} d(\Gamma) = \int_\Gamma \psi^2 (d\Gamma)$$

by construction, and

$$\int_\Gamma \psi^2 d(\Gamma) = \int_\Gamma (\frac{\partial\phi}{\partial n})^2 d(\Gamma) \geqq \alpha \|\varphi\|_{H^1(\ )} \text{ as } \phi = \varphi \text{ on } \Gamma,$$

thus we obtain

$$(\Lambda^*\Lambda\varphi,\varphi) \geqq \alpha \|\varphi\|_{H^1(\Gamma)}$$

and $\Lambda^*\Lambda$ is an isomorphism from $H^s(\Gamma)$ to $H^{s-2}(\Gamma)$ for $s = 1$. Now, the proof for $s > 1$ is clear.

We obtain some interesting regularity results. Let $\varphi^o \in H^o(\partial\Omega) = L^2(\partial\Omega)$, then $\Lambda^*\Lambda\varphi^o \in H^{o-2} = H^{-2}(\partial\Omega)$. As $g_2 \in L^2(\partial\Omega)$ we have on page 157, $\Lambda^*\Lambda\varphi^{-2} = g_2$ then $\varphi^{-2} \in H^2(\Gamma)$, if $\varphi^{-1} = g_1 - \Lambda^*\Lambda\, \varphi^{-2}$, then $\Lambda^*\Lambda\varphi^{-2} \in H^o(\Gamma) = L^2(\Gamma)$, thus $\varphi^{-1} \in H^o(\partial\Omega)$, then $\Lambda^*\Lambda\varphi^{-1} \in H^{-2}$ and by construction $\varphi^o \in H^{-2}$; and by induction we obtain finally

$$\varphi^j \in H^{-2(j+1)} = H^{-2j-2}$$

$$\varphi^{j+1} \in H^{-2(j+2)} = H^{-2j-4}.$$

With the help of these results we are able to estimate

$$\|\varphi_\varepsilon - (\frac{1}{\varepsilon^2}\cdot\varphi^{-2} + \frac{1}{\varepsilon}\cdot\varphi^{-1} + \varphi^o + \dots + \varepsilon^j\varphi^j)\|_{H^{-2j-4}} < C\cdot\varepsilon^{j+1}.$$

Defining

$$\Xi = \varphi_\varepsilon - (\frac{1}{\varepsilon^2}\cdot\varphi^{-2} + \frac{1}{\varepsilon}\varphi^{-1} + \varphi^o + \dots + \varepsilon^j\varphi^j + \varepsilon^{j+1}\varphi^{j+1}),$$

we obtain

$$a_o(\Xi,\varphi) + \varepsilon a_1(\Xi,\varphi) + \varepsilon^2 a_2(\Xi,\varphi) = -\varepsilon^{j+3} a_2(\varphi^{j+1},\varphi) \text{ for } \varphi \in \mathcal{D}(\partial\Omega),$$

and we can write

$$(k+\varepsilon^2\Lambda^*\Lambda)\Xi = -\varepsilon^{j+3}\Lambda^*\Lambda\varphi^{j+1},$$

hence

$$\Xi = -\varepsilon^{j+3}(\varepsilon^2\Lambda^*\Lambda-k)^{-1}\Lambda^*\Lambda\varphi^{j+1} = -\varepsilon^{j+3}\frac{\Lambda^*\Lambda\varphi^{j+1}}{(\Lambda^*\Lambda\varepsilon^2+\frac{\varepsilon^2}{\varepsilon^2}k)} = -\varepsilon^{j+1}(\Lambda^*\Lambda+\frac{k}{\varepsilon^2})^{-1}\Lambda^*\Lambda\varphi^{j+1}.$$

As $\Lambda^*\Lambda : H^s \to H^{s-2}$, then $(\Lambda^*\Lambda)^{-1} : H^{s-2} \to H^s$, therefore $(\Lambda^*\Lambda + \frac{k}{\varepsilon^2})^{-1}\Lambda^*\Lambda$ is a mapping $H^s(\partial\Omega) \to H^s(\partial\Omega)$ for all s. Since $\varphi^{j+1} \in H^{-2j-4}$ we finally have $\Xi \in H^{-2j-4}$ and obtain the estimate.

II. 8.4 The algorithm of the asymptotic solution

Applying the optimal control u_ε to the state equation, then the modified state equation reads

$$Ay_\varepsilon = f \text{ in } \Omega$$

$$\varepsilon^2 \frac{\partial y_\varepsilon}{\partial n} + \varepsilon \cdot p_\varepsilon = 0 \text{ on } \partial\Omega$$

and the adjoint state equation is given (see page 153)

$$A^*p_\varepsilon = 0 \text{ in } \Omega$$

$$\frac{\partial p_\varepsilon}{\partial n^*} = y_\varepsilon - z_{B_0} \text{ on } \partial B_0$$

$$\frac{\partial p_\varepsilon}{\partial n^*} = y_\varepsilon - Z_{B_1} - g_1 \text{ on } \partial B_1$$

$$\frac{\partial p_\varepsilon}{\partial n^*} = -g_2 \text{ on } \Gamma$$

Let us assume that the asymptotic developments of y_ε and p_ε be given by

$$y_\varepsilon = \frac{1}{\varepsilon^2}\cdot y^{-2} + \frac{1}{\varepsilon}\cdot y^{-1} + y^o + \varepsilon y^1 + \ldots,$$

$$p_\varepsilon = \frac{1}{\varepsilon}\cdot p_\varepsilon^{-1} + p^o + \varepsilon p^1 + \ldots .$$

The optimal control u_ε can be written asymptotically

$$u = -\frac{1}{\varepsilon}\cdot p_\varepsilon = -\frac{1}{\varepsilon}(\frac{1}{\varepsilon}p_\varepsilon^{-1} + p^o + \varepsilon p^1 + \ldots),$$

using the asymptotic expansion of the adjoint state equation

$$A^*(\frac{1}{\varepsilon}p^{-1}) + A^*(p^o) + A^*(\varepsilon p^1) + \ldots = 0 \text{ in } \Omega,$$

and the boundary conditions:

$$\frac{\partial p_\varepsilon}{\partial n} = \frac{1}{\varepsilon}\cdot\frac{\partial p^{-1}}{\partial n} + \frac{\partial p^o}{\partial n} + \varepsilon\frac{\partial p^{-1}}{\partial n} + \dots = y_\varepsilon - z_{B_o} \text{ on } \partial B_o$$

$$= \frac{1}{\varepsilon^2} y^{-2} - \frac{1}{\varepsilon} y^{-1} + y^o + \varepsilon\cdot y^1 + \dots - z_{B_o},$$

$$\frac{\partial p_\varepsilon}{\partial n} = \frac{1}{\varepsilon}\cdot\frac{\partial p^{-1}}{\partial n} + \frac{\partial p^o}{\partial n} + \varepsilon\frac{\partial p^1}{\partial n} + \dots = y_\varepsilon - z_{B_1} - g_1 \text{ on } \partial B_1$$

$$= \frac{1}{\varepsilon^2} y^{-2} + \frac{1}{\varepsilon} y^{-1} + y^o + \varepsilon\cdot y^1 + \dots - z_{B_1} - g_1$$

$$\frac{\partial p_\varepsilon}{\partial n} = \frac{1}{\varepsilon}\cdot\frac{\partial p^{-1}}{\partial n} + \frac{\partial p^o}{\partial n} + \varepsilon\frac{\partial p^1}{\partial n} + \dots = -g_2 \text{ on } \Gamma.$$

We obtain:

	on ∂B_o	on ∂B_1	on Γ
ε^o-term			
$A^*p^o = 0$	$\frac{\partial p^o}{\partial n} = y^o - z_{B_o}$	$\frac{\partial p^o}{\partial n} = y^o - z_{B_1} - g_1$	$\frac{\partial p^o}{\partial n} = -g_2$
ε^1-term			
$A^*p^1 = 0$	$\frac{\partial p^1}{\partial n} = y^1$	$\frac{\partial p^1}{\partial n} = y^1$	$\frac{\partial p^1}{\partial n} = 0$
$\frac{1}{\varepsilon}$ - term			
$A^*p^{-1} = 0$	$\frac{\partial p^{-1}}{\partial n} = y^{-1}$	$\frac{\partial p^{-1}}{\partial n} = y^{-1}$	$\frac{\partial p^{-1}}{\partial n} = 0$

The asymptotic expansion of the state equation reads

$$Ay_\varepsilon = \frac{1}{\varepsilon^2} Ay^{-2} + \frac{1}{\varepsilon}\cdot Ay^{-1} + Ay^o + \varepsilon Ay^1 + \dots = f \text{ in } \Omega$$

and the boundary condition

$$\varepsilon^2\left(-\frac{1}{\varepsilon^2}\cdot\frac{\partial y^{-2}}{\partial n} + \frac{1}{\varepsilon}\cdot\frac{\partial y^{-1}}{\partial n} + \frac{\partial y^o}{\partial n} + \varepsilon\frac{\partial y^1}{\partial n}\right) + \varepsilon\left(\frac{1}{\varepsilon}\cdot p^1 + p^o + \varepsilon p^1 + \dots\right) = 0 \text{ on } \partial\Omega.$$

We obtain the ε^o-term

$$Ay^o = 0 \text{ and } \frac{\partial y^{-2}}{\partial n} + p^1 = 0,$$

the ε^1-term

$$Ay^1 = f \text{ and } \frac{\partial y^{-1}}{\partial n} + p^o = 0,$$

the $\frac{1}{\varepsilon}$ - term

$$Ay^{-1} = 0,$$

the ε^2-term $\frac{\partial y^o}{\partial n} + p^1 = 0$, and the ε^3-term $\frac{\partial y^1}{\partial n} = 0$.

The solution u_ε of the optimal control problem

$$J_\varepsilon(u_\varepsilon) = \inf_{v \in L^2(\partial\Omega)} J_\varepsilon(v)$$

is given by the algorithm (1,2,3,4,5,6):

1) $Ay^1 = f$ in Ω

$$\frac{\partial y^1}{\partial n} = 0 \text{ on } \partial\Omega$$

2) $A^*p^1 = 0$ in Ω

$$\frac{\partial p^1}{\partial n} = y^1 \text{ on } \partial B_o$$

$$\frac{\partial p^1}{\partial n} = y^1 \text{ on } \partial B_1$$

$$\frac{\partial p^1}{\partial n} = 0 \text{ on } \Gamma$$

3) $Ay^o = 0$ in Ω

$$\frac{\partial y^o}{\partial n} = -p^1 \text{ on } \partial\Omega$$

4) $A^*p^o = 0$ in Ω

$$\frac{\partial p^o}{\partial n} = y^o - z_{B_o} \text{ on } \partial B_o$$

$$\frac{\partial p^o}{\partial n} = y^o - z_{B_1} - g_1 \text{ on } \partial B_1$$

$$\frac{\partial p^o}{\partial n} = -g_2 \text{ on } \Gamma$$

5) $Ay^{-1} = 0$ in Ω

$$\frac{\partial y^{-1}}{\partial n} = -p^o \text{ on } \partial\Omega$$

6) $A^*p^{-1} = 0$ in Ω

$$\frac{\partial p^{-1}}{\partial n} = y^{-1} \text{ on } \partial B_o$$

$$\frac{\partial p^{-1}}{\partial n} = y^{-1} \text{ on } \partial B_1$$

$$\frac{\partial p^{-1}}{\partial n} = 0 \text{ on } \Gamma.$$

The solution of the optimal control process

$$u_\varepsilon = -\frac{1}{\varepsilon}\cdot p_\varepsilon = -\frac{1}{\varepsilon}\left(\frac{1}{\varepsilon}\cdot p^{-1} + p^o + \varepsilon p^1 + \dots\right).$$

II. 8.5 Summary:

The solution of the optimal control process

$$J_\varepsilon(u_\varepsilon) = \inf_{v \in L^2(\partial\Omega)} J_\varepsilon(v),$$

with the cost function

$$J_\varepsilon(v) = \int_{\partial B_o} |y(v) - z_{B_o}|^2 d(\partial B_o) + \varepsilon \int_{\partial B_1} |y(v) - z_{B_1}|^2 d(\partial B_1) + \varepsilon^2 \int_{\partial\Omega} v^2 d(\partial\Omega)$$

$$-2 \int_{\partial B_1} g_1 \cdot y(v) d(\partial B_1) - 2 \int_\Gamma g_2 \cdot y(v) d\Gamma, \; v \in L^2(\partial\Omega),$$

where y(v) is the solution of the state equation

$$A(y(v)) = f \text{ in } \Omega$$

$$\frac{\partial y(v)}{\partial n} = v \text{ on the boundary } \partial\Omega ,$$

where $v \in L^2(\partial\Omega)$ is the control variable.

The solution of the optimal control process is given by the following algorithm

$$u_\varepsilon = -\frac{1}{\varepsilon} p_\varepsilon = -\frac{1}{\varepsilon}\left(\frac{1}{\varepsilon} \cdot p^{-1} + p^o + \varepsilon p^1 + \ldots\right),$$

where p^{-1}, p^o, and p^1 are solutions of the boundary-value problem of the three-connected domain Ω.

The solutions can be given in closed, analytic form by methods developed in (Grusa (140, 141) for multiconnected regions).

II. 8.6 Evolution of the optimal controlled diffusion process in a three-connected domain

Now, the optimal control process given in closed form, can be simulated by the variation of the

i) domains B_o, B_1 with their boundaries ∂B_o, ∂B_1

ii) 3-connected domain Ω with the outer boundary Γ

iii) fixed cost parameters z_{B_o} on ∂B_o

z_{B_1}, g_1 on ∂B_1

g_2 on Γ

iv) coefficients a_{ij}, a_o of the state equation

v) initial distribution f in Ω.

We obtain some insight into certain complex optimal control processes. The asymptotic method can be interpreted as the evolution of the control process.

For applications it is very interesting that the solution u_ε of the optimal control process is given by the algorithm (1,2,3,4,5,6):

1) $Ay^1 = f$ in Ω

$$\frac{\partial y^1}{\partial n} = 0 \text{ on } \partial\Omega$$

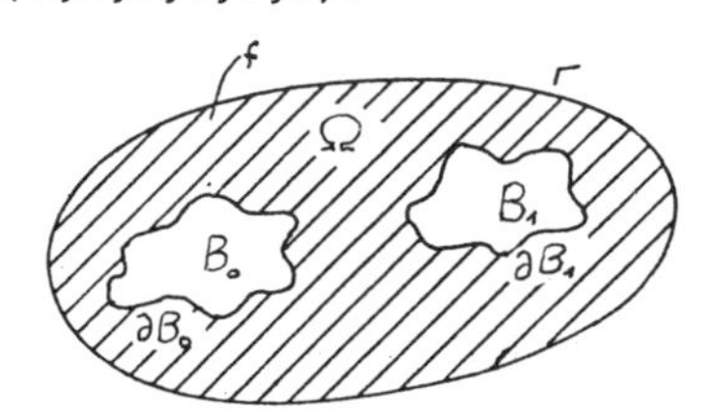

the solution y^1 depends on the initial distribution f, thus $y^1 = y^1(f)$.

2) $A^*p^1 = 0$ in Ω

$$\frac{\partial p^1}{\partial n} = y^1 \text{ on } \partial B_o$$

$$\frac{\partial p^1}{\partial n} = y^1 \text{ on } \partial B_1$$

$$\frac{\partial p^1}{\partial n} = 0 \text{ on } \Gamma$$

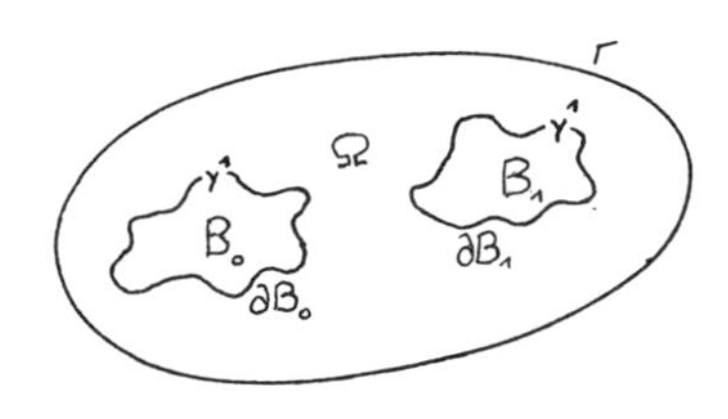

the solution p^1 depends on y^1, thus on f, $p^1 = p^1(y^1,f)$;

3) $Ay^o = 0$ in Ω

$$\frac{\partial y^o}{\partial n} = -p^1 \text{ on } \partial\Omega,$$

the solution y^o depends on the solution p^1, thus $y^o = y^o(p^1,y^1,f)$.

4) $A^*p^o = 0$ in Ω

$$\frac{\partial p^o}{\partial n} = y^o - z_{B_o} \text{ on } \partial B_o$$

$$\frac{\partial p^o}{\partial n} = y^o - z_{B_1} - g_1 \text{ on } \partial B_1$$

$$\frac{\partial p^o}{\partial n} = -g_2 \text{ on } \Gamma,$$

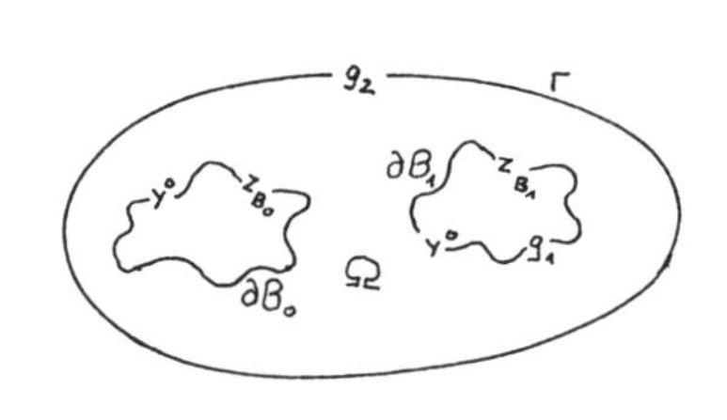

the solution p^o depends on the

i) fixed cost z_{B_o} on ∂B_o

ii) fixed cost z_{B_1} on ∂B_1

iii) cost structure g_1 of the domain B_1

iv) cost structure g_2 of the region outside Ω,

thus $p^o = p^o(z_{B_o}, z_{B_1}, g_1, g_2, y^o)$.

5) $Ay^{-1} = 0$ in Ω

$$\frac{\partial y^{-1}}{\partial n} = -p^o \text{ on } \partial\Omega$$

the solution y^{-1} depends on $y^{-1} = y^{-1}(p^o)$,

6) $A^*p^{-1} = 0$ in Ω

$$\frac{\partial p^{-1}}{\partial n} = y^{-1} \text{ on } \partial B_o$$

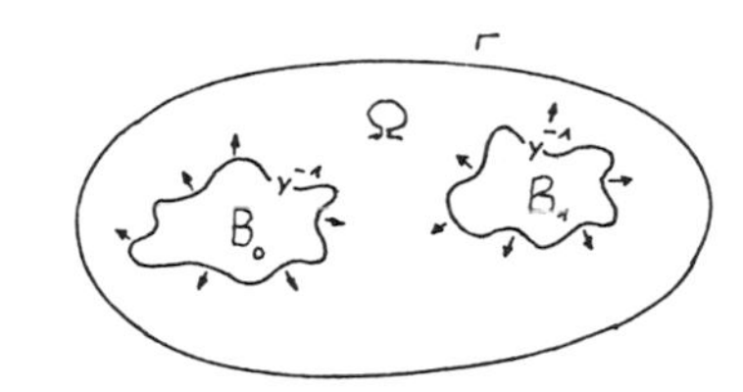

$$\frac{\partial p^{-1}}{\partial n} = y^{-1} \text{ on } \partial B_1$$

$$\frac{\partial p^{-1}}{\partial n} = 0 \text{ on } \Gamma$$

the solution p^{-1} depends on y^{-1} and thus on $p^o = p^o(z_{B_o}, z_{B_1}, g_1, g_2, f)$, and on the coefficients of the elliptic differential operator A^*; and the boundary value y^{-1} depends on $y = y(z_{B_o}, z_{B_1}, g_1, g_2, f,)$.

Some ideas of the chapter are adapted from the work of Lions (214-219, 220a,b,c,d).

III Dynamic reaction-diffusion processes

III.1 Some remarks on population dynamics

III. 1.1 On conditions of migration processes

On the level of populations, individuals interact and move about, and so it is not surprising that there, again, the simplest continuous space-time interaction-migration models have the same general appearance as those for diffusing and reacting chemical systems (R. Aris (12)). The principal ingredients of these models are equations of the form

$$\frac{\partial u}{\partial t} = D \cdot \Delta u + F(u),$$

where u is a vector of diffusing quantities, D is a diffusion-matrix, and F describes all reactions and interactions. Some modelling considerations. The equations of reaction and diffusion are used to model a multispecies population, whose individuals interact in some way to produce other individuals (or eliminate individuals), and also move about in some random manner.

Every mathematical model has numerical parameters (rates, probabilities of various occurrences, coefficients in differential equations, dimension of space,...). Even functional, rather than numerical, parameters are common. Our basic hypothesis of a possible model-building process:
A population is a collection of individuals, each occupying a position in space which is a function of time.

But for the simplicity we shall always speak of geographical space over which the individuals are distributed, at any instant of time. We shall take here only about one-dimensional space, so that the population lies on a line. Many dimensional reaction-diffusion equations are important, and can be derived by a generalization of the process we describe.

The first approach we take is to assume, that the population distribution can be approximated by a smooth density function.

I. To each species i, there corresponds a spatial density function $\rho_i(x,t)$, so that the number of individuals of the species located in the interval $a < x < b$ at time t is approximately

$$\int_a^b \rho_i(x,t)dx.$$

These density functions ρ_i are twice continuously differentiable in x, and once in t.

II. The time evolution of the ρ_i proceeds by a determinisitic process, in the sense that, if any initial distribution in the form of bounded functions $\rho_i^o(x) \in C^2$ is given at any time t_o, then there is a uniquely determined set of density functions $\rho_i(x,t)$ defined for $t \geqq t_o$, with the initial condition

$$\rho_i(x,t_o) = \rho_i^o(x).$$

These assumptions are most reasonable when the population is large. Under II, the density function for a given species, devided by the total population of the species, assuming it is finite, is the probability distribution for the position of a random individual of that species. But we are assuming that the distribution evolves deterministically.

Even if one conceived of the real world as deterministic, the behaviour of physical, chemical or biological processes is subjected to so many influences, that it should be considered largely a matter of chance.

The influences may be reduced to main influences, which can be treated by mathematics.

We reduce the influence to

i) initial-boundary-value problems in simple and multiconnected regions (Chapter I)

ii) control problems in simple and multiconnected regions (Chapter II)

iii) global control processes in space (Chapters VI and VII)

where the control is concentrated on a finite number of small balls, given at internal points of a prescribed domain in $\mathbb{R}^3$. According to the approximation process we take the radius of the balls and tend them to zero, so we obtain for the finite number of points pointwise control processes.

We consider redistribution processes with respect to interior and boundary points.

On a higher level of approximation we consider diffusion processes in multiconnected regions governed by von Stackelberg strategies and differential games.

We derive in the following, under suitable conditions, partial differential equations of parabolic type, describing migration processes and their interactions.

We are considering migration processes, when only one species is present. We assume that the domain of the population is $\mathbb{R}$, the whole line. The extension to $\mathbb{R}^2$ or $\mathbb{R}^n$ is straightforward. Let $\rho(x,t)$ be the density function of the species, assuming there exists a nonnegative function $f(x,t,y,s)$, for $t > s$, and f should be integrable

$$\int_{-\infty}^{+\infty} f(x,t,y,s)dy < \infty.$$

The population density satisfies

$$\text{III.} \quad \rho(x,t) = \int_{-\infty}^{+\infty} f(x,t,y,s)\rho(y,s)dy.$$

We can assume, that the migration process is linear, and the densities $\rho(x,t)$ depend continuously on $\rho(\cdot,s)$.

Small changes in the value of $\rho(\cdot,s)$ at any time s induce small changes in $\rho(\cdot,t)$ at any fixed later time t. Thus, for any $\varepsilon > 0$ there exists a δ, depending on x, t, s, such that if ρ_1 and ρ_2 are two denstity functions, we have

$$\sup_y |\rho_1(y,s) - \rho_2(y,s)| < \delta \text{ implies } |\rho_1(x,t)-\rho_2(x,t)| < \varepsilon .$$

In words: small uniform changes in $\rho(\cdot,s)$ mean small changes in the number of individuals effectively influencing the changes at (x,t).

$$\text{IV. Assuming } \int_{-\infty}^{-\infty} f(x,t,y,s)dx = 1 \text{ for } t > s,$$

we can say that, for finite populations, the population remains constant. Now, $f(x,t,y,s)$ can be interpreted as the probability distribution of the population for the position x at time t of an individual located at y at time s.

We assume another hypothesis, that the individuals cannot migrate finite distances in very short time periods, described by

V. $\lim_{t \searrow s} \frac{1}{t-s} \int_{|x-y|>a} f(x,t,y,s)dx = 0$ for every $a > 0$,

this can be interpreted as the probability of individuals migrating a distance larger than a in a time interval $h = t - s$ approaching zero, as $h \to 0$, at a rate faster than h. If the integral vanish for small (t-s) and with the other assumptions, every individual at (y,s) must move in a predetermined path.

All the above assumptions constitute a model analysing the dynamics of the density ρ, and these dynamics are governed by a parabolic differential equation. We obtain:

III. 1.2 Derivation of the equation describing randomly migrating populations

Theorem:

To each evolution process described by I, II,...,V, there exist functions $\mathcal{D}(x,t) \geqq 0$ and $\mathcal{E}(x,t) \geqq 0$, such that the density function ρ of the process satisfies

$$\frac{\partial \rho}{\partial t} = \frac{\partial}{\partial x} (\mathcal{D}(x,t) \cdot \frac{\partial \rho}{\partial x} + \mathcal{E}(x,t)\rho).$$

Let $\rho(x,t)$ be the density function of the migration-process. The number N(t) of individuals of the species located in the interval $a < x < b$ at time t is given approximately by II

$$N(t) = \int_a^b \rho(x,t)dx.$$

The population density satisfies (III)

$$\rho(x,t) = \int_{-\infty}^{+\infty} f(x,t,y,s)\rho(y,s)dy,$$

hence

$$\rho(x,t+\nu) = \int f(x,t+\nu,y,s)\rho(y,s)dy = \int f(y+x-y,t+\nu,y,s)\rho(y,s)dy, \text{ put } \xi = x-y$$

$$= \int f(y+\xi,t+\nu,y,t)\rho(y,t)dy,$$

defining

$$f(y+\xi,t+\nu,y,t) = L(\xi,y,t,\nu),$$

we have

$$\rho(x,t+\nu) = \int L(\xi,y,t,\nu)\rho(y,t)dy = \int L(\xi,x-\xi,t,\nu)\rho(y,t)dy, \text{ as } y = x-\xi,$$
$$= \int L(\xi,x-\xi,t,\nu)\rho(x-\xi,t)d\xi.$$

We obtain by assumption IV

$$\int_{-\infty}^{+\infty} L(\xi,y,t,\nu)d\xi = 1.$$

Now we consider the change of the number of individuals with respect to the time t

$$\frac{1}{\nu}\{N(t+\nu)-N(t)\} = \frac{1}{\nu}\{\int_a^b (\rho(x,t+\nu)-\rho(x,t))dx\}$$

$$\frac{1}{\nu}\{N(t+\nu)-N(t)\} = \frac{1}{\nu}\int_a^b dx \int L(\xi,x-\xi,t,\nu)\rho(x-\xi,t)d\xi - 1\cdot\rho(x,t)$$
$$= \frac{1}{\nu}\{\int_a^b dx \int[L(\xi,x-\xi,t,\nu)\rho(x-\xi,t)-L(\xi,x,t,\nu)\rho(x,t)]d\xi$$

by IV.

Introducing the kernel

$$K(\xi,x,t,\nu) = L(\xi,x-\xi,t,\nu)\rho(x-\xi,t) - L(\xi,x,t,\nu)\rho(x,t),$$

we obtain

$$\frac{1}{\nu}\{N(t+\nu)-N(t)\} = \frac{1}{\nu}\int_a^b dx \int_{|\xi|<\varepsilon} K(\xi,x,t,\nu)d\xi dx \ +$$
$$+ \frac{1}{\nu}\int_a^b \int_{|\xi|\geqq\varepsilon} K(\xi,x,t,\nu)d\xi dx \underset{\text{def.}}{=} \mathcal{D}_\nu(t,\nu)+C\mathcal{D}_\nu(t,\nu)$$

It follows by V, that the second expression satisfies

$$\lim_{\nu\to 0} C\mathcal{D}_\nu(t,\nu) = 0.$$

As $N(t) = \int_a^b \rho(x,t)\cdot dx$ and by I. ρ_t is continuous, thus $\frac{\partial N}{\partial t} = \int_b^a \rho_t(x,t)dx$

and finally we have

$$\lim_{\nu\to 0} \mathcal{D}_\nu(t,\nu) = \frac{dN}{dt} = \int_a^b \rho_t(x,t)dx.$$

$\mathcal{D}_\nu(t,\nu)$ can be written as follows

$$\mathcal{D}_\nu(t,\nu) = \frac{1}{\nu}\int_{|\xi|<\varepsilon} d\xi \int_a^b K(\xi,x,t,\nu)dx = \frac{1}{\nu}\int_{-\varepsilon}^{+\varepsilon}\int_a^b (L(\xi,x-\xi,t,\nu)\rho(x-\xi,t) -$$

$$L(\xi,x,t,\nu)\rho(x,t))d\xi dx$$

$$= \frac{1}{\nu}\int_{-\varepsilon}^{+\varepsilon}\left(\int_{a-\varepsilon}^{b-\varepsilon} - \int_a^b\right)L(\xi,z,t,\nu)\rho(z,t)dzd\xi, \text{ as } \int_{a-\xi}^{b-\xi} = \int_{a-\xi}^{a} + \int_a^b + \int_b^{b-\xi},$$

$$= \frac{1}{\nu}\int_{-\varepsilon}^{-\varepsilon}\left(\int_{a-\xi}^{a} - \int_{b-\xi}^{b}\right)L(\xi,z,t,\nu)\rho(z,t)dzd\xi \underset{\text{def.}}{=} M(a,t,\nu)-M(b,t,\nu).$$

We obtain

$$M(x,t,\nu) = \frac{1}{\nu}\int_{-\varepsilon}^{+\varepsilon}\int_{x-B}^{x} L(\xi,z,t,\nu)\rho(z,t)dzd\xi$$

by comparison and introducing the transformation of the integration variable $z = x-s$, then $\frac{dz}{ds} = -1$, we have

$$\int_{x-\xi}^{x} (\quad)dz \rightarrow - \int_{\xi}^{0}(\ldots)ds,$$

hence

$$M(x,t,\nu) = \frac{1}{\nu}\int_{-\varepsilon}^{+\varepsilon}\int_0^{\xi} L(\xi,x-s,t,\nu)\rho(x-s,t)dsd\xi.$$

Applying the Taylor expansion

$$\rho(x-s,t) = \rho(x,t) - s\left(\frac{\partial\rho}{\partial x}(x,t) + \frac{\partial\rho}{\partial x}(\tilde{x},t) - \frac{\partial\rho}{\partial x}(x,t)\right),$$

where the value of $\tilde{x}$ is between x and x-s, setting

$$\Delta\frac{\partial\rho}{\partial x} = \frac{\partial\rho}{\partial x}(\tilde{x},t) - \frac{\partial\rho}{\partial x}(x,t),$$

we obtain

$$M(x,t,\nu) = \frac{1}{\nu}\int_{-\varepsilon}^{+\varepsilon}\int_0^{\xi} L(\xi,x-s,t,\nu)\{\rho(x,t)-s(\frac{\partial\rho}{\partial x}(x,t)+\Delta\frac{\partial\rho}{\partial x})\}\, dsd\xi$$

$$= \mathbb{A}(x,t,\nu)\rho(x,t) + \mathbb{B}(x,t,\nu)\frac{\partial\rho}{\partial x}(x,t) + \mathbb{E}(x,t,\nu),$$

$$\mathbb{A}((x,t,\nu)\rho(x,t) = \frac{1}{\nu}\int_{-\varepsilon}^{+\varepsilon}\int_0^{\xi} L(\xi,x-s,t,\nu)\rho(x,t)dsd\xi$$

$$-\ \mathbb{B}(x,t,\nu)\frac{\partial\rho}{\partial x}(x,t) = \frac{1}{\nu}\int_{-\varepsilon}^{\varepsilon}\int_0^{\xi} L(\xi,x-s,t,\nu)s\cdot\frac{\partial\rho}{\partial x}\, dsd\xi$$

$$-\ \mathbb{E}(x,t,\nu) = \frac{1}{\nu}\int_{-\varepsilon}^{+\varepsilon}\int_0^{\xi} L(\xi,x-s,t,\nu)\cdot s\cdot\Delta\frac{\partial\rho}{\partial x}\, dsd\xi.$$

We have

$$\mathcal{D}_\nu(t,\nu) = M(a,t,\nu) - M(b,t,\nu)$$

$$= \mathbb{A}(a,t,\nu)\rho(a,t) + \mathbb{B}(a,t,y)\frac{\partial\rho}{\partial x}(a,t) + \mathbb{E}(a,t,\nu) - (\mathbb{A}(b,t,\nu)\rho(b,t) +$$

$$\mathbb{B}(b,t,\nu)\cdot\frac{\partial\rho}{\partial x}(b,t) + \mathbb{E}(b,t,\nu)).$$

We consider by II, that in a neighbourhood of a, $\rho(x,t) = \rho(a,t) = 0$, and in a neighbourhood of b, $\rho(x,t) = \rho(b,t) = \text{constant}$, thus

$$\mathcal{D}_\nu(t,\nu) = - M(b,t,\nu) = - \mathbb{A}(b,t,\nu)\rho_c,$$

and

$$\lim_{\nu\to 0} \mathcal{D}_\nu(t,\nu) = - \lim_{\nu\to 0} \mathbb{A}(b,t,\nu)\rho_c \underset{\text{def.}}{=} G(x,t).$$

If $\rho(x,t)$ is linear in a neighbourhood of b, we have $\Delta\frac{\partial\rho}{\partial x} = 0$, thus

$$\mathcal{D}_\nu(t,\nu) = - \mathbb{A}(b,t,\nu)\rho - \mathbb{B}(b,t,\nu)\frac{\partial\rho}{\partial x} \text{ and } -\lim_{\nu\to 0} \mathbb{B}(x,t,\nu)\frac{\partial\rho}{\partial x} \text{ exists}$$

Defining

$$-D(x,t) = \lim_{\nu\to 0} \mathbb{B}(x,t,\nu),$$

then

$$\lim_{\nu\to 0} M(x,t,\nu) = \lim_{\nu\to 0} \mathbb{A}(x,t,\nu)\rho(x,t) + \lim_{\nu\to 0} \mathbb{B}(x,t,\nu)\frac{\partial\rho}{\partial x} + \lim_{\nu\to 0} \mathbb{E}(x,t,\nu)$$

$$= -G(x,t)\rho(x,t) - D(x,t)\frac{\partial\rho}{\partial x} + \lim_{\nu\to 0} \mathbb{E}(x,t,\nu),$$

and

$$|\mathbb{E}(x,t,\nu)| \leqq \frac{1}{\nu}\int_{-\varepsilon}^{+\varepsilon}\int_0^\xi |L(\xi,x-s,t,\nu)|\cdot s.\Delta\frac{\partial\rho}{\partial x}\cdot ds d\xi \leqq$$

$$\sup_{|s|<\varepsilon} |\Delta \frac{\partial\rho}{\partial x}|\cdot\frac{1}{\nu}\cdot\int_{-\varepsilon}^{+\varepsilon}\int_0^\xi |L(..)sdsd\xi = \delta(\varepsilon)\cdot D(x,t);$$

taking the limit $v \to 0$, we obtain

$$\limsup_{\nu\to 0} |\mathbb{E}(x,t,\nu)| \leqq \delta(\varepsilon)\cdot D(x,t), \quad \delta(\varepsilon) = \sup_{|x|<\varepsilon} |\Delta\frac{\partial\rho}{\partial x}|,$$

as $\delta(\varepsilon) \to 0$ for $\varepsilon \to 0$, then

$$\lim_{\nu\to 0} \mathbb{E}(x,t,\nu) = 0.$$

We derived above, that

$$\int_a^b \frac{\partial\rho}{\partial t}\,dx = \frac{\partial N}{\partial t} = M(a,t,\nu) - M(b,t,\nu),$$

and

$$\lim_{\nu\to 0} M(x,t,\nu) = - G(x,t)\rho(x,t) - D(x,t)\cdot\frac{\partial\rho}{\partial x},$$

differentiating with respect to the upper limit b, we obtain finally

$$\frac{\partial\rho}{\partial t} = \frac{\partial}{\partial x}\left(G(x,t)\cdot\rho(x,t) + D(x,t)\cdot\frac{\partial\rho}{\partial x}\right).$$

In the theory of time-continuous stochastic processes it is known that the transition probability density function is governed by a parabolic differential equation (see Feller 1966, first obtained by Kolmogoroff (1931) and Feller (1936)).

Now, we derive Kolmogoroff's forward equation:
The coefficient $\mathcal{B}(x,t,\nu)$ is given on page 171

$$-\mathcal{B}(x,t,\nu) = \frac{1}{\nu}\int_{-\varepsilon}^{+\varepsilon}\int_0^{\xi} L(\xi,x-k,t,\nu)k\,dk\,d\xi.$$

We assume that the kernel L is continuous in its second argument, and can be written as

$$L(\xi,x-k,t,\nu) = L(\xi,x,t,\nu)(1 + s),$$

where $s \to 0$ as $k \to 0$, then

$$\mathcal{B}(x,t,\nu) = \frac{1}{\nu}\int_{-\varepsilon}^{+\varepsilon}\int_0^{\xi} k\cdot L(\xi,x,t,\nu)(1+s)\,dk\,d\xi - \frac{1}{8\nu}\cdot\int_{-\varepsilon}^{+\varepsilon}\frac{\xi^2}{2}L(\xi,x,t,\nu)\,d\xi(1+s).$$

Introducing the expression

$$V(x,t) = \lim_{\nu\to 0}\frac{1}{\nu}\int_{-\infty}^{+\infty}\xi^2 L(\xi,x,t,\nu)\,d\xi,$$

where the integral

$$\int = \int_{|\xi|<\varepsilon} + \int_{|\xi|>\varepsilon},$$

and applying condition V , we obtain

$$D(x,t) = -\lim_{\nu\to 0}\mathcal{B}(x,t,\nu) = \frac{1}{2}V(x,t).$$

In an analogous way we have

$$\frac{\partial}{\partial x} D(x,t) = - \lim_{\nu \to 0} \frac{\partial \mathbb{B}}{\partial x}(x,t,\nu) = \frac{1}{2} \frac{\partial}{\partial x} V(x,t).$$

As

$$\lim_{\nu \to 0} M(x,t,\nu) = \lim_{\nu \to 0} \mathbb{A}(x,t,\nu)\rho(x,t) + \lim_{\nu \to 0} \mathbb{B}(x,t,\nu)\frac{\partial \rho}{\partial x} ,$$

and introducing

$$\mathrm{M}(x,t) = \lim_{\nu \to 0} \frac{1}{\nu} \cdot \int_{-\infty}^{+\infty} k \cdot L(\xi, x-k, t, \nu) dk,$$

we obtain

$$G(x,t) = -\mathrm{M}(x,t) + \frac{1}{2} \frac{\partial}{\partial x} V(x,t);$$

using this in

$$\frac{\partial \rho}{\partial t} = \frac{\partial}{\partial x} (D(x,t)\frac{\partial \rho}{\partial x} + G(x,t)\rho),$$

we have

$$G\rho = - \mathrm{M} \cdot \rho + \frac{1}{2} \cdot \frac{\partial V}{\partial x} \cdot \rho$$

and finally

$$\frac{\partial \rho}{\partial t} = \frac{\partial}{\partial x} (\frac{1}{2} \frac{\partial}{\partial x} (V \cdot \rho) - \mathrm{M} \cdot \rho),$$

Kolmogoroff's forward partial differential equation.

This equation describes a large single species population of randomly migrating individuals, whose density ρ is continuous in space and time. The space is one-dimensional and the argument can be readily extended to two-dimensional problems.

The following continuum model includes the creation and annihilation of individuals. Some possible applications are:

i) in biological populations with reproduction and natural death,
ii) in checmically reacting mixtures, where new molecules appear at the expense of others; which may have combined or decomposed to produce new ones.

The equation is given by

$$\frac{\partial \rho}{\partial t} = \frac{\partial}{\partial x} (\frac{1}{2} \frac{\partial}{\partial x} (V \cdot \rho) - \mathrm{M} \cdot \rho) + F(x,t,\rho),$$

where the functional F describes creation and annihilation of individuals. The source function F can be written as follows in terms of the birth rates r_b and the death rates r_d, then the basic dynamic equation is given

$$\frac{\partial \rho}{\partial t} = \frac{\partial}{\partial x} \left(\frac{1}{2} \cdot \frac{\partial}{\partial x}(V \cdot \rho) - M\rho\right) + \rho(r_b - r_d),$$

where the rates r_d, r_b depend on x, t, and ρ.

III. 2 Birth, death, and selection processes

III. 2.1 A biological model

Selection processes are described by the specification of the way in which the rates r_b and r_d may be allowed to depend on the density ρ (93, 95, 96, 101, 102). In biology, weak selection may act through mating preferences, fertility rates and death rates; they may be conditioned by crowding and competition. Thus, we can say that selection is density dependent.

We suppose in the following biological model that only two variant forms A_1, A_2 of the population are available. Denoting by A_i A_k for $i,k = 1,2$ the genotypes and their densities by ρ_{ik}, then we have

$$\sum_{i,k} \rho_{ik} = \rho.$$

By the notation of Crow and Kimura (1970) let

$$P_i = \frac{\sum_k \rho_{ik}}{\rho}$$

denote the frequency of the variant form A_i in the population. Then the fraction of genotype A_i A_k among the total births will be equal to $P_i \cdot P_k$, the fraction of genotype A_i A_k among the total deaths will be equal to $P_{ik} = \frac{\rho_{ik}}{\rho}$. The rate of production of ρ_{ik} is given by

$$\rho(r_b P_i P_k - r_d P_{ik})$$

and this is the source function F.

Thus, the above equation is replaced by the equation-system for the density functions

$$\frac{\partial}{\partial t} \rho_{ik} - \frac{\partial}{\partial x} \left\{\frac{1}{2} \frac{\partial}{\partial x} (V \cdot \rho_{ik}) - M\rho_{ik}\right\} = \rho(r_b P_i P_k - r_d P_{ik}).$$

Now, we insert the selection process into the model-system by varying the rates r. The fraction of genotype A_iA_k among the total births will be

$$P_i \cdot P_k(1 + s\eta_{ik}),$$

and the fraction of genotype A_iA_k among the total deaths will be

$$P_{ik}(1 + s\gamma_{ik}).$$

III. 2.2 Nonlinear migration-selection process

In the previous equation, s is the small parameter of selection (and for s = 0 we have the above conditions) and η_{ik}, γ_{ik} are source functions of (x,t,ρ_{ik}). So we have derived the nonlinear diffusion equation model for the migration-selection process given by

$$\frac{\partial}{\partial t}\rho_{ik} - \frac{\partial}{\partial x}\left(\frac{1}{2}\cdot\frac{\partial}{\partial x}(V\cdot\rho_{ik})-M\rho_{ik}\right) = \rho\{r_bP_iP_k(1+s\cdot\eta_{ik})-r_dP_{ik}(1+s\gamma_{ik})\}.$$

This differential equation system can be reduced by asymptotic methods, based on the smallness of the selection parameter s, to a single nonlinear diffusion equation.

The following reduction, by neglecting influences which are small, is an essential process of the model-building. Considering the above equation, summing on i and k, where

$$\rho = \sum_{i,k}\rho_{ik},\quad P_i = \frac{\sum_k \rho_{ik}}{\rho} \text{ and } P_{ik} = \frac{\rho_{ik}}{\rho},$$

we have

$$\sum_{i,k} P_iP_k = \sum_{i,k} P_{ik} = 1,$$

then

$$r_b\sum_{i,k} P_iP_k - r_d\sum_{i,k} P_{ik} = r_b-r_d,$$

and finally

$$\frac{\partial}{\partial t}\rho - \frac{\partial}{\partial x}\left(\frac{1}{2}\cdot\frac{\partial}{\partial x}(V\cdot\rho)-M\cdot\rho\right) = \rho\{r_b-r_d+s\cdot r_b\sum_{i,k}P_iP_k\,\eta_{ik}-sr_d\sum_{i,k}P_{ik}\gamma_{ik}\}.$$

We derive a dynamic equation: as $P_{ik} = \frac{\rho_{ik}}{\rho}$, then

$$\frac{\partial}{\partial t}P_{ik} = \frac{1}{\rho^2}\left(\frac{\partial}{\partial t}\rho_{ik}\rho - \rho_{ik}\frac{\partial}{\partial t}\rho\right) = \frac{1}{\rho}\cdot\frac{\partial}{\partial t}\rho_{ik} - \frac{\rho_{ik}}{\rho^2}\cdot\frac{\partial\rho}{\partial t},$$

using the nonlinear diffusion equations given above

$$\frac{\partial P_{ik}}{\partial t} = \frac{1}{\rho} \cdot \frac{\partial \rho_{ik}}{\partial t} - \frac{P_{ik}}{\rho}\frac{\partial \rho}{\partial t}$$

$$= \frac{1}{\rho}\left(\frac{\partial}{\partial x}\left(\frac{1}{2}\frac{\partial}{\partial x}(V\cdot\rho_{ik}) - M_{\rho ik}\right)\right) + \{r_b P_i P_k(1+s\eta_{ik}) - r_d P_{ik}(1+s\gamma_{ik})\} -$$

$$- \frac{P_{ik}}{\rho}\left(\frac{\partial}{\partial x}\left(\frac{1}{2}(V\cdot\rho) - M\rho\right)\right) - P_{ik}\{r_b - r_d + s\cdot r_b \sum_{i,k} P_i P_k \eta_{ik} - sr_d \sum_{i,k} P_{ik}\gamma_{ik}\}^o,$$

differentiating

$$\frac{\partial P_{ik}}{\partial t} = \frac{1}{\rho}\cdot\left(\frac{1}{2}\cdot\frac{\partial^2}{\partial x^2}(V\cdot\rho_{ik}) - M\cdot\frac{\partial \rho_{ik}}{\partial x}\right) + \{\ .\ .\ .\ \}$$

$$- \frac{P_{ik}}{\rho}\left(\frac{1}{2}\frac{\partial^2}{\partial x^2}(V\cdot\rho) - M\frac{\partial \rho}{\partial x}\right) - P_{ik}\{\ .\ .\ .\}^o,$$

As $\rho_{ik} = P_{ik}\cdot\rho$, there follows $\frac{\partial \rho_{ik}}{\partial x} = \frac{\partial P_{ik}}{\partial x}\cdot\rho + \frac{\partial \rho}{\partial x}\cdot P_{ik}$, then

$$\frac{\partial}{\partial x}((V\cdot\rho)\cdot P_{ik}) = \frac{\partial}{\partial x}(V\rho)\cdot P_{ik} + V\cdot\rho\cdot\frac{\partial P_{ik}}{\partial x},$$

$$\frac{\partial^2}{\partial x^2}(V\cdot\rho\cdot P_{ik}) = \frac{\partial^2}{\partial x^2}V\rho\cdot P_{ik} + 2\cdot\frac{\partial}{\partial x}V\rho\cdot\frac{\partial P_{ik}}{\partial x} + V\cdot\rho\cdot\frac{\partial^2 P_{ik}}{\partial x^2},$$

and

$$-M\frac{\partial \rho_{ik}}{\partial x} = -M\cdot\rho\cdot\frac{\partial P_{ik}}{\partial x} - MP_{ik}\frac{\partial \rho}{\partial x}$$

$$-M\frac{\partial \rho}{\partial x}P_{ik} = \qquad -MP_{ik}\frac{\partial \rho}{\partial x}; \text{ as } \frac{1}{2\rho^2 V}\frac{\partial}{\partial x}\left(V^2\rho^2\frac{\partial P_{ik}}{\partial x}\right) =$$

$$= \frac{1}{\rho}\frac{\partial}{\partial x}(V\rho)\cdot\frac{\partial P_{ik}}{\partial x} + \frac{1}{2}V\cdot\frac{\partial^2 P_{ik}}{\partial x};$$

the first part of the first expression reads

$$\frac{1}{\rho}\cdot\frac{1}{2}\frac{\partial^2}{\partial x^2}(V\cdot\rho P_{ik}) = \frac{1}{2\rho}\left(\frac{\partial^2}{\partial x^2}(V\cdot\rho)\cdot P_{ik} + 2\cdot\frac{\partial}{\partial x}(V\cdot\rho)\cdot\frac{\partial P_{ik}}{\partial x} + V\cdot\rho\cdot\frac{\partial^2 P_{ik}}{\partial x^2}\right)$$

$$- P_{ik}\cdot\frac{1}{2}\cdot\frac{\partial^2}{\partial x^2}(V\cdot\rho) = -\frac{1}{2\rho}\frac{\partial^2}{\partial x^2}(V\cdot\rho)P_{ik},$$

$$\frac{1}{2\rho}\cdot\frac{\partial^2}{\partial x^2}(V\cdot\rho\cdot P_{ik}) - P_{ik}\cdot\frac{1}{2}\cdot\frac{\partial^2}{\partial x^2}(V\cdot\rho) = \frac{1}{\rho}\cdot\frac{\partial}{\partial x}(V\rho)\cdot\frac{\partial P_{ik}}{\partial x} + \frac{V}{2}\frac{\partial^2 P_{ik}}{\partial x} = \frac{1}{2\rho^2 V}\frac{\partial}{\partial x}\left(V^2\rho^2\cdot\frac{\partial P_{ik}}{\partial x}\right)$$

and finally we obtain

$$\frac{\partial P_{ik}}{\partial t} = \frac{1}{2\rho^2 V}\frac{\partial}{\partial x}(V^2\rho^2\cdot\frac{\partial P_{ik}}{\partial x}) - M\cdot\frac{\partial P_{ik}}{\partial x} + r_b P_i P_k - r_d P_{ik} - P_{ik} r_b + P_{ik} r_d$$

$$s\{r_b P_i\cdot P_k \eta_{ik} - r_d P_{ik}\gamma_{ik}\} - s\{P_{ik} r_b \sum_{i,k} P_i P_k \eta_{ik} - P_{ik} r_d \sum_{i,k} P_{ik}\gamma_{ik}\};$$

defining

$$\bar{\eta} = \sum_{i,k} P_i P_k \eta_{ik}, \quad \bar{\bar{\gamma}} = \sum_{i,k} P_{ik}\gamma_{ik} \text{ and using } \sum_{i,k} P_i P_k = \sum_{i,k} P_{ik} = 1,$$

we obtain one part of the nonlinear diffusion equation describing migration-selection processes

$$\frac{\partial P_{ik}}{\partial t} = \frac{1}{2\rho^2 V}\frac{\partial}{\partial x}(V^2\cdot\rho^2\cdot\frac{\partial P_{ik}}{\partial x}) - M\cdot\frac{\partial P_{ik}}{\partial x} + r_b(P_i P_k - P_{ik}) +$$

$$s\{r_b(P_i P_k \eta_{ik} - P_{ik}\bar{\eta})\} - r_d P_{ik}(\gamma_{ik} - \bar{\bar{\gamma}}).$$

As $P_i = \frac{1}{\rho}\sum_k \rho_{ik}$ and $\rho_{ik} = P_{ik}\cdot\rho$, then $P_i = \sum_k P_{ik}$ and by summing over k we obtain

$$\sum_k \frac{\partial P_{ik}}{\partial t} = \frac{1}{2\rho^2 V}\cdot\frac{\partial}{\partial x}(V^2\cdot\rho^2 \sum_k \frac{\partial P_{ik}}{\partial x} - \sum_k M\cdot\frac{\partial P_{ik}}{\partial x} + \sum_k r_b(P_i\cdot P_k - P_{ik}) +$$

$$s\cdot\{r_b P_i \sum_k P_k\gamma_{ik} - \sum_k P_{ik}\bar{\eta}\} - r_d \sum_k P_{ik}(\gamma_{ik} - \bar{\bar{\gamma}}) + r_d(\sum_k P_{ik}\gamma_{ik} - P_i\bar{\bar{\gamma}}),$$

hence

$$\frac{\partial P_i}{\partial t} = \frac{1}{2\rho^2 V}\frac{\partial}{\partial x}(V^2\cdot\rho^2 \frac{\partial}{\partial x_i} P_i) - M\cdot\frac{\partial P_i}{\partial x} + s r_b P_i(\bar{\eta}_i - \eta) - r_d(\sum_k P_{ik}\gamma_{ik} - P_i\bar{\bar{\gamma}});$$

setting i = 1 and defining $p_1 = p$, the reduced selection-migration model reads

$$\frac{\partial p}{\partial t} - \frac{1}{2\rho^2 V}\cdot\frac{\partial}{\partial x}(V^2\cdot\rho^2\cdot\frac{\partial P}{\partial x}) - M\cdot\frac{\partial P}{\partial x} = s r_b P(\bar{\eta}_1 - \eta) - r_d(\bar{\gamma}_1 - \bar{\bar{\gamma}}), \quad \bar{\gamma}_1 = \sum_k P_{1k}\gamma_{1k}.$$

Let r_m and V_m be the maximum of r_b and V, then we make everything non-dimensional:

$$\xi = (\frac{s\cdot r_m}{V_m})^{1/2}\cdot x \quad \tau = s r_m t; \; \alpha_{d,b} = \frac{r_{d,b}}{r_m}, \; \theta = \frac{V}{r_m}; \; \chi = \frac{M}{(s\cdot r_m V_m)^{\frac{1}{2}}},$$

the equation above reduces to

$$\frac{\partial P}{\partial \tau} - \frac{1}{2\rho^2\theta}\cdot\frac{\partial}{\partial x}(\rho^2\cdot\theta^2\cdot\frac{\partial P}{\partial \xi}) + \chi\cdot\frac{\partial P}{\partial \xi} = \alpha_b P(\bar{\eta}_1 - \eta) - \alpha_d(\bar{\gamma}_1 - \bar{\bar{\gamma}}) = (F(x,t,P),$$

taking (i,j) = (1,2) in the equation on page 178 , we obtain

$$s\left(\frac{\partial P_{12}}{\partial \tau} - \frac{1}{2\rho^2\theta}\frac{\partial}{\partial \xi}\left(\rho^2 \cdot \theta^2 \cdot \frac{\partial P_{12}}{\partial \xi}\right) + x \cdot \frac{\partial P_{12}}{\partial \xi}\right) = \alpha_d (P_1 \cdot P_2 - P_{12}) + s.K_2 ,$$

where K_2 is a function of x,t,P_{ik} and ρ.

The differential equation system can be rewritten as

$$\frac{\partial P}{\partial \tau} - \frac{1}{2\rho^2\theta} \cdot \frac{\partial}{\partial \xi}\left(\rho^2 \cdot \theta^2 \cdot \frac{\partial P}{\partial \xi}\right) + x \cdot \frac{\partial P}{\partial \xi} = \alpha_b P(\bar{\eta}_1 - \eta) - \alpha_d(\bar{\gamma}_1 - \bar{\bar{\gamma}})$$

$$s\left(\frac{\partial P_{12}}{\partial \tau} - \frac{1}{2\rho^2\theta} \cdot \frac{\partial}{\partial \xi}\left(\rho^2 \cdot \theta^2 \frac{\partial P_{12}}{\partial \xi}\right) + x\frac{\partial P_{12}}{\partial \xi}\right) = \alpha_d (P_1 P_2 - P_{12}) + sK_2$$

$$s\left(\frac{\partial \rho}{\partial \tau} - \frac{1}{2}\frac{\partial^2}{\partial \xi^2}(\theta \rho) + \frac{\partial}{\partial \xi}(x \cdot \rho)\right) = \rho(\alpha_b - \alpha_d) + s \cdot K_1 ,$$

where K_i for i = 1,2 are functions of x,t,P_{ik} and ρ.

III. 2.3 Reduced nonlinear differential equation system

On the level of first approximation we obtain by setting the selection parameter s = 0 from the second equation $P_{12} = P_1 \cdot P_2$, and from the third $\alpha_b = \alpha_d$. The differential equation system reduces to a single nonlinear diffusion equation

$$\frac{\partial P}{\partial \tau} - \frac{1}{2\rho^2\theta^2} \cdot \frac{\partial}{\partial \xi}\left(\rho^2 \cdot \theta^2 \cdot \frac{\partial P}{\partial \xi}\right) + x \cdot \frac{\partial P}{\partial \xi} = \alpha_b \cdot P\{(\bar{\eta}_1 - \eta) - (\bar{\gamma}_1 - \bar{\bar{\gamma}})\}.$$

It can be shown, that the single equation is a correct approximation of the equation system above. Moreover, the existence of a wave front for the single equation implies one for the system. Conley and Fife (65) have established the existence of a wave front solution of the single nonlinear diffusion equation, which approximates, in some sense, the former.

If $\alpha_d, \alpha_b, \theta, \eta_{ij}, \gamma_{ij}$ are independent of (x,t), we obtain

$$\frac{\partial P}{\partial \tau} - \frac{\theta}{2}\frac{\partial^2 P}{\partial \xi^2} + x \cdot \frac{\partial P}{\partial \xi} = f(x,t,P),$$

where $f = \alpha_b P(1-P)$.

We have constructed the migration-selection model according to the conditions I,...V; thus, each term of the equation is understood, but the solution is too complicated.

So we have reduced this equation by asymptotic methods based on the smallness of the selection s and obtained the nonlinear equation.

One rule of asymptotics in model-building is to reduce complex models to simpler ones, or to combine simpler ones to complex systems. This is accomplished by identifying terms in equations, or other influences figuring the more complex model, which are significantly smaller than others, and neglecting them.

Among the many rate constants and other parameters in the kinetics for a given reaction mechanism, it is found, that some are of different orders of magnitude than others.

The methods of singular perturbation allow an effective reduction in the order of the system of dynamic equations ...

The "pseudo-steady state" hypothesis, which is standard, is commonly employed to the modelling stage, introduced by Bowen, Acrivos, Oppenheim (43), see also Aris (12), Heineken, Tsuchiya, Aris (154), Lin Seger (213), Murray (253).

The source function F(. .) may be interpreted as the rate of production due to reaction. Since chemical reactions produce or absorb heat, the time rate of change in temperature depends on the rates of the various chemical reactions. The rates depend on temperature and on the concentrations of the reacting species. Temperature and concentration can be described by ρ in the framework of the above equations. Also it is possible to describe reactions which occur among individuals of a population. The reader is referred to the work done by Feinberg, Horn (98), Feinberg (100) and the development of the theory by Aris (11), Feinberg (1972-77), Gavalas (128), Horn (1977), Vol'pert, Hudyaev (349). Biological populations are sometimes segregated into homogeneous geographical groups. For determinisitc analysis of migration, selection, and mutation in colony models, see Nagylaki ((255), (258),...). Colony models incorporate stochastic effects, too, see e.g. Mollison (250) Sawyer (301, 300, 302, 303), Nagylaki (257, 259).

Situations, inwhich the population is intrinsically very mobile, may be described by homogeneous systems. This means, that the average time it takes an individual to diffuse or otherwise travel from one end of the habitat to the other will be small compared to the time scale in which reaction effects (chemical reactions, death, births, disease infections,...) serve to change the composition of the population.

III. 3 On travelling waves and front

III. 3.1 Some remarks on travelling waves

It is possible to analyse multidimensional nonlinear diffusion phenomena (on a first level of approximation) by semilinear parabolic partial differential equations of the type

$$\frac{\partial u}{\partial t} = \Delta u + F(u),$$

where $\Delta = \sum_{j=1}^{n} \frac{\partial^2}{\partial x_j^2}$ is the Laplace operator in $\mathbf{R}^n$, In this part we take the restriction, that $F(0) = 0 = F(1)$ and the solutions $u(x,t)$ have values in $[0,1]$.

It can be shown that essential features of one-dimensional diffusion phenomena are also present in the multi-dimensional case. It is known, that there exist plane wave solutions of the semilinear equation of the form

$$u(x,t) = q(x \cdot \nu - C \cdot t) \text{ for } C \in \mathbf{R},$$

and $\nu \in \mathbf{R}^n$ as an arbitrary unit vector, C is interpreted as the velocity of the travelling wave. One can show that there exists a minimal wave speed C_{min}, which is determined by the forcing term $F(u)$. C_{min} can be seen as an asymptotic speed of propagation of disturbances.

In the one-dimensional case we have the following:

We consider the existence of travelling wave solutions of

$$\frac{\partial u}{\partial t} = \frac{\partial^2 u}{\partial x^2} + F(u),$$

in the form $u(x,t) = U(x-Ct)$.

The only stable travelling waves are the wave fronts, characterized by solutions $U(z)$ approaching distinct values as $z \to \pm\infty$. It is necessary that the distinct values be the zeros of $F(u)$. As $F(0) = 0 = F(1)$, by assumption, a wave front does exist, such that

$$\lim_{z\to\infty} U(z) = 1 \text{ and } \lim_{z\to-\infty} U(z) = 0.$$

Now we show how the existence and development of the wave fronts may be obtained.

Substituting the ansatz

$$u(x,t) = U(x-Ct)$$

into

$$\frac{\partial u}{\partial t} = \frac{\partial^2 u}{\partial x^2} + F(u),$$

we have

$$\frac{d^2U}{dz^2} + C\,\frac{dU}{dz} + F(U) = 0,$$

introducing the (U,V) phase plane, we obtain the equivalent differential equation system

$$\frac{dU}{dz} = V; \quad \frac{dV}{dz} = -C\cdot V-F(U).$$

Now, the wave front corresponds to the trajectory, described by the system, and connecting the stationary points (0,0) and (1,0). The equation

$$\frac{d^2U}{dz^2} + C\cdot\frac{dU}{dz} + F(U) = 0$$

can be written as

$$\frac{dV}{dz} + C\cdot V + F(U) = 0,$$

and to this equation corresponds the boundary condition

$$V(0) = 0 = V(1).$$

Thus, to any travelling wave front there corresponds the function $V(U)$. Integrating the equation

$$V(1)-V(0) = \int_0^1 \frac{\partial V}{du}\,du, \text{ and } \int_0^1 \frac{dV}{dU}\,dU + \int_0^1 F(U)dU + C\cdot\int_0^1 V(U)dU = 0,$$

we obtain

$$C \int_0^1 V(U)dU = -\int_0^1 F(U)dU$$

by the boundary condition. For positive solutions V of the above differential equation, we have the relation

$$\int_0^1 F(U)\,dU \begin{cases} \leqq 0 \text{ if } C > 0 \\ = 0 \text{ if } C = 0 \\ \geqq 0 \text{ if } C < 0 \end{cases}$$

III. 3.2 Construction of a travelling front

To construct a travelling front, we have to find an orbit in the (U,V) phase plane connecting the critical points (0,0) and (1,0). The existence and the behaviour of the trajectory depends on the types of the critical points. The system

$$\frac{dU}{dz} = V$$

$$\frac{dV}{dz} = -C\cdot V-F(U),$$

can be written

$$\frac{dQ}{dz} = A\cdot Q, \text{ where } Q = \binom{U}{V} \text{ and } A = \begin{pmatrix} 0 & 1 \\ -F(U) & -C \end{pmatrix}.$$

By linearization at the critical point (0,0), the matrix is given by

$$A = \begin{pmatrix} 0 & 1 \\ -F'(0) & -C \end{pmatrix},$$

hence

$$\begin{pmatrix} \frac{dU}{dz} \\ \frac{dV}{dz} \end{pmatrix} = \begin{pmatrix} 0 & 1 \\ -F'(0) & -C \end{pmatrix} \cdot \begin{pmatrix} U \\ V \end{pmatrix} = \begin{pmatrix} V \\ -F'(0)\cdot U-C\cdot U \end{pmatrix}.$$

The eigenvalue problem $AQ = \lambda\cdot Q$ is given as

$$\begin{pmatrix} V \\ -F'(0)U-CV \end{pmatrix} = \lambda \begin{pmatrix} U \\ V \end{pmatrix},$$

thus we have

$$V = \lambda U$$

$$\lambda V = -F'(0)U-C.V = -F'(0)\,\frac{V}{\lambda} - C\cdot V,$$

hence

$$\lambda^2 - C\cdot\lambda + F'(0) = 0$$

with the solution

$$\lambda_{1,2} = \frac{C}{2} \pm \sqrt{\frac{C^2}{4} - F'(0)}.$$

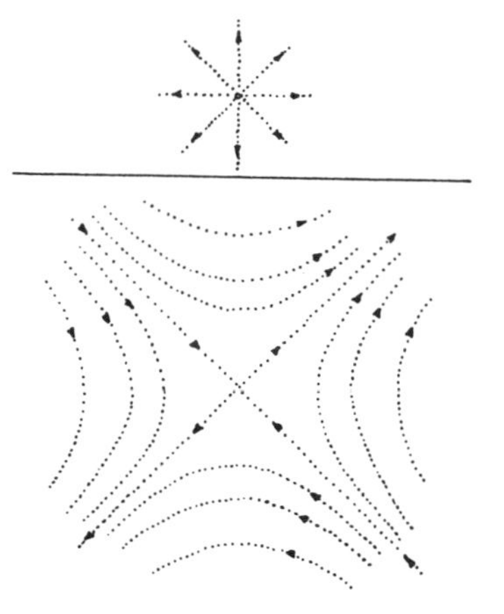

I. For $F'(0) > 0$ and $C^2 \geq 4F'(0)$, both eigenvalues λ_1, λ_2 are real and the critical point (0,0) represents a node.

II. In the case $F'(0) < 0$, the eigenvalues are real too, but of opposite sign and the critical point is a saddle.

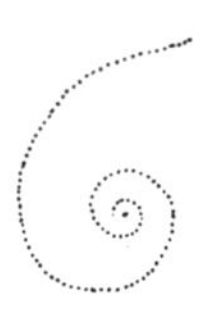

III. In the extreme case $C^2 < 4F'(0)$, all eigenvalues are complex and the trajectory shows a spiral.

Quite analogously the critical point (1,0) may be characterized. So we consider the trajectories describing the transitions

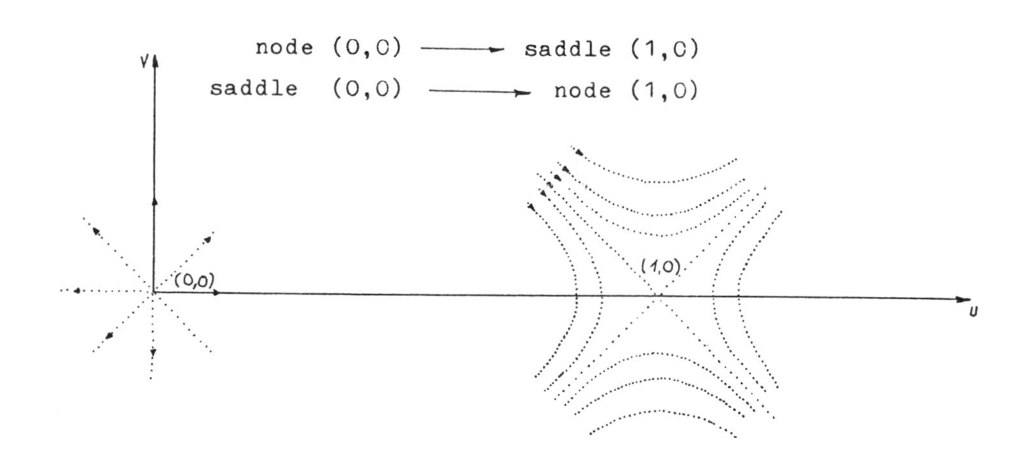

As (1.0) is a saddle, there exists a trajectory coming from the left and above.

For further considerations we analyse the trajectories in the region of a triangle:

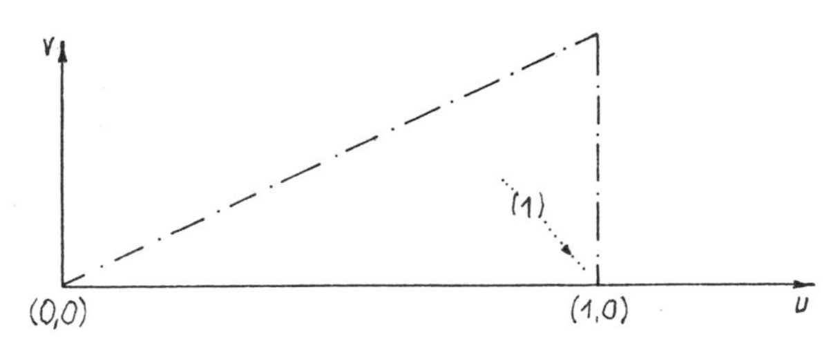

the trajectory (1) comes from a critical point, or crosses the boundary of the triangle.

As U is strictly increasing on the trajectory, the trajectory cannot cross the boundary U = 1. Now we assume, that V > 0 but small, as

$$\frac{dU}{dz} = V; \; \frac{dV}{dz} = -C \cdot V - F(U),$$

we have

$$\frac{dU}{dV} = \frac{dU}{dz} \cdot \frac{dz}{dV} = \frac{dU}{dz} \cdot \frac{1}{\frac{dV}{dz}} = \frac{V}{-C \cdot V - F(U)} < 0,$$

as $F(U) > 0$ for $U \in [0,1]$ the trajectories are given by $\frac{\partial V}{\partial U} < 0$. Let the diagonal be given by $V = \mu U$ for $\mu > 0$, now we estimate $\frac{dV}{dU}$ on this line.

We have

$$\frac{dV}{dU} = -C - \frac{F(U)}{V},$$

defining

$$\gamma = \sup \frac{F(U)}{U},$$

then $\frac{F(U)}{U} < \gamma$ and $-\gamma < -\frac{F(U)}{U}$, thus we have

$$\frac{dV}{dU} = -C - \frac{F(U) \cdot U}{V \cdot U} \geqq -C - \frac{\gamma \cdot U}{V} = -C - \frac{\gamma}{\mu};$$

let the positive number μ satisfy $-C - \frac{\gamma}{\mu} \geqq \mu$, then $\frac{dV}{dU} \geqq \mu$. This means, that the trajectory leaves the triangle.

The relation $-c - \frac{\gamma}{\mu} \geqq \mu$ may be written as $\mu^2 + C\mu + \gamma \leqq 0$; the equation $\mu^2 + C\mu + \gamma = 0$ has the solution

$$\mu_{1,2} = -\frac{C}{2} \pm \sqrt{\frac{C^2}{4} - \gamma} = -\frac{C}{2} \pm \frac{1}{2}\sqrt{C^2 - 4\gamma},$$

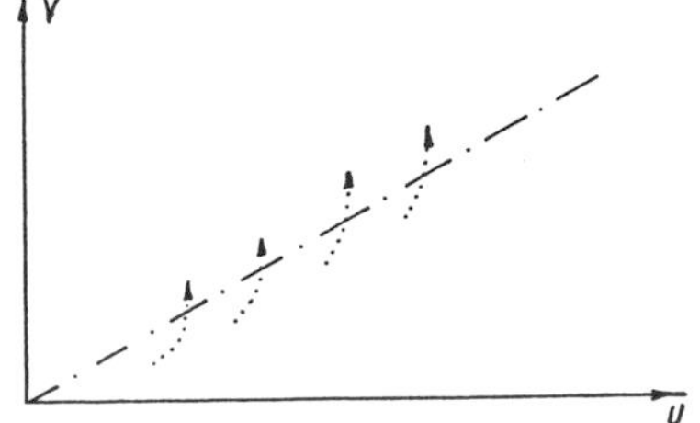

therefore the solution μ lies between the real solutions

$$-C \pm \frac{1}{2}\sqrt{C^2-4\gamma} \quad \text{if c satisfies } C \leqq -\sqrt{4\gamma} < 0.$$

We obtain by the theorem of means $F'(0) \leqq \sup \frac{F(U)}{U} \underset{\text{def.}}{=} \gamma$, hence $4F'(0) < 4\gamma$ and finally $-\sqrt{4\gamma} < -\sqrt{4F'(0)}$. There exists a maximal velocity C_{max} in the interval

$$-\sqrt{4\gamma} < C_{max} \leqq -\sqrt{4 \cdot F'(0)},$$

such that for every $C \leqq C_{max}$ there exists a wave front.

Thus, we have proved

Theorem I

Assuming $F(0) = 0 = F(1)$, and $F'(0) < 0$, $F'(1) < 0$, and F has only one zero in $(0,1)$, there exists a unique wave front solution U of the semilinear equation with $\lim_{z \to -\infty} U(z) = 0$, but $\lim_{z \to +\infty} U(z) = 1$ and

Theorem II

For $F'(0) > 0$ (node) and $F'(1) < 0$ (saddle), $f(u) > 0$ for $u \in (0,1)$, there exists a number $C_{max} < 0$ and the wave front solution

$$\lim_{z \to -\infty} U(z) = 0, \quad \lim_{z \to +\infty} U(z) = 1,$$

if and only if

$$C \leqq C_{max}.$$

In the well known work (194) of Kolmogoroff et al (1937), the condition $F'(u) \leqq F'(0)$ is taken and the existence of a family of wave fronts is proved. Special solutions of the semilinear equation start as a step function and converge in wave form to the front U with speed C_{max}.

Let $F(u) > 0$, $F'(u) < F'(0)$ for $u \in (0,1)$, and $F'(0) > 0$ with the step function

$$\phi(x) = \begin{cases} 0 \text{ for } x < 0 \\ 1 \text{ for } x > 0, \end{cases}$$

as initial conditions.

Then there exists a maximal wave velocity C_{max} with the associated wave front U and a function $\psi(t)$, such that $\lim_{t\to\infty} \psi'(t) = 0$, and

$$|u(x,t) - U(x-C_{max}\cdot t-\psi(t))| \to 0 \text{ for } t \to \infty$$

uniformly in x.

Other convergence results are given by McKean (236), Hoppensteadt (167) Rothe (293), (295), Stokes (329), Kametaka (180), Moet (249), Uchiyama (340), Bramson (45). Kanel' (181), (182) gave applications in combustion theory and Larson gave generalizations in (207).

Fife, McLeod (105) and McLeod, Fife (237) have proved the uniform convergence to the wave front (with ψ = constant) for a wide class of initial data. If F(...) depends on $\frac{\partial u}{\partial x}$, Chueh (59) gave interesting results.

Let F have only one zero at the point $\alpha \in (0,1)$ and assuming

$$F'(0) < 0,\ F'(1) < 0 \text{ and } F(u) < 0 \text{ for } u \in (0,\alpha),\ F(u) > 0 \text{ for } u \in (\alpha,1).$$

If the initial data satisfy

$$\lim_{x\to-\infty} \sup \phi(x) < \alpha \quad \lim_{x\to+\infty} \inf \phi(x) > \alpha,$$

then for some constants z_0, c, and w, c > 0, w > 0, we obtain

$$|u(x,t) - U(x-C\cdot t-z_0| < Ke^{-wt};$$

this is a result of global stability of travelling waves.

In his monograph (44) Bramson analysed the solution of the Kolmogoroff equation under arbitrary initial conditions. Criteria will be given for convergence to the travelling waves, as well as a formula for the value of the centring term $\psi(t)$.

III. 4 Mathematical approach to synergetics

We deal with equations of motion of continuously extended systems containing fluctuations described by equations of the type

$$\frac{\partial u_i}{\partial t} = D_i \Delta u_i + G_i(u, \text{grad } u) + F_i(t) \text{ for } i = 1,\dots,n,$$

where $u = (u_1,u_2,\dots,u_n)$ are physical quantities.

The G_i are nonlinear functions of u and perhaps of a gradient (page 19) and the $F_i(t)'$ are fluctuating forces, which are caused by external reservoirs and internal dissipation.

We consider on pages 20f, 134f, 205f, nonlinear parabolic equations in multiconnected domains.

In most application like lasers or hydrodynamic, G is a linear or bilinear function of u, in chemical reaction models G is a cubic coupling term, in population genetics D satisfies inequalities (see Aronson, Weinberger (15)). For D_i real, the equation describes diffusion phenomena. For D_i imaginary we have to replace the first order time derivative by the second order time derivative, and the equation describes wave phenomena. We consider hyperbolic initial-boundary-value problems in two-connected domains on page 27f and wave phenomena governed by differential games on page 131f.

Haken (143) considers laser phenomena; then u stands for the electric field strength, for the polarization of the medium, and for the inversion density of laser active atoms.

In nonlinear optics, u stands for the strength of several interacting modes. See, on page 385f, the very interesting model of image transmission in optical wave-guides and applications to image reconstruction in multi-connected domains.

In hydrodynamics u stands for the temperature, density, velocity field, ... In chemical reactions, u stands for the numbers of molecules, respectively for the densities, participating in the chemical reaction. Haken (144f.) derived from the parabolic equation above, equations for the undamped modes, which acquire a macroscopic size and determine the dynamics of the system in the vicinity of the instability point. These modes form a mode skeleton which grows out from fluctuations above the instability and thus describes the embryonic state of the evolving spatio-temporal structure.

The methods and models derived by Haken (148) can be applied in our model-building processes.

Haken obtained generalized Ginzburg-Landau equations for nonequilibrium phase transitions.

The method of generalized Ginzburg-Landau equations allows Haken (144) to treat the buildup of ultrashort laser pulses (270).

Instabilities present fascinating problems in fluid dynamics. Haken (148) considers an infinitely extended horizontal fluid layer, which is heated from below so that a temperature gradient is maintained. This gradient is called the Rayleigh number R. As long as the Rayleigh number is not too large, the fluid remains quiescent and heat is transported by conduction. If the number exceeds a certain value, however, suddenly the fluid starts to convect. What is most surprising, is that the convection pattern is very regular and may either show rolls or hexagons. This mechanism is treated theoretically by Haken (148), where the hexagons are explained.

Morphogenetic models given by Gierer, Meinhardt (133, 134) may be described by analytical methods derived in Haken ((148), p. 302f.).

The theory of the 2-dimensional splines in (139) and the global differential-splines in (141) can be applied to morphogenetic models.

III. 5 Systems of reaction-diffusion equations

III. 5.1 Invariant bounded regions

Systems of reaction-diffusion equations received great attention in recent years, in the simplext form they are described by

$$\frac{\partial u}{\partial t} = C\cdot\Delta u + M\,\frac{\partial u}{\partial x} + f(u) \text{ in a bounded domains } D \text{ in } \mathbb{R}^n$$

$$\frac{\partial u}{\partial n} = 0 \text{ on the smooth boundary } \partial D$$

$$u(x,0) = u_o(x) \text{ the initial condition in } D,$$

where C is an $n \times n$ matrix, the diffusion matrix, and $f(u)$ is a smooth non-linear function.

The diffusion terms $C\cdot\Delta u$ interact with the nonlinear term $f(u)$. $C\Delta u$ acts in such a way as to dampen u and $f(u)$ tends to produce dissipation. This leads, under suitable boundary and initial conditions, to the possibility of threshold phenomena (see Aronson, Weinberger (15)) or of solitary waves (see page 308) and many other interesting features.

Introducing the mean value

$$\bar{u}(t) = \frac{1}{\text{Volume } D} \cdot \int_D u(x,t)dx,$$

where $u(x,t)$ is the spatial density as a solution of the above system, we can say that $u(t)$ approximately (see (68)) satisfies the ordinary differential equation

$$\frac{du}{dt} = f(u).$$

Under suitable conditions we have, that the solution of the partial differentia equation gets exponentially close to its average $\bar{u}$, as $t \to \infty$. For each $t > 0$ the solution u of the ordinary differential equation is an element of a Banach space $\mathfrak{B}$, which represents a curve in $\mathfrak{B}$ and describes the evolution of u.

When modelling chemical and biological processes, the asymptotic representation means: all solutions of the partial differential equation decay as $t \to \infty$ to spatially homogeneous functions, i.e. to the solution of the associated "kinetic" equation $\frac{du}{dt} = f(u)$.

To obtain qualitative information on nonlinear diffusio-reaction processes, we have to introduce the notion of invariant bounded regions, i.e. bounded regions Ξ in phase space (u-space).

A bounded region Ξ is said to be invariant for the above initial-boundary-value problem, if, whenever the initial condition $u(x,0) \in \Xi$ and the boundary condition $\in \Xi$, it follows, that the solution $u(x,t) \in \Xi$ for all $x \in D$ and $t > 0$.

Invariant regions Ξ will be made up of the intersection of half-spaces. Let U be an open subset of $\mathbf{R}^n$ and E_i are smooth real valued functions

$$\mathbf{R}^n$$

$$\mathbf{u}$$

$$E_i : U \to \mathbf{R},$$

where the grad E_i never vanishes, then

$$\Xi = \bigcap_{i=1}^{m} \{u \in U | E_i(u) \leqq 0\},$$

and the boundary

$$\partial\Xi = \bigcap_{i=1}^{m} \{u \in U | E_i(u) = 0\}.$$

Let u be a solution of the initial-boundary-value problem, where the boundary and initial values are in Ξ for all $x \in D$.

We now assume that the solution $u(x,t)$ is not in Ξ for $t > 0$ (i.e. the region is not invariant).

This means that there is a function E_i, a time t_1 and a point $x_1 \in \mathbf{R}$, such that for $t \leqq t_1$ and $x \in \mathbf{R}$, $E_i \circ u(x,t) \leqq 0$, and for any $\varepsilon > 0$, there is a t_o, with $t_1 < t_o < t_1 + \varepsilon$, such that

$$E_i \circ u(x_1, t_o) > 0 \quad (\text{i.e. } \frac{\partial E_i \circ u}{\partial t} > 0 \text{ at } (x_1, t_1)).$$

Remark:

With these assumptions in mind, we have that

$$E_i \circ u(x_1, t) < 0 \text{ for } 0 \leqq t < t_1 \text{ and } E_i \circ u(x_1, t_1) = 0$$

imply

$$\frac{\partial E_i \circ u}{\partial t} < 0 \text{ at } (x_1, t_1),$$

then Ξ is an invariant region.

Under these conditions, the following theorem is proved.

Theorem:

Let $\Xi = \bigcap_{i=1}^{m} \{u \in U | E_i \circ u = E_i(u) \leqq 0\}$ be given (an invariant region), we suppose, that for all $t \in \mathbb{R}_+$ and every

$$u_o \in \partial\Xi = \bigcap_{i=1}^{m} \{u \in U | E_i(u) = 0\}$$

and the conditions:

1) $\frac{\partial E_i}{\partial u}$ at u_o is a left eigenvector of C and M at (u_o, x)

2) if $\frac{\partial E_i}{\partial u}(u_o, x) = \mu \frac{\partial E_i}{\partial u}$ with $\mu > 0$, then E_i is quasi convex at u_o, i.e. if whenever $\frac{\partial E}{\partial u}(\eta) = 0$, then $\frac{\partial^2 E}{\partial u^2}(\eta) > 0$.

3) $\frac{\partial E_i}{\partial u} f < 0$ at $u_o \in \partial\Xi$ for $t \in \mathbb{R}_+$,

then Ξ is invariant with respect to the initial-boundary-value problem.

Proof:

Take $E = E_i$ for simplicity, applying the remark, we have to show that

$$\frac{\partial E \circ u}{\partial t} < 0 \text{ at } (x_1, t_1).$$

At (x_1, t_1) we have

$$\frac{\partial E \circ u}{\partial t} = \frac{\partial}{\partial t} E(u) = \frac{\partial E}{\partial u} \cdot \frac{\partial u}{\partial t} = \frac{\partial E}{\partial u} (C \cdot \frac{\partial^2 u}{\partial x^2} + M \cdot \frac{\partial u}{\partial x} + f \cdot u)$$

applying 1) we get

$$\frac{\partial E}{\partial u} C = \mu \cdot \frac{\partial E}{\partial u} \text{ and } \frac{\partial E}{\partial u} M = \tau \cdot \frac{\partial E}{\partial u}$$

with eigenvalues μ and τ, hence

$$\frac{\partial E \circ u}{\partial t} = \mu \frac{\partial E}{\partial u} \cdot \frac{\partial^2 u}{\partial x^2} + \tau \cdot \frac{\partial E}{\partial u} \cdot \frac{\partial u}{\partial x} + \frac{\partial E}{\partial u} \cdot f.$$

Lemma:

We obtain

$$\frac{\partial E}{\partial u} \cdot \frac{\partial u}{\partial x} = 0 \text{ at } (x_1, t_1).$$

Proof:

Defining $g(x) = E \circ u(x,t_1)$ and using the remark, we have $g(x_1) = E \circ u(x_1,t_1) = 0$, then

$$\frac{\partial g(x)}{\partial x} = \frac{\partial}{\partial x} E(u)(x,t_1) = \frac{\partial E}{\partial u} \cdot \frac{\partial u}{\partial x} (x,t_1).$$

Assumign $g'(x_1) > 0$, we obtain

$$0 < \int_{x_2}^{x} g'(x_1,t)dt = g(x_1,t)\Big|_{t=x_2}^{t=x} = g(x_1,x) - g(x_1,x_2) = g(x_1,x)$$

for $|x-x_1|$ sufficient small, thus $g(x) > 0$ for all $x \geqq x_2$, if $|x-x_1|$ is small. By construction

$(E \circ u)(x_1,t_1) > 0$ for $|x-x_1|$ sufficient small, then

$(E \circ u)(x,t) > 0$ for $|t-t_1| < \varepsilon$ and $|x-x_1| < \varepsilon$,

in particular $(E \circ u)(x,t) > 0$ for some x and $t < t_o$, this violates the remark. In a similar way we can show that $g'(x_1) < 0$ is impossible, thus $g'(x) = 0$, and as $\frac{\partial E}{\partial u} \cdot \frac{\partial u}{\partial x} = \frac{\partial g(x)}{\partial x}$, we have $\frac{\partial E}{\partial u} \cdot \frac{\partial u}{\partial x} (x_1,t_1) = g'(x_1) = 0$, thus the lemma is proved.

It is shown quite analogously that

$$\frac{\partial^2 g(x)}{\partial x^2}(x_o) \leqq 0 \; g''(x) \leqq 0.$$

As $g(x) = E(v)$, there follows

$$\frac{\partial g}{\partial x} = \frac{\partial E}{\partial v} \cdot \frac{\partial v}{\partial x}, \; \frac{\partial^2 g}{\partial x^2} = \frac{\partial}{\partial x} \left(\frac{\partial E}{\partial v}\right) \frac{\partial v}{\partial x} + \frac{\partial E}{\partial v} \cdot \frac{\partial^2 v}{\partial x^2},$$

hence

$$0 \geqq g''(x) = \frac{d^2}{dx^2} g(x) = \frac{\partial^2 E}{\partial u^2} \cdot \frac{\partial u}{\partial x} + \frac{\partial E}{\partial u} \cdot \frac{\partial^2 u}{\partial x^2},$$

this can be estimated as follows: applying condition 2), then

$$\frac{\partial E}{\partial u}(u_o, x) = \mu \cdot \frac{\partial E}{\partial u} \text{ with } \mu > 0,$$

and differentiating, we obtain

$$\frac{\partial}{\partial x}\left(\frac{\partial E}{\partial u}\right) \cdot C(u_o, x) + \frac{\partial E}{\partial u} \cdot \frac{\partial u}{\partial x} \cdot \frac{\partial C}{\partial u}(u_o, x) = \mu \cdot \frac{\partial}{\partial x}\left(\frac{\partial E}{\partial u}\right) - \mu \cdot \frac{\partial^2 E}{\partial u^2} \cdot \frac{\partial u}{\partial x}.$$

Using the lemma, we obtain finally

$$\frac{\partial^2 E}{\partial u^2} \cdot \frac{\partial u}{\partial x} \geqq 0 \text{ at } (x_1, t_1),$$

there follows

$$\frac{\partial E}{\partial u} \cdot \frac{\partial^2 u}{\partial x^2} \leqq 0 \text{ at } (x_1, t_1).$$

This result is now applied to

$$\frac{\partial E \circ u}{\partial t} = \mu \cdot \frac{\partial E}{\partial u} \cdot \frac{\partial^2 u}{\partial x^2} + \tau \cdot \frac{\partial E}{\partial x} \cdot \frac{\partial u}{\partial x} + \frac{\partial E}{\partial u} \cdot f$$

and with the lemma

$$\frac{\partial E \circ u}{\partial t} \leqq \frac{\partial E}{\partial u} \cdot f < 0 \text{ by condition 3);}$$

thus,

$$\frac{\partial E \circ u}{\partial t} < 0 \text{ at } (x_1, t_1)$$

and by the remark Ξ is an invariant region.

Remark:

If C and M are diagonal matrices, we take $E_i = u_i - a_i$, where a_i = constant. The invariant regions can be written as

$$\Xi = \bigcap_{i=1}^{m} \{u \mid a_i \leqq u_i \leqq a^i\},$$

where a_i, a^i are constant.

Thus, the conditions 1) and 2) of the theorem are satisfied, then we have the

Corollary:

Let C and M be diagonal matrices and suppose, that for $t \in \mathbb{R}_+$ and for every $v_0 \in \Xi$ the condition $f(\ .\ .\ ) < 0$ at u_0 for all $t \in \mathbb{R}_+$ is satisfied, then the defined set Ξ is an invariant region, called an invariant rectangle.

The notion of invariant regions is introduced by Chueh, Conley and Smoller (60), and Weinberger (355), related papers are Alikakos (5), (6), Amann (9), Bebernes, Chueh and Fulks (20) and Kuiper (202), (203), (204). Now we are interested in the interactive growth of only two species; thus, the system reduces to (a generalization on page 229)

$$\frac{\partial u}{\partial t} = d_1 \Delta u + u \cdot M(u,v) \text{ in D for } t > 0$$

$$\frac{\partial v}{\partial t} = d_2 \Delta v + v \cdot N(u,v) \text{ in D for } t > 0$$

$$\frac{\partial u}{\partial n} = 0 = \frac{\partial v}{\partial n} \quad \text{on } \partial D$$

$$u(x,0) = u_0(x);\ v(x,0) = v_0(x) \text{ in D.}$$

We assume that the initial values u_0 and v_0 are bounded nonnegative smooth functions. The boundary conditions can be interpreted to mean that there is no migration of either species across ∂D; thus D is considered to be the habitat of the species u and v.

III. 5.2 Asymptotic methods

In this section, we introduce the asymptotic behaviour of solutions of initial-boundary-value problems. These solutions can be estimated by the corresponding ordinary differential equations. Until now, we have considered general processes governed by parabolic partial differential equations and their corresponding initial-boundary-value problems. Aronson, Weinberger (15) analysed the asymptotic behaviour of the solutions of parabolic differential equations and obtained comparison theorems as a consequence of general maximum principles. For systems of partial differential equations there are no maximum principles available.

Now, we consider processes described by initial-boundary-value problems of systems of partial parabolic differential equations in bounded domains in $\mathbb{R}^n$. Under suitable conditions it is possible to simplify the descritpion by asymptotic methods. The asymptotic behaviour of the solution of initial-boundary-value problems can be estimated by the corresponding system of ordinary differential equations. The methods obtained can be applied to the analysis of the stability of the corresponding process.

Certain applications can be given in physics, biology and chemistry. Some mathematical models describe physiological phenomena of signal transmission across axons; the Hodgkin-Huxley equation (163) is approximated from the Fitz-Hugh-Nagumo equations (110), (254). It is known that there exist certain analogies. The electric potential of a membrane can diffuse like a chemical, and of course can interact with real chemical species (ions) which are transported through the membrane. These facts gave rise to Hodgkin's and Huxley's celebrated model for the propagation of nerve signals. Our modelling-processes are based on such analogies (Grusa (141)).

Other applications are models for the Belousov-Zhabotinsky reactions in chemical kinetics (153), (168) and the Field-Noyes equations (153), and reaction phenomena in a chemical process (Gel'fand (129), Gardner (126)), problems arising in the theory of combustion describing flame propagation (128), reaction-diffusion equations in chemical reactor theory (10) or in certain biochemical reactions (41).

Let there be a system of partial differential equations of the form

$$\frac{\partial u}{\partial t} = C\cdot\Delta u + f(u,t) \text{ in a bounded domain D in } \mathbb{R}^n,\ t > 0$$

$$\frac{\partial u}{\partial n} = 0 \quad \text{on the smooth boundary } \partial D,\ t > 0$$

$$u(x,0) = (u_1^0,(x),u_2^0(x),\ldots,u_n^0(x)) \text{ the initial condition in D, } t > 0.$$

Let C be a diagonal matrix, then there exists a bounded invariant region (rectangle)

$$\Xi = \prod_{i=1}^{n} [a_i,b_i],$$

by the corollary on page 194 , if the nonlinear term f, as a smooth function

$$f : \underset{\mathbf{u}}{\overset{\mathbb{R}^n}{U}} \to \mathbb{R}^n,$$

satisfies $f(u,t,\ldots) \leqq 0$ on the boundary $\partial\Xi$ of the invariant rectangle. The nonlinear term is given by $(f_1(u,t),f_2(u,t),\ldots,f_n(u,t))$. We take the i-th component $f_i(u,t)$ and define the sup

$$f_i^+(u,t) \sim f_i(\eta_1,\eta_2,\ldots,\eta_{i-1}, \underset{u_i}{\uparrow}, \eta_{i+1},\ldots, \eta_n,t),$$

thus we have a set of natural numbers

$$\{1,2,\ldots,i-1,i,i+1,\ldots,n\} = \mathfrak{m} \cup K \text{ where } \mathfrak{m} \cap K = \emptyset.$$

From the invariant rectangle we obtain

$$f_i(\eta_1,\eta_2,\ldots,\eta_{i-1}, u_i, \eta_{i+1},\ldots,\eta_n,t)$$

with the side conditions

$$a_j \leqq \eta_j \leqq u_j \text{ for } j \in \mathfrak{m},\ j \neq i \text{ and } u_j \leqq \eta_j \leqq b_j \text{ for } j \in K,\ j \neq i,$$

finally we define

$$f_i^+(u,t) = \sup\ (f_i(\eta_1,\ldots,u_i,\ldots,\eta_n,t) \text{ where } a_j \leqq \eta_j \leqq u_j \text{ for } j \in \mathfrak{m} \text{ and } j \neq i)$$
$$u_j \leqq \eta_j \leqq b_j \text{ for } j \in K \text{ and } j \neq i,$$

and

$$f_i^-(u,t) = \inf\ (f_i(\eta_1,\ldots,u_i,\ldots,\eta_n,t), \text{ where } a_j \leqq \eta_j \leqq u_j \text{ for } j \in \mathfrak{m} \text{ and } j \neq i$$
$$u_j \leqq \eta_j \leqq b_j \text{ for } j \in \mathfrak{m} \text{ and } j \neq i.$$

Let

$$a_j \leqq \eta_j \leqq u_j \text{ for } j \in \mathfrak{m},\ j \neq i,$$

then

$$f(a_j) \leqq f(\eta_j) \leqq f(u_j),$$

and f_i^+ is nondecreasing in the j-th component for $j \in \mathfrak{m}$,
but f_i^- is nonincreasing in the j-th component for $j \in K$;

analogously

f_i^+ is nonincreasing in the j-th component for $j \in K$,
f_i^- is nondecreasing in the j-th component for $j \in \mathfrak{m}$.

We set

$$g_i(u,t) = \begin{cases} f_i^+(u,t) \text{ if } i \in \mathfrak{m} \\ f_i^-(u,t) \text{ if } i \in K, \end{cases}$$

and define

$$F_{\mathfrak{m}}(u,t) = (g_1(u,t),\ldots,g_n(u,t)).$$

We take this nonlinear term in the ordinary differential equation

$$\frac{dv}{dt} = F_{\mathfrak{M}}(v,t),$$

with the initial condition $v(0) = v^o$, where

$$v_i^o = \begin{cases} \sup u_i^o(x), \; x \in \bar{D}, \; i \in \mathfrak{M} \\ \\ \inf u_i^o(x), \; x \in \bar{D}, \; i \in K \end{cases}$$

and $v(t)$ is also a solution of

$$\frac{\partial v}{\partial t} = C\Delta v + F_{\mathfrak{M}}(v,t) \text{ in } D, \; v(x,0) = v^o \text{ and } \frac{\partial v}{\partial n} = 0 \text{ on } \partial D.$$

First we show that Ξ is an invariant rectangle for the ordinary differential equation. According to the corollary on page 194, it remains to prove

$$F_{\mathfrak{M}}(v,t) \leqq 0 \text{ on the boundary } \partial\Xi \text{ of the rectangle.}$$

Let ν be an outward pointing normal to $\partial\Xi$, e.g. $\nu = (1,0,0,\ldots,0)$ and $u_1 = b_1$, then

$$F_{\mathfrak{M}}\cdot\nu = (g_1(u,t),\ldots,g_n(u,t))\cdot(1,0,\ldots,0) = g_1(u,t) = g_1(b_1,u_2,u_3,\ldots,u_n) =$$

$$= f_1^+(b_1,u_1,u_2,\ldots,u_n) \text{ for } 1 \in \mathfrak{M}$$

by construction.

Then $f_+^1(b_1,u_1,\ldots,u_n) \leqq 0$, as Ξ is an invariant rectangle for the system of partial differential equation. Analogously we can prove that

$$F_{\mathfrak{M}} \leqq 0 \text{ on } \partial\Xi \text{ if } 1 \notin \mathfrak{M}.$$

Now we are able to prove the comparison theorem. By assumption u_i is a solution of

$$\frac{\partial u_i}{\partial t} = C\Delta u_i + f_i(u,t) \text{ in } D$$

$$\frac{\partial u_i}{\partial n} = 0 \quad \text{on } \partial D$$

$$u_i^o(x,0) = u_i^o(x) \text{ in } D$$

and v_i is a solution of

$$\frac{dv_i}{dt} = g_i(v,t)$$

$$v_i(0) = v_i^o = \begin{cases} \sup u_i^o(x) \text{ for } i \in \mathfrak{M} \\ \inf u_i^o(x) \text{ for } i \in K. \end{cases}$$

It follows that

$$\frac{\partial u_i}{\partial t} - \frac{\partial v_i}{\partial t} = C\Delta u_i - C\Delta v_i + f_i(u_i,t) - g_i(v_i,t),$$

as $v_i = v_i(t)$ then $\Delta v_i = 0$, we obtain by setting $\omega_i = u_i - v_i$, that ω_i satisfies

$$\frac{\partial \omega_i}{\partial t} = C\Delta\omega_i + f_i(u,t) - g_i(v,t) \text{ in D for } t > 0$$

$$\frac{\partial \omega_i}{\partial n} = 0 \text{ on } \partial D \text{ for } t > 0,$$

and the initial condition $\omega(x,0) \leqq 0$ in D. By construction

$$v_i^o = \sup u_i^o(x) \text{ for } i \in \mathfrak{M},$$

thus

$$u_i(x,0) = u_i^o(x) < v_i^o = v_i(x,0),$$

hence

$$u_i(x,0) - v_i(x,0) \leqq 0,$$

thus

$$\omega(x,0) \leqq 0 \text{ in D}.$$

Finally, we obtain that there exists a $\omega = (\omega_1,\dots,\omega_n)$ such that

$$\omega_i = \begin{cases} u_i - v_i & i \in \mathfrak{M} \\ v_i - u_i & i \in K \end{cases}$$

and the nonlinear term

$$L(\omega,t) = (\ell_1(\omega,t),\dots,\ell_n(\omega,t)),$$

where

$$\ell_i(\omega,t) = \begin{cases} f_i(u,t) - g_i(v,t), & i \in \mathfrak{M} \\ g_i(v,t) - f_i(u,t), & i \in K, \end{cases}$$

then ω satisfies the system of partial differential equations

$$\frac{\partial \omega}{\partial t} = C\Delta\omega + L(\omega,t).$$

As the initial condition $\omega(x,0) \leqq 0$ for $x \in D$, we have only to prove, that

$$\omega(x,t) \leqq 0 \text{ in } D, \text{ for } t > 0.$$

Remembering the corollary on page 194 , it remains to prove that the nonlinear term

$$L(\omega,t) \leqq 0 \text{ on the boundary, the set } \omega_i \leqq 0 \text{ for } i = 1,2,\dots,n.$$

We must show that

$$\ell_i(\omega,t) \leqq 0 \text{ on } \omega_j \leqq 0 \text{ for } j\ i.$$

Let $i \in \mathfrak{M}$, then

$$\ell_i(\omega,t) = f_i(u,t) - g_i(v,t) = f_i(u,t) - f_i^+(v,t)$$

by construction, taking the sup $f_i(u,t)$, we have

$$\ell_i(\omega,t) \leqq f_i^+(u_1,\dots,u_i,\dots,u_n,t) - f_i^+(v_1,\dots,v_i,\dots,v_n,t).$$

We assume $1 \in \mathfrak{M}$, $1 \neq i$, as $\omega_j < 0$ for $j \neq i \in \mathfrak{M}$ then

$$\omega_1 \leqq 0 \text{ for } i \in \mathfrak{M}, \text{ e.g. } \omega_1 = u_1 - v_1 \leqq 0,$$

thus

$$u_1 \leqq v_1.$$

As f_i^+ is nondecreasing in all components, special in $1 \in \mathfrak{M}$, then, if $u_1 \leqq v_1$, it follows $f_i^+(u_1,u_2,\dots,u_n,t) \leqq f_i^+(v_1,u_2,\dots,u_n,t)$.

Let $1 \in K$, then $\omega_i = v_i - u_i \leqq 0$, thus $u_1 \geqq v_1$, as f_j^+ is nonincreasing for $j \in K$ (p.196) then, if $u_1 \geqq v_1$ it follows $f_1^+(u_1,u_2,\dots,u_n,t) \leqq f_i^+(v_1,u_2,\dots u_n)$. This process of construction can be continued, such that for $j \neq i$ if $u_j < v_j$ it follows

$$f_i^+(u_1,\dots,u_{j-1},\overset{j}{\overset{\downarrow}{u_j}},u_{j+1},\dots,u_n,t) \leqq f_i^+(u_1,\dots,u_{j-1},\overset{j}{\overset{\downarrow}{v_j}},\dots,u_n,t)$$

and finally $\ell_i(\omega,t) \leqq 0$.

We have proved the comparison

III. 5.3 <u>Theorem:</u>

Let $u_i(t,x)$ be a solution of the system of partial differential equations in D, for $t > 0$, and let $v_i(t)$ be a solution of the corresponding system of ordinary differential equations for $t > 0$. If the initial conditions

$$u_i(0,x) \leqq v_i(0)$$

are satisfied for $x \in D$, then

$$u_i(x,t) \leqq v_i(t) \text{ for } x \in D \text{ and all } t > 0, \text{ if } i \in \mathfrak{M}.$$

For $i \in K$ we have

$$u_i(x,t) \leqq v_i(t) \text{ for } x \in D,\ t > 0. \quad (125)$$

We are now interested in the interactive growth of only two species; thus the system reduces to

$$\frac{\partial u}{\partial t} = d_1 \Delta u + uM(u,v) \text{ in } D, \text{ for } t > 0$$

$$\frac{\partial v}{\partial t} = d_2 \Delta v + vN(u,v) \text{ in } D, \text{ for } t > 0$$

$$\frac{\partial u}{\partial n} = 0 = \frac{\partial v}{\partial n} \text{ on } \partial D, \text{ for } t > 0$$

$$u(x,0) = u_o(x) \text{ and } v(x,0) = v_o(x) \text{ in } D, \text{ for } t > 0.$$

We assume that the initial values u_o and v_o are bounded nonnegative, smooth functions. We apply the proof of the comparison theorem in the case of the nonlinear term

$$f = (f_1, f_2) = (uM(u,v), vN(u,v)).$$

We obtain the rectangle

$$\Xi = \{(u,v) \mid a_1 \leqq u \leqq b_1,\ a_2 \leqq v \leqq b_2\}.$$

III. 5.4 Interactive growth of two species

Then Ξ is an invariant rectangle if and only if the vector field

$$(uM(u,v), vN(u,v))$$

satisfies the conditions:

$$\begin{array}{llll} \text{if } u = b_1, & \text{then } uM(u,v) \leqq 0 & \text{for } a_2 < v < b_2 \\ v = b_2, & vN(u,v) \leqq 0 & a_1 < u < b_1 \\ u = a_1, & uM(u,v) \geqq 0 & a_2 < v < b_2 \\ v = a_2, & vN(u,v) \geqq 0 & a_1 < u < b_1. \end{array}$$

Given the nonlinear term

$$f = (f_1, f_2) = (f_1(u,v),\ f_2(u,v)),$$

we take $f_1(u,v)$ and quite analogously (page 196), obtain the set of natural numbers and the partition

$$\{1,2\} = \mathfrak{M} \cup K, \text{ e.g. } \mathfrak{M} = \{1\} \text{ and } K = \{2\}.$$

We want to determine

$f_i^+(u,t) = \sup f_i(u,\eta)$ under the side conditions

$a_1 \leqq \eta \leqq u$ for $1 \in \mathfrak{M}$, $1 \neq 2$ and $v \leqq \eta \leqq b_2$ for $2 \in K$, $2 \neq 1$.

Hence

$$f_1^+(u,t) = \sup_{v \leqq \eta \leqq b_2} f_1(u,\eta) = u \cdot \sup_{v \leqq \eta \leqq b_2} M(u,\eta) \text{ by definition of } f_1 \text{ and}$$

$$f_2^+(u,t) = \sup_{a_1 \leqq \eta \leqq u} f_2(\eta,v) = v \cdot \sup_{a_1 \leqq \eta \leqq u} N(\eta,v) \text{ by definition of } f_2,$$

quite analogously

$$f_1^-(u,t) = \inf_{v \leqq \eta \leqq b_2} f_1(u,\eta) = u \cdot \inf_{v \leqq \eta \leqq b_2} M(u,\eta) \quad \text{and}$$

$$f_2^-(u,t) = \inf_{a_1 \leqq \eta \leqq u} f_2(\eta,v) = v \cdot \inf_{a_1 \leqq \eta \leqq u} N(\eta,v),$$

and defining

$$f_2^+(u,t) = v \cdot \sup_{a_1 \leqq \eta \leqq u} N(\eta,v) \underset{\text{def.}}{=} v \cdot N^*(u,v),$$

$$f_1^-(u,t) = u \cdot \inf_{v \leqq \eta \leqq b_2} M(u,\eta) \underset{\text{def.}}{=} u \cdot \bar{M}(u,v), \text{ and}$$

$$M^*(u,v) = \sup_{b_1 \leqq \eta \leqq v} M(u,\eta) \text{ and } \bar{N}(u,v) = \inf_{u \leqq \eta \leqq a_2} N(\eta,v).$$

III. 5.5 Modelling predator-prey-phenomena

Let

$$\Xi = \{(u,v) \mid 0 \leqq u \leqq a,\ 0 \leqq v \leqq b\}$$

be an invariant rectangle; then we can show that M^* and N^* are locally Lipschitz continuous in

$$\{(u,v) \mid u > 0,\ v > 0\}.$$

Therefore, the system of ordinary differential equations

$$\frac{du}{dt} = u \cdot M^*(u,v) \quad \text{and} \quad \frac{dv}{dt} = v \cdot N^*(u,v)$$

$$u(0) = U^o = \sup_{\bar{D}} u_o(x) \qquad v(0) = v^o = \sup_{\bar{D}} v_o(x)$$

have unique solutions u(t) and v(t). The comparison theorem yields

$$0 \leq u(x,t) \leq u(t) \quad \text{and} \quad 0 \leq v(x,t) \leq v(t).$$

Now, we obtain very interesting ecological interaction models (further references are, for example, Fife (106), May (234), Maynard Smith (235)).

The interactions of two species are determined by the signs of the partial derivatives:

Modelling

i) predator-prey-phenomena, we take $\frac{\partial M}{\partial v} < 0$, $\frac{\partial N}{\partial u} > 0$, for $u > 0$, $v > 0$

ii) competition phenomena, we take $\frac{\partial M}{\partial v} < 0$, $\frac{\partial N}{\partial u} < 0$, for $u > 0$, $v > 0$

iii) symbiosis interactions, we take $\frac{\partial M}{\partial v} > 0$, $\frac{\partial N}{\partial u} > 0$, for $u > 0$, $v > 0$.

Now some remarks on predator-prey-phenomena. In the case of predator-prey interaction, the conditions can be interpreted:

the prey growth rate M ($\frac{\partial M}{\partial v} < 0$) decreases as the predator population increases ($\frac{\partial N}{\partial u} > 0$). Thus, in other words, an increase in prey is favourable for the growth of the predators.

We take specific growth rates by choosing γ, δ, μ as positive constants, where

$$0 < \beta < 1, \beta < \frac{\mu}{\delta} < 1,$$

then

$$M(u,v) = -(u-\beta)(u-1)-\gamma v \quad \text{and} \quad N(u,v) = -\mu - \delta v + \gamma u.$$

The zero set of M, $M(u,v) = 0$, gives

$$v = \frac{1}{\gamma} \cdot (-u^2+(1+\beta)u-\beta);$$

for $u = \beta$ we obtain $M(\beta,v) = -\gamma v$, and, thus $M(\beta,v) = 0$ for $v = 0$,

$M(1,v) = 0$ for $v = 0$,

represented in the phase plane

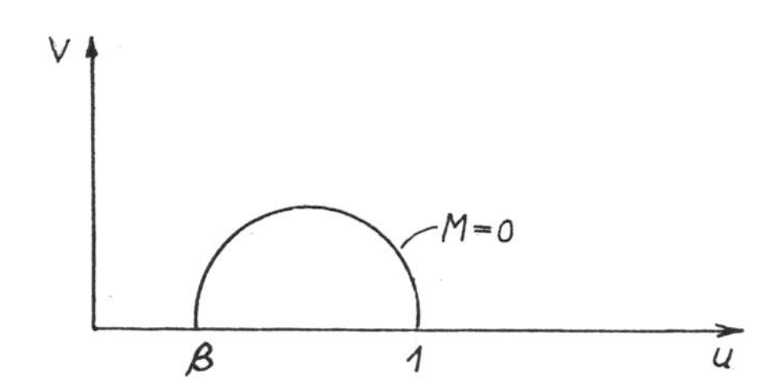

We have

$$\frac{\partial M}{\partial v} = -\gamma, \text{ as } \gamma > 0, \text{ we obtain } \frac{\partial M}{\partial v} < 0.$$

The zero set of N, N(u,v) = 0, gives

$$v = \frac{1}{\delta}(\gamma u - \mu),$$

and the representation in the phase plane

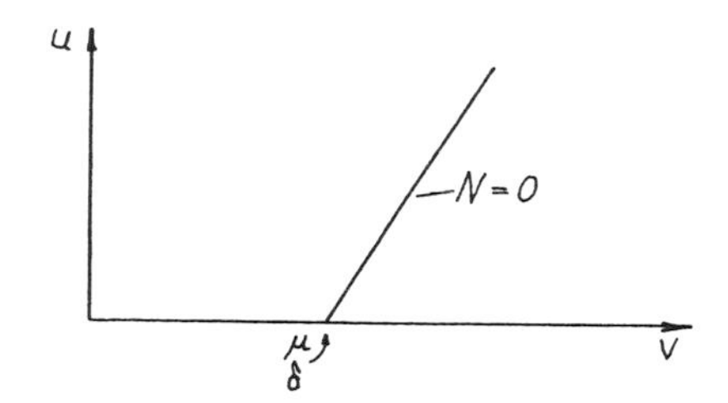

We have

$$\frac{\partial N}{\partial u} = \gamma, \text{ as } \gamma > 0, \text{ we obtain } \frac{\partial N}{\partial u} > 0.$$

Combining these results

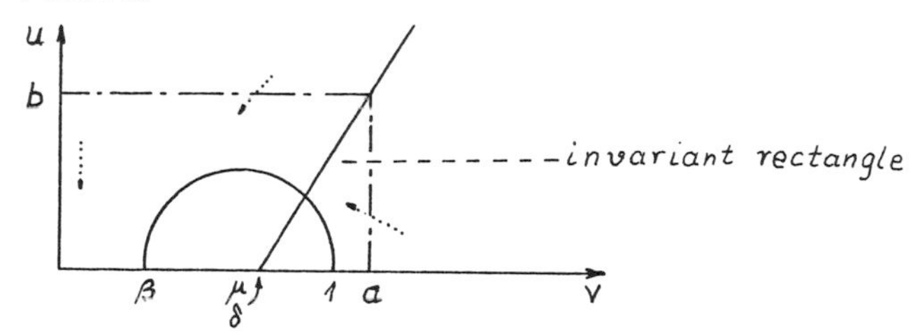

We obtain for the system of ordinary differential equations the representation in the phase plane

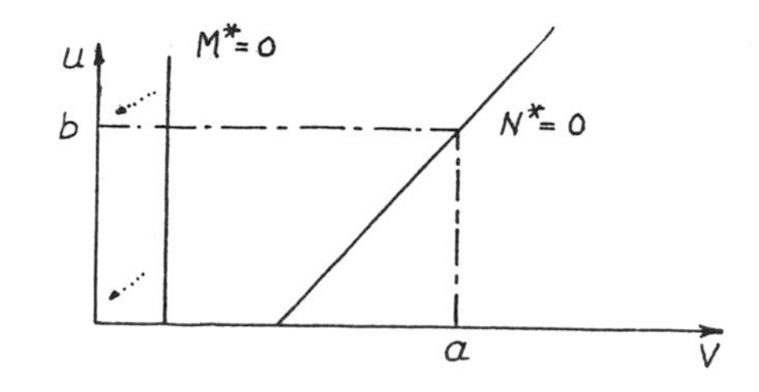

The invariant rectangle can be seen as a global attractor. The comparison theorem gives

$$0 \leqq u(x,t) \leqq u(t) \text{ and } 0 \leqq v(x,t) \leqq v(t) \text{ for } x \in D \text{ and } t > 0.$$

If the initial condition is given by $u_0(x) < \beta$, for all $x \in D$, then the diagram shows (see the vector) that

$$(u(t),v(t)) \to (0,0) \text{ for } t \to \infty,$$

and applying the comparison theorem

$$0 \leqq u(x,t) \leqq \underset{\substack{\downarrow\, t\to\infty \\ 0}}{u(t)} \;; \qquad 0 \leqq v(x,t) \leqq \underset{\substack{\downarrow\, t\to\infty \\ 0}}{v(t)}$$

we obtain

$$u(x,t) = 0 \text{ and } v(x,t) = 0, \text{ as } t \text{ tends to } \infty.$$

Thus we obtain a very interesting interpretation. If the initial density $u_0(x)$ of the prey is too low, e.g. $u_0(x) < \beta$, then the asymptotic model for $t \to \infty$ shows, that

the density of the predators $v(x,t) \underset{t\to\infty}{\to} 0$,

the density of the prey $u(x,t) \underset{t\to\infty}{\to} 0$,

hence, both u and v must become extinct.

III. 5.6 Modelling competition phenomena

Considering the case of competition, where the interaction is described by

$$\frac{\partial M}{\partial v} < 0 \quad \text{and} \quad \frac{\partial N}{\partial u} < 0;$$

we obtain the phase plane of the general system:

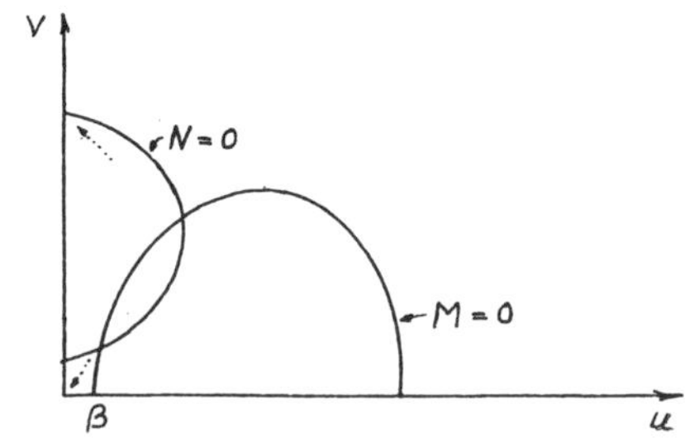

the phase plane of the ordinary differential equation

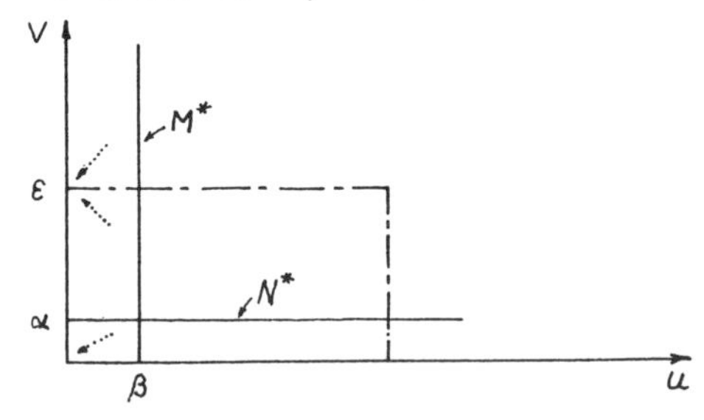

Let $u(t)$ and $v(t)$ be the solutions of the system of ordinary differential equations. If the initial densities of both species u and v are low, e.g.

$$u_o(x) < \beta \quad \text{and} \quad v_o(x) < \alpha,$$

the phase plane shows, that

$$u(t) \to 0 \quad \text{and} \quad v(t) \to 0 \quad \text{for} \quad t \to \infty.$$

Applying the comparison theorem, the densities

$$u(x,t) \to 0 \quad \text{and} \quad v(x,t) \to 0 \quad \text{for} \quad t \to \infty.$$

Consider the following specific case.
Let the initial density be $u_o(x) < \beta$, but the initial density of the species v may be higher, $v_o(x) < \varepsilon$. The phase plane shows that the solutions of the ordinary differential equation system satisfy

$$u(t) \to 0 \text{ for } t \to \infty \text{ and } v(t) \to \varepsilon \text{ for } t \to \infty.$$

Applying the comparison theorem, it follows

$$u(x,t) \to 0 \quad \text{for} \quad t \to \infty, \quad \text{and} \quad v(x,t) \to \varepsilon \text{ for } \quad t \to \infty.$$

As the initial density of the species u is too low, the species u becomes extinct, but the higher initial density of v means that v survives. This shows a certain exclusion principle, also applicable in economics. The results can be applied to a finite number of species interacting in space.

III. 5.7 Limitation of the growth (322)

The interacting species in a bounded domain are determined by their growth rates, but we have also to consider the environment with their specific resource limitation influencing the growth rates. Now, we restrict our attention to the predator-prey interactions. Let u and v be the population densities of the prey and predator, and M, N are their corresponding growth rates, satisfying

CI. $\frac{\partial M}{\partial v} < 0$ and $\frac{\partial N}{\partial u} > 0$ for $u > 0$, $v > 0$.

Imposing strict limits to the growth, we assume that there exists an $x > 0$, such that

CII. $M(u,0) < 0$ for all $u > x$.

This means, that even when there is no predator (i.e. $v = 0$), i.e. under the most favourable conditions for the prey u, the environment does not allow growth once the density exceeds a critical value x. The condition CI. is now restricted by the condition:

CIII. There exists a function h such that

$$N(u,v) < 0 \text{ for all } u > 0 \text{ and } v > h(u).$$

CI. says, that an increase of the prey u represents an enrichment of the environment for the predators v. CIII. insures that no matter what the value $(u > 0)$ of the prey, the growth rate for the predators v becomes negative once v becomes large $(h(u))$ enough.

In the case of competition, the growth rates satisfy

$$\frac{\partial M}{\partial v} < 0 \text{ and } \frac{\partial N}{\partial u} < 0 \text{ for } u > 0,\ v > 0.$$

Imposing strict limits to the growth, we assume that there exists a $\mu > 0$, such that $M(u,0) < 0$ for all $u > \mu$. We obtain the condition:

CIV. There is a function q, such that

$$M(u,v) < 0 \text{ for } v > 0 \text{ and } u > q(v).$$

CIV. insures that no matter what the value $(v > 0)$ of the v, the growth rate for u becomes negative once u becomes large enough $(q(v))$. These growth conditions are directly applicable to various problems in economics.

If the growth rate N of the predators is smooth, there exists a function $v = h(u)$ such that

$$N(u,h(u)) = 0 \text{ for all } u > 0 \text{ and } h(u) > 0.$$

Differentiating, we obtain

$$0 = \frac{\partial N}{\partial u}\cdot\frac{\partial u}{\partial u} + \frac{\partial N}{\partial v}\cdot\frac{\partial v}{\partial u} = N_u + N_v\cdot\frac{\partial v}{\partial u} = \frac{\partial}{\partial u} N(u,h(u)) + \frac{\partial}{\partial v} N(u,h(u))\frac{dh(u)}{\partial u}.$$

As $\frac{\partial N}{\partial u} > 0$ by condition CI., and $\frac{\partial}{\partial v} N(u,h(u)) < 0$ by CIII, it follows that $\frac{\partial h(u)}{\partial u} = \frac{\partial v}{\partial u} > 0$. This means that an increase in u is advantageous for the growth of v.

In a similar way we can show that $q'(v) > 0$. Combining these results, we have the following growth limitation. There exist nondecreasing nonnegative functions $h(u)$ and $q(v)$ such that if $v > h(u)$, then $N(u,v) < 0$, and

if $u > q(v)$, then $M(u,v) < 0$.

The growth functions satisfy the asymptotic growth condition

$$h(u) = O(u) \text{ if } u \to \infty \text{ and } q(v) = O(v) \text{ if } v \to \infty .$$

The curves $v = h(u)$ and $u = q(v)$ can intersect more than once; a possible graph is shown.

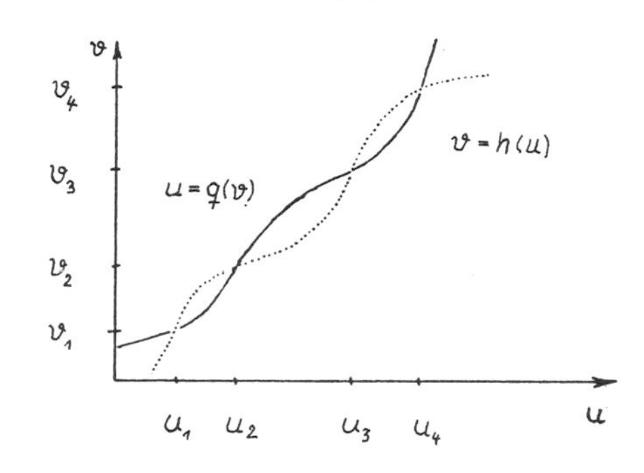

It is natural to introduce an unbounded set:

$$U = \{(u,v) | u \geqq q(v) \text{ and } v \geqq h(u)\}$$

is not connected. Denoting the interaction points of the curves by

$$\{(u_\alpha, v_\alpha)\}_\alpha,$$

and defining max $u = S$ and max $v = T$, then the unbounded component of u is given by

$$U_{unb.} = U \cap \{(u,v) | u \geqq s,\ v \geqq T\}.$$

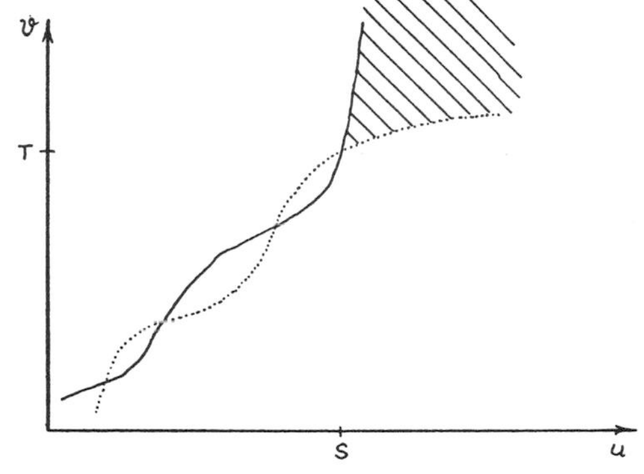

Lemma:

The interactive growth of two species is influenced by the parabolic partial differential equation system

$$\frac{\partial u}{\partial t} = d_1 \Delta u + uM(u,v) \text{ in } D, \text{ for } t > 0$$

$$\frac{\partial v}{\partial t} = d_2 \Delta v + vN(u,v) \text{ in } D, \text{ for } t > 0$$

$$\frac{\partial u}{\partial n} = 0 = \frac{\partial v}{\partial n} \qquad \text{on } \partial D, \text{ for } t > 0$$

$$u(x,0) = u_o(x);\ v(x,0) = v_o(x) \text{ in } D, \text{ for } t > 0,$$

assuming that the growth limitation is satisfied, and let (a,b) be in the interior of the set U, then

$$\Sigma(a,b) = \{(u,v) | 0 \leqq u \leqq a,\ 0 \leqq v \leqq b\}$$

is an invariant rectangle.

Let (a,b) be in the interior of the set U, then $a > q(b)$ and $b > h(a)$, as q, h are nondecreasing, nonnegative functions, then $b \geqq v$ implies $q(b) \geqq q(v)$, thus $a > q(v)$. As $a > q(v)$ the growth limitation implies $M(a,v) < 0$; as $a > 0$, we have $a \cdot M(a,v) < 0$ for $0 \leqq v \leqq b$.

In a quite analogous way, we obtain $b \cdot N(u,b) < 0$ if $0 \leqq u \leqq a$, thus the nonlinear term satisfies

$$f(a,b) = (aM(a,v), b \cdot N(b \cdot v)) < 0 \text{ for } (a,b) \in \Sigma$$

and by the corollary on page 194, Σ is an invariant rectangle.

Corollary:

Let the initial conditions u_0 and v_0 be bounded; we can choose (a,b) in the interior of U, such that $0 \leqq u_0(x) \leqq a$ and $0 \leqq v_0(x) \leqq b$ for $x \in D$. Applying the theorem, then the solution of the system is in the invariant rectangle Σ, thus

$$0 \leqq u(x,t) \leqq a \text{ and } 0 \leqq v(x,t) \leqq b \text{ for } x \in D \text{ and } t > 0,$$

Remark 1:

We assume that the growth limitation holds; then the maximal growth rates M^* and N^* (defined on page 201) are negative in the interior of U.

Let (u,v) be in the interior of U, then $u > q(v)$ (and $v > h(u)$), let v be $v \geqq \delta \geqq 0$ and as q is nondecreasing, then $u > q(v) > q(\delta)$ and by the growth limitation $u > q(\delta)$ implies

$$M(u,\delta) < 0 \text{ if } 0 \leqq \delta \leqq v.$$

By definition

$$M^*(u,v) = \sup_{0 \leqq \delta \leqq v} M(u,\delta) < 0$$

and analogously we obtain $N^*(u,v) < 0$.

Remark 2:

U is an invariant region for the maximal vector field

$$(uM^*(u,v),vN^*(u,v)).$$

Proof:

The set

$$U = \{(u,v)|u \geqq q(v) \text{ and } v \geqq h(u),\ u > 0,\ v > 0\}$$

has the boundary ∂U given by the segments

1) $u = q(v)$, $q'(v) \geqq 0$
2) $v = h(u)$, $h'(u) \geqq 0$
3) $u = 0$
4) $v = 0$.

The trajectories of the system

$$\frac{du}{dt} = u \cdot M^*(u,v) \qquad \frac{dv}{dt} = v \cdot N^*(u,v)$$

$$u(0) = a, \qquad v(0) = b, \qquad \text{as} \quad \begin{matrix} 0 \leq u_0(x) < a \\ 0 \leq v_0(x) < b \end{matrix},$$

cannot cross the boundary of type 3) or 4). In the interior of U, $v > h(u)$ and $N^*(u,v) < 0$ by Remark 1. As h is chosen to be the smallest function satisfying the growth limitation, it follows $N^*(u,h(u)) = 0$.

Thus a trajectory of the above system cannot pass out of U across the boundary of the type 2); similarly they cannot cross the boundary of type 1).

Thus U is invariant for the maximal vector field

$$(uM^*(u,v),vN^*(u,v)).$$

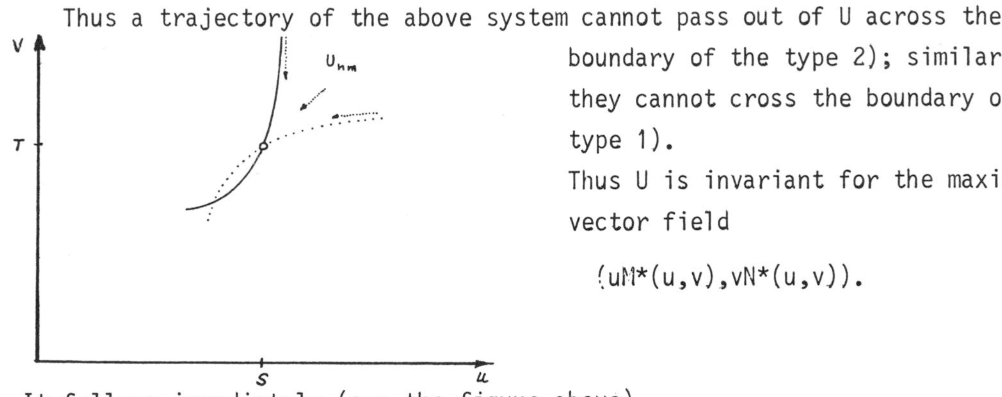

It follows immediately (see the figure above).

Remark 3:

Every trajectory of the maximal vector field, beginning at $t = 0$ in U_{nm} converges to the point (S,T) for $t \to \infty$.

Let $u^*(t)$ and $v^*(t)$ be solutions of the above system, then Remark 3 reads

$$\lim_{t\to\infty} (u^*(t),v^*(t)) = (S,T).$$

We are now able to apply the comparison theorem (on page 199)

$$u(x,t) \leqq u^*(t) \text{ and } v(x,t) \leqq v^*(t)$$

and finally

$$\limsup_{t\to\infty} u(x,t) \leqq S \quad \text{and} \quad \limsup_{t\to\infty} v(x,t) \leqq T$$

uniformly for $x \in D$.

Thus we have proved a growth limitation (see further investigations in (142)).

Theorem:

Let D be a bounded domain in $\mathbb{R}^n$ with sufficient smooth boundary ∂D. Let the growth rates M and N of the nonlinear term satisfy the growth limitation. Then every solution of the initial-boundary-value problem

$$\frac{\partial u}{\partial t} = d_1 \Delta u + uM(u,v) \text{ in D, for } t > 0$$

$$\frac{\partial v}{\partial t} = d_2 \Delta v + vN(u,v) \text{ in D, for } t > 0$$

$$\frac{\partial u}{\partial n} = 0 = \frac{\partial v}{\partial n} \qquad \text{on } \partial D, \text{ for } t > 0$$

$$u(x,0) = u_0(x) \text{ and } v(x,0) = v_0(x) \text{ in D, for } t > 0$$

(where the initial conditions u_0, v_0 are nonnegative) is nonnegative for $t > 0$ and satisfies

$$\limsup_{t\to\infty} u(x,t) \leqq S \text{ and } \limsup_{t\to\infty} v(x,t) \leqq T \quad (322).$$

Other aspects are treated by Conway, Hoff, and Smoller (68), Alikakos (5), (6), Kuiper (203), (204), Mimura et al. (242), (243), (244), (245), (246), Williams and Chow (357).

III. 6 Stability and bifurcation in certain reaction-diffusion systems

III. 6.1 Lotka-Volterra model

The Lotka-Volterra model was devised to explain temporal oscillations in the occurrence of fish in the Adriatic sea. Two kinds of fish are considered, the predator fishes and the prey fishes. The rate equations of the prey and the predators are given by

$$\frac{d\ n(prey)}{dt} = gain(prey)-loss(prey) \quad \frac{d\ n(pred.)}{dt} = gain(pred.)-loss(pred.)$$

The prey will multiply according to gain(prey) = $\gamma \cdot n(prey)$, where γ is the growth rate, if there are no predator fishes. The prey fishes have losses of their individuals by being eaten by the predators, thus the loss is proportional to the number of the prey and predators

$$Loss(prey) = \delta \cdot n(prey) \cdot n \cdot (predators).$$

Turning to the predators, we obtain their growth-rate

$$gain(predator) = \alpha \cdot n(prey) \cdot n(predators),$$

this is clear, as the multiplication rate of the predators is proportional (α) to the prey fishes and their own number. The loss of the predators is given by their death, then

$$loss(predator) = 2 \cdot \varepsilon \cdot n(predator),$$

where ε may be proporitional to the escape time of the prey (determined by methods given in Grusa (141), p. 165ff).

For instance, the prey wander to other regions, where the predators do not follow so quickly, or they can hide in certain places not accessible to the predators.

Denoting the prey by u and the predators by v, we obtain the Lotka-Volterra equations

$$\frac{du}{dt} = \gamma \cdot u - \delta\ u \cdot v$$

$$\frac{dv}{dt} = \alpha \cdot uv - 2\varepsilon v.$$

Two typical trajectories in the (u,v)-phase space for fixed parameters are given by

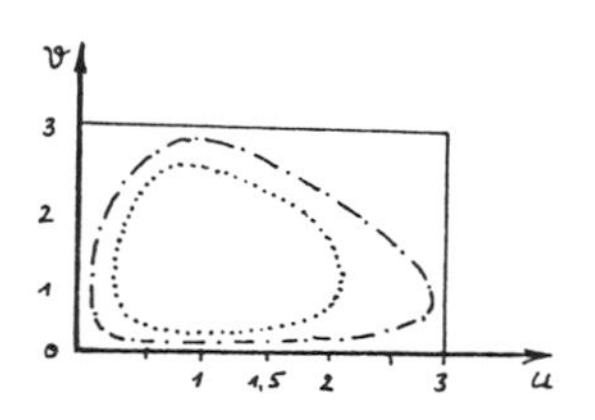

The variation of u with respect to v for various values of the parameters and the initial conditions u(0) and v(0) with the values

0.2 and 0.8 for the curve — · — · — · —

2.0 and 0.5 for the curve . . . .

The time variation of the two populations u, v corresponding to the above trajectories, where the initial values are the values for t = 0, by setting $u = f_1$ and $v = f_2$.

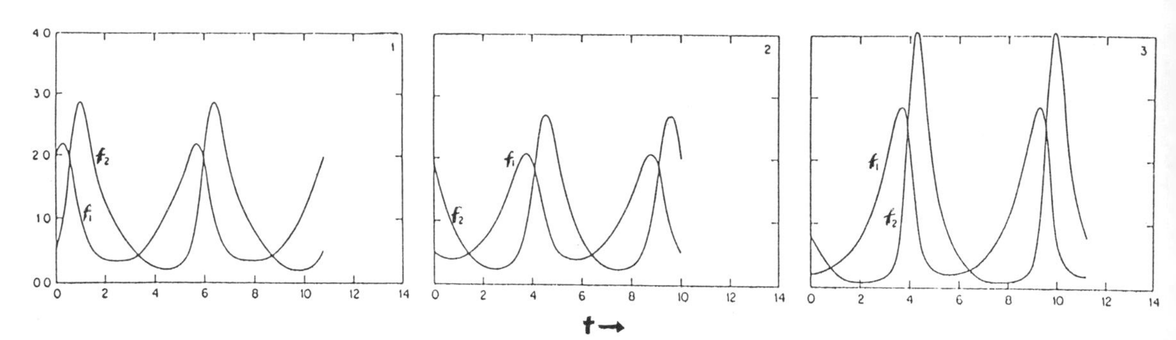

(from: (137): Reviews of modern Physics, 43, (1971), p. 235)

The periodic solution can be interpreted as follows. If the number of the predators become too great, the prey is eaten too quickly; thus the food supply of the predators decreases and therefore their population decreases. This allows for an increase of the number of prey animals, so that a greater food supply becomes available for the predators whose number increases again.

Treating this problem stochastically, we can show that both populations die out. We must take into account effects like the enviornment, time lags, different death rates, different reaction behaviour. Furthemore the predator may live on two or several kinds of prey, thus the oscillations become smaller and the whole system more stable.

III. 6.2 Interactive growth of two species

The interactive growth of the two species (prey u, predators v) can be given by the generalized Lotka-Volterra equation

$$\frac{du}{dt} = u \cdot f(u) - u \cdot v \cdot \psi(u)$$

$$\frac{dv}{dt} = -v \cdot g(v) + u \cdot v \cdot \psi(u),$$

where $f(u)$, $\{g(v)\}$ is the normalized growth rate of the prey u {predator v} in the absence of the predators {the prey}. We assume that

$$f(1) = 0,\ f(u) < 0 \text{ for } u > 1 \text{ and } g(0) = \mu \in \mathbb{R},\ g'(v) \geqq 0 \text{ for } v > 0.$$

Common examples of ψ are

$$\psi(u) = \alpha \text{ and } \psi(u) = \frac{\alpha}{1+\beta u},$$

where α and β are constants. The solution can be described in the phase plane (u,v) and the location of the equilibrium points is given by

$$\frac{du}{dt} = 0 \qquad\qquad u = 0 \qquad\qquad v\cdot\psi(u) = f(u)$$

and equivalent   or

$$\frac{dv}{dt} = 0 \qquad\qquad v = 0 \qquad\qquad u\cdot\psi(u) = g(v),$$

thus

$$v = F(u) = \frac{f(u)}{\psi(u)}$$ can be interpreted geometrically

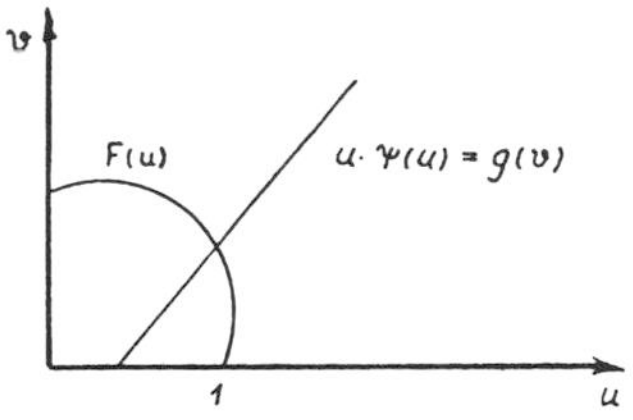

The stability character of the various rest points is determined by the linearization of the vector field

$$(u\cdot f(u) - u\cdot v\,\psi(u),\ -vg(v) + u\cdot v\psi(u))).$$

Differentiating the first (second) expression with respect to u (v), we obtain

$$(f(u) + u\cdot f'(u) - v(u\cdot\psi(u))',\ -g(v) - v\cdot g'(v) + u\cdot\psi(u)),$$

then, at a point (u,v) we obtain the functional matrix

$$\mathcal{F}(u,v) = \begin{pmatrix} f(u)+u\cdot f'(u)-v(u\cdot\psi(u))' & -u\cdot\psi(u) \\ v(u.\psi(u))' & -g(v)-v.g'(v) + u\cdot\psi(u) \end{pmatrix}$$

In the case

$$\mathcal{F}(u,0) = \begin{pmatrix} f(u)+uf'(u) & -u\cdot\psi(u) \\ 0 & -g(0)+u\cdot\psi(u) \end{pmatrix}$$ the diagonal matrix $$\mathcal{F}(0,0) = \begin{pmatrix} f(0) & 0 \\ 0 & -g(0) \end{pmatrix},$$ follows

as $g(0) = \mu > 0$, we obtain, that the point $(0,0)$ is a saddle if $f(0) > 0$; $f(u)$ may be interpreted as the growth rate of the prey u in the absence of the predators v.

Possible distributions of f:

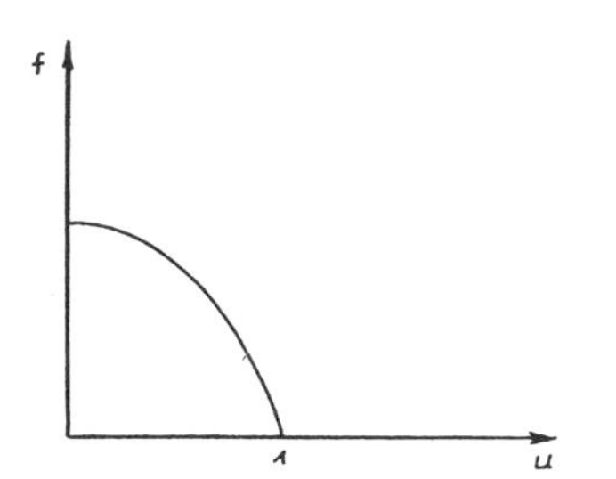

A logistic S-shaped curve of saturated growth

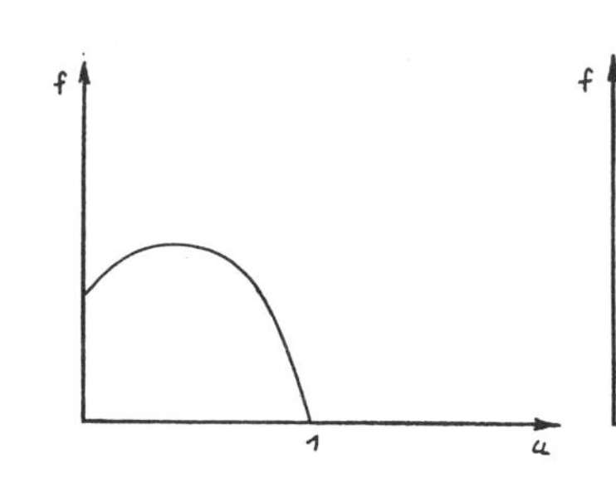

There are allee effects at low population densities

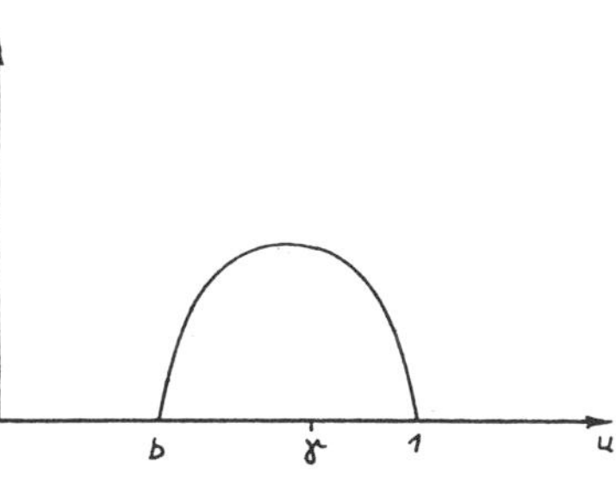

A fall off in the growth rate, known as asocial growth by Phillip (1957, (280)).

The functional matrix at the point (1,0) reads

$$\mathcal{F}(1,0) \begin{pmatrix} f(1) + f'(1) & -\psi(1) \\ 0 & -g(0)+\psi(1) \end{pmatrix}$$

thus,

$$\det \mathcal{F}(1,0) = (f(1)+f'(1))(-g(0)+\psi(1)), \text{ as } f(1) = 0 \text{ and } g(0) = \mu$$

$$= f'(1)\cdot(-\mu+\psi(1)) < 0, \text{ as } f'(1) < 0$$

hence, the point (1,0) is a saddle in all three cases. We show that (b,0) is a saddle; therefore, we take

$$\mathcal{F}(b,0) = \begin{pmatrix} f(b)+bf'(b) & -b\psi(b) \\ 0 & -g(0) + b\psi(b) \end{pmatrix}$$

then,

$$\det \mathcal{F}(b,0) = bf'(b)(b\psi(b)-g(0)), \text{ as } f(b) = 0, \text{ let } \gamma \in (b,1),$$

$$\text{then, } b < \gamma \text{ and } b\psi(b) < \gamma\cdot\psi(\gamma) \text{ and } b\psi(b)-g(0) < \gamma\cdot\psi(\gamma)-g(0).$$

We derived the equation system (see the figure on page 213)

$$v = F(u) \text{ and } F(u) = \frac{f(u)}{\psi(u)} \quad \text{and}$$

$$u\cdot\psi(u) = g(v),$$

for the point $(\gamma,0)$ we can write

$$\gamma\cdot\psi(\gamma) = g(0),$$

so we obtain

$$b\cdot\psi(b) - g(0) < 0, \text{ thus } \det \mathcal{F}(b,0) < 0,$$

and $(b,0)$ is a saddle, too.

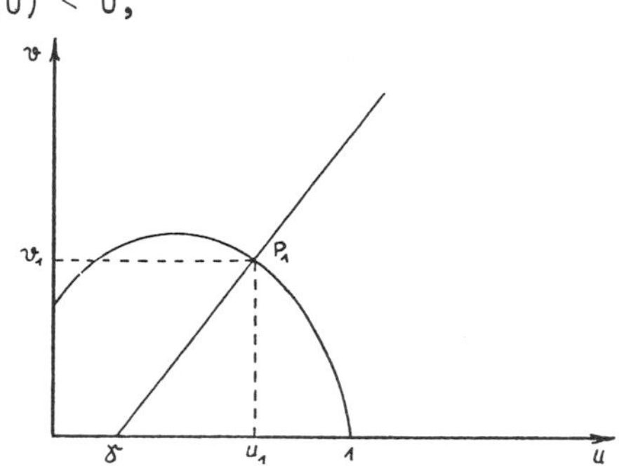

III. 6.3 Functional matrix

The point $P_1 = (u_1,v_1)$ satisfies the equation

$$v_1 = F(u_1) \text{ and } u_1\psi(u_1) = g(v_1),$$

and we obtain the functional matrix

$$\mathcal{F}(u_1,v_1) = \begin{pmatrix} f(u_1)+u_1f'(u_1)-v_1(u_1\cdot\psi(u_1))' & -u_1\cdot\psi(u_1) \\ v_1(u_1\cdot\psi(u_1))' & -g(v_1)-v_1(g'(v_1)+u_1\cdot\psi(u_1) \end{pmatrix}$$

By setting

$$C = f(u_1)+u_1f'(u_1),\ A = v_1(u_1\cdot\psi(u_1))',\ D = -g(v_1)-v_1\cdot g'(v_1) \text{ and}$$

$$B = u_1\cdot\psi(u_1),$$

we have

$$\mathcal{F}(u_1,v_1) = \begin{pmatrix} C-A & -B \\ A & D+B \end{pmatrix},$$

and

$$\det \mathcal{F}(u_1,v_1) = (C-A)\cdot(D+B) + AB = CD + CB - AD$$

hence

$$\det \mathcal{F}(u_1,v_1) = C(D+B) - AB$$

$$= \{f(u_1)+u_1\cdot f'(u_1)\}\cdot(-g(v_1)-v_1g'(v_1)+u_1\psi(u_1)) -$$

$$\{v_1(u_1\psi(u_1))'\cdot(-g(v_1)-v_1(g'(v_1)\}$$

$$\text{as } u_1\psi(u_1) = g(v_1) \text{ and } f(u_1) = F(u_1)\cdot\psi(u_1)$$

$$\psi(u_1)+u_1\cdot\psi'(u_1) = g'(v_1) \quad \text{and } f'(u_1) = F'(u_1)\cdot\psi(u_1) + F(u_1)\cdot\psi'(u_1)$$

$$= \{F(u_1)\cdot\psi(u_1) + u_1F'(u_1)\cdot\psi(u_1)+u_1F(u_1)\cdot\psi'(u_1)\}\cdot(-v_1\cdot g'(v_1)) - AD,$$

where $AD = v_1(\psi_1(u_1) + u_1\psi'(u_1))\cdot(-u_1\psi(u_1)-v_1g'(v_1))$

$$= v_1\{-u_1\psi^2(u_1)-v_1g'(v_1)\cdot\psi(u_1)-u_1^2\psi'(u_1)\psi(u_1)-u_1\cdot v_1\psi'(u_1)\cdot g'(v_1)\}$$

$$= v_1u_1\psi(u_1)\cdot(-\psi(u_1)-u_1\psi'(u_1))-v_1^2g'(v_1)\psi(u_1)-u_1v_1^2\psi'(u_1)\cdot g'(v_1),$$

then

$$\det \mathcal{F}(u_1,v_1) = v_1u_1\psi(u_1)\cdot(-F'(u_1)\cdot g'(v_1) + \psi(u_1)+u_1\cdot\psi'(u_1)) +$$

$$+ v_1^2g'(v_1)\cdot\psi(u_1)+u_1\cdot v_1^2\psi'(u_1)\cdot g'(v_1)-v_1g'(v_1)\cdot F(u_1)\cdot\psi(u_1)-$$

$$- v_1u_1g'(v_1)\cdot F(u_1)\cdot\psi'(u_1)$$

$$v_1g'(v_1)\cdot\psi(u_1)\cdot(v_1-F(u_1)) + u_1v_1\psi'(u_1)\cdot g'(v_1)\cdot(v_1-F(u_1)), \text{ as } v_1 = F(u_1).$$

finally we obtain

$$\det \mathcal{F}(u_1,v_1) = v_1u_1\cdot\psi(u_1)\ \{-F'(u_1)\cdot g'(v_1) + \psi(u_1)+u_1\cdot\psi'(u_1)\}.$$

The trace of the functional matrix is given by

$$\operatorname{trace} \mathcal{F}(u_1,v_1) = C-A + D + B$$

$$= f(u_1) + u_1f'(u_1)-v_1(\psi(u_1)+u_1\psi'(u_1)) + (-g(v_1)-v_1g'(v_1))+u_1\psi(u_1)$$

$$= F(u_1)\psi(u_1)+u_1F'(u_1)\cdot\psi(u_1)+u_1F(u_1)\cdot\psi'(u_1)-v_1\psi(u_1)-v_1u_1\psi'(u_1) -$$

$$- g(v_1)-v_1g'(v_1) + u_1\psi(u_1)$$

$$= u_1\psi(u_1)F'(u_1)-v_1\cdot g'(v_1)+\psi(u_1)\cdot(F(u_1)-v_1)+u_1\psi'(u_1)\cdot(F(u_1)-v_1)$$

$$= u_1\psi(u_1)F'(u_1) - v_1\cdot g'(v_1).$$

We call the point P_1 an attractor, if the trace $\mathcal{F}(u_1,v_1) < 0$, and the det $\mathcal{F}(u_1,v_1) > 0$, therefore, the point $P_1 = (u_1,v_1)$ is an attractor, if

$$F'(u_1) < 0 \text{ and } g'(v_1) = 0, \text{ then } F'(v_1) < 0.$$

The equilibrium point P satisfies the equation

$$v = F(u),\ F(u) = \frac{f(u)}{\psi(u)},$$

and

$$u\psi(u) = g(v)$$

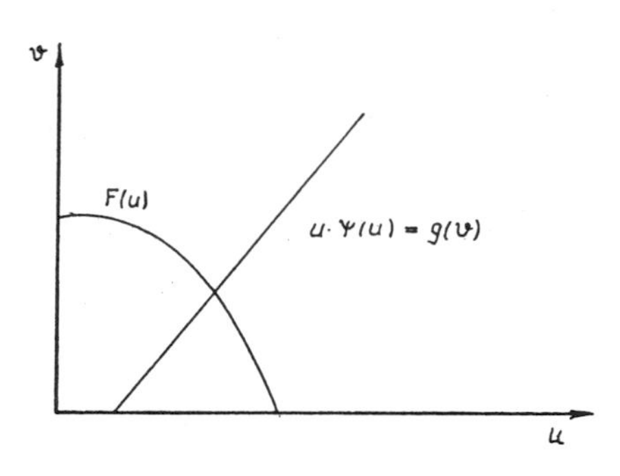

III. 6.4 Equilibrium point

The point $P_1 = (u_1, v_1)$ is an attractor, if

$$F'(u) < 0.$$

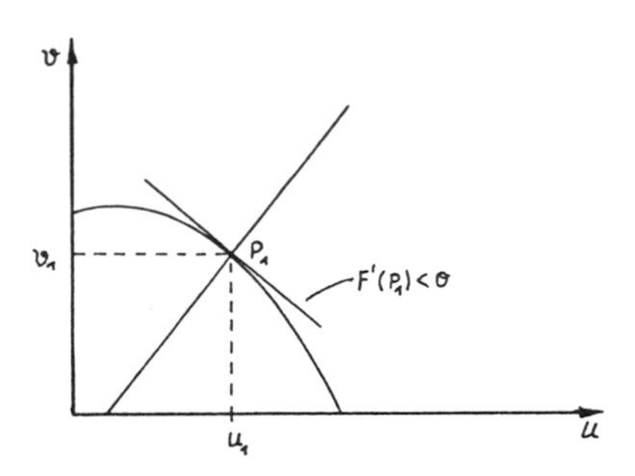

By the Poincaré-Bendixon theorem the equilibrium point P_1 is a global attractor, if the initial values satisfy

$$u(0) > 0,\ v(0) > 0$$

and

$$(u(t), v(t) \underset{t\to\infty}{\to} P_1.$$

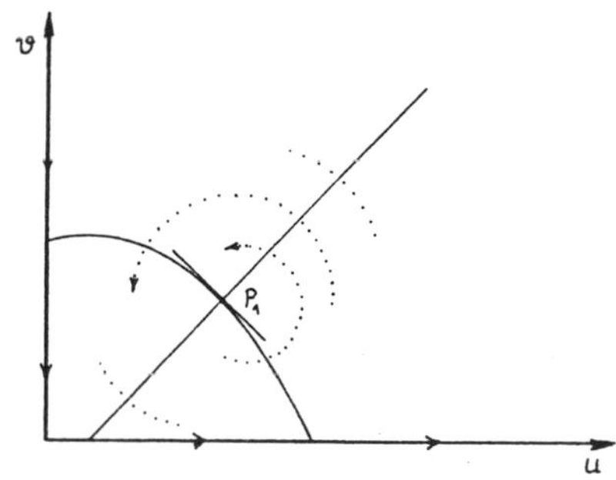

The transition from the stable point P_1 to an unstable point looks like

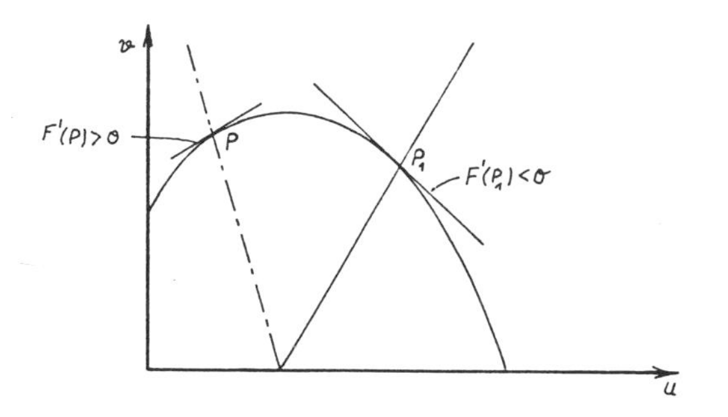

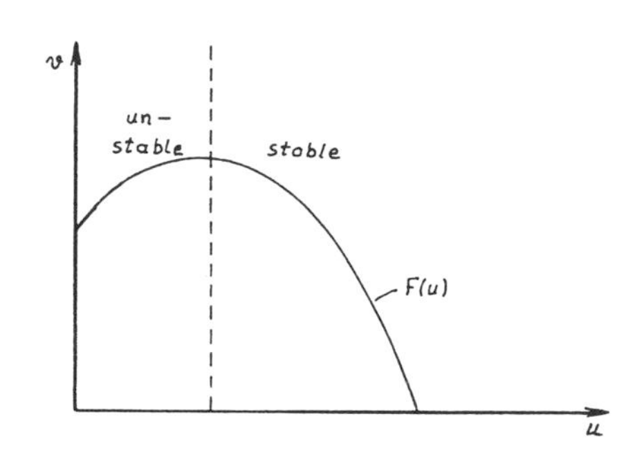

Let f and ψ be fixed and g varies by translation (see the notation above), then there appears a limit cycle through a Hopf bifurcation (see Hassard (151))

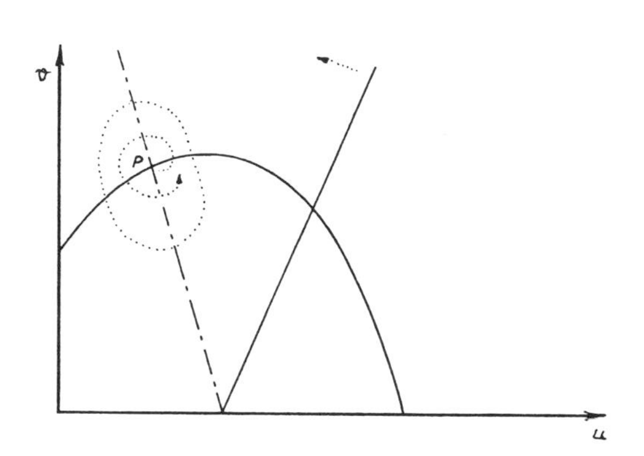

As P is unstable, then there exists an attracting limit cycle, which is unique and contains the point P.

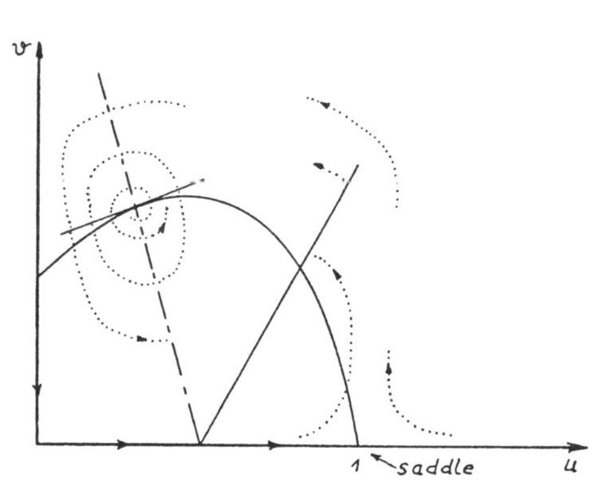

The transition from the stable to the unstable point can be explained by the complex eigenvalues of the functional matrix $\mathcal{F}(u,v)$, where at transition the eigenvalues cross the imaginary axis. When they do so, the limit cycle appears through a Hopf bifurcation (Hassard (151)).

III. 6.5 Some aspects of bifurcation

We generalize our model by describing the classical two species interactive growth

$$\frac{du}{dt} = u \cdot M(u,v)$$

$$\frac{dv}{dt} = v \cdot N(u,v),$$

where the variables u, v denote measures of the total population as:

the number of individuals (in economics, sociology or biology),

or the density functions (in physics, economics, biology, and in ecology (233, 234, 235)).

The interaction of the population is determined by the functions M and N, considered as growth rates for u and v.

In this volume, we shall assume, that u and v are continuously distributed throughout a spatial domain D.

We concentrate our attention on spatial densities u(t,x) and v(t,x) such that

$$\bar{u}(t) = \frac{1}{\text{Volume } D} \int_D u(x,t)dx \quad \text{and} \quad \bar{v}(t) = \frac{1}{\text{Volume } D} \int_D v(x,t)dx.$$

We suppose, that there are two populations undergoing certain simple diffusion processes given by the system of partial differential equations

$$\frac{\partial u}{\partial t} = d_1 \Delta u + uM(u,v)$$

$$\frac{\partial v}{\partial t} = d_2 \Delta v + vN(u,v).$$

The time dependent diffusion phenomena can be described in bounded domains D in $\mathbb{R}^n$ with sufficient smooth boundary ∂D. The boundary conditions

$$\frac{\partial u}{\partial n} = 0 = \frac{\partial v}{\partial n} \text{ on } \partial D$$

are interpreted as an assumption that both populations are confined to the domain D, thus there is no migration across the boundary. This no flux condition is quite reasonable from an ecological point of view. One imagines that the species are, say, on an island, or in a valley surrounded by mountains from which they cannot escape. These phenomena can be treated by our modelling methods. We refer the reader to Okubo (271) for general continuous models in population dynamics, where general population models are described by Maynard-Smith (1974, (235)), and May (1976, (234)).

The transition from the system of partial differential equations to the system of ordinary differential equations, by methods of the mean values $\bar{u}$, $\bar{v}$, can be made precise. This transition process is often referred to in the biological and chemical literature as the "lumped parameter assumption". But this assumption cannot always be valid. Heuristically under suitable circumstances, it might be reasonable to assume that u and v do not vary too much from point to point, so that possible transport processes can be ignored. In fact, let f be the map: $(u,v) \to (uM(u,v), vN(u,v))$ and denoting df as its differential, then

$$\Theta = \max \{df_n | n \in \text{invariant rectangle}\}$$

is a measure of the strength of the nonlinear term. Let λ denote the smallest, positive eigenvalue of $-\Delta$ in D with respect to the homogeneous boundary conditions

$$\frac{\partial u}{\partial n} = 0, \frac{\partial v}{\partial n} = 0, \text{ and } C = \min(d_1, d_2),$$

where the d_i are the diffusion coefficients, then we define

$$\mu = \lambda \cdot C - \Theta.$$

Introducing the mean values

$$\bar{u}(t) = \frac{1}{\text{Vol.}} \int_D u(x,t)dx \text{ and } \bar{v}(t) = \frac{1}{\text{Vol.}} \int_D v(x,t)dx,$$

we obtain under the condition $\mu > 0$ that every solution of the system of the partial differential equations $(u(x,t),v(x,t))$, which is confined to an invariant bounded region, decays exponentially fast to the spatial averages $\bar{\mu}$, $\bar{v}$; hence

$$\sup_{x \in D}|u(x,t)-\bar{u}(t)| + \sup_{x \in D}|v(x,t)-\bar{v}(t)| < C_i \cdot \text{constant} \cdot \exp(-\mu t).$$

The condition $\mu > 0$ can be interpreted as

$$\lambda > \frac{\Theta}{C}.$$

If λ is large, we can say that the spatial region is small; or, we have

$$C > \frac{\Theta}{\lambda}.$$

This means that the diffusion C is strong relative to the nonlinear reaction term Θ.

According to Courant-Hilbert (69) the eigenvalues depend on the size of the domain D, i.e. λ is inversely proportional to the squared diameter of D. This theorem gives us a sufficient condition for the damping of the spatial variations. Thus, spatial inhomogenities become quickly damped out.

III. 6.6 Stability of bifurcating solutions

We shall restrict our attention to the case of diffusion and reaction processes governed by the system

$$\text{I.}\quad \begin{aligned} \frac{\partial u}{\partial t} &= d_1 \Delta u + u \cdot f(u) - u \cdot v \cdot \psi(u) \\ \frac{\partial v}{\partial t} &= d_2 \Delta v - v \cdot g(v) + u \cdot v \cdot \psi(u) \end{aligned}$$

in a bounded spatial domain D, subject to the homogeneous boundary conditions

$$\frac{\partial u}{\partial n} = 0 \text{ and } \frac{\partial v}{\partial n} = 0 \text{ on } \partial D,$$

thus, we have a "closed" system, which allows no flux through the boundary.

Introducing the spatial average, we obtain a system of ordinary differential equations for $t \to \infty$ approximately given by

$$\text{II.}\quad \begin{aligned} \frac{du}{dt} &= u \cdot f(u) - u \cdot v \cdot \psi(u) \\ \frac{dv}{dt} &= -v \cdot g(v) + u \cdot v \cdot \psi(u). \end{aligned}$$

If the diffusion coefficients are large enough, or if the spatial domain is small enough, then every solution of the partial differential equation system will converge to a spatially homogeneous solution of the ordinary differential equation system..

We consider now the bifurcation of a spatially inhomogeneous equilibrium solution from an equilibrium solution which is constant in space. We are looking for a critical point that is stable as an equilibrium solution of the ordinary differential equation II, but is unstable as a solution of the system I. The problem of this type of instability goes back to Turing (338), his main idea was that the diffusion acts as a destabilizing factor. For various other contexts, see Prigogine and Nicolis (286). Othmer and Scriven (277); for the problem with respect to population dynamics see, for example, Segel and Levin (318), Levin and Segel (212), Segel and Jackson (317). Brown determined the stability of bifurcating solutions in (49).

The discussion of predator-prey systems is given in Mimura et al (242-246), and of the global behaviour of the bifurcating branch of the solution in Nishiura (267).

The stability of system II in the point (u,v) is given by the functional matrix $\mathcal{F}(\tilde{u},\tilde{v})$, where the trace $\mathcal{F}(\tilde{u},\tilde{v}) < 0$, and det $\mathcal{F}(\tilde{u},\tilde{v}) > 0$.

We determine the stability analysis of the system I (Casten and Holland 54,5) by introducing the linearized system I (taking for the nonlinear term the functional matrix $\mathcal{F}(\tilde{u},\tilde{v})$)

$$\frac{\partial}{\partial t}\,\omega(x,t) = C\cdot\Delta\omega(x,t) + \mathcal{F}(\tilde{u},\tilde{v})\omega(x,t) \text{ in } D \text{ for } t > 0$$

$$\frac{\partial}{\partial n}\,\omega(x,t) = 0 \text{ on } \partial D, \text{ for } t > 0, \text{ where } C = \operatorname{diag}(d_1,d_2).$$

Now, $\omega(x,t)$ can be thought of as an approximation to

$$(u(x,t)-\tilde{u},\ v(x,t)-\tilde{v}).$$

The constant solution $(\tilde{u},\tilde{v})$ is an asymptotically stable solution of system I if, and only if, every solution of the linearized system I converges to zero for $t\to\infty$. We treat the following criterion for stability.

The solution $\omega(x,t)$ of the linearized system I can be expanded by eigenfunctions $\varphi_i(x)$ of $-\Delta$, i.e.

$$\begin{aligned} &\Delta\varphi_i + \lambda\varphi_i = 0 \text{ in } D \\ &\frac{\partial\varphi_i}{\partial n} = 0 \quad \text{on } \partial D \end{aligned} \qquad \text{and } \omega(x,t) = \sum_{i=0}^{\infty} z_i(t)\cdot\varphi_i(x).$$

The eigenfunctions satisfy

$$0 \leqq \lambda_o \leqq \lambda_1 \leqq \lambda_2 \leqq \ldots \leqq \lambda_n \to \infty,$$

by normalization of φ_i we obtain an orthonormal basis in $L^2(D)$.

Using the ansatz $\omega(x,t)$ in the linearized system, then each z_i has to satisfy the ordinary differential equation

$$\frac{dz_i}{dt} = (\mathcal{F}(\tilde{u},\tilde{v}) - \lambda C)z_i.$$

If there exist for each matrix $\mathcal{F}(\tilde{u},\tilde{v})-\lambda_i C$, for $i = 0,1,2,...$ two eigenvalues with negative real parts, then $(\tilde{u},\tilde{v})$ is an asymptotically stable equilibrium solution of the system of partial differential equations.

If there are eigenvalues with positive real part for any matrix $\mathcal{F}(\tilde{u},\tilde{v})-\lambda_i C$, then $(\tilde{u},\tilde{v})$ is unstable. Defining

$$\mathcal{F}(\lambda) = \mathcal{F}(\tilde{u},\tilde{v})-\lambda\cdot C,$$

we obtain that the

$$\text{trace } \mathcal{F}(\lambda) = \text{trace } \mathcal{F}(\tilde{u},\tilde{v}) - \lambda \text{ trace}(\text{diag.}(d_1,d_2)) = \text{trace } \mathcal{F}(\tilde{u},\tilde{v})-\lambda(d_1+d_2),$$

by methods of linear algebra see Greub ((138), p. 125). The characteristic polynomial of a linear transformation φ of a 2-dimensional space can be written as

$$f(\lambda) = \lambda^2 - \lambda(\text{trace } \varphi) + \det \varphi$$

applied to $\mathcal{F}$, follows

$$\det(\mathcal{F}(\lambda)) = \lambda^2 - \lambda(\text{trace } \mathcal{F}) + \det \mathcal{F},$$

we have on page 216:

$$\text{trace } \mathcal{F}(u_1,v_1) = u_1\psi(u_1)\cdot F'(u_1)-v_1\cdot g'(v_1),$$

hence

$$\det(\mathcal{F}(\lambda)) = d_1\cdot d_2\cdot\lambda^2 + (d_1\cdot\tilde{v}\cdot g'(\tilde{v}) - d_2\tilde{u}\cdot\psi(\tilde{u})\cdot F'(\tilde{u}))\cdot\lambda + \det \mathcal{F}.$$

The stability forces of the ordinary differential system II are

$$\text{trace } \mathcal{F} < 0 \text{ and } \det \mathcal{F} > 0.$$

III. 6.7 Instability and eigenvalues

In order to have instability for the system I, we must consider

$$\det |\mathcal{F}(\lambda)| < 0,$$

for λ equal to one of the λ_j's, then

$$\text{trace } \mathcal{F}(\lambda) < 0$$

for $\lambda > 0$ by construction, as $d_1 + d_2 > 0$. We obtain by construction

$$\det \mathcal{F}(0) = \det \mathcal{F} > 0,$$

thus we have to consider

$$\det (\mathcal{F}(\lambda)) < 0 \text{ for } \lambda = \lambda_j \text{ where } j \geqq 1.$$

As $d_1 \cdot d_2$ is positive and if we assume that

$$d_1 \tilde{v}\, g'(\tilde{v}) - d_2 \tilde{u} \cdot \psi(\tilde{u}) \cdot F'(\tilde{u})$$

is positive or zero, then

$$\det \mathcal{F}(\lambda) > 0$$

for all $\lambda > 0$, this means that $(\tilde{u},\tilde{v})$ is stable for the system I. If we have

$$\frac{d}{d\lambda}\det(\mathcal{F}(\lambda))\Big|_{\lambda=0} = d_1 \tilde{v} g'(\tilde{v}) - d_2 \tilde{u} \psi(\tilde{u}) F'(\tilde{u}) < 0,$$

then

$$\det \mathcal{F}(\lambda) < 0 \text{ and } (\tilde{u},\tilde{v}) \text{ is unstable.}$$

By assumption we have $g'(v) \geqq 0$, and if

$$F'(u) > 0,$$

we obtain

$$\det \mathcal{F}(\lambda) < 0.$$

Let $g'(\tilde{v}) = 0$, then $(\tilde{u},\tilde{v})$ is asymptotically stable only if

$$F'(\tilde{u}) < 0.$$

We need the graph of $\det \mathcal{F}(\lambda)$ for further considerations:

$$\det \mathcal{F}(\lambda) = d_1 \cdot d_2 \cdot \lambda^2 + (d_1 \tilde{v} g'(\tilde{v}) - d_2 \tilde{u} \psi(\tilde{u}) F'(\tilde{u}))\lambda + \det \mathcal{F},$$

taking $A_2 = \tilde{v} g'(\tilde{v})$ and $B_1 = \tilde{u} \cdot \psi(\tilde{u}) \cdot F'(\tilde{u})$, we have

$$\det \mathcal{F}(\lambda) = d_1 \cdot d_2 \lambda^2 + (d_1 A_2 - d_2 B_1)\lambda + \det \mathcal{F},$$

and

$$\frac{d}{d\lambda} \det \mathcal{F}(\lambda) = 2\lambda d_1 \cdot d_2 + (d_1 A_2 - d_2 B_1) = 0,$$

then

$$\lambda = \frac{1}{2}\left(\frac{A_2}{d_1} - \frac{B_1}{d_2}\right).$$

The minimum value of det $\mathcal{F}(\lambda)$ at $\lambda = \lambda_{min}$ is

$$\lambda_{min} = \frac{1}{2}\left(\frac{A_2}{d_1} - \frac{B_1}{d_2}\right) \text{ and } \det \mathcal{F}(\lambda_{min})$$

is given by

$$\det \mathcal{F}(\lambda_{min}) = -\frac{d_2}{d_1}\frac{A_2^2}{4} - \frac{d_1}{d_2}\frac{B_1^2}{4} + \frac{1}{2} A_2 B_1 + \det \mathcal{F}.$$

Let there be $\varepsilon > 1$, such that

$$\frac{d_2}{d_1} > \varepsilon, \text{ then } \lambda_{min} > 0,$$

and

$$\det \mathcal{F}(\lambda_{min}) < 0.$$

We have instability, if one of the eigenvalues λ_i is in the interval (λ_1, λ_2) where

$$\det \mathcal{F}(\lambda) < 0.$$

As the eigenvalue depend upon the size of the domain D, we are able to vary $\lambda_1, \lambda_2, \ldots, \lambda_n$ by varying the domain or one of the parameters $d_1, d_2, A_2, B_1, \psi, \ldots$ Now, these variations lead to bifurcation. We can say that bifurcation takes place at least when λ_i is a simple eigenvalue of $-\Delta$, when it crosses the graph of det $\mathcal{F}(\lambda)$ and all other λ_i's are strictly below the graph of det $\mathcal{F}(\lambda)$.

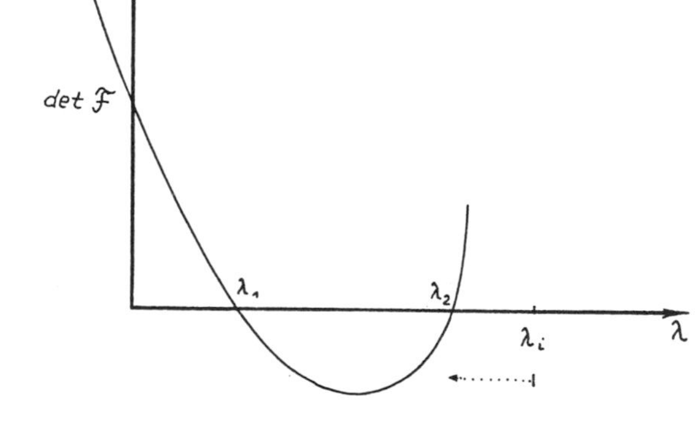

Let the diffusion matrix C and $\mathcal{F}(\tilde{u}, \tilde{v})$ be fixed, such that det $\mathcal{F}(\lambda(s))$ has two positive roots λ_1 and λ_2. Let there be a variation of the domain by a parameter s, thus we have a one parameter family of domains D(s).

As s increases (past s) the eigenvalues $\lambda_i(s)$ are very smooth and one eigenvalue $\tilde{\lambda}_j$ moves exactly into the interval, while the other λ_i's remain outside.

The partial differential equation system

$$\frac{\partial u}{\partial t} = d_1\Delta u + u\cdot f(u) - u\cdot v\psi(u)$$

$$\frac{\partial v}{\partial t} = d_2\Delta v - v\cdot g(v) + u\cdot v\psi(u)$$

can be written as

$$1)\quad \frac{\partial p}{\partial t} = C\cdot\Delta P + K(P)$$

where

$$P = (u,v),\ C = \mathrm{diag}(d_1,d_2),\ \text{and } K(P) = (f(u)\cdot u - u\cdot v\cdot\psi(u),\quad -v\cdot g(v)+u.v.\psi(u)),$$

and the linearized system

$$\frac{\partial u}{\partial t} = d_1\Delta u + \mathcal{F}(\tilde{u},\tilde{v})\cdot u$$

$$\frac{\partial v}{\partial t} = d_2\Delta v + \mathcal{F}(\tilde{u},\tilde{v})v$$

can be rewritten

$$2)\quad \frac{\partial P}{\partial t} = C\cdot\Delta P + \mathcal{F}(\tilde{u},\tilde{v})P.$$

Denoting the right-hand side of 1) by the operator

$$O(P) = C\cdot\Delta P + K(P).$$

Let $\tilde{P} = (\tilde{u},\tilde{v})$ be the equilibrium point (see page 217), we have $O(\tilde{P}) = 0$. The right-hand side of the linearized equation 2) can be written as operator

$$O(P,\tilde{P}) = C\cdot\ P + \mathcal{F}(\tilde{u},\tilde{v})P.$$

The spectrum of this operator consists only of eigenvalues μ_{ij} for $i = 1,2$ and $j = 0,1,\ldots$ with the corresponding eigenfunctions

$$\varphi_j(x)\cdot\xi_{ij} \text{ for } i = 1,2;\ j = 0,1,2,\ldots \text{ i.e. } O(P,\tilde{P})(\varphi_j\ \xi_{1j}) =$$

$$(F+C\Delta)(\varphi_j\xi_{1j}) = \mu_{ij}\ \varphi_i\ \xi_{1j}.$$

The eigenfunctions φ_j of $-\Delta u$ satisfy $-\Delta\varphi_j = \lambda_j\cdot\varphi_j$, where μ_{1j}, μ_{2j} are two eigenvalues of the 2×2 matrix $(\mathcal{F}-\lambda_j C)$. Since we have by construction

$$\text{trace } \mathcal{F}(\tilde{\lambda}_j) < 0 \text{ and det } \mathcal{F}(\tilde{\lambda}_j),$$

then one of $\mu_{1\tilde{j}}$ or $\mu_{2\tilde{j}}$ is zero, while the other is negative; we assume $\mu_{1\tilde{j}} = 0$.

We have, $\tilde{\lambda}_j$ is a simple eigenvalue of $-\Delta$ by assumption, then $\mu_{1\tilde{j}} = 0$ is a simple eigenvalue of $O(P,\tilde{P})$. This means that the kernel of $O(P,\tilde{P})$ is one dimensional.

III. 6.8 Bifurcation

As the eigenfunctions $\varphi_j\xi_{ij}$ for $i = 1,2$ and $j = 0,1,2,\ldots$ generate a complete system, one can show that the range of $O(P,\tilde{P})$ is one-dimensional, and we apply a result of Crandall and Rabinowitz (70); there will be bifurcation, if

$$\frac{\partial}{\partial s} O(P,\tilde{P})\ \varphi_{\tilde{j}}\xi_{1\tilde{j}} \notin \text{range } O(P,\tilde{P}).$$

As $\varphi_j\xi_{ij}$ are the eigenfunctions of $O(P,\tilde{P})$, we have

$$O(P,\tilde{P})\varphi_j\xi_{ij} = (\mathcal{F}+C\cdot\Delta)\cdot(\varphi_j\xi_{ij}),$$

and as C and $\mathcal{F}$ are fixed with respect to s, we obtain by differentiating

$$\frac{\partial}{\partial s} O(P,\tilde{P})(\varphi_j\xi_{ij}) = C\,\frac{\partial}{\partial s}\,\Delta\varphi_j\xi_{ij} = \Delta'\varphi_j C\cdot\xi_{ij}.$$

Let ψ_j be the eigenfunctions of $-\Delta$, then

$$-\Delta\varphi_j = \lambda_j\varphi_j \text{ and } -\frac{\partial}{\partial s}(\Delta\varphi_j) = \frac{\partial}{\partial s}(\lambda_j\varphi_j) = \lambda'_j\varphi_j + \lambda_j\cdot\varphi'_j;$$

on the other hand

$$\frac{\partial}{\partial s}\Delta\cdot\varphi_j = \Delta'\varphi_j + \Delta\varphi'_j,$$

so we have

$$\Delta'\varphi_j + \Delta\varphi'_j = -\lambda'_j\,\varphi_j - \lambda_j\varphi'_j.$$

thus

$$\begin{aligned}
\Delta'\varphi_j &= -\Delta\varphi'_j - \lambda_j\varphi'_j - \lambda'_j\varphi_j \\
&= -(\Delta+\lambda_j)\varphi'_j - \lambda'_j\varphi_j \\
&= -(\Delta+\lambda_j)\sum_k a_k\varphi_k - \lambda'_j\varphi_j, \quad \text{taking the expansion of the eigenfunction } \varphi'_j = \sum_k a_k\varphi_k, \\
&= \sum_{k\neq j}(\lambda_k-\lambda_j)a_k\varphi_k - \lambda'_j\varphi_j
\end{aligned}$$

Using this, we obtain

$$\frac{\partial}{\partial s} O(\varphi_j \xi_{1\tilde{j}}) = \Delta' \varphi_{\tilde{j}}(C \cdot \xi_{1\tilde{j}}) = \sum_{k \neq j} (\lambda_k - \lambda_{\tilde{j}}) a_k \varphi_k ((C \cdot \xi_{1\tilde{j}}) - \lambda'_{\tilde{j}} \varphi_j (C \xi_{1\tilde{j}})) \notin \text{range}\, O(P, \tilde{P})$$

Thus, the bifurcation to a new solution occurs as $\lambda_{\tilde{j}}$ crosses the graph of the det $\mathcal{F}(\lambda)$. The trace derived on page 223 satisfies

$$\text{trace} \quad \mathcal{F}(\tilde{u}, \tilde{v}) = \tilde{u}\psi(\tilde{u}) \cdot F'(\tilde{u}) - \tilde{v} g'(\tilde{v}) < 0.$$

On page 224 we assumed that $g'(\tilde{u}) > 0$ and if $F'(\tilde{u}) \geq 0$, we obtain

$$\tilde{v} g'(\tilde{v}) > \tilde{u}\psi(\tilde{u}) \cdot F'(\tilde{u}) > 0.$$

Under these conditions we obtained on page 224 that det $\mathcal{F}(\lambda) < 0$. Let $(\tilde{u}, \tilde{v})$ be an equilibrium point and let λ_i be a simple eigenvalue of $-\Delta$, where λ_i crosses the graph of

$$\det \mathcal{F}(\lambda) \text{ for } i \geq 1,$$

then a non constant equilibrium bifurcates from the equilibrium point $(\tilde{u}, \tilde{v})$. If the inequality

$$\tilde{v} g'(\tilde{v}) > \tilde{u}\psi(\tilde{u}) F'(\tilde{u}) > 0$$

is satisfied, and

$$\det \mathcal{F}(\lambda) = d_1 \cdot d_2 \lambda^2 + (d_1 \tilde{v} g'(\tilde{v}) - d_2 \tilde{u}\psi(\tilde{u}) \cdot F'(\tilde{u}))\lambda + \det \mathcal{F},$$

the bifurcation for any simple eigenvalue λ_i can be influenced by variation of the diffusion coefficients d_1 and d_2.

In general reaction-diffusion processes, the diffusion phenomena are acting as a destabilizing factor.

III.7 Symmetric reaction-diffusion systems

For now, we present only one model in two spatial dimensions of a system of nonlinear partial differential equations (further treatment will be given in a later work). We consider one possible application in morphogenesis (Thom (334)), describing a process of formation of microstructures. The shapes of the microstructures may be formed by two species u and v acting on each other according to the system

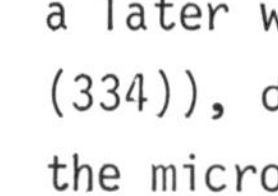

$$\frac{\partial u}{\partial t} + \Delta u = (\beta v^2 + \gamma(u^2+v^2-1))\cdot u$$

$$\frac{\partial v}{\partial t} + \Delta v = (\beta u^2 + \gamma(u^2+v^2-1))v.$$

The modelling of morphogenetic processes is based on the idea, that certain diffusion-reaction processes form patterns. The production of different patterns depends on the:

boundary conditions, initial conditions, fluctuations, ...
and various other parameters,

which can be treated by our modelling methods.

We construct periodic and nonperiodic, analytic solutions, describing the evolution of patterns.

The corresponding density distributions are given in explicit form and can be illustrated. We have chosen this approach, which would yield as a probability distribution of the patterns of the whole system.

Our methods of modelling complex processes can also be seen under the general view point of synergetics. In synergetics, profound analogies are considered between quite different systems such as biological, physical or chemical ones. It seems, that biological systems combine dissipative and non dissipative structures. Some of these processes and phenomena can be described by our modelling methods.

III. 7.1 Asymptotic method

The system of nonlinear parabolic differential equations

$$\frac{\partial u}{\partial t} = D\cdot\Delta u - \{\beta v^2 + \gamma(u^2+v^2-1)\}u$$

$$\frac{\partial v}{\partial t} = D\cdot\Delta v - \{\beta u^2+\gamma(u^2+v^2-1)\}v,$$

is symmetric with respect to u and v, thus we are able to construct analytic solutions by introducing the ansätze

$$u(x,y,t) = R(x,y,t)\cdot\sin\theta(x,y,t),$$

$$v(x,y,t) = R(x,y,t)\cdot\cos\theta(x,y,t).$$

Then we obtain finally

$$\frac{\partial R}{\partial t} = D\ \{\Delta R - R\cdot(\text{grad}\theta)^2\} - \frac{\beta\cdot R^3}{4}(1-\cos(4\theta)) - \gamma R(R^2-1)$$

$$R\cdot\frac{\partial\theta}{\partial t} = D\ \{2(\text{grad}R)(\text{grad}\theta) + R\cdot\Delta\theta\} - \frac{\beta\cdot R^4}{4}\cdot\sin(4\theta).$$

We take the asymptotic expansion of R and θ in series of $\varepsilon = \frac{\beta}{\gamma} << 1$, then we have $R_o = 1$, and thus

$$R = R_o + \varepsilon R_1 + {}^2R_2 + \ldots \text{ and } \theta = \theta_o + \varepsilon\theta_1 + \varepsilon^2\theta_2 + \ldots$$

The zero order relation satisfies

$$\frac{\partial\theta_o}{\partial t} = D\cdot\Delta\theta_o - \frac{\beta}{4}\sin(4\cdot\theta_o).$$

First we consider the time independent equation

$$\frac{\partial^2\theta_0}{\partial x^2} + \frac{\partial^2\theta_0}{\partial y^2} = \frac{1}{2}\sin\theta_o,$$

and obtain the solution

$$\theta(x,y) = \pm\ 4\ \text{arc tan}\ (\varphi(x)\cdot\varepsilon(y)),$$

where $\varphi(x)$, $\varepsilon(y)$ are linear combinations of the Jacobian elliptic function cn..
The solution of the time dependent problem

$$\frac{\partial\theta}{\partial t} - \Delta\theta = \frac{1}{4}\sin(4\theta)$$

is given in closed, analytic form

$$\theta(x,y,t) = \pm\ \text{arc tan}\ (\varphi(x)\cdot\varepsilon(y))\text{-constant}\ \Sigma(x,y)\cdot e^{-k\cdot\beta t},$$

where

$$\Sigma(x,y) = \frac{s(x)+s(y)}{1+s(x)\cdot s(y)} \text{ and } s(x) = k\cdot\text{sn}\ (\frac{x}{\sqrt{1+k^2}},k)$$

the Jacobian elliptic function.
Introducing the density distribution

$$\rho_n(x,y) = \frac{1}{T}\int_0^T \theta(x,y,t)\cdot\theta(x,y,T-t)dt,$$

we obtain closed, analytic expressions in compact form.
The expression

$$\frac{1}{2}\cdot(1 - \Sigma(x,y))$$

can be given explicitly as a surface over the (x,y)-plane: in the perspective

for the parameter $k^2 = 0.7$

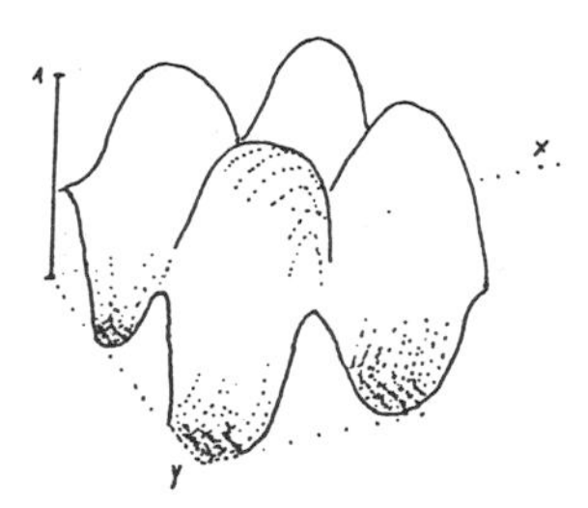

and the projection of the surface gives the contour-lines

then + and - refer to the domain where the value of the surface assumes its maximum (= + 1), and its minimum (= 0).

III. 7.2 Analytic solution

We are looking for a symmetric, periodic solution of

$$\frac{\partial^2\theta}{\partial x^2} + \frac{\partial^2\theta}{\partial y^2} = \frac{1}{4}\sin\theta.$$

It is known by methods from the theory of solitons (131), (268) that in the one-dimensional case

$$\frac{\partial^2\varphi}{\partial x^2} = \sin\varphi$$

has a solution

$$\varphi = 4 \text{ arc tan } (f(x),$$

where f satisfies the partial differential equation

$$(1+f^2)\frac{\partial^2 f}{\partial x^2} - 2f\left(\frac{\partial f}{\partial x}\right)^2 - f(1-f^2) = 0.$$

Therefore, we try the ansatz

$$\theta(x,y) = 4\cdot\text{arc tan}(\varphi(x)\cdot\varepsilon(y)).$$

We obtain by differentiating

$$\frac{\partial\theta}{\partial x} = 4\,\frac{1}{1+(\varphi(x)\cdot\varepsilon(y))^2}\,\varphi'(x)\cdot\varepsilon(y),$$

and

$$\frac{\partial^2\theta}{\partial x^2} = 4(-1)(1+(\varphi(x)\cdot\varepsilon(y))^2)^{-2}2\cdot(\varphi(x)\cdot\varepsilon(y))\,\varphi'(x)\varepsilon(y)\cdot(\varphi'(x)\cdot\varepsilon(y)) +$$

$$4(1+(\varphi(x)\cdot\varepsilon(y))^2)^{-1}\varepsilon(y)\cdot\varphi''(x)$$

$$\frac{\partial\theta}{\partial y} = 4\,\frac{1}{1+(\varphi(x)\cdot\varepsilon(y))^2}\,\varepsilon'(y)\cdot\varphi(x)$$

$$\frac{\partial^2\theta}{\partial y^2} = 4(-1)(1+(\varphi(x)\cdot\varepsilon(y))^2)^{-2}2\cdot(\varphi(x)\cdot\varepsilon(y))\varphi(x)\varepsilon'(y)\cdot(\varepsilon'(y)\cdot\varphi(x)) +$$

$$4(1+(\varphi(x)\cdot\varepsilon(y))^2)^{-1}\varphi(x)\varepsilon''(y)$$

then

$$(1+(\varphi(x)\cdot\varepsilon(y))^2)^2\,\frac{\partial^2\theta}{\partial x^2} = -4\cdot 2\cdot\varphi(x)\cdot(\varphi'(x))^2\varepsilon^2(y)+4(1+(\varphi(x)\cdot\varepsilon(y))^2)\varphi''(x)\varepsilon(y$$

$$(1+(\varphi(x)\cdot\varepsilon(y))^2)^2\,\frac{\partial^2\theta}{\partial y^2} = +4\cdot 2\;\varepsilon(y)\cdot(\varepsilon'(y))^2\varphi^2(x)+4(1+(\varphi(x)\cdot\varepsilon(y))^2)\varepsilon''(y)\varphi(x$$

$$-(1+(\varphi(x)\cdot\varepsilon(y))^2)^2\,\sin\,(\theta(x,y)) = -(1+(\varphi(x)\cdot\varepsilon(y))^2)\varphi(x)\cdot\varepsilon(y).$$

Thus we obtain the equation system

$$-2\varphi(x)\cdot\varepsilon(y)[(\varphi'(x))^2\varepsilon^2(y) + (\varepsilon'(y))^2\varphi^2(x)]$$

$$+\ (1+(\varphi(x)\varepsilon(y))^2)\cdot(\varphi''(x)\cdot\varepsilon(y)+\varepsilon''(y)\cdot\varphi(x))-\varphi(x)\varepsilon(y)-\varphi^3(x)\varepsilon^3(y) = 0.$$

Remembering Zagrodzinski (362), we introduce the ordinary differential equation system

$$(X')^2 = AX^4 + \frac{1}{2}X^2 + A$$

$$(Y')^2 = AY^4 + \frac{1}{2}Y^2 + A$$

and show that the solution $X = \varphi(x)$, $Y = \varepsilon(y)$ satisfies our equation system. Therefore, we differentiate and obtain

$$2X'\,X'' = 4AX^3 + \frac{1}{2}\,2X,\quad 2Y'Y'' = 4AY^3 + \frac{1}{2}\,2Y,$$

then we use

$$X'' = 2AX^3 + \frac{1}{2}X \text{ and } Y'' = 2AY^3 + \frac{1}{2}Y$$

to show that

$$(1+X^2Y^2)\cdot(X''\cdot Y+Y''X)-2XY(X'^2\cdot Y^2+Y'^2X^2)-X\cdot Y-X^3Y^3 = 0$$

is satisfied, we have

$$(1+X^2Y^2)\cdot(X''\cdot Y+Y''X) = X''Y+YX''+Y^3X^2X'' + X^3Y^2\cdot Y'' =$$

$$= 2AX^3Y + \frac{1}{2}X\cdot Y + 2AY^3X + \frac{1}{2}Y\cdot X + 2AX^5Y^3\cdot\frac{1}{2}\cdot Y^3X^3+2AY^5X^3 + \frac{1}{2}X^3Y^3$$

$$= Y\cdot X + X^3Y^3 + 2AX^3\cdot Y + 2AY^3X + 2AX^5Y^3 + 2AY^5X^3.$$

$$\text{as } -2XY(X'^2Y^2+Y'^2X^2) = -2AX^3Y^3 - X^3Y^3-2AX\cdot Y^3-2AY^5X^3-X^3Y^3-2AX^3Y.$$

Now we take the ordinary differential equation

$$\left(\frac{d\varphi}{dx}\right)^2 = (\varphi')^2 = A\varphi^4 + (2A+1)\varphi^2 + A \neq 0$$

and try to solve it by methods of elliptic integrals. Let $A > 0$ be given, then

$$\frac{1}{\frac{\partial\varphi}{\partial x}} = \frac{1}{\sqrt{\left(\frac{\partial\varphi}{\partial x}\right)^2}} = \frac{1}{\sqrt{A\varphi^4+(2A+1)\varphi^2+A}},$$

integrating

$$\int \frac{1}{\sqrt{A\varphi^4+(2A+1)\varphi^2+A}}\, d\varphi = \int \frac{dx}{d\varphi}\cdot d\varphi = x + \text{constant}.$$

We consider the zeros of the expression

$$A\varphi^4 + (2A+1)\varphi^2 + A.$$

Setting $\varphi^2 = x$, we obtain the quadratic equation

$$Ax^2 + (2A+1)x + A = 0,$$

where the solution is given by

$$x_{1,2} = -\frac{1}{2}\frac{2A+1}{A} \pm \sqrt{\left(\frac{2A+1}{A}\right)^2-1} = \frac{1}{2A}(2A+1) \pm \sqrt{4A+1},$$

denoting $x_{1,2}$ by $x_{1,2} = -\mathfrak{A} + \mathfrak{B}$, then the equation

$$A\varphi^4 + (2A+1)\varphi^2 + A = 0$$

has the solution

$$\varphi_{1,2,3,4} = \pm\sqrt{-\mathbb{A} + \mathbb{B}}, \text{ explicitly } \begin{array}{l} \varphi_1 = \sqrt{-\mathbb{A}+\mathbb{B}} \\ \varphi_2 = -\sqrt{-\mathbb{A}+\mathbb{B}} \\ \varphi_3 = \sqrt{-\mathbb{A}-\mathbb{B}} \\ \varphi_4 = \sqrt{-\mathbb{A}-\mathbb{B}} \end{array} \quad \text{denoting} \quad \begin{array}{l} \varphi_1 = a \\ \varphi_2 = -a \\ \varphi_3 = b \\ \varphi_2 = -b. \end{array}$$

The expression above can be written

$$A\varphi^4 + (2A+1)\varphi^2 + A = (\varphi_1-\varphi)(\varphi_2-\varphi)(\varphi_3-\varphi)(\varphi_4-\varphi)$$
$$= (a-\varphi)(-a-\varphi)\cdot(b-\varphi)\cdot(-b-\varphi)$$
$$= (a-\varphi)(a+\varphi)(b-\varphi)(b+\varphi)$$
$$= (a^2-\varphi^2)\cdot(b^2-\varphi^2)$$
$$= (a^2 - \frac{a^2}{a^2}\cdot\varphi^2)\cdot(b^2 - \frac{b^2}{b^2}\cdot\varphi^2) = a^2b^2(1-(\frac{\varphi}{a})^2)\cdot(1-(\frac{\varphi}{b})^2).$$

Using this in the integral

$$\int \frac{d\varphi}{\sqrt{A\varphi^4+(2A+1)\varphi^2+A}} = \int \frac{d\varphi}{a.b\sqrt{(1-(\frac{\varphi}{a})^2)\cdot(1-(\frac{\varphi}{b})^2)}} =$$

by introducing the transformation

$$\frac{\varphi}{a} = \sin\psi, \text{ then } (\frac{\varphi}{a})^2 = \sin^2\psi \text{ and } \sqrt{(1-\sin^2\psi)} = \cos\psi, \text{ as } \frac{d\varphi}{d\psi} = a\cdot\cos\psi,$$

then

$$= \int \frac{a\cdot\cos\psi\, d\psi}{a.b\cdot\cos\psi\cdot\sqrt{(1-(\frac{a}{b})^2\sin^2\psi}} = \frac{1}{b}\int_0^\psi \frac{d\psi}{\sqrt{1-(\frac{a}{b})^2\sin^2\psi}} .$$

Now, the complete elliptic integral of the first kind with modulus k

$$F(\psi,k) = \int_0^\psi \frac{d\psi}{\sqrt{1-k^2\cdot\sin^2\psi}} \quad \text{with } k^2 = (\frac{a}{b})^2.$$

The Jacobian elliptic function is defined by

$$\frac{\varphi}{a} = \text{sn } F(\psi,k).$$

As arc sin $(\frac{\varphi}{a}) = \psi$ = the amplitude of $F(\psi,k)$, then

$$\frac{\varphi}{a} = \sin \text{arc} \sin \frac{\varphi}{a} = \sin \psi = \sin(\text{amplitude of } F),$$

this is the "sinus amplitudinis" of F. Thus

$$\varphi(x) = \frac{a}{b} \text{ sn } F(\psi,k) = k\cdot\text{sn } F(\psi,k) = k\cdot\text{sn } (\frac{1}{\sqrt{1+k^2}} (x-x_o),k)$$

and $\varepsilon(y)$ can be derived as

$$\varepsilon(y) = k\cdot\text{sn } (\frac{1}{\sqrt{1+k^2}} (y-y_o),k)$$

It can be shown (1), (89) that

$$\text{sn}^2(\frac{t}{2}) = \frac{1-\text{cn}t}{1+\text{dn}t} \quad \text{and } \text{cn}(t,k) = \text{dn}(k\cdot t, \frac{1}{k}),$$

where cn, sn, dn are the Jacobian elliptic functions, which can be expressed by Taylor expansions:

$$\text{sn}(t,k) = t - (1+k^2) \frac{t^3}{3!} + (1 + 14k^2 + k^4) \frac{t^5}{5!} + \ldots$$

$$\text{cn}(t,k) = 1 - \frac{t^2}{2!} + (1+4k^2) \frac{t^4}{4!} - (1 + 44k^2 + 16k^4) \frac{t^6}{6!} + \ldots$$

$$\text{dn}(t,k) = 1 - k^2 \frac{t^2}{2!} + k^2(4 + k^2)\frac{t^4}{4!} - k^2(16+44k^2+k^4)\frac{t^6}{6!} + \ldots$$

Thus, we have

$$\text{sn } \frac{t}{2} = \sqrt{\frac{1-\text{cn}t}{1+\text{dn}t}} \, .$$

Applying the Landen's transformation, given by

$$F(\psi,k) = \frac{2}{1+k'} F(\frac{1-k'}{1+k'}),$$

where $k' = \sqrt{1-k^2}$ is the complementary modulus, to

$$\varphi(x) = \sqrt{k}\cdot\text{sn}(\frac{x-x_o}{2\cdot\sqrt{1+k^2}} ,k),$$

we obtain

$$\varphi(x) = \left(\frac{1-\mathrm{cn}\left(\frac{1}{\sqrt{1+\tilde{k}^2}}\,(x-x_o),\tilde{k}\right)}{1+\mathrm{cn}\left(\frac{1}{\sqrt{1+\tilde{k}^2}}\,(x-x_o)\tilde{k}\right)} \right)^{\frac{1}{2}}, \text{ where } \tilde{k} = \frac{1-k}{1+k} \quad ,$$

and quite analogously

$$\varepsilon(y) = \left(\frac{1-\mathrm{cn}\left(\frac{1}{\sqrt{1+\tilde{k}^2}}\,(y-y_o),\tilde{k}\right)}{1+\mathrm{cn}\left(\frac{1}{\sqrt{1+\tilde{k}^2}}\,(y-y_o),\tilde{k}\right)} \right)^{\frac{1}{2}}$$

III. 7.2.1 Time-independent solution

We have derived that the equation

$$\frac{\partial^2\theta}{\partial x^2} + \frac{\partial^2\theta}{\partial y^2} = \frac{1}{4}\cdot\sin\theta$$

has a solution

$$\theta(x,y) = 4\cdot\mathrm{arc\ tan}\ (\varphi(x)\cdot\varepsilon(y)),$$

with $\varphi(x)$, $\varepsilon(y)$ given in the closed form.

The ordinary differential equation

$$(\varphi')^2 = A\varphi^4 + (2A+1)\varphi^2 + A$$

reduces in the case $A = 0$ to $(\varphi')^2 = \varphi^2$, and the integral representation follows

$$\pm x = \int \frac{\partial x}{\partial\varphi}\,d\varphi = \int \frac{d\varphi}{\sqrt{(\frac{d\varphi}{dx})^2}} = \int \frac{d\varphi}{\sqrt{\varphi^2}} = \int \frac{d\varphi}{\varphi} = \ln\varphi,$$

now the solution is given by

$$\varphi = \exp\ (\pm\ x).$$

Remembering the identity

$$\frac{1-\tanh(x)}{1+\tanh(x)} = e^{-2x}, \text{ as } \tanh(x) = \frac{\sinh(x)}{\cosh(x)} = \frac{e^x-e^{-x}}{e^x+e^{-x}} ,$$

then

$$\varphi(x) = \exp(-x) = (\frac{1-\tanh(x)}{1+\tanh(x)})^{\frac{1}{2}},$$

this may be regarded as the limit, when k tends to one. It is interesting to compare the structure of the solutions. One can show analogously, that for $A < 0$

$$\varphi(x) = \left(\frac{1-\tilde{k}\mathrm{cn}(\frac{1}{\sqrt{1+\tilde{k}^2}}(x-x_0),\tilde{k})}{1+\tilde{k}\mathrm{cn}(\frac{1}{\sqrt{1+\tilde{k}^2}}(x-x_0),\tilde{k})} \right)^{\frac{1}{2}} .$$

III. 7.2.2 Time-dependent solution

The time dependent model is governed by the nonlinear parabolic partial differential equation

$$\frac{\partial\theta}{\partial t} - \Delta\theta = \frac{1}{4}\cdot\sin(4\cdot\theta).$$

We construct a two-dimensional solution

$$\theta(x,y,t) = \pm \text{ arc tan } (\varphi(x)\cdot\varepsilon(y)) + \text{constant}\cdot z(x,y)\cdot e^{-\lambda t}$$

for large t, where

$$\varphi(x) = \left(\frac{1-s(x)}{1+s(x)}\right)^{\frac{1}{2}} \text{ and } \varepsilon(y) = \left(\frac{1-s(y)}{1+s(y)}\right)^{\frac{1}{2}},$$

with the Jacobian elliptic functions

$$s(x) = s(x,k) = k\cdot\mathrm{sn}\left(\frac{x}{\sqrt{1+k^2}},k\right) \text{ and } s(y) = s(y,k) = k\cdot\mathrm{sn}\left(\frac{y}{\sqrt{1+k^2}},k\right);$$

and $z(x,y)$ satisfies

$$\frac{d^2z}{dx^2} + \frac{d^2z}{dy^2} + \{\lambda + 1-2\cdot\Sigma(x,y)\}\, z = 0,$$

where

$$\Sigma(x,y) = \frac{s(x) + s(y)}{1 + s(x)\cdot s(y)} .$$

The solution can be given in closed form (Grusa (141)).
We refer to page 236, that in the case $k = 1$, $s(x)$ can be written as $s(x) = \tan h\ (\frac{x}{\sqrt{2}})$, then

$$\Sigma(x,y) = \frac{s(x)+s(y)}{1+s(x)\cdot s(y)} = \frac{\tan h(\frac{x}{\sqrt{2}}) + \tanh(\frac{y}{\sqrt{2}})}{1+\tanh(\frac{x}{\sqrt{2}})\cdot\tanh(\frac{y}{\sqrt{2}})} = \tanh(\frac{x+y}{\sqrt{2}})$$

and $\Sigma(.)$ satisfies the partial differential equation.

If we take the above solution θ as an approximation for the case $k \approx 1$, we can write

$$\theta(x,y,t) = \pm \text{ arc tan } (\varphi(x)\cdot\varepsilon(y)) + c_3 \ \ \Sigma(x,y)e^{-k\cdot\beta t}$$

Studying density profiles of the two species A and B we introduce normalized density distributions

$$\rho_A(x,y) = \frac{1}{N}\int_0^N u(x,y,t)\cdot u(x,y,N-t)dt \qquad \text{and}$$

$$\rho_B(x,y) = \frac{1}{N}\int_0^N v(x,y,t)\cdot v(x,y,N-t)dt,$$

where u and v are explicit solutions of the system of nonlinear partial differential equations.

Applying the solution θ, we obtain

$$\rho(x,y) = \frac{1}{N}\int_0^N \theta(x,y,t)\cdot\theta(x,y,N-t)dt = \frac{1}{N}\int_0^N (\pm \text{ arc tan } (\varphi(x)\cdot\varepsilon(y) +$$

$$C_3 \ \Sigma(x,y)e^{-k\beta t})\cdot(\pm \text{ arc tan } (\varphi(x)\cdot\varepsilon(y)) + C_3 \ \Sigma(x,y)e^{-k\beta(N-t)})dt.$$

The integrand can be written as

$$(\pm \text{ arc tan}(\varphi(x)\cdot\varepsilon(y)))^2 + \text{arc tan } (\varphi\cdot\varepsilon)C_3 \ \Sigma(x,y)e^{-k\beta N}e^{k\beta t} +$$

$$\text{arc tan}(\phi\cdot\varepsilon) \ \Sigma(x,y)e^{-k\cdot\beta t} + C_3^2(\Sigma(x,y))^2 e^{-k\beta t} \ e^{k\beta t} \ e^{-\beta\cdot kN}.$$

As $\frac{1}{N}\int_0^N e^{k\beta t}dt = \frac{1}{k\cdot\beta N}(e^{k\beta N}-1)$ and $\frac{1}{N}\int_0^N e^{-k\beta t}dt = \frac{1}{k\beta N}\cdot(1-e^{-k\beta N})$

$$\frac{d}{dx}\text{ arc tan } (\varphi(x)\varepsilon(y)) = \frac{1}{2}(\Sigma(x,y) + 1)$$

we obtain approximately

$$\rho(x,y) \approx \frac{1}{2}(1-\Sigma(x,y)) + 2\,\frac{1}{N\beta k}(1-e^{-k\beta N})C_3\Sigma(x,y)\cdot(1-\Sigma^2(x,y))^{\frac{1}{2}}.$$

The functions $\frac{1}{2}(1-\Sigma(x,y))$ and $\Sigma(x,y)\ (1-\Sigma^2(x,y))^{\frac{1}{2}}$ can be given in explicit form by varying k from 0,5 to 1,5 (see page 231).

III. 8 A biological global model

We are interested in the interactive growth of two species in the bounded domains D_1 and D_2, where the species are influenced by linear diffusion processes in the 4-connected domain D_0 (93)-(96), (101), (102), (283)-(286), (149))

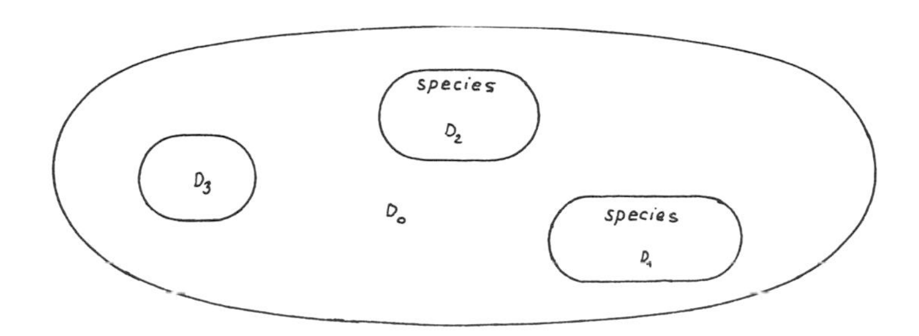

The interactive growth process of the species is governed by systems of nonlinear differential equations (see this chapter). Thus, we obtain that the whole process in the multiconnected domain can be described by

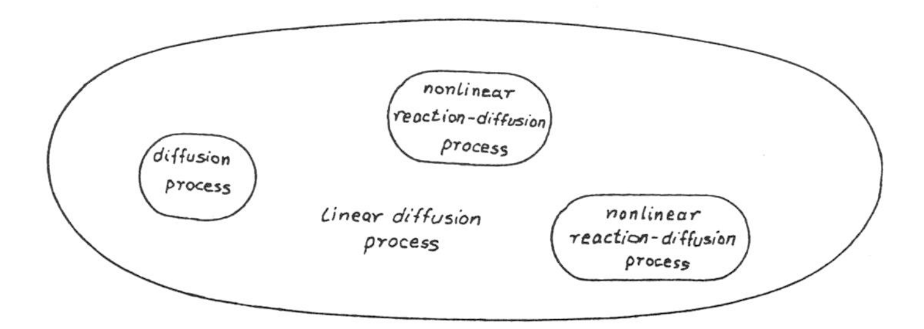

The interaction of the species may be modelled by competition phenomena (page 204) and predator-prey phenomena (page 201)

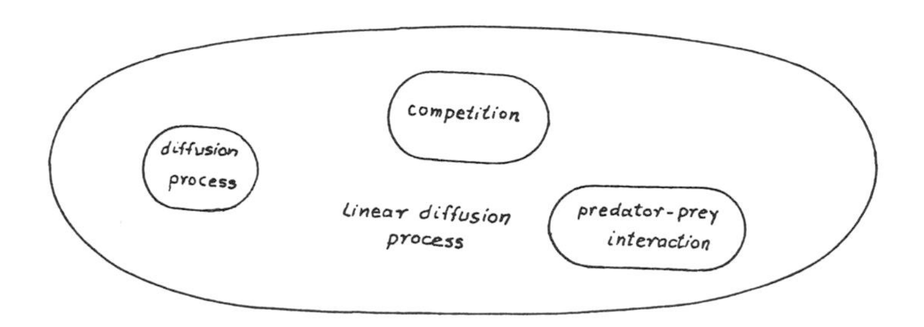

the specified biological global model reads

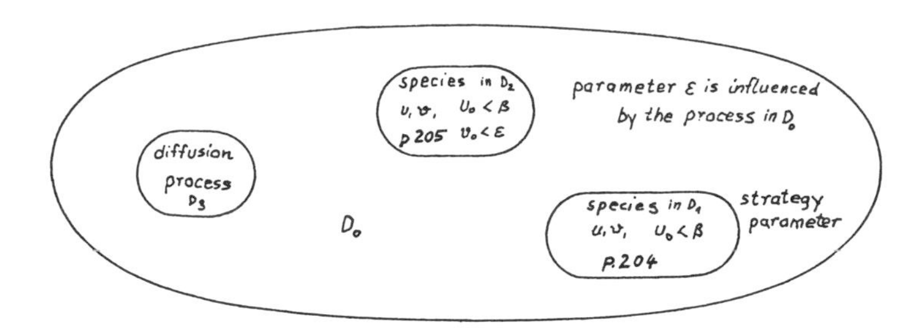

We consider first predator-prey phenomena in D_1. Let $u_0(x)$ be the initial density of the prey, but u_0 is too low, e.g. $u_0(x) < \beta$, where β can be interpreted as a strategy parameter derived (see page 148) from the controlled diffusion process in the multiconnected domain D_0.

The asymptotic model for $t \to \infty$ shows that the density of the predators v and the density of the prey u in D_1 become extinct.

We consider in the second case competition phenomena in D_2. Let $u_0(x)$ and $v_0(x)$ be the initial densities of the species u and v, such that u_0, v_0 are low, e.g.

$$u_0(x) < \beta \quad \text{and} \quad v_0(x) < \alpha .$$

Then the asymptotic model for $t \to \infty$ shows that the species u and v become extinct. If the initial density of u is low, e.g. $u_0(x) < \beta$, but the initial density of the species v is higher, e.g. $v_0(x) < \varepsilon$, then we find that v survives. Thus, we have a certain exclusion principle.

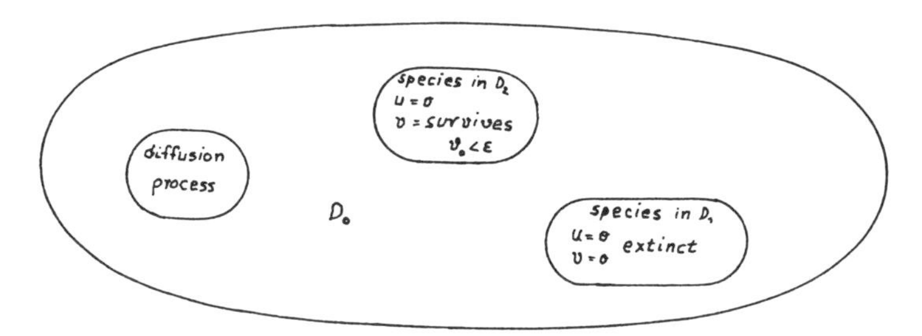

The parameter ε can be controlled by strategies derived from differential game theory (see page 119f), and the von Stackelberg conditions (see page 134f). The strategy parameter ε can be derived from the global solution (see page 148) of the multiconnected diffusion process in D_0. Thus, the exclusion principle in the domain D_2 is governed by the diffusion process in D_0. Further biological applications are left to the reader (especially in economics).

III. 9 Nonlinear model for a chemical reaction

Let y be the state of a concentration in a chemical reaction, which takes place at $x \in (0,1)$; let $i(x,t)$ be the concentration of a product which is slowing down the reaction. The model can be described by

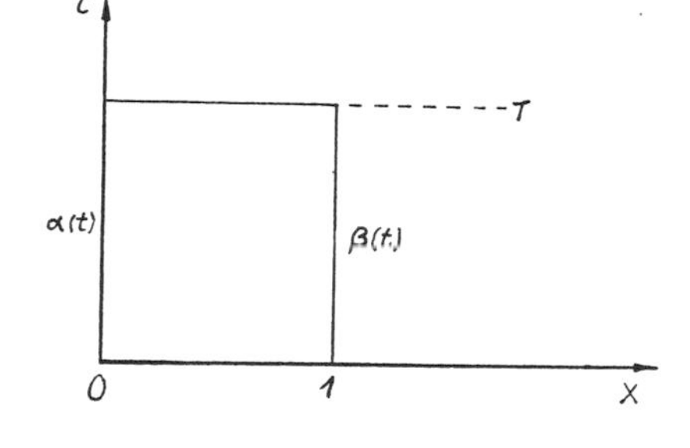

$$\frac{\partial y}{\partial t} - \frac{\partial^2 y}{\partial x^2} + F(y,i) = 0 \text{ in } 0 < x < 1$$
$$0 < t < T$$

$$\frac{\partial i}{\partial t} - \frac{\partial^2 i}{\partial x^2} = 0$$

with the boundary conditions

$$y(0,t) = \alpha(t), \; y(1,t) = \beta(t)$$

where α, β are positive, and the control variable

$$i(0,t) = v(t), \; i(1,t) = v(t),$$

where v satisfies enzymes reaction

$$0 \leq v(t) \leq M,$$

the initial condition satisfies

$$y(x,0) = 0, \; i(x,0) = 0.$$

We obtain the initial boundary value problem

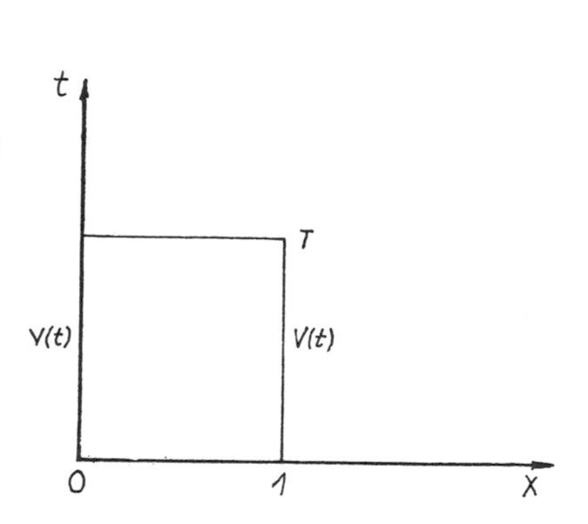

$$\frac{\partial i}{\partial t} - \frac{\partial^2 i}{\partial x^2} = 0$$

with

$$i(0,t) = v(t), \; i(1,t) = v(t)$$

on the boundary and $i(x,0) = 0$, then there exists a unique solution.
The analytic solution can be given in closed form

$$\sup_{(x,t)\in(0,1)\times(0,T)} \left| i(x,t) - \sum_{n}^{\text{finite}} c_n i_n(x,t) \right| < \varepsilon,$$

where $i_n(x,t)$ are the Hermitian polynomials

$$i_n(x,t) = n! \sum_{k=0}^{[\frac{n}{2}]} \frac{x^{n-2k}t^k}{(n-2k)!k!} ,$$

where c_n can be derived explicitly and the c_n are functions of the control variable v. Thus we obtain the solution $i = i(v)$. Now we have to solve the initial boundary value problem

$$\frac{\partial y}{\partial t} - \frac{\partial^2 y}{\partial x^2} + F(y,i(v)) = 0 \text{ in } (0,1) \times (0,T)$$

$$y(0,t) = \alpha(t),\ y(1,t) = \beta(t) \text{ on the boundary}$$

$$y(x,0) = 0.$$

For Hölder continuous right-hand side and continuous boundary values, there exists a solution of the linearized problem

$$\frac{\partial y}{\partial t} - \frac{\partial^2 y}{\partial x^2} + H(i(v))\cdot y = 0,$$

where the coefficient H does not depend on y, but H may be a function of x and t. The analytic solution can be given in closed form (141). Thus the solution can be written as

$$y = y(x,\alpha(t),\beta(t),i(v)).$$

Now we observe the process at $x = \frac{1}{2}$, hence, we obtain the solution

$$y(\tfrac{1}{2},t,v)$$

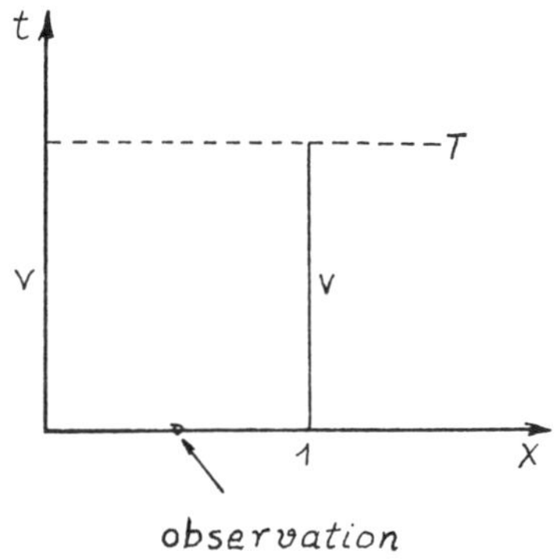

and we can introduce the cost function

$$J(v) = \int_0^T |y(\tfrac{1}{2},t,v)-z_d(t)|^2 dt,$$

where $z_d(t)$ can be seen as the time-dependent fixed cost. Let

$$U_{cc} = \{v | 0 \leq v(t) \leq M\},$$

then we consider the optimal control problem

$$\inf_{v \in U_{cc}} J(v).$$

We prove that there exists a solution that is not necessarily unique. As the control vari able v spans U_{cc} we can show that the solution of

$$\frac{\partial i}{\partial t} - \Delta i = 0$$

$$i(0,t) = v = i(1,t)$$

$$i(x,0) = 0$$

remains in a relative compact subset of $L^2(D \times (0,T))$.

Proof:

Defining the initial boundary value problem by

$$-\frac{\partial \psi}{\partial t} - \frac{\partial^2 \psi}{\partial x^2} = \phi$$

$$\psi(0,t) = 0 = \psi(q,t)$$

$$\psi(x,T) = 0,$$

if $\phi \in L^2(D \times (0,T))$ then there exists a solution. Multiplying the differential equation by ψ we obtain

$$\int_0^1 \int_0^T \frac{\partial i}{\partial t} \psi \, dxdt - \int_0^1 \int_0^T \Delta i \cdot \psi dxdt = 0.$$

Then

$$\int_0^T \frac{\partial i}{\partial t} \psi(x,t) dt = i(x,T) \cdot \psi(x,T) - i(x,0) \cdot \psi(x,0) + \int_0^T \left(\frac{\partial^2 \psi}{\partial x^2} + \phi\right) i \, dt,$$

as $\psi(x,T) = 0$, $i(x,0) = 0$ we obtain

$$\int_0^1 \int_0^T \frac{\partial i}{\partial t} \psi(x,t) dtdx = \iint \left(\frac{\partial^2 \psi}{\partial x^2} + \phi\right) \cdot i \, dxdt.$$

Integration by parts yields

$$\int_0^1 \frac{\partial^2 \psi}{\partial x^2} \cdot i dx = \frac{\partial \psi}{\partial x}(1,t) \cdot i(1,t) - \frac{\partial \psi}{\partial x}(0,t) \cdot i(0,t) - \int_0^1 \frac{\partial \psi}{\partial x} \cdot \frac{\partial i}{\partial x} \, dx$$

$$= \left(\frac{\partial \psi}{\partial x}(1,t) - \frac{\partial \psi}{\partial x}(0,t)\right) v(t) - \int_0^1 \frac{\partial \psi}{\partial x} \cdot \frac{\partial i}{\partial x} \, dx$$

and

$$\int_0^1 \frac{\partial\psi}{\partial x}\frac{\partial i}{\partial x}\,dx = \psi(1,t)\cdot i(1,t) - \psi(0,t)\cdot i(0,t) - \int_0^1 \psi\cdot \frac{\partial^2 i}{\partial x^2}\,dx =$$

$$- \int_0^1 \psi\cdot \frac{\partial^2 i}{\partial x^2}\,dx, \text{ setting } D = (0,1),$$

thus

$$\int_{Dx(0,T)} \frac{\partial^2\psi}{\partial x^2}\, i(x,t)dxdt = \int_0^T v(t)\cdot(\frac{\partial\psi}{\partial x}(1,t) - \frac{\partial\psi}{\partial x}(0,t))dt$$

$$+ \int_{Dx(0,T)} \psi\cdot \frac{\partial^2 i}{\partial x^2}\,dx.$$

Using

$$0 = \int_{Dx(0,T)} \frac{\partial i}{\partial t}\,\psi dxdt - \int_{Dx(0,T)} \Delta i\cdot\psi dxdt,$$

and

$$\int_{Dx(0,T)} \frac{\partial i}{\partial t}\,\psi dxdt = \int_{Dx(0,T)} (\frac{\partial^2\psi}{\partial x^2} + \phi) i\, dxdt,$$

$$\int_{Dx(0,T)} \frac{\partial^2\psi}{\partial x^2}\, i(x,t)dxdt = \int_0^T v(t)(\frac{\partial\psi}{\partial x}(1,t) - \frac{\partial\psi}{\partial x}(0,t))dt + \int_{Dx(0,T)} \psi\cdot\frac{\partial^2 i}{\partial x^2}dx,$$

we obtain

$$\int_{Dx(0,T)} \phi(x,t)i(x,t)dxdt = - \int_0^T (\frac{\partial\psi}{\partial x}\,(1,t) - \frac{\partial\psi}{\partial x}\,(0,t))v(t)dt,$$

as $v \in U_{cc}$, we have for $v(t) \leqq M$ the following estimation

$$|\int_0^T \frac{\partial\psi}{\partial x}\,(\cdot,t)\cdot v(t)dt| \leqq (\int_0^T |\frac{\partial\psi}{\partial x}(\cdot,t)|^2 dt)^{\frac{1}{2}}\cdot(\int_0^T |v(t)|^2 dt)^{\frac{1}{2}} < M\cdot K,$$

where $K = \text{const.}\ (\int_0^T |\frac{\partial\psi}{\partial x}(\cdot,t)|^2 dt)^{\frac{1}{2}}$.

As

$$\int_{Dx(0,T)} |\phi(x,t)\cdot i(x,t)|dxdt \leqq (\int_{Dx(0,T)} |\phi(x,t)|^2)^{\frac{1}{2}}\cdot(\int_{Dx(0,T)} |i(x,t)|^2)^{\frac{1}{2}}$$

and

$$- \int_{Dx(0,T)} \phi(x,t)\cdot i(x,t)dxdt = \int_0^T (\frac{\partial\psi}{\partial x}\,(1,t) - \frac{\partial\psi}{\partial x}(0,t))v(t) < M\cdot k'$$

we have

$$-(\int |\phi|^2)^{\frac{1}{2}}\cdot(\int |i|^2)^{\frac{1}{2}} \leqq - \int_{Dx(0,T)} \phi\cdot i \leqq M\cdot k',$$

thus i(v) remains in a relatively compact subset of $L^2(D \times (0,T))$. This is clear, as ϕ remains in a bounded set of $L^2(D \times (0,T))$, then there follows by regularity

$$- \frac{\partial\psi}{\partial t} - \frac{\partial^2\psi}{\partial x^2} = \phi$$

and ψ remains in a bounded subset of $L^2(0,T),H^2(D))$, $\int_0^T \|\psi\|_{H^2(Dx(0,T)}^2 dt < \infty$ and $\frac{\partial\psi}{\partial x}$ remains in a bounded set of $L^2((0,T),H^{-1}(D))$.

The solution y(v) of the initial boundary value problem

$$\frac{\partial y}{\partial t} - \frac{\partial^2 y}{\partial x^2} + F(y,i(v)) = 0 \text{ in } D \times (0,T)$$

$$y(0,t) = \alpha(t),\ y(1,t) = \beta(t)$$

$$y(x,0) = 0$$

remains in a bounded set of $L^2(0,T),H^1)$. Then $\frac{\partial y(v)}{\partial t}$ remains in a bounded set of $L^2(0,T),H^{-1})$. Applying the imbedding theorem

$$H^1(D) \subset H^2(D) \subset C(0,1),$$

then

$$y(\tfrac{1}{2},t,v) \in L^2(0,T),$$

and the cost function is defined

$$J(v) = \int_0^T |y(\tfrac{1}{2},t,v)-z_d(t)|^2 dt.$$

As

$$U_{cc} = \{v|0 \leq v(t) \leq M\}$$

is a bounded and closed set in $L^2(0,T)$, if v_n is a minimizing sequence, then there exists a subsequence, denoted again by v_n, such that

$$v_n \to u \text{ in } L^2(0,T) \text{ weakly.}$$

Taking v_n as a control-variable, then, as i(v) remains in a relatively compact subset of $L^2(D \times (0,T))$, there exists a subsequence

$$i(v_n) \underset{\text{def.}}{=} i_n \to i \text{ strongly in } L^2(D \times (0,T)).$$

Taking $i_n = i(v_n)$ in F(y,i(v)), then the solution $y(v) = y(i(v_n)) = y(v_n) \underset{\text{def.}}{=} y_n$

and $\frac{\partial y_n}{\partial t}$ remains in a bounded set of $L^2(0,T),H^1(D))$, $L^2(0,T),H^{-1}(D))$, then there exist subsequences

$$y_n \to y; \frac{\partial y_n}{\partial t} \to \frac{\partial y}{\partial t} \text{ in } L^2(0,T),H^1(D)); L^2((0,T), H^{-1}(D)) \text{ weakly,}$$

As $v_n \to u$, we have $i(v_n) \to i(u)$ and then $y(u)$. As $y_n \to y$ in $L^2(0,T), H^1(D))$

$$y_n(\tfrac{1}{2},t) \to y(\tfrac{1}{2},t) \text{ in } L^2(0,T) \text{ weakly,}$$

we obtain

$$y_n(\tfrac{1}{2},t,\cdot)-z_d(t) \to y(\tfrac{1}{2},t,\cdot)-z_d(t) \text{ in } L^2(0,T) \text{ weakly}$$

thus, we have

$$J(v_n) = \int_0^T |y(\tfrac{1}{2}\cdot t,\underset{\substack{\downarrow \\ u}}{v_n}) - z_d(t)|^2 dt$$

and finally

$$\liminf_{\substack{n\to\infty \\ v_n}} J(v_n) \geq J(u)$$

we have:

$$u \in U_{cc}, \text{ where } J(u) = \inf J(v) \text{ for } v \in U_{cc}.$$

Now we obtain a very interesting characterization of the optimal control u. Remembering Chapter II, we get

$$\frac{1}{2}(J'(u),\phi) = \frac{1}{2}\frac{d}{d\lambda}(J(u+\lambda\phi))\Big|_{\lambda=0} = \frac{1}{2}\frac{d}{d\lambda}\left(\int_0^T |y(\tfrac{1}{2},t,u+\lambda\phi)-z_d(t)|^2 dt\right)\Big|_{\lambda-0}$$

$$= \frac{1}{2}2\cdot\int_0^T |y(\tfrac{1}{2},t,u)-z_d(t)|\cdot\frac{\partial}{\partial\lambda} y(\tfrac{1}{2},t,u+\lambda\phi)\Big|_{\lambda=0} dt,$$

for $u, \phi \in U_{cc}$ $\lambda \to y(\frac{1}{2},t,u+\lambda\phi)$ and $\lambda \to i(\frac{1}{2},t,u+\lambda\phi)$ are differentiable, defining

$$\frac{d}{d\lambda} y(\tfrac{1}{2},t,u+\lambda\phi) = \tilde{y} \text{ and } \frac{d}{d\lambda} i(\tfrac{1}{2},t,u+\lambda\phi) = \tilde{i},$$

then we obtain

$$\frac{1}{2}(J'(u),v) = \int_0^T |y(\tfrac{1}{2},t,u)-z_d(t)|\cdot\tilde{y}(\tfrac{1}{2},t)dt.$$

Applying the differential operator $\frac{\partial}{\partial\lambda}$ to

$$\frac{\partial y}{\partial t} - \frac{\partial^2 y}{\partial x^2} + F(y,i) = 0,$$

we obtain

$$\frac{\partial}{\partial t}\frac{\partial y}{\partial \lambda} + \frac{\partial^2}{\partial x^2}\frac{\partial y}{\partial \lambda} + \frac{\partial F}{\partial \lambda} \cdot \frac{\partial y}{\partial \lambda} + \frac{\partial F}{\partial i}\frac{\partial i}{\partial \lambda} = 0,$$

hence

$$\frac{\partial \tilde{y}}{\partial t} - \frac{\partial^2 \tilde{y}}{\partial x^2} + \frac{\partial}{\partial y} F(y,i)\cdot\tilde{y} + \frac{\partial}{\partial i} F(y,i)\tilde{i} = 0$$

$$\frac{\partial \tilde{i}}{\partial t} - \frac{\partial^2 \tilde{i}}{\partial x^2} = 0$$

with the boundary conditions

$$\tilde{y}(0,t) = 0,\ \tilde{y}(1,t) = 0 \text{ and } \tilde{i}(0,t) = \phi(t),\ \tilde{i}(1,t) = \phi(t)$$

and the initial conditions

$$\tilde{y}(x,0) = 0 \text{ and } \tilde{i}(x,0) = 0.$$

Introducing the corresponding adjoint state of the system

$$-\frac{\partial p}{\partial t} - \frac{\partial^2 p}{\partial x^2} + \frac{\partial}{\partial y} F(y,i)p = y(\tfrac{1}{2},t,u)-z_d(t)\delta(x-\tfrac{1}{2}) \quad \text{and}$$

$$-\frac{\partial q}{\partial t} - \frac{\partial^2 q}{\partial y^2} + \frac{\partial}{\partial i} F(y,i)\cdot q = 0$$

$$p(0,t) = 0,\ p(1,t) = 0 \qquad\qquad q(0,t) = 0,\ q(1,t) = 0$$

$$p(x,t) = 0 \qquad\qquad q(x,T) = 0.$$

Multiplying the equation with respect to $\tilde{y}(x,t)$ and $\tilde{i}(x,t)$, then

$$\frac{\partial p}{\partial t}\tilde{y}(x,t) - \frac{\partial^2 p}{dx^2}\tilde{y}(x,t) + \frac{\partial}{\partial y} F(y,i)\cdot p\cdot\tilde{y}(x,t) = (y(\tfrac{1}{2},t,u)-z_d)\cdot\tilde{y}(x,t)\cdot\delta(x-\tfrac{1}{2})$$

and $$-\frac{\partial q}{\partial t}\tilde{i}(x,t) - \frac{\partial^2 q}{\partial x^2}\tilde{i}(x,t) + \frac{\partial F}{\partial i} q\, \tilde{i}(x,t) = 0$$

and by integration over $D \times (0,T)$, we obtain

$$\int_0^T \frac{\partial p}{\partial t}\tilde{y}(x,t)dt = p(x,T)\tilde{y}(x,T)-p(x,0)\tilde{y}(x,0) - \int_0^T p\cdot\frac{\partial \tilde{y}}{\partial t}\,dt = -\int_0^T p\cdot\frac{\partial \tilde{y}}{\partial t}\,dt,$$

remembering the boundary conditions, we have

$$\int_0^1 \frac{\partial^2 p}{\partial x^2} \tilde{y}(x,t)dx = \frac{\partial p}{\partial x}(1,t)\cdot\tilde{y}(1,t) - \frac{\partial p}{\partial x}(0,t)\cdot\tilde{y}(0,t) -$$

$$- \int_0^1 \frac{\partial p}{\partial x} \cdot \frac{\partial \tilde{y}}{\partial x}(x,t)dx = - \int_0^1 \frac{\partial p}{\partial x} \cdot \frac{\partial \tilde{y}}{\partial x} dx$$

and

$$\int_0^1 \frac{\partial p}{\partial x}\cdot \frac{\partial}{\partial x} \tilde{y}(x,t)dx = p(1,t)\cdot\frac{\partial}{\partial x}\tilde{y}(1,t)-p(0,t)\frac{\partial}{\partial x} \tilde{y}(0,t) -$$

$$- \int_0^1 p\cdot\frac{\partial^2 \tilde{y}}{\partial x^2} dx = - \int_0^1 p\cdot\frac{\partial^2 \tilde{y}}{\partial x^2} dx.$$

Thus we obtain

$$\int_0^1 \frac{\partial^2 p}{\partial x^2}\cdot\tilde{y}(x,t)dx = \int_0^1 p\cdot\frac{\partial^2 \tilde{y}}{\partial x^2} (x,t)dx,$$

and by definition

$$\int_D \tilde{y}(x,t)\delta(x-\tfrac{1}{2})dx = \tilde{y}(\tfrac{1}{2},t),$$

then

$$\int_{Dx(0,T)} (y(\tfrac{1}{2},t,u)-z_d(t))\cdot\tilde{y}(\tfrac{1}{2},t)dxdt = \int_0^T (y(\tfrac{1}{2},t,u)-z_d(t))\tilde{y}(\tfrac{1}{2},t)dt =$$

$$= \tfrac{1}{2}(J'(u),\phi)$$

by construction.

Multiplying the adjoint state with respect to $\tilde{y}(x,t)$ and integrating, we obtain

$$- \int_{Dx(0,T)} \frac{\partial p}{\partial t} \tilde{y}(x,t)dtdx - \int_{Dx(0,T)} \frac{\partial^2 p}{\partial x^2} \tilde{y}(x,t)dxdt \quad +$$

$$+ \int_{Dx(0,T)} \frac{\partial F}{\partial y} p\cdot\tilde{y}(x,t)dxdt = \int_{Dx(0,T)} (y(\tfrac{1}{2},t,u)-z_d(t))\tilde{y}(x,t)\cdot\delta(x-\tfrac{1}{2}) dxdt$$

applying the above considerations

$$\int_{Dx(0,T)} p\cdot\frac{\partial \tilde{y}}{\partial t} dtdx - \int_{Dx(0,T)} p\cdot\frac{\partial^2 \tilde{y}}{\partial x^2}(x,t)dxdt \quad +$$

$$+ \int_{Dx(0,T)} \frac{\partial F}{\partial y} p\cdot\tilde{y}(x,t)dxdt = \tfrac{1}{2}(J'(u),\phi),$$

we obtain finally

$$\int_{D\times(0,T)} p\cdot\left(\frac{\partial\tilde{y}}{\partial t} - \frac{\partial^2 y}{\partial x^2} + \frac{\partial}{\partial y} F(y,i)\cdot\tilde{y}(x,t)\right)dxdt = \frac{1}{2}(J'(u),v).$$

Analogously we obtain for the second adjoint state equation

$$\int_0^T \frac{\partial q}{\partial t}\,\tilde{i}(x,t)dt = q(x,T)\cdot\tilde{i}(x,T)-q(x,0)\cdot\tilde{i}(x,0) -$$

$$\int_0^T q\cdot\frac{\partial}{\partial t}\,\tilde{i}(x,t)dt = -\int_0^T q\cdot\frac{\partial}{\partial t}\,\tilde{i}(x,t)dt,$$

with the boundary conditions

$$\int_0^1 \frac{\partial^2 q}{\partial x^2}\,\tilde{i}(x,t)dx = \left(\frac{\partial q}{\partial x}(1,t) - \frac{\partial q}{\partial x}(0,T)\right)\phi(t) - \int_0^1 \frac{\partial q}{\partial x}\cdot\frac{\partial\tilde{i}}{\partial x}\,dx$$

and

$$\int_0^1 \frac{\partial q}{\partial x}\cdot\frac{\partial\tilde{i}}{\partial x}\,dx = q(x,t)\cdot\frac{\partial\tilde{i}}{\partial x}(x,t)\Big|_{x=0}^{x=1} - \int q\,\frac{\partial^2\tilde{i}}{\partial x^2}\,dx,$$

thus

$$\int_{D\times(0,T)} \frac{\partial^2 q}{\partial x^2}\,\tilde{i}(x,t)dxdt = \int_0^T \phi(t)\cdot\left(\frac{\partial q}{\partial x}(1,t)- \frac{\partial q}{\partial x}(0,t)\right)dt$$

$$+ \int_{D\times(0,T)} q\cdot\frac{\partial^2\tilde{i}}{\partial x^2}\,dx$$

and finally

$$\int_{D\times(0,T)} \frac{\partial q}{\partial t}\,\tilde{i}(x,t)dtdx = -\int_{D\times(0,T)} q\,\frac{\partial\tilde{i}}{\partial t}(x,t)dxdt.$$

Using the adjoint system, multiplying by $\tilde{i}(x,t)$, then

$$-\frac{\partial q}{\partial t}\,\tilde{i}(x,t) - \frac{\partial^2 q}{\partial x^2}\,\tilde{i}(x,t) + \frac{\partial F}{\partial i}\,q.\tilde{i}(x,t) = 0,$$

integrating

$$-\int_{D\times(0,T)} \frac{\partial q}{\partial t}\,\tilde{i}(x,t)dxdt - \int_{D\times(0,T)}\frac{\partial^2 q}{\partial x^2}\,\tilde{i}(x,t)dxdt +$$

$$+ \int_{D\times(0,T)} \frac{\partial F}{\partial i}\,q\cdot\tilde{i}(x,t)dxdt = 0.$$

Integration by parts yields

$$\int_{Dx(0,T)} q\cdot\frac{\partial \tilde{i}}{\partial t}(x,t)dxdt - \int_0^T \phi(t)\cdot(\frac{\partial q}{\partial x}(1,t) - \frac{\partial q}{\partial x}(0,t))dt -$$

$$- \int_{Dx(0,T)} q\cdot\frac{\partial^2 \tilde{i}}{\partial x^2}dxdt + \int_{Dx(0,T)} q\cdot\frac{\partial F}{\partial i}\tilde{i}(x,t)dxdt = 0,$$

then $\int_{Dx(0,T)} q(\frac{\partial \tilde{i}}{\partial t} - \frac{\partial^2 \tilde{i}}{\partial x^2})dxdt + \int_0^T \phi(t)\cdot(\frac{\partial q}{\partial x}(0,t) - \frac{\partial q}{\partial x}(1,t)dt +$

$$+ \int_{Dx(0,T)} q\cdot\frac{\partial F}{\partial i}\tilde{i}(x,t)dxdt = 0$$

and finally we obtain

$$\frac{1}{2}(J'(u),\phi) = \int_0^T \phi(t)(\frac{\partial q}{\partial x}(0,t) - \frac{\partial q}{\partial x}(1,t))dt + \int_{Dx(0,T)} q\cdot(\frac{\partial \tilde{i}}{\partial t} - \frac{\partial^2 \tilde{i}}{\partial x^2})dxdt$$

$$+ \int_0^T p(\frac{\partial \tilde{y}}{\partial t} - \frac{\partial^2 \tilde{y}}{\partial x^2} + \frac{\partial F}{\partial y}(y,i)\cdot\tilde{y} + \frac{\partial F}{\partial i}(y,i)\cdot\tilde{i})dxdt.$$

According to the construction of $\tilde{i}$ and $\tilde{y}$, we obtain

$$\frac{1}{2}(J'(u),\phi) = \int_0^T \phi(t)\cdot(\frac{\partial q}{\partial x}(0,t) - \frac{\partial q}{\partial x}(1,t))dt.$$

The optimality system is given by

$$\frac{\partial y}{\partial t} - \frac{\partial^2 y}{\partial x^2} + F(y,i(u)) = 0 \text{ in } D \times (0,T) \text{ and } \frac{\partial i}{\partial t} - \frac{\partial^2 i}{\partial x^2} = 0 \text{ in } D \times (0,T)$$

$$y(0,t) = \alpha(t),\ y(1,t) = \beta(t) \qquad i(0,t) = u,\ i(1,t) = u$$

$$y(x,0) = 0 \qquad i(x,0) = 0,$$

and the adjoint equations can be written as

$$-\frac{\partial p}{\partial t} - \frac{\partial^2 p}{\partial x^2} + \frac{\partial F}{\partial y}(y,i)p = (y(\tfrac{1}{2},t,u) - z_d(t))\delta(x - \tfrac{1}{2}) \text{ and}$$

$$-\frac{\partial q}{\partial t} - \frac{\partial^2 q}{\partial x^2} + \frac{\partial F}{\partial i}(y,i)p = 0 \text{ in } D \times (0,T)$$

$$p(0,t) = 0,\ p(1,t) = 0 \qquad q(0,t) = 0,\ q(1,t) = 0$$

$$p(x,T) = 0, \qquad q(x,T) = 0;$$

and the integral inequality reads

$$\int_0^T \left(\frac{\partial q}{\partial x}(0,t) - \frac{\partial q}{\partial x}(1,t)\right)(v-u)dt = \frac{1}{2}\,(J'(u),\phi) \geqq 0$$

for $v \in U_{cc}$ and $u \in U_{cc}$, where $\phi = v-u$. In the unconstrained case, we obtain

$$\left(\frac{\partial q}{\partial x}(0,t) - \frac{\partial q}{\partial x}(1,t)\right)\cdot(v-u) \geqq 0, \text{ as } v \in U_{cc}, \text{ then } v \geqq 0,$$

this can be written

$$\left(\frac{\partial q}{\partial x}(0,t) - \frac{\partial q}{\partial x}(1,t)\right)v \geqq \left(\frac{\partial q}{\partial x}(0,t) - \frac{\partial q}{\partial x}(1,t)\right)u \geqq 0$$

as

$$\left(\frac{\partial q}{\partial x}(0,t) - \frac{\partial q}{\partial x}(1,t)\right) > 0$$

and by setting

$$\left(\frac{\partial q}{\partial x}(0,t) - \frac{\partial q}{\partial x}\,(1,t)\right)u(t) = 1,$$

we obtain one solution of the optimal control

$$u(t) = \frac{1}{\frac{\partial q}{\partial x}(0,t) - \frac{\partial q}{\partial x}\,(1,t)} \quad .$$

IV Nonlinear wave phenomena

We consider an inviscid incompressible fluid in a constant gravitational field. Denoting the space coordiantes by (x,y,z) and the components of the velocity vector u by (u_1,u_2,v). Let $u(x,t)$ be the velocity of a fluid particle at the position x at time t, and let $\rho(x,t)$ and $p(x,t)$ denote the density and pressure of the fluid particle at x at time t.

Let V be a volume element enclosed in a smooth surface S in the fluid, and the unit normal to the surface S is denoted by u.

The gravitational acceleration g is in the negative z direction, and the external force $F = -\rho\cdot g\cdot j$ acts on V, where j is the unit vector in the z direction.

The equation of conservation of mass is

$$\frac{d}{dt}\int_V \rho\, dxdydz = -\int_S \rho\cdot n_j\cdot u_j\cdot ds.$$

Assuming all quantities are sufficiently smooth, we can apply the divergence theorem

$$\int_V \mathrm{div}(\rho\cdot u_i)dV = \int_S \rho\cdot n_j u_j ds.$$

Using this theorem, we obtain

$$\int_V \frac{\partial q}{\partial t}\, dV + \int_V \frac{\partial}{\partial x_i}\,(\rho\cdot u_j)dV = 0,$$

where the summation is taken for repeated indices. This equation is true for all volume elements V, then

$$\frac{\partial \rho}{\partial t} + \frac{\partial}{\partial x_j}(\rho u_j) = 0$$

is the known continuity equation.

Let F be the body force per unit mass, then the conservation of the momentum in the i-th direction:

$$\frac{d}{dt}\int_V \rho\cdot u\, dV = \text{transport of momentum} - \text{surface force} - \text{body force}$$

$$= -\int_S \rho u_i u_j \cdot n_j ds - \int_S n_i p\, ds - \int_V F_i \cdot \rho\, dV$$

$$= -\int_S (\rho u_i u_j + p\cdot\delta_{ij}) n_j ds + \int_V F_i \cdot \rho dV$$

$$= - \int_V \{\frac{\partial}{\partial x_j}(\rho u_i u_j + p\delta_{ij}) + \rho F_i\} dV, \text{ by the divergence theorem.}$$

Since this is true for all volume elements Y, we finally have

$$\frac{\partial}{\partial t}(\rho\cdot u_i) + \frac{\partial}{\partial x_j}(\rho u_i u_j) + \frac{\partial p}{\partial x_i} = \rho F_i \ .$$

Considering water waves, we assume ρ = constant and by the continuity equation it follows,

$$\frac{\partial u_j}{\partial x_j} = 0, \text{ in vector notation, div } u = 0.$$

The momentum equation is given by

$$\frac{\partial u_i}{\partial t} + u_j \frac{\partial u_i}{\partial x_j} = -\frac{1}{\rho}\cdot\frac{\partial p}{\partial x_i} + F_i$$

and in vector notation

$$\frac{\partial u}{\partial t} + (u\cdot \text{grad})u = -\frac{1}{\rho}\text{gradp} - g\cdot j.$$

We assume that curl u = 0, i.e. the flow is irrotational. Applying methods of vector analysis, u can be written as

$$u = \text{grad } \phi \ ,$$

where ϕ is some scalar field, called: velocity potential.

By assumption, ρ = constant, we derived div u = 0, then div grad ϕ = 0; this means, that

$$\Delta\phi = 0$$

the velocity potential is a solution of the Laplace equation.

IV. 1 Nonlinear waves: Breaking and run-up effects

IV. 1.1 Equation of the free surface

Now, the momentum equation

$$\frac{\partial u}{\partial t} + (u \cdot \text{grad})u + \text{grad}(\frac{p}{\rho}) + g \cdot j = 0$$

can be written as

$$\text{grad}(\frac{\partial \phi}{\partial t}) + \text{grad}(\frac{1}{2} u^2) + \text{grad}\ (\frac{p}{\rho}) + \text{grad}(g \cdot z) = 0$$

this means that

$$(\frac{\partial \phi}{\partial t} + \frac{1}{2}\ (\text{grad}\ \phi)^2 + \frac{p-p_o}{\rho} + g \cdot z)$$

is a function of t and can be set to be zero. We have finally

$$\frac{p-p_o}{\rho} = \frac{\partial \phi}{\partial t} + \frac{1}{2}\ (\text{grad}\ \phi)^2 + g \cdot z,$$

thus the pressure p can be expressed by the potential ϕ.

Taking $z = \eta(x_1, x_2, t)$ as the free surface of the fluid (water) and $p = p_o$, we obtain the boundary condition at the free surface

$$\frac{\partial \phi}{\partial t} + \frac{1}{2}\ (\text{grad}\ \phi)^2 + g \cdot \eta = 0 \text{ at } z = \eta(x_1, x_2, t).$$

We consider the behaviour of the free surface. Let the surface at time t be given by

$$f(x_1, x_2, x_3, t) = 0.$$

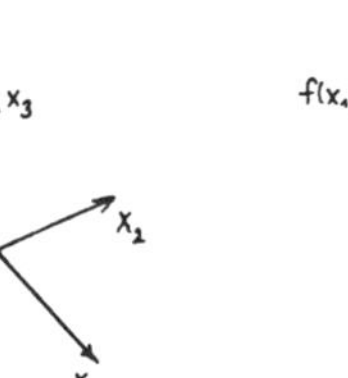

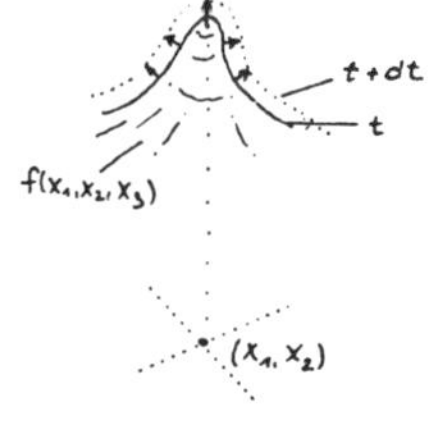

We consider positions at time t and t + dt of the surface. Let the points x on the first surface and x + n·ds on the second one be separated by a distance ds along the normal n, where the unit normal

$$n = \frac{\text{grad} f}{|\text{grad} f|}\ .$$

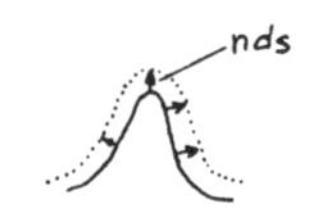

The first surface is written as $f(x,t) = 0$, the second surface

$$f(x + nds, t+dt) = 0$$

can be expressed by Taylor

$$f(x+nds, t+dt) = f(x,t) + (n\cdot \text{grad} f)ds + f_t dt + \ldots$$

written as differentials

$$(n\cdot \text{grad} f)ds = - f_t dt,$$

and finally

$$\frac{ds}{dt} = - \frac{f_t}{|\text{grad } f|} .$$

The velocity of the surface in the direction of n is given by

$$\frac{ds}{dt} = - \frac{f_t}{|\text{grad} f|} .$$

The velocity u of the fluid in the direction of n is

$$u\cdot n = u\cdot \frac{\text{grad } f}{|\text{grad} f|} .$$

Now, one defining property of the free surface can be written:

the normal velocity of the surface = the normal velocity of the fluid, hence

$$- \frac{f_t}{|\text{grad } f|} = u\cdot \frac{\text{grad } f}{|\text{grad } f|} .$$

So we have

$$u\cdot \text{grad } f + f_t = 0,$$

where $u\cdot \text{grad } f$ is expressed by the total differential

$$\frac{df(x_1,x_2,x_3)}{dt} = \frac{\partial f}{\partial x_1}\frac{\partial x_1}{\partial t} + \frac{\partial f}{\partial x_2}\frac{\partial x_2}{\partial t} + \frac{\partial f}{\partial x_3}\frac{\partial x_3}{\partial t} = f_{x_1}\cdot u_1 + f_{x_2}\cdot u_2 + f_{x_3}\cdot v =$$

$$= u\cdot \text{ grad } f,$$

as $u = \text{grad } \phi$, we obtain finally

$$f_t + f_{x_1}\phi_{x_1} + f_{x_2}\cdot\phi_{x_2} + f_{x_3}\cdot v = 0.$$

Let $x_1 = x$, $x_2 = y$, and $x_3 = z$, specializing f to

$$f(x,y,z,t) = \eta(x,y,t) - z = 0,$$

this means, that the free surface is represented by $z = \eta(x,y,t)$, we obtain

$$\eta_t + \eta_x \phi_x + \eta_y \phi_y = v = \phi_z.$$

Thus, the surface $z = \eta(x,y,t)$ is determined by

$$\eta_t + \eta_x \phi_x + \eta_y \phi_y = \phi_z.$$

We derive the momentum equation

$$\frac{p-p_o}{\rho} = \frac{\partial\phi}{\partial t} + \frac{1}{2}(\text{grad } \phi)^2 + g \cdot z,$$

where $z = \eta(x,y,t)$.

Now we consider pressure phenomena:

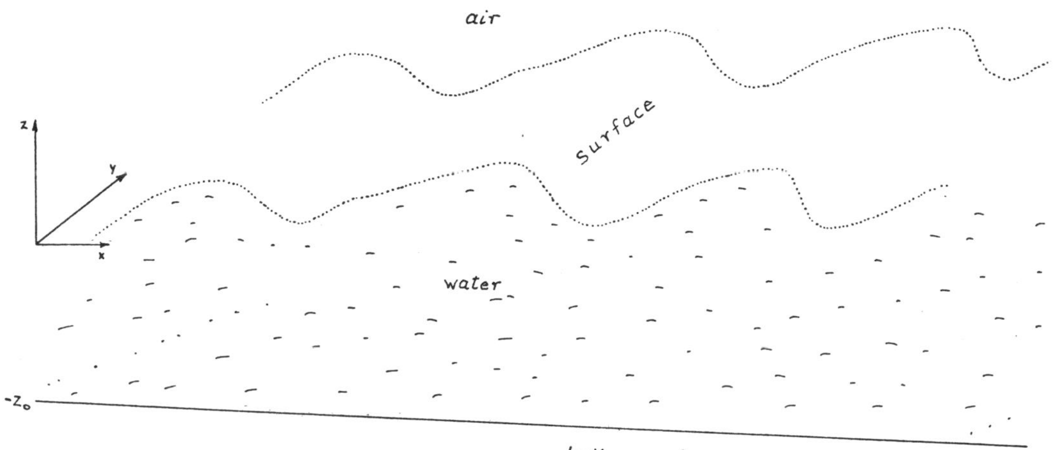

Since the interface between air and water has no mass, the forces in the fluid on both sides must be equal. Neglecting the surface tension, the pressure in the air and the pressure in the water must be equal at the surface. Any perturbation of the surface implies some motion of the air. But on a first level of approximation, we assume that the pressure in the air generated by the motion of the surface is negligible. We approximate the air pressure by the undisturbed constant value p. This is also clear, as the density of the air is very small compared with that of the water.

Neglecting the motion of the air, we assume that p_o is the pressure of the air, and p is the pressure in the water, given by $p = p_o$; thus, the momentum equation reduces to

$$\frac{\partial\phi}{\partial t} + \frac{1}{2}(\text{grad }\phi)^2 + g.\eta = 0.$$

Finally we can formulate the free boundary value problem:

$\Delta\phi = 0$ in the volume generated by the bottom surface and the interface $z = \eta(x,y,t)$.

$\frac{\partial\phi}{\partial n} = 0$ at the bottom surface

$\frac{\partial\phi}{\partial t} + \frac{1}{2}(\text{grad }\phi)^2 + g\cdot z = 0$ at the surface $z = \eta(x,y,t)$, which is determined by

$$\eta_t + \eta_x\phi_x + \eta_y\phi_y = v = \phi_z.$$

Considering small perturbations, we obtain a linearized theory, where the expressions η, ϕ, grad ϕ, ϕ_t are small.

The boundary conditions at the free surface

$$\eta_t + \eta_x\phi_x + \eta_y\phi_y = \phi_z$$

$$\phi_t + \frac{1}{2}(\text{grad }\phi)^2 + g\cdot\eta = 0$$

reduce to

$$\eta_t = \phi_z$$

$$\phi_t + g\cdot\eta = 0,$$

differentiating the second equation, we have

$\phi_{tt} + g\phi_z = 0$ at the free surface, and

$\phi_z = 0$ at the bottom surface $z = -z_o$.

IV. 1.2 Analytic solution of the water surface elevation

The linearized free boundary value problem is given by

$\Delta\phi = 0$ in the volume generated by the bottom surface and the free surface

$\phi_z = 0$ at the bottom $z = -z_o$

$\phi_{tt}+g\phi_y =$ at the free surface.

Taking the ansatz for a one-dimensional wave

$$\phi = \phi(z)e^{ikx}e^{-i\omega t},$$

then the Laplace equation reduces to

$$\phi_{zz} - K\phi = 0,$$

and the boundary value at the free surface

$$\phi_z - \frac{\omega^2}{g}\phi = 0,$$

the boundary value at the bottom

$$\phi_z = 0.$$

The solution of the reduced Laplace equation is given by

$$\phi = \cosh(k\cdot(z_o + z))$$

and satisfies $\phi_z = 0$ at the bottom $z = -z_o$, and the free surface condition

$$\phi_z - \frac{\omega^2}{g}\phi = 0$$

is satisfied, provided

$$\omega^2 = g\cdot k \tanh(k\cdot z_o).$$

As k denotes the wave number, ω the frequency of the wave, then $c = \frac{\omega}{k}$ is defined as the phase velocity of the wave and explicitly

$$c^2 = \frac{g}{k}\tanh(k\cdot z_o).$$

This formula shows us that the phase velocity depends on the wave number k, and thus expresses the fact, that for general disturbance the waves will disperse.

The water surface $\eta(x,y,t)$ can be derived from $\phi_t + g\cdot\eta = 0$, and the velocity potential is derived as

$$\phi = \cosh(k(z_o+z))e^{-kx}e^{-i\omega t},$$

we obtain

$$\eta = -\frac{1}{g}\phi_t \text{ at } z = 0.$$

The water surface elevation is given by

$$\eta = -\frac{1}{g}\phi_t\Big|_{z=0} = \frac{i\omega}{g}\cosh(k\cdot z_o)e^{ikx}e^{-i\omega t},$$

hence

$$\eta(x,t) = A(k)\cdot e^{i(k\cdot x-\omega(k)\cdot t)}$$

where the amplitude

$$A(k) = \frac{i\omega}{g}\cosh(k\cdot z_o) \text{ and } \omega(k) = (g\cdot k\cdot \tanh(k\cdot z_o))^{1/2}.$$

Now, we solve the initial value problem

$$\eta(x,t)\Big|_{t=0} = s_o(x)$$

$$\frac{\partial\eta(x,t)}{\partial t}\Big|_{t=0} = s_1(x),$$

where s_o, s_1 can be given as one dimensional splines. The general solution of the water surface elevation represented as

$$\eta(x,t) = \int_{-\infty}^{+\infty} A_1(k)e^{i(kx-\omega(k)\cdot t)}dk + \int_{-\infty}^{+\infty} A_2(k)e^{i(kx-\omega(k)t)}dk,$$

the initial values satisfy

$$\eta(x,t)\Big|_{t=0} = \int_{-\infty}^{+\infty}(A_1(k) + A_2(k))e^{ikx}dk = s_o(x)$$

$$\frac{\partial\eta(x,t)}{\partial t}\Big|_{t=0} = \int_{-\infty}^{+\infty} i\omega(k)(A_2(k) - A_1(k))e^{ikx}dk = s_1(x).$$

Applying the Fourier inversion theorem, the coefficients are expressed by the initial values

$$\hat{A}_1(k) = \frac{1}{2}\int_{-\infty}^{+\infty} \{s_0(x) - \frac{i}{\omega(k)} s_1(x)\}e^{-ikx}dx, \quad \text{and}$$

$$\hat{A}_2(k) = \frac{1}{2}\int_{-\infty}^{+\infty} \{s_0(x) + \frac{i}{\omega(k)} s_1(x)\}e^{-ikx}dx.$$

Thus, the solution of the initial-boundary-value problem of the water surface elevation is derived in closed form.

IV. 1.3 System of nonlinear partial differential equations

To illustrate some model-building processes, we go on to construct the two-dimensional water surface elevation with respect to given initial data.

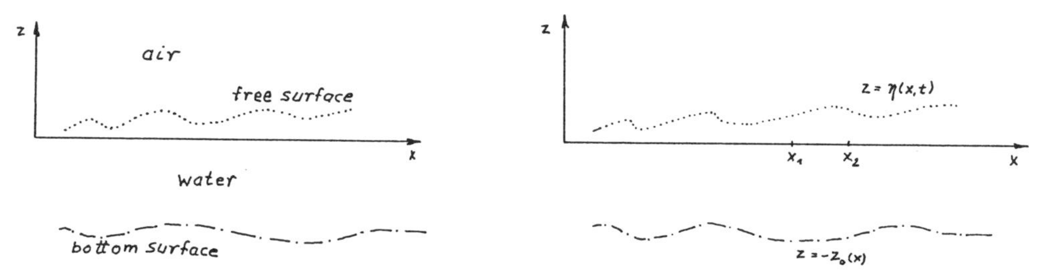

The depth $d(x,t)$ is given by

$$d(x,t) = z_0(t) + \eta(x,t).$$

The equation of continuity or conservation of mass can be written (page 252)

$$\frac{d}{dt}\int_{x_1}^{x} d(x,t)dx + u\cdot d \Big|_{x_1}^{x_2} = 0.$$

If all quantities are smooth, we take the limit $x_1 \to x_2$ and obtain

$$\frac{\partial d(x,t)}{\partial t} + \frac{\partial}{\partial x}(u(x,t)\cdot d(x,t)) = 0.$$

We derive from the momentum equation (page 256) that the condition at the free surface is

$$\frac{\partial\phi}{\partial t} + \frac{1}{2}(\text{grad } \phi)^2 + g\cdot\eta = 0.$$

Applying the grad-operator to this equation and remembering that $u = \text{grad } \phi$, then we obtain

$$\frac{\partial u}{\partial t} + u\cdot\frac{\partial u}{\partial x} + g\cdot\frac{\partial\eta}{\partial x} = 0.$$

Thus, the differential equation system reads

$$d_t + ud_x + d \cdot u_x = 0$$

$$u_t + u \cdot u_x + g \cdot \eta_x = 0.$$

Now we consider the behaviour of nonlinear waves on a sloping beach

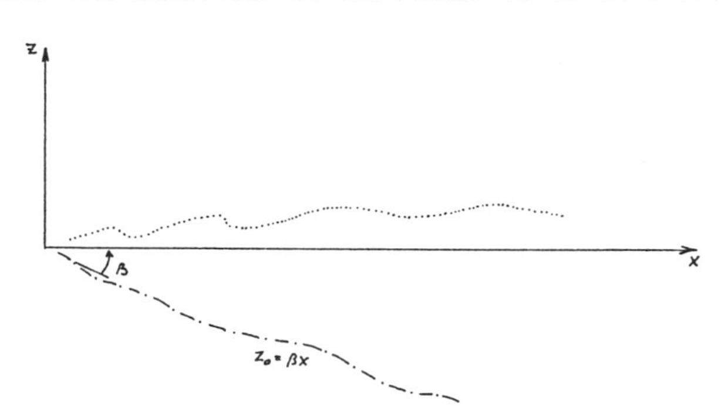

The depth is given by

$$d(x,t) = \beta x + \eta(x,t),$$

then

$$d_x = \beta + \eta_x \text{ and } g \cdot \eta_x = d_x \cdot g - \beta g.$$

Thus, the reduced form of the nonlinear differential equation system is given by

$$d_t + u \cdot d_x + du_x = 0$$

$$u_t + u \cdot u_x + d_x g - \beta g = 0.$$

This equation system is able to describe breaking phenomena observed at the seashore, bores on beach, and run-up phenomena.

Linearizing the above differential equation system, we derive the differential equation

$$\eta_{tt} = g \cdot \beta x \cdot \eta_{xx} + g\beta\eta_x \quad \text{(page 268)},$$

with the complete solution

$$\eta(x,t) = \{A . J_0(2\omega \cdot \sqrt{\frac{x}{g\beta}}) - iC \cdot Y_0(2\omega)\sqrt{\frac{x}{g \cdot \beta}})\}e^{-i\omega t}.$$

At the seashore $x = 0$, $Y_0(\xi)$ is singular and this singularity describes the breaking of the wave.

Introducing the y-coordinate along the shore, the linearized equation

$$\eta_{tt} = g\cdot\beta\cdot x(\eta_{xx}+\eta_{yy}) + g\beta\eta_x$$

with the solution

$$\eta(x,y,t) = e^{-kx}L_n(2kx)e^{\pm iky}e^{\pm i\omega t}$$

where $L_n(..)$ are the Laguerre polynomials. The initial value problem can be solved in closed form.

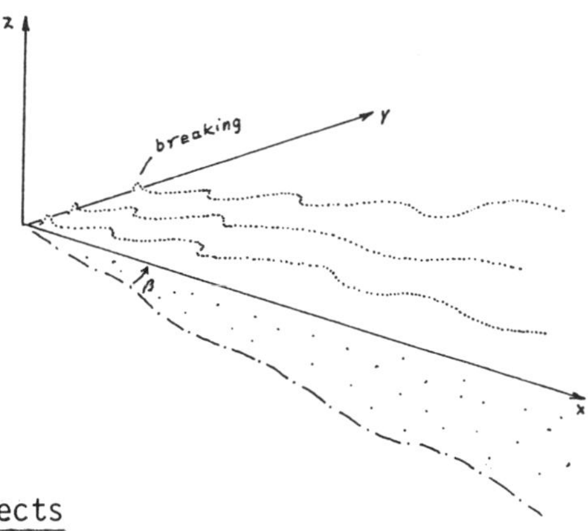

Nonlinear waves: Breaking and run-up effects

Introducing the depth

$$h = \beta\cdot x + \eta(x,t),$$

(see the figure on page 261), and let g be the acceleration of gravity, then we obtain a nonlinear system of differential equations (like that on page 261)

$$h_t + uh_x + hu_x = 0$$

$$u_t + u\cdot u_x + g\cdot h_x - g\beta = 0.$$

Introducing the variable $c = \sqrt{g\cdot h}$ we obtain the equation system (Carrier and Greenspan (53))

$$2c_t + 2uc_x + cu_x = 0$$

$$u_t + uu_x + 2cc_x - g\beta = 0,$$

the equivalent system in characteristic form is given by

$$(u+2c)_t + (u+c)\cdot(u+2c)_x - g\beta = 0$$

$$(u-2c)_t + (u-c)\cdot(u-2c)_x - g\beta = 0.$$

This can be rewritten, as $(g\beta t)_t = g\beta$, $t_x = 0$,

$$(u+2c-g\beta t)_t + (u+c)\cdot(u+2c-g\beta t)_x = 0$$

$$(u-2c-g\beta t)_t + (u+c)\cdot(u-2c-g\beta t)_x = 0.$$

Now we define the characteristic curves

$$C_+ : \frac{dx}{dt} = u + c, \text{ where } u + 2c - g\beta t = \text{constant}$$

$$C_- : \frac{dx}{dt} = u - c, \text{ where } u - 2c - g\beta t = \text{constant},$$

and introduce the characteristic variables

$$p = u+2c-g\beta t \text{ and } q = u-2c-g\beta t.$$

Now $\frac{dx}{dt} = u + c$, where p = constant, thus $x(p,q)$ and

$$\frac{dx}{dt} = \frac{\partial x}{\partial q} \cdot \frac{\partial q}{\partial t} + \frac{\partial x}{\partial p} \cdot \frac{\partial p}{\partial t} = \frac{\partial x}{\partial q} \cdot \frac{\partial q}{\partial t} .$$

It can be written

$$\frac{\partial x}{\partial q} = (u+c) \frac{\partial t}{\partial q} .$$

Then

$$x_q = (u+c)t_q \quad \text{and} \quad x_p = (u-c)t_p,$$

solving the equation of p and q, for u and c

$$u = \frac{1}{2} (p+q), \; c = \frac{1}{2}(p-q),$$

we have

$$X_q = (\frac{3p+q}{4} + g\beta t)t_q, \quad X_p = (\frac{p+3q}{4} + g\beta t)t_p.$$

Differentiating the first equation with respect to p and the second with respect to q, we obtain the linear equation observed by Carrier and Greenspan (53)

$$2(p-q)t_{pq} + 3(t_q - t_p) = 0.$$

Applying the transformation

$$\sigma = p-q, \; \lambda = -p-q,$$

we obtain a cylindrical wave equation, like

$$t_{\lambda\lambda} = t_{\sigma\sigma} + \frac{3}{\sigma} \cdot t_\sigma .$$

Simplifying by introducing the transformation

$$g\beta t = \frac{\lambda}{2} - \frac{\phi_\sigma}{\sigma} ,$$

we finally obtain

$$\phi_{\lambda\lambda} = \phi_{\sigma\sigma} + \frac{1}{\sigma} \cdot \phi_{\sigma}.$$

We have $c = \frac{\sigma}{4}$ and the particle velocity

$$u = -\frac{\phi_{\sigma}}{\sigma}.$$

The simplest solution of the wave equation is given

$$\phi = N(\lambda)\cdot\cos(\alpha\cdot\lambda),$$

where λ is an arbitrary separation constant, and $N(\sigma)$ satisfies

$$N'' + \frac{1}{\sigma} \cdot N' + \alpha^2 N = 0$$

with the solution $N = A\cdot J_0(\alpha\cdot 0)$, hence

$$\phi = A\cdot J_0(\alpha\cdot\sigma)\cdot\cos(\alpha\cdot\lambda).$$

In the case of linear approximation we obtain, if u is small, then ϕ is small. From the known relation ((359a),p. 70)

$$g\beta x = -\frac{1}{4}\phi_{\lambda} + \frac{1}{2}\left(\frac{\phi_{\sigma}}{\sigma}\right)^2 + \frac{\sigma^2}{16}$$

we obtain

$$g\beta x \approx \frac{\sigma^2}{16}$$

and the transformation reduces to $g\beta t \approx \frac{1}{2}\lambda$.

By taking $\alpha = \frac{\omega}{2g\beta}$ we have the solution

$$\phi = A\cdot J_0\left(2\omega\sqrt{\frac{x}{g\beta}}\right)\cdot \cos\omega t$$

and the particle velocity can be written

$$u = -\frac{\phi_{\sigma}}{\sigma} = \frac{2\omega}{\sigma} A\cdot \frac{J_0'\left(2\omega\sqrt{\frac{x}{g\beta}}\right)}{2\omega\sqrt{\frac{x}{g\cdot\beta}}} \cdot \cos\omega t.$$

The shore-line is characterized by $\sigma = 0$ as $z = \sigma\cdot\alpha$, consider

$$\lim_{z\to 0} \frac{J_0'(z)}{z} = -\frac{1}{2}.$$

From

$$g\beta x = -\frac{1}{4}\phi_\lambda + \frac{1}{2}\cdot\left(-\frac{\phi_\sigma}{\sigma}\right)^2 + \frac{\sigma^2}{16}$$

follows

$$g\beta x = \frac{1}{4}\cdot\alpha A\sin\alpha\cdot\lambda + \frac{1}{8}\alpha^4 A^2\cos^2\alpha\lambda.$$

Now the particle velocity = 0 at the maximum or minimum run-up, thus $u = -\frac{\phi_\sigma}{\sigma} = 0$ shows $\cos a\cdot\lambda = 0$, then we have $\sin a\lambda = \pm 1$, using this in $g\beta x = \ldots$ we obtain the

maximum run-up $g\beta x = \frac{1}{4}\alpha\cdot A$

minimum run-up $g\beta x = -\frac{1}{4}\cdot\alpha\cdot A$

, then $-\frac{\alpha\cdot A}{4g\beta} \leq x \leq \frac{\alpha\cdot A}{4g\beta}$.

The geometrical interpretation

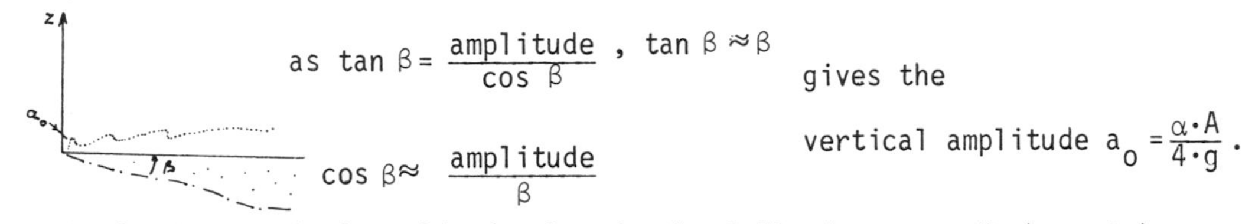

as $\tan\beta = \frac{\text{amplitude}}{\cos\beta}$, $\tan\beta \approx \beta$

$\cos\beta \approx \frac{\text{amplitude}}{\beta}$

gives the vertical amplitude $a_o = \frac{\alpha\cdot A}{4\cdot g}$.

We obtain the vertical amplitude also in the following way. We have taken $\alpha = \frac{\omega}{2g\beta}$ and $\sigma \sim 4\cdot\sqrt{g\beta x}$, then

$$\alpha\cdot\sigma = 2\omega\frac{\sqrt{x}\,\sqrt{g\beta}}{g\beta} = 2\omega\cdot\frac{\sqrt{x}}{\sqrt{g\cdot\beta}} .$$

As $\lambda = 2g\beta\cdot t$, then $\alpha\cdot\lambda = 2g\beta t\cdot\frac{\omega}{2g\beta} = \omega\cdot t$ and the velocity of a particle

$$u = -\frac{\phi_\sigma}{\sigma} = -\alpha^2 A\cdot\frac{J_o'(\alpha\cdot\sigma)}{\alpha\,\sigma}\cos(\alpha\cdot\lambda) = -\frac{2\omega}{\beta}\cdot\frac{\beta}{2\omega}\alpha^2\cdot A\frac{J_o'(\alpha\cdot\sigma)}{\alpha\,\sigma}\cos\omega t =$$

$$= -\frac{2\omega}{\beta}a_o\frac{J_o'\left(2\omega\sqrt{\frac{x}{g\beta}}\right)}{2\omega\sqrt{\frac{x}{g\beta}}}\cos\omega t,$$

where

$$a_o = \frac{\beta}{2\omega}\alpha\cdot\alpha\cdot A = \frac{\beta}{2\omega}\cdot\frac{\omega}{2g\beta}\alpha A = \frac{\alpha.A}{2g} .$$

IV. 1.4 Breaking phenomena

When the Jacobian (J) of the transformation $(\lambda,\sigma) \to (x,t)$ becomes zero, the solution in the (x,t)-plane will be multivalued, i.e. breaking will occur:

$$J = x_\lambda \cdot t_\sigma - x_\sigma \cdot t_\lambda = (ut_\lambda + c \cdot t_\sigma)t_\sigma - (ct_\lambda + ut_\sigma)t_\lambda = c(t_\sigma^2 - t_\lambda^2) = c(t_\sigma - t_\lambda)\cdot(t_\sigma + t_\lambda).$$

As $g\beta t = \frac{\lambda}{2} - \frac{\phi_\sigma}{\sigma}$, we have

$$g\beta t_\sigma = \frac{A\alpha^3}{z}\left(J_0'' + \frac{2}{z}J_0'\right)\cdot \cos(\alpha\cdot\lambda),$$

where $z = \alpha\cdot\sigma$ and

$$g\beta t_\lambda = \frac{1}{2} + \frac{A\alpha^3}{z} J_0' \cdot \sin(\alpha\cdot\lambda),$$

and finally

$$g\beta(t_\lambda - t_\sigma) = \frac{1}{2} - A\alpha^3 \frac{J_1 \sin(\alpha\cdot\lambda) - J_2\cos(\alpha\cdot\lambda)}{z} =$$

$$= \frac{1}{2} - A\alpha^3 \frac{J_1^2 + J_2^2}{z^2}\sin(\alpha\cdot\lambda + \eta), \text{ as } \frac{J_1^2 + J_2^2}{z^2} \xrightarrow[z\to 0]{} \left(\frac{1}{2}\right)^2 + \left(\frac{1}{2}\right)^2 = \frac{1}{2}.$$

If the Jacobian becomes zero, the expression above must be zero, thus, we have to take

$$A\alpha^3 \geqq 1,$$

such that

$$A\alpha^3 . \sin(\alpha . \lambda + \eta) \approx 1.$$

The breaking condition can be written as $A\alpha^3 \geqq 1$. We have chosen $\alpha = \frac{\omega}{2g\beta}$, then

$$1 \leqq A\alpha^3 = A\alpha . \alpha^2 = \frac{A\alpha\cdot\omega^2}{4g^2\cdot\beta^2} = \frac{\omega^2}{g\beta^2} a_0,$$

where a_0 is the vertical amplitude given by $a_0 = \frac{A\cdot\alpha}{4g}$. In observations made by Galvin (121), the quantity $Q = \frac{\omega^2}{g\beta^2}\cdot a_0$, plays an important role.

The behaviour of the wave amplitudes in Galvin's experiments are given by:

collapsing plunging spilling

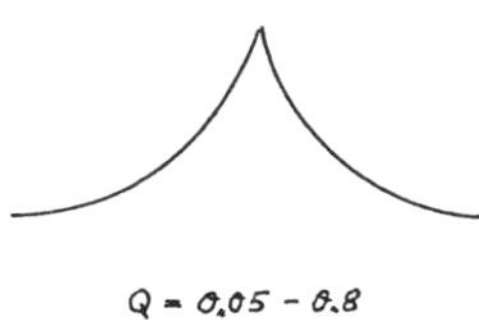

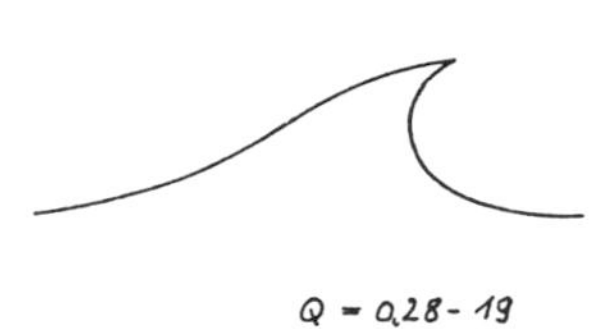

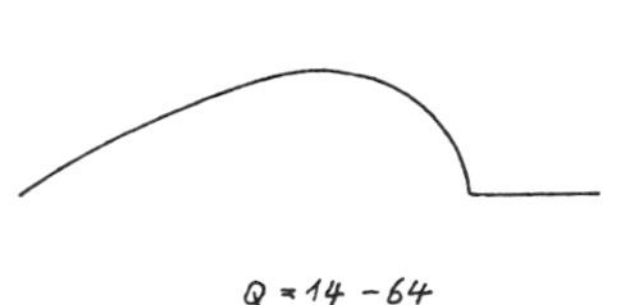

IV. 1.5 Modelling by linearizing

Linearizing the known equation system

$$h_t + uh_x + hu_x = 0$$

$$u_t + u \cdot u_x + g\eta_x = 0,$$

where the depth

$$h(x,t) - h_o(x) + \eta(x,t),$$

we obtain

$$0 = \eta_t + u(h_o' + \eta_x) + (h_o + \eta)u_x = \eta_t + uh_o' + h_o u_x + u\eta_x + \eta u_x,$$

assuming small disturbances $\varepsilon << 1$, such that

$$\frac{\eta}{h_o} = O(\varepsilon) \quad \text{and} \quad \frac{u}{\sqrt{g \cdot h_o}} = O(\varepsilon),$$

then $u \cdot h_o' = O(\varepsilon)$, $h_o u_x = O(\varepsilon)$, but $u\eta_x = O(\varepsilon^2)$, $\eta \cdot u_x = O(\varepsilon^2)$ and $u \cdot u_x = O(\varepsilon^2)$. On a first level of approximation, we have

$$\eta_t + h_o u_x + h_o' u = 0$$

$$u_t + g\eta_x = 0,$$

differentiating

$$\eta_{tt} + h_o u_{xt} + h_o' u_t = 0, \text{ as } u_t = -g\eta_x; \text{ and } u_{xt} = -g\eta_{xx},$$

we obtain finally

$$\eta_{tt} - h_o g\eta_{xx} - gh_o'\eta_x = 0.$$

The solution with $h_o = x \cdot \tan\beta$ (see the figure on page 261) gives $h_o' = \tan\beta \approx \beta$

for small β, hence we obtain a linear equation

$$\eta_{tt} = g\beta x\eta_{xx} + g\beta\eta_x.$$

We make the ansatz $\eta(x,t) = B(x)e^{-i\omega t}$, where B(x) is a solution of

$$B'' + \frac{1}{x}\cdot B' + \frac{\omega^2}{g\beta}\frac{1}{x} B = 0,$$

introducing the transformation

$$x = \frac{g\beta}{\omega^2} \cdot \frac{\xi^2}{4},$$

we obtain the Bessel equation of order zero

$$\frac{d^2B}{d\xi^2} + \frac{1}{\xi}\frac{dB}{d\xi} + B = 0.$$

The general solution is given by (359a)

$$B = A\cdot J_0(\xi) - iCY_0(\xi),$$

where A, C are constants, and $J_0(\xi)$ is regular at the beach $x = 0$, but $Y_0(\xi)$ is singular for $x = 0$, with logarithmic singularity. The most obvious phenomenon we observe at the seashore is breaking. On the level of the linerarized equation it is an attempt to represent the breaking of the wave by the singularity. As C increases, more energy goes into the singularity (seen as breaking) and less is reflected.

IV. 1.6 Two-dimensional water surface elevation

We introduce on a second level of approixmation the y-coordinate along the shore (see the figure on page 262). The linearized equation is given by

$$\eta_{tt} = g\beta x(\eta_{xx} + \eta_{yy}) + g\beta\eta_x,$$

separating the variables

$$\eta = B(x)e^{\pm iky}\, e^{\pm i\omega t},$$

then B satisfies

$$B'' + \frac{1}{x} B' + (\frac{\omega^2}{g\beta x} - k^2)B = 0.$$

For $x \to \infty$ the equation is reduced to

$$B'' - k^2 B = 0,$$

with the solution $B \sim e^{-kx}$.

Choosing the new ansatz

$$\eta = G(x) \cdot e^{-kx} \, e^{\pm iky} \, e^{\pm i\omega t},$$

where $x = 2kx$ for $k > 0$, then $G(x)$ satisfies

$$xG_{xx} + (1-x)G_x + \frac{1}{2}\left(\frac{\omega^2}{g\beta k} - 1\right)G = 0,$$

and the solutions are the known Laguerre polynomials $L_n(\xi) = e^{\xi} \frac{d^n}{dx^n}(\xi^n e^{-\xi})$ and $\omega^2 = gk(2n+1)$ for n = positive integer. We find the final solution (359a)

$$\eta(x,y,t) = e^{-kx} \, L_n(2kx) e^{\pm iky} \, e^{\pm i\omega t},$$

and for negative values of k, we obtain a solution

$$\eta(x,y,t) = \int_{-\infty}^{+\infty} e^{-|k|x} \, L_n(2(k)x) e^{\pm iky} \cos(g(2n+1) \; |k|)^{1/2} t \; dk.$$

As the solution $e^{-kx} \, L_n(2kx)$ (for integers $n > 0$) form a complete set, we are able to represent any square integrable function of x as a series in the corresponding Hilbert space. Therefore, we can solve the initial value problem of

$$\eta_{tt} = g\beta x(\eta_{xx} - \eta_{yy}) + g\beta\eta_x$$

with respect to the initial data

$$\eta(x,y,0) = Sp(x,y),$$

$$\left.\frac{\partial\eta(x,y,t)}{\partial t}\right|_{t=0} = 0.$$

The solution can be represented as

$$\eta(x,y,t) = \sum_{n=0}^{\infty} \int_{-\infty}^{+\infty} \mathcal{D}_n(k) \; \{e^{-|k|x} L_n(2|k|x) e^{iky} \cos(t(g(2n+1) \cdot |k|)^{\frac{1}{2}}) \; dk,$$

where the coefficients are chosen corresponding to the initial data, given by the known inversion formula

$$\mathcal{D}_n(k) = \frac{1}{\pi}\int_0^{\infty}\int_{-\infty}^{+\infty} S_p(\tau_1,\tau_2)e^{-|k|\tau_1}L_n(2|k|\tau_1).\ \frac{|k|}{(n!)^2}\ e^{-ik\tau_2}\ d\tau_1\ d\tau_2.$$

The initial condition $S(\tau_1,\tau_2)$ can be given by two-dimensional interpolating Lg-splines (139).

The initial disturbance may be given in the form

$$S_p(x,y) = S_1(x)\cdot S_2(y),$$

then, the inversion formula reads

$$\mathcal{D}_n(k) = \int_0^{\infty} S_1(\tau_1)e^{-|k|\tau_1}\ \frac{L_n(2|k|\tau_1)}{n!}\ \frac{2|k|}{n!}d\tau_1\ \frac{1}{2\pi}\int_{-\infty}^{+\infty} S_2(\tau_2)e^{-ik\tau_2}\ d\tau_2.$$

Defining the operator

$$K = \frac{1}{2\pi}\int_{-\infty}^{+\infty} S_2(\tau_2)e^{-ik\tau_2}d\tau_2.$$

Let f be continuous in $x = 0$, then (see page 70)

$$\int_{-\infty}^{+\infty} \delta(x)f(x)dx = f(0),$$

where $\delta(x)$ is the Dirac-distribution; applied to the operator K, when $S_2(\cdot) \to \delta(\cdot)$, we obtain

$$\int_{-\infty}^{+\infty} \delta(x)e^{-ik\tau_2}\ d\tau_2 = e^{-ik0} = 1,$$

and $\mathcal{D}_n(k)$ can be written as

$$\mathcal{D}_n(k) = K\int_0^{\infty} S_1(\tau_1)e^{-|k|\tau_1}.\ \frac{L_n(2|k|\tau_1)}{n!}\ \frac{2|k|}{n!}\ d\tau_1.$$

Using this, the solution of the initial value problem reads

$$\eta(x,y,t) = \sum_{n=0}^{\infty}\int_{-\infty}^{+\infty}\left(\int_0^{\infty} S_1(\tau)e^{-|k|\tau_1}.\frac{L_n(2|k|\tau_1)}{n!}\right).\frac{L_n(2|k|x)}{n!}\cdot e^{-|k|x}e^{iky}.$$

$$2|k|\cdot\cos(\sqrt{g(2n+1)\ k}\ \cdot t)dk.$$

It is known that if $n \to \infty$

$\dfrac{L_n(2|k|\cdot\tau_1)}{n!}$ can be approximated by $J_o(2\cdot\sqrt{2n|k|\cdot\tau_1}$

$\dfrac{L_n(2|k|\cdot x)}{n!}$ can be approximated by $J_o(2\cdot\sqrt{2n|k|\cdot x})$.

Now, by taking $k \to 0$ and introducing (page 264): $2|k|n = \frac{\omega^2 n}{g\cdot\beta}$; $2|k| = \psi(\frac{\omega^2 n}{g\cdot\beta})$, we obtain the closed solution

$$\eta(x,y,t) = \int_{-\infty}^{+\infty}\left(\int_0^{\infty} S_1(\tau_1)\cdot J_o(2\omega\cdot\sqrt{\frac{\tau_1}{g\beta}})d\tau_1\right)\cdot J_o(2\omega\sqrt{\frac{x}{g\beta}})\cdot\ \psi(\frac{\omega^2}{g\cdot\beta})\cdot\cos(\omega t).$$

IV. 2 Modelling influence phenomena: Influence models in economics

IV. 2.1 Crowd movements characterized by friction phenomena and random influence impulses

The speed at which individuals move along a busy street changes due to two factors. First there are friction phenomena within the movement and secondly there are random influence-impulses.

The friction is proportional to the speed v, but the direction is opposite to the speed of the individual. Denoting the friction constant by β, we have

$$f_v = -\beta\cdot v.$$

This can be described by movement processes of the pedestrian ∘

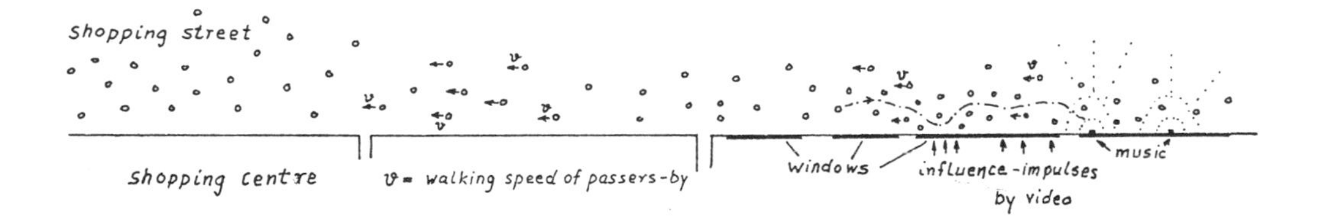

A possible path of an individual is denoted by — · — · — · — (and the corresponding friction phenomenon).

The individuals are influenced by music coming from nearby shops, or listening to others talking, or by looking at advertisements (fashion). Thus, there are various impulses influencing the pedestrian. These influence impulses last for only a very short time, thus, we represent the influence by a δ-function of strength s, as

$$I = s\cdot\delta(t-t_j)$$

where t_j is the moment the influence occurs.

As the individuals are constantly lured by optical (see Chapter VI) and acoustic advertising, they change their behaviour (e.g. their speed (velocity) v) accordingly

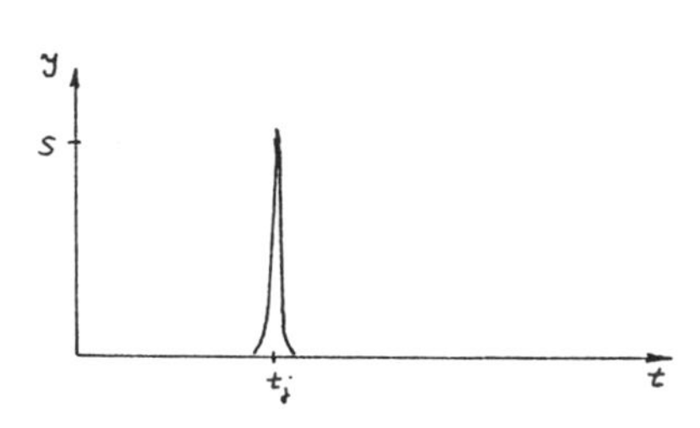

$$b\,\frac{\partial v}{\partial t} = I = s\cdot\delta(t-t_j),$$

where we have introduced b as a constant of the behaviour. This equation may be integrated over a short time interval around $t = t_j$

$$\int_{t_{j-0}}^{t_{j+0}} b.\frac{\partial v}{\partial t}\,dt = \int_{t_{j-0}}^{t_{j+0}} s\delta(t-t_j)dt = s\int_{t_{j-0}}^{t_{j+0}} \delta(t-t_j)dt = s,$$

thus

$$b(v(t_{j+0})-v(t_{j-0})) = b\cdot\Delta v = s,$$

and $\Delta v = \frac{s}{b}$, in words:

at time t_j the velocity v of the individual is increased by the amount

$$\frac{s}{b} = \frac{\text{strength of the influence impulse}}{\text{behaviour constant}}$$

During the pedestrian's walk —·—·— down the street, he is influenced by a sequence of j influence-impulses, as

$$I(t) = s\cdot\sum_j \delta(t-t_j).$$

To come to a more realistic model we must consider that the influence-impulses are acting not only in one direction but randomly also in the reverse direction; hence, introducing the random sequence $(\pm 1)\cdot j$ of plus and minus signs

$$I(t) = s\cdot\sum_j \delta(t-t_j)(\pm 1)j.$$

Individuals change their behaviour, e.g. their velocity v, described by

$$b\cdot\frac{\partial v}{\partial t} = \text{random influence-impulses + friction phenomena}$$

$$= s\cdot\sum_j \delta(t-t_j)\cdot(\pm 1)j + (-1)\beta\cdot v$$

dividing gives

$$\frac{\partial v}{\partial t} = - \frac{\beta}{b} v + \frac{s}{b} \sum_i \delta(t-t_i)\cdot(\pm 1)i,$$

defining

$$A(t) = \frac{s}{b} \sum_j \delta(t-t_i)\cdot(\pm 1)i,$$

then

$$\frac{\partial v}{\partial t} = - \frac{\beta}{b} v + A(t).$$

The speed of the pedestrian on a shopping trip changes due to random influence-impulses:

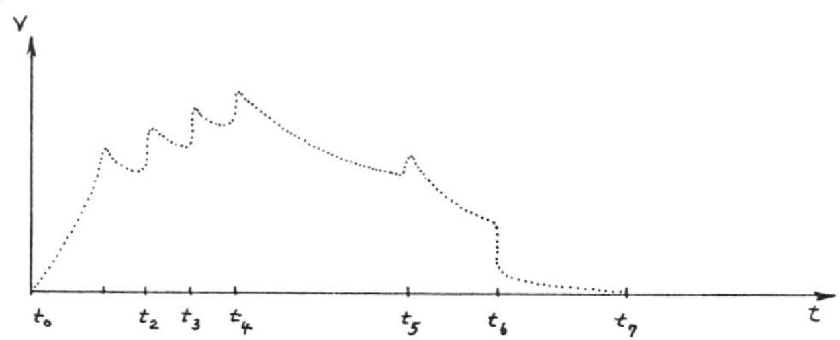

At time t_0 the individual is starting the shopping trip. At t_1 he greets a friend; they are talking until t_2. Thus the speed of the individual may be reduced. At t_2 and t_3 the individual listens to discussions in passing, and this reduces his speed. At t_4 and t_5 the individual meets friends and is talking until t_6, where they arrive at a shopping centre and are influenced by video-advertising, thus they reduce their speed while they look at the windows. Then there follows a "creeping motion" in front of the windows. The individual is influenced by advertising and at t_7 he visited the centre .

On the whole we can say that the sequence of times and directions are random events. We cannot predict the single path but only averages. We are considering averages over different time sequences and directions of the influence-impulses.

Denoting the average A over the random sequence of plus and minus signs by $\langle A \rangle$, as the random sequence occurs with equal probability, we have

$$\langle A \rangle = 0.$$

Let $\delta(t)$ be the random influence-impulses at time t,

$$A(t) = \frac{s}{b} \sum_i \delta(t-t_i)(\pm 1)i$$

and at time t*

$$A(t^*) = \frac{s}{b} \cdot \sum_i \delta(t^*-t_i)\cdot(\pm 1)i.$$

By forming the product, we have to take the average over the times of the impulses and their directions

$$\langle A(t), A(t^*)\rangle = (\frac{s}{b})^2 \sum_j \langle\delta(t-t_j)\delta(t^*-t_j)\rangle(\pm 1)j,$$

assuming that t_j is a Poisson process, we have

$$\langle\delta(t-t_j)\delta(t^*-t_j)\rangle = \frac{1}{t_j} \delta(t-t^*),$$

and the correlation function is given by

$$\langle A(t), A(t^*)\rangle = (\frac{s}{b})^2 \frac{1}{t_1} \delta(t-t^*).$$

The velocity of the individual satisfies

$$b\cdot\frac{\partial v}{\partial t} = b\cdot A(t) - \beta\cdot v,$$

defining P = b·v, the equation can be rewritten as

$$\frac{dP}{\partial t} = Q(p(t)),$$

and we describe the motion of individuals by probability functions. The solutions of this equation are the paths of individuals and may be given in the P-t-plane.

Let t be a fixed time, we ask for the probability of finding the individuals at the point p. If this point p is not on p(t), then the probability is zero. But if the point p is on p(t) (the solution of the above equation), the probability is one.

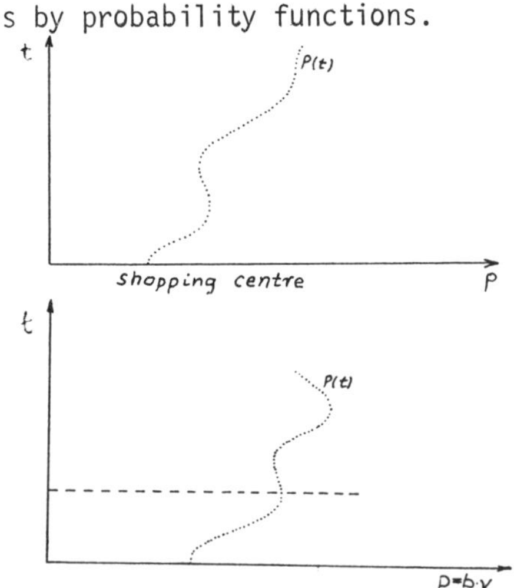

Now we ask for probability functions which are

$$\begin{cases} 1 \text{ if } p = p(t) \\ 0 \text{ if } p \neq p(t). \end{cases}$$

Thus the δ-function can be interpreted as the probability density

$$P(p,t) = \delta(p-p(t)).$$

As the integral over the δ-function $\delta(p-p_0)$ vanishes, if the integration interval does not contain p_0, and is one, if the integration interval contains a neighbourhood of p_0:

$$\int_{p_0-\varepsilon}^{p_0+\varepsilon} \delta(p-p_0)dp = \{{}^1_0 \; .$$

We are able to derive an equation for the probability density P. Differentiating P(p,t) with respect to t

$$\frac{\partial P(p,t)}{\partial t} = \frac{d}{dp(t)} (\delta(p-p(t))\cdot\frac{dp(t)}{dt}$$

by the chain rule.

As $\frac{dp}{dt} = Q(p(t))$, the derivative of the Heaviside function is the Dirac distribution. (Yosida (367), p. 50).

The derivative of the Dirac δ-function

$$- \frac{d}{dp} (\delta(p-p(t))\cdot\frac{\partial p(t)}{\partial t}$$

can be rewritten as

$$- \frac{d}{dp} (\delta(p-p(t))\cdot Q(p(t)).$$

Let $\varphi(p)$ be an arbitrary test function, then

$$\int \varphi(p)\cdot\frac{\partial P(p,t)}{\partial t} dp = - \int_{p(t)-\varepsilon}^{p(t)+\varepsilon} \varphi(p)\cdot\frac{d}{dp} (\delta(p-p(t))\cdot Q(p(t))dp =$$

$$= - \int_{p(t)-\varepsilon}^{p(t)+\varepsilon} - \frac{\partial\varphi}{\partial p} \cdot\delta(p-p(t))Q(p(t))dp$$

we obtain this expression by integration by parts, as

$$\int \varphi(p) \frac{d}{dp} \{\delta(p-p(t))\cdot Q(p(t))\}dp = - \int \frac{\partial\varphi(p)}{\partial p} \{\delta(p-p(t))\cdot Q(p(t))\}dp;$$

finally we have

$$\frac{\partial P(p,t)}{\partial t} = - \frac{\partial}{\partial p} (\delta(p-p(t))\cdot Q(p)) = - \frac{\partial}{\partial p} (P(p,t)\cdot Q(p)),$$

and this is interpretable as a continuity equation.

IV. 2.2 Probability distributions of the paths

The solutions of the equation

$$\frac{\partial P(p,t)}{\partial t} = Q(p(t))$$

characterize the paths of the individuals.

The probability distribution of the path $p_1(t)$ is $P_1(p,t) = \delta(p-p_1(t))$, ...
" " " " $p_2(t)$ is $P_2(p,t) = \delta(p-p_2(t))$, - · - · -
" " " " $p_3(t)$ is $P_3(p,t) = \delta(p-p_3(t))$, ——·——

Let q_i be the probability of the occurrence of a path p_i, then the average over all paths is given by

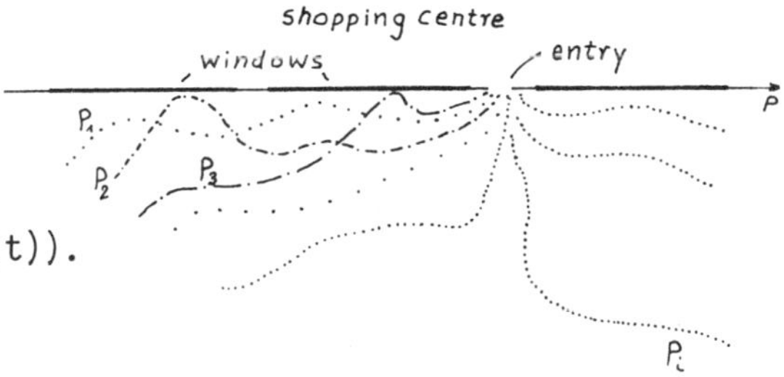

$$u(p,t) = \sum_i q_i P_i(p,t) = \sum_i q_i \delta(p-p_i(t)).$$

Thus $u(p,t)$ can be written in an abbreviated form

$$u(p,t) = \langle\delta(p-p(t)\rangle,$$

and $u\cdot dp$ is interpretable as the probability of finding an individual at position p in the interval dp at time t.

Now we derive a differential equation for u. Introducing the change of u in a time interval Δt

$$\Delta u(p,t) = u(p,t+\Delta t) - u(p,t);$$

this can be written as

$$\langle\delta(p-p(t+\Delta t))\rangle - \langle\delta(p-p(t))\rangle,$$

setting $p(t+\Delta t) = \Delta p(t) + \Delta p(t)$,

$$\Delta u(p,t) = \langle\delta(p-p(t) - \Delta p(t))\rangle - \langle\delta(p-p(t))\rangle.$$

Expanding the δ-function with respect to power of Δp by Taylor, up to quadratic order in Δp, then

$$\Delta u(p,t) = \langle -\frac{d}{dp}(\delta(p-p(t)))\Delta p(t)\rangle + \frac{1}{2}\langle\frac{d^2}{dp^2}(\delta(p-p(t)))\cdot(\Delta^2 p(t)\rangle.$$

First we express $\Delta p(t)$ by known facts. The individuals are changing their behaviour, for example, their velocity according to

$$b\cdot\frac{\partial v}{\partial t} = \text{random influence-impulses} + \text{friction phenomena}$$

$$= s\cdot\sum_i \delta(t-t_i)\cdot(\pm 1)_i - \beta\cdot v,$$

as we have defined $p = b\cdot v$; $\mu = \frac{\beta}{b}$, $s\sum_i \delta(t-t_i)(\pm 1)_i = A(t)$, thus we obtain

$$\frac{\partial p}{\partial t} = -\mu\cdot p(t) + A(t).$$

We find Δp by integrating over the time interval Δt and applying the differential equation

$$\Delta p = p(t+\Delta t)-p(t) = \int_t^{t+\Delta t} \frac{\partial p(\tau)}{\partial \tau}\, d\tau = -\int_t^{t+\Delta t} \mu p(\tau)d\tau + \int_t^{t+\Delta t} A(\tau)d\tau =$$

$$= -\mu p(t)\cdot\Delta t + \Delta A(t),$$

hence, the first part of the expansion can be written as

$$\frac{d}{dp}(\delta(p-p(t))\cdot\Delta p(t)) = \frac{d}{dp}(\delta(p-p(t))\cdot(-\mu p(t)\cdot\Delta t) + \frac{d}{dp}(\delta(p-p(t))\cdot\Delta A(t),$$

taking the average

$$\langle \quad " \quad \rangle = \langle \quad " \quad \rangle + \langle \quad " \quad \rangle.$$

$A(t)$ are the influence-impulses acting randomly in all directions after the time t, whereas $p(t)$ is determined prior to this time. By independence of these effects

$$\langle\delta(p-p(t))\,\Delta A(t)\rangle = \langle\delta(p-p(t))\rangle\cdot\langle\Delta A(t)\rangle,$$

and due to page 273

$$<A(t)> = 0, \text{ thus } <\Delta A(t)> = 0.$$

There remains

$$<-\frac{d}{dp}(\delta(p-p(t))\cdot\Delta p(t)> = -\frac{d}{dp}<\delta(p-p(t))\cdot(-\mu p(t)\Delta t> =$$

$$= \frac{d}{dp}<\delta(p-p(t))\cdot\mu\cdot p(t)>\Delta t = \frac{d}{dp}<\delta(p-p(t))>\mu\cdot p\cdot\Delta t = \frac{d}{dp}(u\cdot\mu\cdot p)\Delta t.$$

Now we determine

$$<\frac{d^2}{dp^2}(\delta(p-p(t))\cdot(\Delta p(t))^2>,$$

this can be expressed by the above argument, as

$$\frac{d^2}{dp^2}<\delta(p-p(t)>\cdot<(\Delta p(t)^2>.$$

As $\Delta p = -\mu p(t)\cdot\Delta t + \Delta A(t)$, then

$$(\Delta p)^2 = \Delta^2 A(t) - 2\mu\cdot p\cdot\Delta t\cdot\Delta A + \mu^2\cdot p^2\cdot\Delta^2 t,$$

taking the average and remembering that $<\Delta A(t)> = 0$, it follows that

$$<\Delta^2 p(t)> = <\Delta^2 A(t)>.$$

Assuming that t_j is a Poisson process, we have

$$<A(t)\cdot A(t^*)> = (\frac{s}{b})^2 \frac{1}{t_1}\delta(t-t^*),$$

and therefore

$$<\Delta A(t)\Delta A(t)> = \int_t^{t+\Delta t}\int_t^{t+\Delta t} <A(t)\cdot A(t^*)>dtdt^* =$$

$$= \int_{\cdot}^{\cdots}\int_{\cdot}^{\cdots} (\frac{s}{b})^2 \frac{1}{t_1}\delta(t-t^*)dtdt^* = \int_{\cdot}^{\cdots}\int_{\cdot}^{\cdots} D\cdot\delta(t-t^*)dtdt^* = D\cdot\Delta t,$$

defining $D = (\frac{s}{b})^2 \frac{1}{t_1}$, we have finally

$$\frac{d^2}{dp^2}<\delta(p-p(t)>\cdot<\Delta^2 p(t)> = \frac{d^2}{dp^2}(u)\ D\cdot\Delta t.$$

We obtained the expansion of the δ-function with respect to power Δp

$$u(p,t+\Delta t)-u(p,t) = <-\frac{d}{dp}(\delta(p-p(t))\cdot\Delta p(t)> + \frac{1}{2}<\frac{d^2}{dp^2}(\delta(p-p(t))\cdot(\Delta p(t))^2>$$

$$= \frac{d}{dp}(\mu\cdot p\cdot u)\Delta t + \frac{1}{2}D\cdot\frac{d^2}{dp^2}u\cdot\Delta t,$$

dividing by Δt and taking the limit $\Delta t \to 0$, we obtain finally

$$\frac{\partial u}{\partial t} = \frac{d}{dp}(\mu\cdot p\cdot u) + \frac{1}{2}\cdot D\cdot\frac{d^2}{dp^2}u.$$

In the following section, we solve the partial differential equation, which is of Kolmogoroff-Focker-Planck type. It can be rewritten as

$$\frac{\partial u}{\partial t} + \frac{d}{dp}(-\mu pu - \frac{1}{2}D\frac{\partial u}{\partial p}) = 0.$$

The stationary solution is defined by $\frac{du}{dt} = 0$, thus, u is time-indepdent. We obtain

$$\frac{d}{dp}(-\mu pu - \frac{1}{2}D\cdot\frac{\partial u}{\partial p}) = 0;$$

as we impose the condition that u vanishes for $p \to \pm\infty$, we have

$$\frac{1}{2}D\cdot\frac{\partial u}{\partial p} = -\mu p\cdot u.$$

Separating the variables, we can write the differential form

$$\frac{1}{2}D\cdot\frac{du}{u} = -\mu p\,dp = Ldp,$$

and defining $L = -\mu p$.

Integrating

$$\frac{1}{2}\int D\frac{du}{u} = \int L\,dp,$$

we obtain the stationary solution

$$u = C\exp(\frac{2}{D}\cdot\int L\,dp),$$

where C is a constant, thus explicitly

$$u(p) = C\cdot\exp(-\mu\frac{p^2}{D}),$$

this is the form of a Gaussian distribution.

Now we try the ansatz of a Gaussian distribution for the time dependent solution in the form

$$u(p,t) = M(t)\cdot\exp\left(-\frac{p^2}{a(t)} + \frac{2b(t)}{a(t)}\cdot p\right)$$

and insert it into

$$\frac{\partial f}{\partial t} + \frac{\partial}{\partial p}\left(-\mu pf - \frac{1}{2}D\frac{\partial f}{\partial p}\right) = 0.$$

Differentiating the ansatz

$$\frac{\partial u}{\partial t} = \frac{\partial M(t)}{\partial t}\cdot\exp(\cdot) + M(t)\cdot\exp(\cdot)\ \{-(-1)a(t)^{-2}\ \dot{a}(t)\cdot p^2 + 2p(\frac{\dot{b}\cdot a - b\cdot\dot{a}}{a^2})\}$$

$$\frac{\partial u}{\partial p} = M(t)\exp(\cdot)\cdot(\frac{2}{a(t)}\ p + 2\ \frac{b(t)}{a(t)})$$

$$\frac{\partial^2 u}{\partial p^2} = M(t)\ \{\exp(\cdot)\cdot(\frac{2}{a(t)}p + 2\ \frac{b(t)}{a(t)})^2 + \exp(\cdot)\cdot(-\ \frac{2}{a(t)}\}$$

and inserting into

$$\frac{\partial u}{\partial t} - \mu\frac{d}{dp}(p.u) - \frac{1}{2}D\cdot\frac{\partial^2 u}{\partial p^2} = 0,$$

we have

$$\exp(\cdot)(\dot{M} + M\{\frac{1}{a^2}.\dot{a}\cdot p^2 + 2p(\frac{\dot{b}\cdot a - b\cdot\dot{a}}{a^2})\}) - \mu\{(M\cdot\exp(\cdot) + p\cdot M\exp(\cdot)\cdot(\frac{2}{a}\ p + 2\ \frac{b}{a})\}$$

$$-\ \frac{1}{2}D\{M\cdot\exp(\cdot)\cdot(-\ \frac{2}{a}p + 2\frac{b}{a})^2 + \exp(\cdot)\cdot(-\ \frac{2}{a})\},\ \text{as}\ (-\ \frac{2}{a}p + 2\ \frac{b}{a})^2 = 2(2\frac{b^2}{a^2} - 4\frac{b}{a^2}p + 2\ \frac{p^2}{a});$$

we obtain

$$\frac{\dot{M}}{M} + (\frac{\dot{a}}{a^2}\ p^2 + 2p(\frac{\dot{b}a - b\dot{a}}{a^2}) - \mu - \mu p\cdot(-\ \frac{2}{a}p + 2\frac{b}{a}) - \ \frac{1}{2}D\{2(2\frac{b^2}{a^2} - 4\frac{6}{a^2}p + 2\frac{p^2}{a} - \ \frac{2}{a}\}.$$

Now we collect the expressions with respect to p^0

$$\frac{\dot{M}}{M} - \mu - D2\ \frac{b^2}{a^2} + \frac{D}{a} = 0,$$

the expressions with p

$$2p(\frac{\dot{b}a - b\cdot\dot{a}}{a^2}) - \mu p\cdot 2\frac{b}{a} + \frac{1}{2}D\ \frac{4}{2}\cdot\frac{b}{a^2}p = 0,$$

and with p^2

$$\frac{\dot{a}}{a^2} + \mu\frac{2}{a}\frac{a}{a} - \frac{1}{2}D(2\cdot 2\cdot \frac{1}{a} \cdot \frac{a}{a}) = 0.$$

Hence

$$\dot{a} + 2\mu a - 2D = 0,$$

thus

$$\dot{a} = -2\mu\cdot a + 2D.$$

We obtain from the second equation by rearrangements $\dot{b} = -\mu\cdot b$, and finally the equation system

$$\dot{a} = -2\mu a - 2D \qquad a(t) = \frac{D}{\mu}(1-\exp(-2\mu t) + \mu_o \exp(-2\mu t))$$

and the solutions

$$\dot{b} = -\mu\cdot b \qquad b(t) = b_o\cdot \exp(-\mu t)$$

$$\frac{\dot{M}}{M} = \mu + D2\cdot(\frac{b}{a})^2 - \frac{D}{a} \qquad M(t) = \frac{1}{\sqrt{\pi\cdot a}}\exp(-\frac{b^2}{a}).$$

As the ansatz is a Gaussian distribution, we interpret

a(t) = the width of the Gaussian distribution

b(t) = the displacement of the Gaussian distribution.

The explicit solution of

$$D\cdot \frac{\partial^2 u}{\partial p^2} = \frac{\partial u}{\partial t} - 2\mu p\frac{\partial u}{\partial p} - 2\mu\cdot u$$

can be given as

$$u(p,t) = M(t)\cdot \exp(-\frac{p^2}{a} + 2\frac{b}{a}p) = \frac{1}{\sqrt{\pi\cdot a}}\exp(-\frac{b^2}{a} - \frac{p^2}{a} + 2\frac{b}{a}p) =$$

$$= \frac{1}{\sqrt{\pi\cdot a}}\exp(-\frac{1}{a}(p-b)^2);$$

and finally the closed form can be written:

$$u(p,t) = \frac{1}{\sqrt{\pi a(t)}}\exp(-\frac{(p-b(t))^2}{a(t)}),$$

where the width a and the displacement b satisfy

$$\frac{da}{dt} = -2\mu a + 2D \quad \text{and} \quad \frac{db}{dt} = -\mu b.$$

We derived $\mu = \frac{\beta}{D}$, where β is the friction coefficient. In (141), p. 191 we introduced friction phenomena generated in boundary layers, given by

$$\tau = \varkappa \frac{\partial u}{\partial y} ,$$

where $\varkappa$ is the "Zähigkeit" of the population movement. At the boundaries, the equation of motion reduces to

$$v_r \frac{\partial u}{\partial y} = \frac{\varkappa}{\rho} \frac{\partial^2 u}{\partial y^2} .$$

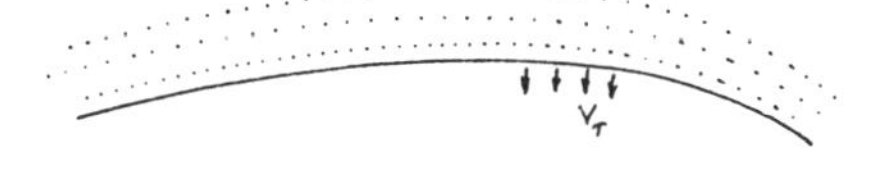

The friction phenomena are described by

$$\tau = \varkappa \frac{\partial u}{\partial y} = \frac{\varkappa^2}{\rho \cdot v_r} \frac{\partial^2 u}{\partial y^2} = - \mu .$$

thus we obtain

$$\frac{da}{dt} = 2D + \frac{\varkappa^2}{\rho \cdot v_r} \frac{\partial^2 u}{\partial y^2}$$

and consider regions, where

$$\frac{\partial^2 u}{\partial y^2} > 0, \frac{\partial^2 u}{\partial y^2} < 0 \text{ and at } \frac{\partial^2 u}{\partial y^2} = 0,$$

there is a transition from laminar to turbulent phenomena ((141), p. 193). Therefore the width of the Gaussian distribution is stabilized in regions, where $\frac{\partial^2 u}{\partial y^2} < 0$ as

$$a(t) = \frac{D}{\mu} (1-\exp(\frac{2\varkappa^2}{\rho \cdot v_r} \cdot t \cdot \frac{\partial^2 u}{\partial y^2}) + \alpha_0 \cdot \exp(\frac{2\varkappa^2}{\rho \cdot v_r} \cdot t \cdot \frac{\partial^2 u}{\partial y^2}),$$

but the width of the Gaussian distribution is unstable in regions, where

$$\frac{\partial^2 u}{\partial y^2} > 0.$$

This can be applied to very interesting socio-economic phenomena ((141, p. 134, pp. 198f).

The solution $u(p,t)$ can be interpreted as the average over all paths of the individuals, thus, as the probability distribution of the paths; $u(p,t)dp$ gives the probability of finding an individual at the position p in the interval dp at time t.

The displacement of the distribution is shown in the figure. The line ..... shows the most probable path of the individuals.

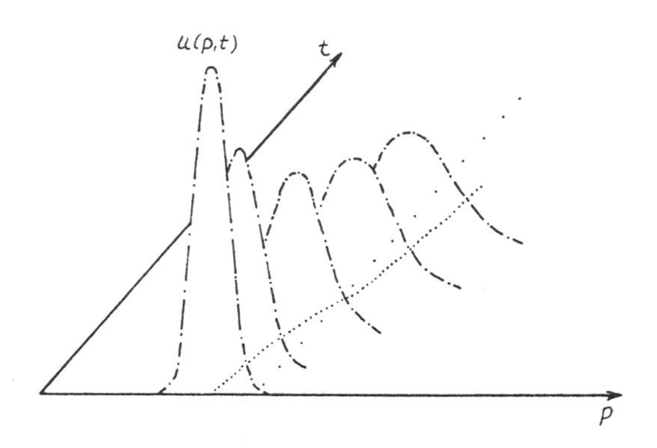

IV. 3 Prediction of certain irreversible processes

The following simple model gives a good intuitive understanding of an irreversible process. We consider the effect of entropy on macroscopic predictions (in the sense of Jaynes ((132), (173)-(177)). Introducing a macrostate $(A_1,\ldots,A_m)$ and the time-dependent probability $p(A_1,\ldots,A_m,t)$ which represents a "bubble" of probability in the macroscopic state space, which expands due to diffusion and satisfies a phenomenological equation of bubble dynamics

$$\frac{\partial P}{\partial t} + \frac{D}{k} \operatorname{div} (P \operatorname{grad} S) = D \cdot \Delta P,$$

where $S(A) = k \cdot \ln W(A)$ is the ordinary entropy of thermodynamics (339). A given macrostate $(A_1,\ldots,A_m)$ in physics can be realized by an enormous number $W(A_1,\ldots,A_m)$ of microstates, which would lead to slightly different macrostates a short time later.

The result is, that after an initial transient period in which a non-Gaussian bubble becomes Gaussian, at as good a level of accuracy as one could want, the solution is a Gaussian bubble moving along a trajectory in A-space and readjusting its size in accordance with the local entropy curvature.

In the one dimensional case, putting $A_1 = x$, we obtain

$$P(x,t) = \frac{1}{\sqrt{2\pi a(t)}} \exp \left(-\frac{(x-b(t))^2}{2a(t)}\right),$$

where

1) $\dot{b} = K^{-1}D \cdot S'(b)$ and

2) $\dot{a} = 2D(1 + K^{-1}S''(b) \cdot a)$.

Equation 1) is of Onsager form (275), (276); the system is moving along a trajectory defined by the local entropy gradient. But 2) shows a welcome new feature: the curvature of the entropy function stabilizes the bubble if it is convex ($S' < 0$). A strongly convex (i.e. downward curving $S'' < 0$) entropy function is strongly stabilizing, leading to a small bubble. When the curvature of the entropy function decreases, stabilizing forces are weaker and the bubble enlarges, representing more "random" behaviour. When the entropy curvature is zero, the restoring forces are zero, and the bubble spreads following Einstein's law of Brownian motion.

IV. 4.1 Crowd movement described by stochastic methods

Let x stand for discrete points, the lattice points

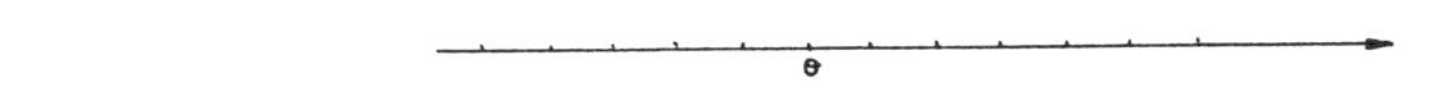

An individual starts from the origin $x = 0$ and moves with speed v from lattice point to lattice point step by step. He can move either in a positive or in the negative direction. Each step is of duration Δt and covers a distance Δx, thus we have $\Delta x = \Delta v \cdot t$.

Each time the individual arrives at a lattice point there is a probability $p \cdot \Delta t$ of reversal of direction, then

$$1 - p \cdot \Delta t$$

is the probability that the direction of motion will be maintained.

What is wanted is the probability that after a certain time t the individual is at a certain interval. If the individual starts from the origin, the displacement after n steps, denoted by $\mathcal{D}_n$, it is the displacement after time $n \cdot t$. Introducing the random variable

$$\mathfrak{X} = \begin{cases} 1 & \text{with probability } 1-p\cdot\Delta t \text{ (the direction will be maintained)} \\ -1 & \text{"} \quad \text{"} \quad p\cdot\Delta t \text{ (reversal of direction)} \end{cases}$$

and considering a sequence of such independent random variables $\mathfrak{X}_1, \mathfrak{X}_2, \ldots, \mathfrak{X}_{n-1}$. Now the individual starts in the positive direction from the origin. The first step will take it a distance $v \cdot \Delta t$ in positive direction, then the individual tosses a coin and finds his velocity will change from v to $x_1 v$. This means that it will be maintained or reverse according to the outcome of the toss.

In the next step the individual will move an additional distance

$$x_1 v\Delta t.$$

So the displacement after n-steps will be

$$\begin{aligned} \mathcal{D}_n &= v\cdot\Delta t + x_1 v\Delta t + x_2 x_1 v\Delta t + x_3 x_2 x_1 v\Delta t + \ldots \\ &= v\cdot\Delta t\ (1 + x_1 + x_1 x_2 + x_1 x_2 x_3 + \ldots + x_1 x_2 \ldots x_n). \end{aligned}$$

If the individual started in the negative direction, the displacement would have been

$$\tilde{\mathcal{D}}_n = -\ v\Delta t\ (1 + x_1 + x_1 x_2 + x_1 x_2 x_3 + \ldots + x_1 x_2 \ldots x_n) = -\ \mathcal{D}_n.$$

Now we introduce an arbitrary function f(x) and ask for the average

$$\langle f(x + \mathcal{D}_n)\rangle.$$

If f(x) is the characteristic function of an interval, then the average will be the probability of finding the individual in that interval after n steps, if he started at the point x. Now let us consider the two functions

$$F_n^+(x) = \langle f(x+\mathcal{D}_n)\rangle,\ F_n^-(x) = \langle f(x-\mathcal{D}_n)\rangle,$$

using $\mathcal{D}_n$ and factoring x_1 out, we obtain

$$F_n^+(x) = \langle f(x+v\cdot\Delta t+x_1 v\cdot\Delta t(1+x_2+x_2x_3+\ldots+x_2x_3\ldots x_{n-1}))\rangle.$$

Now we first take the average on x_1 and then on all the other x_i's. Averaging on x_1, this variable can assume the value -1 with probability $p\cdot\Delta t$, and it can assume the value +1 with probability $1 - p\cdot\Delta t$, so we obtain

$$\begin{aligned} F_n^+(x) &= p\cdot\Delta t\langle f(x+v\Delta t-v\Delta t(1+x_2+x_2x_3+\ldots))\rangle + \\ &+ (1-p\Delta t)\langle f(x+v\Delta t(1+x_2+x_2x_3 + \ldots))\rangle. \end{aligned}$$

If we replace x by $x + v\cdot\Delta t$ and n by n-1, then we obtain

$$F_n^+(x) = p\cdot\Delta t\cdot F_{n-1}^-(x+v\Delta t) + (1-p\Delta t)F_{n-1}^+(x+v\Delta t).$$

In a quite analogous way, we derive

$$F_n^-(x) = p\cdot\Delta t F_{n-1}^+(x-v\Delta t) + (1-p\Delta t)F_{n-1}^-(x-v\Delta t).$$

We can pass formally from the difference equation to the differential equation by taking $\Delta t \to 0$. If we pass from the discrete to the continuous, we notice that n measures time, as n is the number of steps of the individual and $n\cdot\Delta t$ the time. Rewriting the above relations, we obtain by Kac (179)

$$\frac{F_n^+(x)-F_{n-1}^+(x)}{\Delta t} = \frac{F_{n-1}^+(x+v\Delta t)-F_{n-1}^+(x)}{\Delta t} - p\cdot F_{n-1}^+(x+v\Delta t) + pF_{n-1}^-(x+v\Delta t),$$

and passing to the limit we get

$$\frac{\partial F^+}{\partial t} = \frac{\partial F^+}{\partial x}\cdot v - p\cdot F^+ + p\cdot F^-,$$

in an analogous way

$$\frac{\partial F^-}{\partial t} = -\frac{\partial F^-}{\partial x}v + pF^+ - pF^-.$$

Introducing the functions

$$F = \frac{1}{2}(F^+ + F^-) \text{ and } H = \frac{1}{2}(F^+ - F^-),$$

and adding the above equations we get

$$\frac{\partial F}{\partial t} = v\cdot\frac{\partial H}{\partial x}.$$

Subtracting $\frac{\partial F^+}{\partial t}$ from $\frac{\partial F^-}{\partial t}$ we have

$$\frac{\partial H}{\partial t} = v\frac{\partial F}{\partial x} - 2p\cdot H.$$

We eliminate H by differentiating $\frac{\partial F}{\partial t} = v.\frac{\partial H}{\partial x}$ with respect to t and $\frac{\partial H}{\partial t} = v.\frac{\partial F}{\partial x}-2pH$ with respect to x, and we obtain finally the partial differential equation

$$\frac{1}{v}\frac{\partial^2 F}{\partial t^2} = v\frac{\partial^2 F}{\partial x^2} - \frac{2p}{v}\frac{\partial F}{\partial t}.$$

Now we assume that the motion of the individuals is continuous. During the time interval dt there is a probability pdt of changing direction (the probability of not changing direction is then 1-pdt).

We suppose that R(t) is a random variable (distribution), that is a

measurable function, and $R(t)$ can assume only integral values 0,1,2,...
The probability that $R(t)$ is equal to k at time t is given by the Poisson formula

$$\text{Prob.}\{R(t) = k\} = \frac{(a\cdot t)^k}{k!} e^{-at}.$$

Now take a finite number of increasing time points

$$t_1 < t_2 < t_3 < \dots < t_n$$

then the increments

$$R(t_2)-R(t_1),\ R(t_3)-R(t_2),\dots,R(t_n)-R(t_{n-1})$$

are independent, and $R(t)$ is a Poisson process.

We think of $R(t)$ as the number of individuals going out a shopping-centre up to time t. Then $R(t_2) - R(t_1)$ represents the number of individuals going out in the time interval (t_1,t_2). If we are looking at another time interval, which does not overlap the first one, then the number of individuals going out during this interval is independent of the number in the other one.

It is well known in stochastics that, if we assume that we have an event with probability pdt of happening in t+dt and 1-pdt of not happening, then the number of events which occur up to time t is a Poisson process. Thus we can say, the number of meetings our individual undergoes on a shopping trip up to time t is just a Poisson process. Every time our individual meets somebody it reverses the velocity. If the individual (she or he) starts in the positive direction, then for some time the velocity will be v; if he or she meets somebody, then he or she changes to -v. It remains the same until the individual meets another person and so forth.

Now we can relate the velocity to the Poisson process. Remembering that the number of meetings up to time t is just $R(t)$, and the velocity changes sign at each meeting, thus we can write

$$v(t) = v\cdot(-1)^{R(t)},$$

this says that after an even number of meetings the individual has his old velocity; after an odd number, the individual has the negative of it.

Now the displacement $x(t)$ is very simple

$$x(t) = \int_0^t v(\tau)d\tau = v \int_0^t (-1)^{R(\tau)}d\tau,$$

and this is the continuous analogue of $\mathcal{D}_n$. We have a direct analogy with the discrete case.

We have constructed on page 286

$$F(x,t) = \frac{1}{2} \langle f(x+\mathcal{D}_n)\rangle + \frac{1}{2} \langle f(x-\mathcal{D}_n)\rangle,$$

which satisfies the partial differential equation

$$\frac{1}{v} \cdot \frac{\partial^2 F}{\partial t^2} = v \cdot \frac{\partial^2 F}{\partial x^2} - \frac{2p}{v} \cdot \frac{\partial F}{\partial t}$$

with respect to the initial value problem

$$F(x,0) = f(x), \ \left(\frac{\partial F}{\partial t}\right)_{t=0} = 0.$$

If we replace the discrete random walk by the continuous one, we can show that

$$F(x,t) = \frac{1}{2} \langle f(x + v \cdot \int_0^t (-1)^{R(\tau)}d\tau)\rangle + \frac{1}{2}\langle f(x-v \int_0^t (-1)^{R(\tau)}d\tau)\rangle$$

satisfies the partial differential equation (179)

$$\frac{1}{v} \frac{\partial^2 F}{\partial t^2} + \frac{2p}{v^2} \frac{\partial F}{\partial t} = \Delta F,$$

with the initial conditions

$$F(x,y,0) = f(x,y),$$

$$\left(\frac{\partial F}{\partial t}\right)_{t=0} = 0.$$

Thus we have replaced the time t by the randomized time

$$\int_0^t (-1)^{R(\tau)}d\tau,$$

so we can construct the solution of the above partial differential equation of hyperbolic type in terms of the Poisson process.

IV. 4.2 Crowd movement and advertising pulses

Let x be the coordinate measured along the length of the shopping street, and 2a the distance with respect to the shopping fronts.

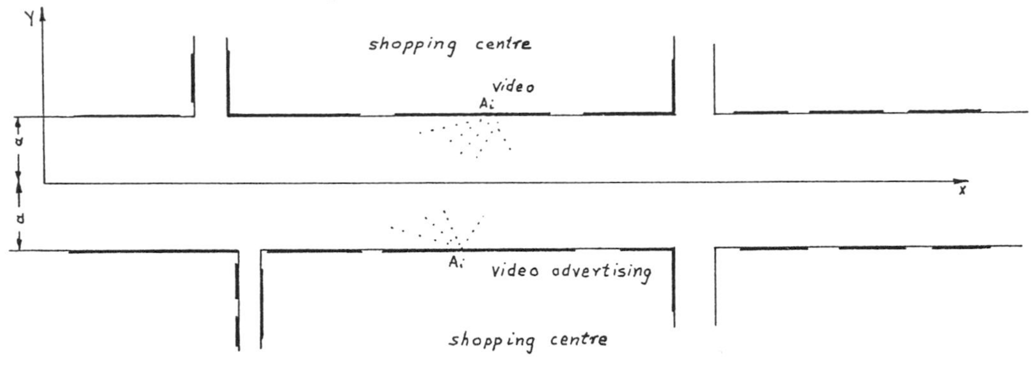

We assume that the movement phenomena in the shopping street are described by the nonlinear 2-dimensional system of partial differential equations ((141), p. 270).

$$\frac{\partial u}{\partial t} + u \cdot \frac{\partial u}{\partial x} + v \cdot \frac{\partial u}{\partial y} = -\frac{1}{\rho} \cdot \frac{\partial p}{\partial x} - \nu \Delta u$$

$$\frac{\partial v}{\partial t} + u \cdot \frac{\partial v}{\partial x} + v \cdot \frac{\partial v}{\partial y} = -\frac{1}{\rho} \cdot \frac{\partial p}{\partial y} + \nu \Delta v.$$

We assume, that the components of the velocity in the y- and z-direction are zero, thus $v = 0$. The equation of continuity $\frac{\partial u}{\partial x} + \frac{\partial u}{\partial y} = 0$ implies that $\frac{\partial u}{\partial x} = 0$; thus, the velocity component u depends on y and t, therefore $u = u(y,t)$. The movement phenomenon is given by

$$\frac{\partial u}{\partial t} = -\frac{1}{\rho} \cdot \frac{\partial p}{\partial x} + \nu \cdot \frac{\partial^2 u}{\partial y^2} \text{ and } 0 = -\frac{\partial p}{\partial y}, \quad 0 = -\frac{\partial p}{\partial z},$$

the two last equations show that p is a function of x and t only, $p = p(x,t)$.

We introduce the dimensionless variables

$$v = a\xi, \quad \tau = \frac{\mu \cdot t}{\rho \cdot a^2}, \quad u = u_0 \cdot v,$$

then we obtain the following transformations

$$\frac{\partial}{\partial \xi} = a \frac{\partial}{\partial y}; \quad \frac{\partial}{\partial t} = \frac{\mu}{\rho \cdot a^2} \frac{\partial}{\partial \tau} \quad ; \quad \frac{\partial^2}{\partial y^2} = \frac{1}{a^2} \frac{\partial^2}{\partial \xi^2},$$

and

$$-\frac{1}{\rho}\frac{\partial p}{\partial x} = \frac{\partial u}{\partial t} - \nu\cdot\frac{\partial^2 u}{\partial y^2} = \frac{\mu\cdot u_0}{\rho\cdot a^2}\frac{\partial v}{\partial \tau} - \nu\cdot\frac{u_0}{a^2}\frac{\partial^2 v}{\partial \xi^2} = \frac{\mu\cdot u_0}{\rho\cdot a^2}\left(\frac{\partial v}{\partial \tau} - \frac{\partial^2 v}{\partial \xi^2}\right), \text{ as}$$

$$\nu = \frac{\mu}{\rho} \text{ and } \frac{\partial v}{\partial \tau} - \frac{\partial^2 v}{\partial \xi^2} = -\,G(\tau),$$

thus

$$-\frac{1}{\rho}\ \frac{\partial p}{\partial x} = -\frac{\mu\cdot u_0}{\rho\cdot a^2}\cdot G(\tau),\ \frac{\partial p}{\partial x} = \frac{\mu\, u_0}{a^2}\, G(\tau).$$

We have to solve the parabolic differential equation

$$\frac{\partial^2 v}{\partial \xi^2} = \frac{\partial v}{\partial \tau} + G(\tau)$$

with respect to the initial conditions

$$v(0,\tau) = 0 \text{ for } \tau < 0,\ v = 0 \text{ for } \tau = 0$$

and the boundary conditions

$$\frac{\partial v}{\partial \xi},\ v \text{ finite at } \xi = \pm\, 1, \text{ this means } y = \pm\, a, \text{ and}$$

$$v(\pm 1,\tau) = \sum_{i=0}^{N} A_i\delta(\tau-\tau_i) \text{ for } \tau > 0,$$

where A_i are nondimensional random quantities:

A_i = the strength of the i-th pulse

τ_i = the random instant at which the i-th pulse occurs.

Now we consider the influenced 'crowd' movement generated by the advertising pulses A_i. We are able to solve the parabolic differential equation by the methods of finite sine transform, S_ξ, defined (46, 47, 48):

$$S(f(x)) = \int_0^{\ell} f(x)\sin(n\,\frac{\pi}{\ell}\cdot x)dx,$$

and the inverse transform is given by the Fourier series

$$f(x) = \frac{2}{\ell}\cdot \sum_{n=1}^{\infty} F_n\cdot\sin n\,\frac{\pi}{\ell}\cdot x,$$

where $F_n = S(f(x))$ and $(0,\ell)$ is an interval.

We apply the sine transform to the parabolic differential equation

$$S_\xi(\frac{\partial v}{\partial t}) - S_\xi(\frac{\partial^2 v}{\partial \xi^2}) = - S_\xi(F(\tau)).$$

The differentiation with respect to τ commutes with the integral

$$S_\xi(\frac{\partial v}{\partial \tau}) = \frac{\partial}{\partial \tau} (S_\xi v)(\xi,\tau)) \underset{\text{def.}}{=} \frac{\partial}{\partial \tau} V_n(\tau).$$

It is known (193) that

$$S_\xi(v''(\xi)) = - (n \frac{\pi}{\ell})^2 S_\xi(v) - n \frac{\pi}{\ell} \{(-1)^n v(\ell)-v(0)\},$$

where $v(\ell)$ and $v(0)$ are the boundary values,

$$S_\xi(v''(\xi)) = - (n \frac{\pi}{\ell})^2 V_n(\tau) - n\cdot\frac{\pi}{\ell} \{(-1)^n \sum_{i=1}^{N} A_i\delta(\tau-\tau_i)-0\};$$

the transformed differential equation

$$\frac{\partial}{\partial \tau} V_n(\tau) - \{-(n \frac{\pi}{\ell})^2 V_n(\tau)\} = n\cdot\frac{\pi}{\ell} \{(-1)^n \sum_i A_i\delta(\tau-\tau_i)\} - G(\tau)$$

defining

$$\phi(\tau) = n \frac{\pi}{\ell} \{(-1)^n \sum_i A_i\delta(\tau-\tau_i)\} - G(\tau),$$

thus, the differential equation reads

$$V_n'(\tau) + (n.\frac{\pi}{\ell})^2 V_n(\tau) = \phi(\tau)$$

and has the solution

$$V_n(n,\tau) = \int_0^\tau \phi(\tau')\exp(-(n \frac{\pi}{\ell})^2(\tau-\tau'))d\tau';$$

using the right-hand side of the differential equation, we obtain the explicit solution in closed form by

$$V_n(\tau) = \pi\cdot\frac{n}{\ell}\{(-1)^n \sum_{i=1}^{N} \int_0^T A_i\exp(-(n \frac{\pi}{\ell})^2(\tau'-\tau_i)\delta(\tau'-\tau_i)d\tau' \pm$$

$$\pm \int_0^\tau G(\tau')\exp(-(n \frac{\pi}{\ell})^2(\tau-\tau')d\tau'.$$

Introducing the Heavyside-function (Yosida (367), p. 50) $\frac{d}{d\tau'}(H(\tau') = \delta$, then

$$V_n(\tau) = \pi \frac{n}{\ell} \{(-1)^n \sum_{i=1}^{N} A_i \exp(-(n \frac{\pi}{\ell})^2(\tau-\tau_i))H(\tau-\tau_i) \pm$$

$$\pm \int_0^\tau G(\tau')\exp(-(n \frac{\pi}{\ell})^2(\tau-\tau'))d\tau'.$$

Applying the inverse transform

$$v(\xi,\tau) = \frac{2}{\ell} \sum_{m=1}^{\infty} V_m(\tau)\sin(m \frac{\pi}{\ell})\xi,$$

we obtain the closed solution

$$v(\xi,\tau) = \frac{2}{\ell} \sum_{m=1}^{\infty} \{(-1)^m (m \frac{\pi}{\ell}) \sum_{i=1}^{N} A_i \exp(-(m \frac{\pi}{\ell})^2(\tau-\tau_i))H(\tau-\tau_i) \pm$$

$$\pm \int_0^\tau G(\tau')\exp(-(m \frac{\pi}{\ell})^2(\tau-\tau')d\tau'\} \sin(m \frac{\pi}{\ell})\xi.$$

The quantity of the population moving per unit width of the shopping street is

$$M(\tau) = \int_{-1}^{+1} v(\xi,\tau)d\xi = 2\cdot \int_0^1 v(\xi,\tau)d\xi,$$

thus,

$$M(\tau) = \frac{2}{\ell} \sum_{m=1}^{\infty} \{(-1)^m \sum_{i=1}^{N} A_i \exp(-(m \frac{\pi}{\ell})^2(\tau-\tau_i))H(\tau-\tau_i) -$$

$$- \frac{1}{\frac{m\cdot\pi}{\ell}} \int_0^\tau G(\tau')\exp(-(m \frac{\pi}{\ell})^2(\tau-\tau')d\tau'\},$$

written as a stochastic integral, we obtain finally

$$M(\tau) = \frac{2}{\ell} \sum_{m=1}^{\infty} \{(-1)^m \int_0^\tau A(t)\exp(-(m \frac{\pi}{\ell})^2(\tau-t)d(P(t)) -$$

$$- \frac{\ell}{m\cdot\pi} \int_0^\tau G(\tau')\exp(-(m \frac{\pi}{\ell})^2(\tau-\tau'))d\tau'\},$$

where $P(t)$ is represented by the number of pulses in time t, such that $dP(t)$ represents the number of advertising pulses in $(t,t+dt)$.

The points of the time axis τ_i belonging to the times of occurrence of advertising pulses are assumed to be distributed in accordance with the Poisson law with constant average density ρ_p. The product densities in this case are

$$E(dP(t)) = \rho_p$$

and the expectation of the distribution of A(t) is given by

$$E\{A(t)\} = C = \text{constant}.$$

Since the pulses act in a direction opposing to the free stream, C is negative. The condition that the population movement is reversed in an averaged sense, if

$$E(M(\tau)) < 0,$$

can be written as

$$\frac{2}{\ell} \cdot \sum_{m=1}^{\infty} \{(-1)^m \int_0^{\tau} E(A(t))\exp(-(m\,\tfrac{\pi}{\ell})^2(\tau-t))E\,\{dP(t)\}$$

$$< \sum_m \frac{\ell}{m\pi} \int_0^{\tau} E\{G(\tau')\}\exp(-(m\,\tfrac{\pi}{\ell})^2(\tau-\tau'))d\tau'.$$

Assuming that $G(\tau)$ is determinisitic, we take $G(t) = \alpha e^{-\beta t}$ and the integral can be written as

$$\int_0^{\tau} \alpha\cdot\exp(-\beta\tau')\cdot\exp(-(m\,\tfrac{\pi}{\ell})^2(\tau-\tau')d\tau = \int_0^{\tau} \alpha\,\exp(-(m\,\tfrac{\pi}{\ell})^2\tau)\cdot$$

$$\exp(\tau'((m\,\tfrac{\pi}{\ell})^2-\beta))d\tau' = \frac{\alpha.\exp(-((m\cdot\frac{\pi}{\ell})^2\;)}{(m\,\frac{\pi}{\ell})^2 - \beta}\; e^{((m\,\frac{\pi}{\ell})^2-\beta)\tau'}\Big|_0^{\tau} =$$

$$= \frac{\alpha}{(m\,\frac{\pi}{\ell})^2 - \beta}(e^{-\beta\tau}-e^{-(m\,\frac{\pi}{\ell})^2\tau}),$$

the first integral

$$\int_0^{\tau} \exp(-(m\,\tfrac{\pi}{\ell})^2\tau)\exp(m\,\tfrac{\pi}{\ell})^2 t = \frac{\exp(-(m\,\frac{\pi}{\ell})^2)\tau}{(\frac{m\cdot\pi}{\ell})^2}\,(e^{(\frac{m\cdot\pi}{\ell})^2\tau} - 1) =$$

$$\frac{1}{(\frac{m\cdot\pi}{\ell})^2}\,(1 - e^{-(\frac{m\cdot\pi}{\ell})^2\tau}),$$

and finally

$$\frac{2}{\ell}\sum_{m=1}^{\infty} C\cdot x(-1)^m \frac{1}{(\frac{m.\pi}{\ell})^2}\,(1-e^{-(\frac{m\cdot\pi}{\ell})^2\tau)} < \sum_{m=1}^{\infty} \frac{\ell}{m\cdot\pi}\,\frac{\alpha}{(\frac{m\cdot\pi}{\ell})^2 - \beta}\,(e^{-\beta\tau}-e^{-(\frac{m\pi}{\ell})^2\tau)}.$$

As $\frac{\partial p}{\partial x} = \frac{\mu u_0}{\alpha^2} \cdot G(t)$, we may say that G(t) characterizes the pressure gradient in the deterministic sense, if $G(t) = \alpha \cdot e^{-\beta t}$.

Now we consider the random pressure gradient under the condition that there are no pulses of strength A_i influencing the population movement. Thus we have the expected value of the velocity $v(\xi,\tau)$, given by

$$E(v(\xi,\tau)) = \frac{2}{\ell} \sum_{m=1}^{\infty} \sin(\frac{m \cdot \pi}{\ell})\xi \cdot \int_0^{\tau} E(G(\tau'))\exp(-(\frac{m\pi}{\ell})^2(\tau-\tau'))d\tau'.$$

By construction G( ) characterizes the pressure gradient, hence $G(\tau)$ is a random function of τ.

Now we introduce the population movement correlation, defined by

$$E(v(\eta_1,\tau_1) \cdot v(\eta_2,\tau_2)) = \frac{4}{\ell} \sum_{m,n=1}^{\infty} \sin(m \frac{\pi}{\ell})\eta_1 \sin(n \frac{\pi}{\ell}) \eta_2 \cdot$$

$$\int_0^{\tau_1} \int_0^{\tau_2} E\{G(t_1) \cdot G(t_2)\} \exp(-(m \frac{\pi}{\ell})^2(\tau_1-t_1))\exp(-(n \frac{\pi}{\ell})^2(\tau_2-t_2))dt_1 dt_2.$$

This is the correlation at two different points distant η_1 and η_2 from the $\eta = 0$ plane at two different instants of time τ_1 and τ_2.

Thus we obtain, for example, the correlation at different points at the same time, or the same points at different times. We assume that

$$E\{G(t_1) \cdot G(t_2)\} = \sin \beta t_1 \cdot \delta(t_1-t_2),$$

then we have the population movement correlation at the same points at different times τ_1 and τ_2:

$$E(v(\eta,\tau_1) \cdot v(\eta,\tau_2))$$

$$= \frac{4}{\ell} \sum_{m,n=1}^{\infty} \sin(m \frac{\pi}{\ell})\eta \cdot \sin(n \frac{\pi}{\ell})\eta \cdot \int_0^{\tau_1} \int_0^{\tau_2} \sin\beta t_1 \cdot \delta(t_1-t_2)$$

$$\exp(-(m \frac{\pi}{\ell})^2(\tau_1-t_1))\exp(-(n \frac{\pi}{\ell})^2(\tau_2-t_2))dt_1 dt_2.$$

If the process G(t) is stationary, such that the second order correlation of G(t) depends only on the difference

$$\tau_2-\tau_1 = \tau > 0,$$

we may consider

$$\lim_{\substack{\tau_1 \to \infty \\ \tau_2 \to \infty \\ \tau = \tau_2 - \tau_1}} E\{v(\eta,\tau_1)\cdot v(\eta,\tau_2)$$

and obtain an explicit formula, for example, of the power spectrum.

IV. 5 Nonlinear phenomena of influence

IV. 5.1 Threshold-influence

Observing pedestrians in front of shop windows, denoting by $\bar{i}$ the influence rate, then

$$\frac{d\bar{i}}{dt} = \text{gain} - \text{loss}.$$

The gain is proportional to (the present number $\bar{i}$ of influenced individuals) $\times$(the number I of prospective customers). Thus we obtain the

$$\text{gain} = g\cdot I\cdot \bar{i},$$

where g is the gain-constant.

The loss term comes from the escape of individuals from the influence-region.

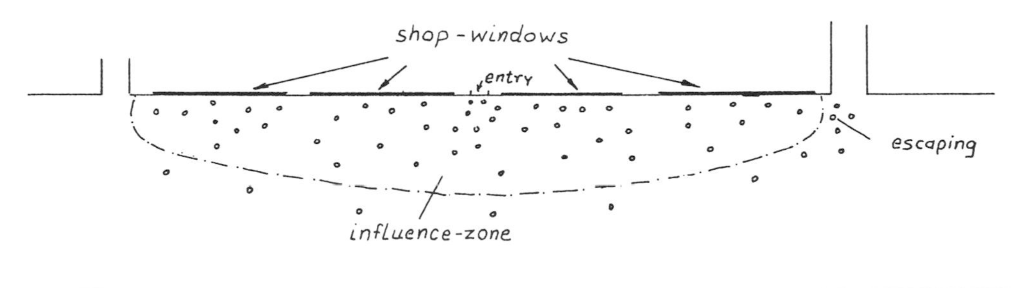

The escape-time is defined as the earliest time at which an individual initially inside the boundaries of the influence-zone, leaves that region (Grusa (141), p. 165f).

Thus the loss is proportional to the present number $\bar{i}$ of influenced individuals $(2\cdot f_{esc.})$, where

$$f_{esc.} = \frac{1}{\text{escape-time}},$$

then the loss $= 2 \cdot f_{esc.} \cdot \bar{i}$.

I is the number of prospective customers, this number decreases as the influenced individuals escape from the influence-zone. The decrease ΔI is proportional to the number $\bar{i}$ of influenced persons, thus $\Delta I = \alpha \cdot \bar{i}$ with α as the proportionality constant.

Let I_o be a fixed number of prospective customers generated by an external innovation field E, given, for example, by the shop-windows or video-display. Thus the number of prospective customers is given by

$$I = I_o - \Delta I.$$

The time-dependent influence-rate can be written as

$$\frac{d\bar{i}}{dt} = \text{gain} - \text{loss}$$

$$= g \cdot I \cdot \bar{i} - 2f_{esc.} \bar{i}$$

$$= -(2f_{esc.} - g \cdot I_o)\bar{i} - g\alpha \cdot \bar{i}^2.$$

Defining the constants

$$C = 2f_{esc.} - gI_o, \quad C_1 = g \cdot \alpha,$$

we obtain the nonlinear differential equation

$$\frac{d\bar{i}}{dt} = - C \cdot \bar{i} - C_1 \bar{i}^2,$$

with the closed solution

$$\bar{i}(t) = - \frac{C}{2C_1} - \frac{|C|}{2} \cdot \frac{A \cdot e^{-|C| \cdot t} - 1}{A \cdot e^{-|C|t} + 1}, \text{ where } A = \frac{\left|\frac{C}{C_1}\right| - \frac{C}{C_1} - 2u_o}{\left|\frac{C}{C_1}\right| - \frac{C}{C_1} + 2u_o},$$

and u_o is the initial condition.

The time dependent influence $\bar{i}(t)$ depends on $C = 2f_{esc.} - gI_o$; if I_o is a small number, this means that the innovation field of the shopping centre is small, then $C > 0$, and by the explicit solution $\bar{i}(t)$, the number of the influenced persons is small.

But if I_0 is high, then $C < 0$, thus a great number of influenced persons may be emitted. This model describes threshold-influence phenomena.

In particular the solution $\bar{i}(t)$ tends for $t \to \infty$ to the following equilibrium value

$$i = \begin{cases} 0 & \text{for } 2 > \aleph g I_0 \\ \frac{C}{C_1} & \text{for } 2 < \aleph g I_0 . \end{cases}$$

The temporal behaviour is expressed in the following figures:

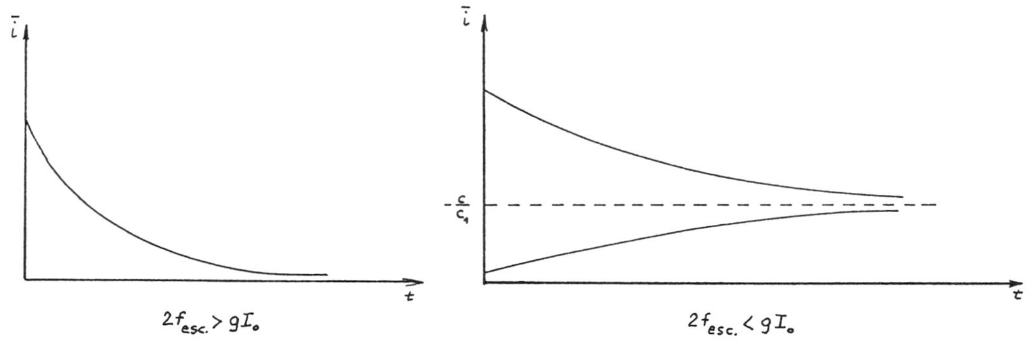

The threshold-influence is given by

$$-\frac{C}{C_1} = \frac{gI_0 - 2f_{esc.}}{g \cdot \alpha} \quad ;$$

thus, the threshold-influence of advertising is proportional to I_0, the innovation field (of the windows) of the shopping centre, and $\bar{i}(t)$ tends to an equilibrium solution without any oscillations.

Now we consider the case

$$2f_{esc.} < g \cdot I_0 ,$$

as

$$f_{esc.} = \frac{1}{\text{escape time}} \quad ,$$

then $f_{esc.}$ can be interpreted as the frequency of escape, and $2f_{esc.} < g \cdot I_0$ means that the frequency of escape is bounded by $g \cdot I_0$, the innovation field of the windows of the shopping centre.

If the individuals satisfy various initial conditions at time $t = 0$ in front of the shop-windows, then for $t \to \infty$ the influence can be written as

$$-\frac{C_1}{C} = \frac{gI_0 - 2f_{esc.}}{g \cdot \alpha} \quad ,$$

this expression is a constant and depends on all parameters, I_o, g, α, f_{esc}. Thus, we have "bleibenden" influence on the individuals independent of their corresponding initial states.

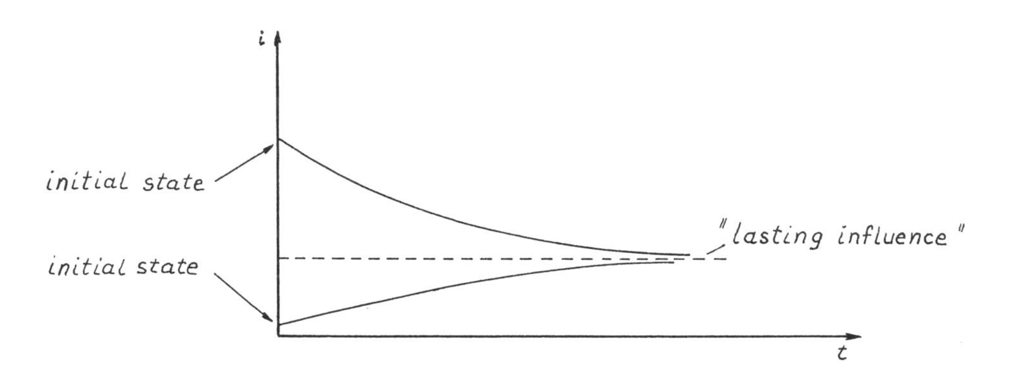

If the frequency of escape from the influence region is not bounded above, e.g.

$$gI_o < 2f_{esc}.$$

then the rate of influence tends to zero for $t \to \infty$.

This simple, but instructive model motivates us to consider influence phenomena under the heading of nonlinear differential equations.

IV. 5.2 Influence phenomena and possible behaviour of individuals

Influence fields which appear in economics, sociology, biology,... can be described by partial differential equations characterizing certain variational principles and generalized variational principles ((139, (141)).

In the following part, we only consider wave phenomena governed by influence potential fields. On a first level of approximation (of real processes in space) we consider the time independent wave equation

$$\sum_\alpha \frac{k^2}{2q} \Delta_\alpha\psi + (I-U)\psi = 0,$$

where Δ_α is the Laplace operator in R^3, U the influence potential field, I the influence coefficient, and k, q_α are coefficients specified by the model-building.

Substituting $\psi = e^{\frac{i}{k}\varphi}$ in the wave equation and remembering the methods of vector analysis, the function φ satisfies the equation

$$\sum_{\alpha} \frac{1}{2q_{\alpha}} (\text{grad}_{\alpha}\varphi)^2 - \sum_{\alpha} \frac{ik}{2q_{\alpha}} \Delta_{\alpha}\varphi = I-U, \ i \in \mathbb{C}.$$

We are seeking φ in the form of a power series

$$\varphi = \varphi_0 + (\frac{k}{i})\varphi_1 + (\frac{k}{i})^2 \varphi_2 + \dots .$$

We are beginning the model-building by considering the simplest case: the one-dimensional motion of individuals under the influence

$$I-U(x)$$

with respect to the x-coordinate

$$\frac{1}{2q} (\frac{\partial\varphi}{\partial x})^2 - \frac{ik}{2q} \frac{d^2\varphi}{dx^2} = I-U(x).$$

Supposing that

$$\frac{i \cdot k}{2 \cdot q} \frac{d^2\varphi}{dx^2} < \frac{1}{2q} (\frac{\partial\varphi}{\partial x})^2 \text{ then } k \left| \frac{\frac{d^2\varphi}{dx^2}}{(\frac{d\varphi}{dx})^2} \right| << 1 \text{ or } \left| \frac{d}{dx} (\frac{k}{\frac{d\varphi}{dx}}) \right| << 1,$$

so we obtain the first approximation

$$\frac{1}{2q} (\frac{d\varphi_0}{dx})^2 = I-U(x), \text{ and finally } \varphi_0 = \pm \int \sqrt{2q(I-U(x))}dx.$$

Defining $h(x) = \sqrt{2q(I-U(x))}$, then $\varphi_0 = \pm \int h(x)dx$.

By the first approximation

$$\frac{d\varphi}{dx} \sim \frac{d\varphi_0}{dx} = h(x), \text{ hence the expression } \frac{d}{dx} (\frac{k}{\frac{d\varphi}{dx}})$$

can be written

$$\frac{d}{dx} (\frac{k}{\frac{d\varphi}{dx}}) = - \frac{k}{h^2} \frac{dh}{dx}, \text{ and as } \frac{dh}{dx} = - \frac{q}{h} \frac{dU}{dx},$$

the derivative of the influence potential U can be interpreted by the action of an economical, sociological or biological influence field.

This influence field is acting on the individuals under the condition of approximation

$$\left| \frac{d}{dx} (\frac{k}{\frac{d\varphi}{dx}}) \right| = \frac{q \cdot k}{h^3} \left| \frac{\partial U}{\partial x} \right| << 1.$$

This means that the approximate solution can be applied only to a short distance of the generated influence field.

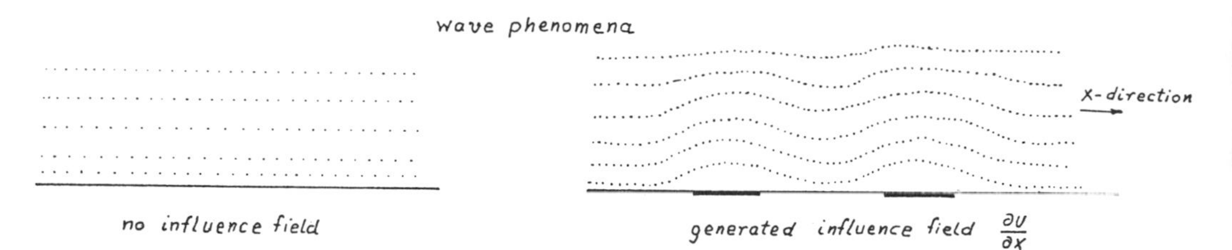

The first term of the approximate solution

$$\varphi = \varphi_0 + (\frac{k}{i})\varphi_1 + (\frac{k}{i})^2 \varphi_2 + \dots$$

is given by

$$\varphi_0 = \pm \int h(x)\cdot dx$$

Now, we calculate the next terms. Using

$$\varphi' = \varphi_0' + \frac{k}{i}\varphi_1' + (\frac{k}{i})^2\varphi_2' \text{ and } \varphi'' = \varphi_0'' + \frac{k}{i}\varphi_1'' + (\frac{k}{i})^2\varphi_2''$$

in

$$\frac{1}{2q}(\frac{\partial\varphi}{\partial x})^2 - \frac{ik}{2q}\frac{d^2\varphi}{dx^2} = I-U(x),$$

we obtain

$$\frac{1}{2q}(\varphi_0'^2 + 2\frac{k}{i}\varphi_0'\varphi_1' + (\frac{k}{i})^2\varphi_1'^2 + 2\varphi_0'\varphi_2'(\frac{k}{i})^2 + 2\cdot\varphi_1'\varphi_2'(\frac{k}{i})^3 +$$

$$(\frac{k}{i})^4\varphi_2'^2) - \frac{ik}{2q}(\varphi_0'' + (\frac{k}{i})\varphi_1'' + (\frac{k}{i})^2\varphi_2'' = I-U(x),$$

the h^0 term

$$\varphi_0' = \sqrt{2q(I-U(x)}$$ (see page 299).

k^1-term

$$\frac{1}{2q}2\frac{k}{i}\varphi_0'\varphi_1' - \frac{ik}{2q}\varphi_0'' = 0 \text{ thus } \varphi_0'\varphi_1' + \frac{1}{2}\varphi_0'' = 0,$$

k^2-term

$$\frac{1}{2q}(\frac{2^2}{i^2}\varphi_1'^2 + 2\varphi_0'\varphi_2'\frac{k^2}{i^2}) - \frac{ik^2}{2qi}\varphi_1'' = 0 \text{ hence } \frac{1}{2}\varphi_1'^2 + \varphi_0'\varphi_2' + \frac{1}{2}\varphi_1'' = 0.$$

The second term can be written as

$$\frac{d\varphi_o}{dx}\frac{d\varphi_1}{dx} + \frac{1}{2}\frac{d^2\varphi_o}{dx^2} = 0, \text{ then } \frac{d\varphi_1}{dx} = \frac{\frac{d^2\varphi_o}{dx^2}}{2\frac{d\varphi_o}{dx}} = -\frac{\frac{dh(x)}{dx}}{h(x)},$$

$$\varphi_1 = -\frac{1}{2}\ln h = \ln\left(\frac{1}{\sqrt{h}}\right);$$

the solution of the influence wave is given by

$$\psi = e^{\frac{i}{k}\varphi} \sim e^{\frac{i}{k}(\varphi_o + \frac{k}{i}\varphi_1)} = e^{\frac{i}{k}\varphi_o} e^{\ln\frac{1}{\sqrt{h}}} = \frac{1}{\sqrt{h}} e^{\frac{i}{k} \pm \int hdx}.$$

Then there exist C_1, C_2 constants, such that

$$\psi = \frac{C_1}{\sqrt{h}}\exp\left(\frac{i}{k}\int h(x)dx\right) + \frac{C_2}{\sqrt{h}}\cdot\exp\left(-\frac{i}{k}\int h(x)dx\right).$$

We treat one possible application in economics. We consider individuals in front of shop-windows. Now let U be the shop-window-influence-potential, then we analyse the influence of advertising on the individuals and give some aspects of their behaviour. We take the influence-wave-solution and consider $\psi\cdot\bar{\psi}$, where $\bar{}$ is the conjugate complex, then

$$|\psi|^2 = \psi\cdot\bar{\psi} \sim \frac{1}{h}.$$

We can interpret h as h = const · (speed of the individuals in front of the shop-windows). Then $|\psi|^2$ = the probability of finding an individual at a point with coordinate between x and x + dx. This probability is proportional to $\frac{1}{h}$, thus the time the individual spends in the window segment dx of the shop front is inversely proportional to the velocity of the individual:

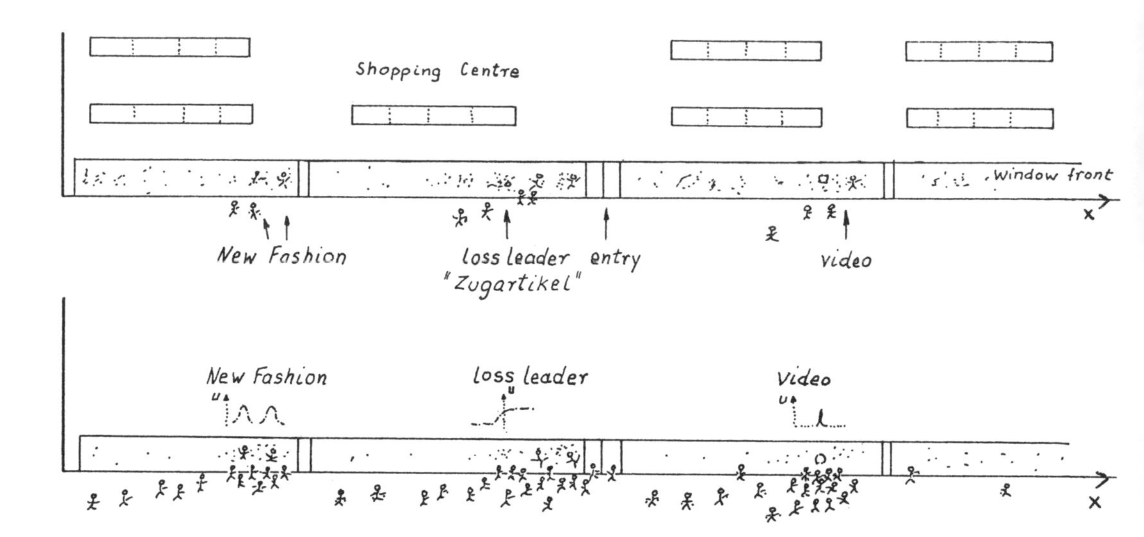

The new fashion, the 'Zugartikel', and video advertising are described by the influence-potential curves given above (169). Now we look from above:

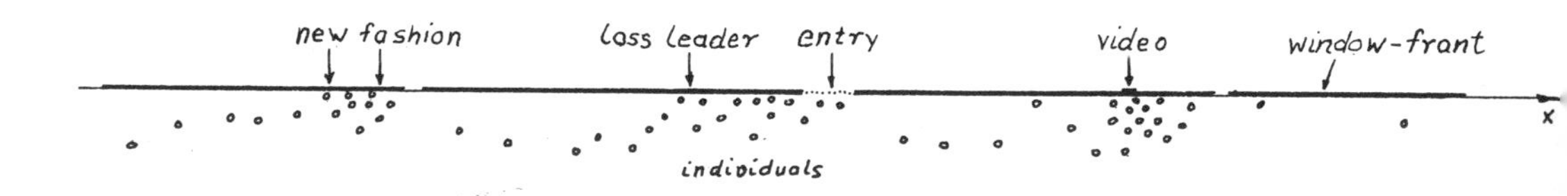

We determine the higher order approximations of the influence wave function. For the term of the order k^2 we have

$$\frac{1}{2}\varphi_1'^2 + \varphi_0' \cdot \varphi_2' + \frac{1}{2}\varphi_1'' = 0, \text{ hence } \varphi_2' = -\frac{1}{2}\frac{(\varphi_1')^2}{\varphi_0'} - \frac{1}{2}\frac{\varphi_1''}{\varphi_0'} \ .$$

By construction we have

$$\varphi_0' = h \text{ and } \varphi_1' = -\frac{h'}{2h}, \text{ thus } \varphi_1'' = \frac{h''2 \cdot h}{4h^2} + \frac{2(h')^2}{4h^2} \ ,$$

and

$$\frac{1}{2}\frac{\varphi_1''}{\varphi_0'} = -\frac{h''}{4h^2} + \frac{(h')^2}{4h^3} \ ,$$

with the expression

$$- \frac{1}{2} \frac{\varphi_1'^2}{\varphi_0'} = - \frac{1}{2} \frac{(h')^2}{4h^2 \cdot h} = - \frac{1}{8} \frac{(h')^2}{h^3} ,$$

we obtain finally

$$\varphi_2' = \frac{h''}{4h^2} - \frac{3}{8} \frac{(h')^2}{h^3} .$$

On page 299 we introduced the influence field $F(x) = - \frac{\partial U}{\partial x}$, where U is the influence-potential and obtained

$$\frac{dh(x)}{dx} = - \frac{q}{h} \frac{\partial U}{\partial x} = \frac{q}{h} F(x),$$

thus we write

$$F(x) = \frac{h \cdot h'}{q} .$$

Integrating φ_2', we have

$$\varphi_2 = \int \frac{h''}{4h^2} dx - \frac{3}{8} \int \frac{(h')^2 \cdot h^2}{h^3 \; h^2} dx,$$

integrating the first term by parts, we can write

$$\varphi_2 = \frac{qF}{4h^3} + q^2 \frac{3}{8} \int \frac{F(x)^2}{h^5} dx.$$

The approximated influence-wave is given by

$$\psi = e^{\frac{i}{k}\varphi} = e^{\frac{i}{k} \varphi_0 + \varphi_1} (1 - ik\varphi_2).$$

and explicitly

$$\psi = \text{constant} \frac{1}{\sqrt{h}} \left(1 - \frac{i \cdot q \cdot k}{4} \frac{F}{h^3} - \frac{ikq^2 \cdot 3}{8} \int \frac{F^2}{h^5} dx\right) \exp\left(\frac{i}{k} \int h(x) dx\right)$$

We assumed on page 299 that

$$\left| \frac{d}{dx} \left(\frac{k}{\frac{\partial \varphi}{\partial x}} \right) \right| = \frac{q \cdot k}{h^3} |F| << 1;$$

now the second term $kq^2 \cdot \int \frac{F^2}{h^2} dx$ may also be $<< 1$. Therefore, we suppose that the influence field, given by F^2, tends quickly to zero over a short distance.

Thus, we obtain the approximated influence-wave

$$\psi = \frac{1}{\sqrt{h}}\,(C_1\cdot\exp(\frac{i}{k}\int h(x)dx) + C_2\cdot\exp(-\frac{i}{k}\int h(x)dt)),$$

as $h(x) = \sqrt{2q(I-U)}$ we obtain that $h(x)$ is imaginary for $I < U$ and

$$\psi = \frac{1}{\sqrt{h}}\,\{C_1\cdot\exp(\frac{1}{k}\int|h(x)|\,dx) + C_2\cdot\exp(-\frac{1}{k}\int|h(x)|dx)\}.$$

Let the influence potential U be given in the form and the influence coefficient I be given by

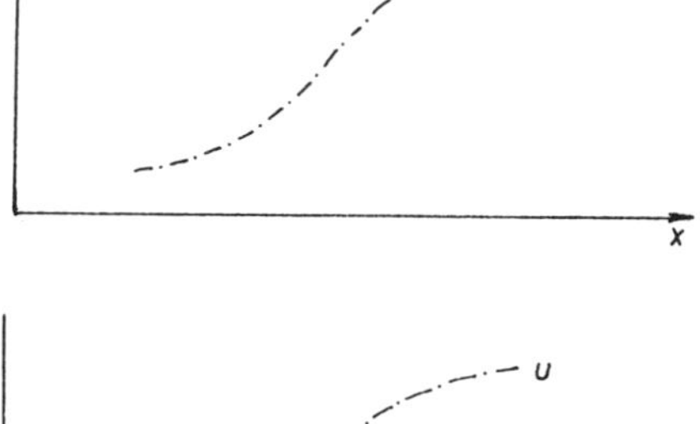

$$I = U(a).$$

We obtain a diagram of the influence potential

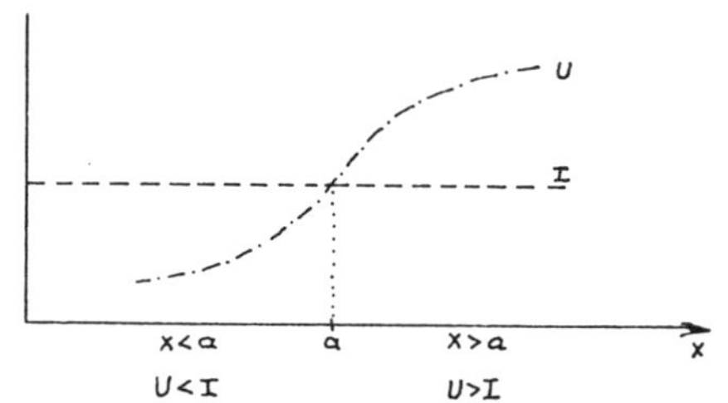

In the region to the right of $x = a$ the influence-wave may be damped and has the form

$$\psi = \frac{C}{\sqrt{h}}\exp(-\frac{1}{k}\int_a^x h(x)dx) \text{ for } x > a,$$

and to the left of $x = a$ the influence-wave is given by the explicit solution above. The coefficients C_1 and C_2 may be determined by methods of the function theory. We are following the variation of the influence-wave from positive x-a to negative x-a. For small values of x-a we have

$$I-U(x) \approx \left.\frac{\partial U}{\partial x}\right|_{x=a}(x-a) = F(a)\cdot(x-a),$$

and we interpret $F(a)$ as the influence field at $x = a$.

A possible application in economics:
We consider a large departmental store with a long window front (see above).

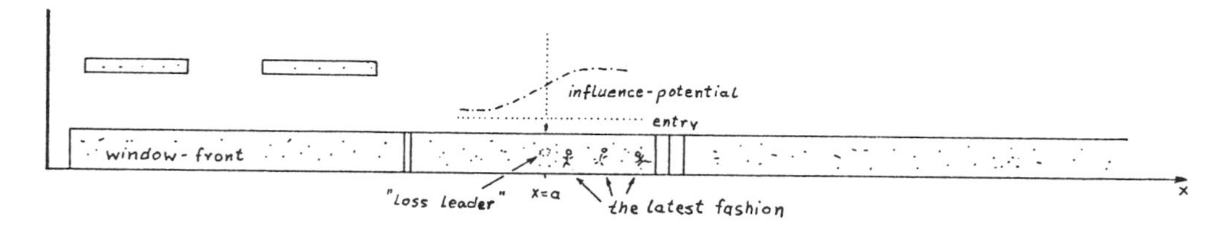

Then $F(a)$ is interpretable as the influence of the "loss leader" (169). Therefore, $h(x)$ takes the form

$$h(x) = \sqrt{2qF(a)(x-a)}$$

and the influence-wave to the right of the shopping front is

$$\psi(x) = \frac{C}{\sqrt{h(x)}} \exp\left(-\frac{1}{k}\int_a^x h(x)dx\right)$$

$$= \frac{C}{|2q\cdot F(a)|^{1/4}|x-a|^{1/4}} \exp\left(-\frac{1}{k}\int_a^x \sqrt{2q\cdot F(a)(x-a)}dx\right).$$

Assuming $\psi(x)$ complex valued, we take a path in the complex plane, which is sufficiently far from $x = a$:

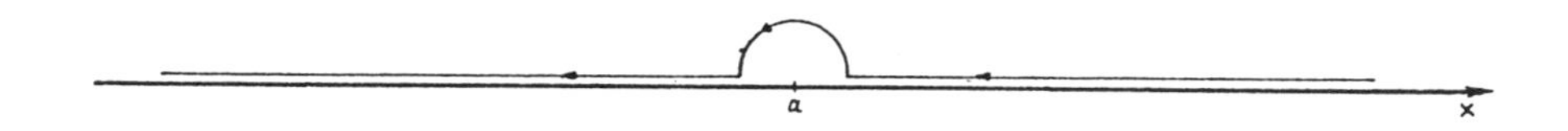

then the arc $(x-a) = \rho\cdot e^{i\varphi}$ gives $\frac{dx}{d\varphi} = i\rho e^{i\varphi}$, and the integral

$$\int \sqrt{x-a}\ dx = \int \rho^{\frac{1}{2}} e^{i\frac{\varphi}{2}} i\cdot\rho\cdot e^{i\varphi}d\varphi = i\int \rho^{3/2}e^{i\frac{3}{2}\varphi}d\varphi = i\rho^{\frac{3}{2}}(\cos\tfrac{3}{2}\varphi + i\sin\tfrac{3}{2}\varphi).$$

The exponential factor increases for $\varphi \in (0, \frac{2}{3}\pi)$ and decreases to 1, and the exponent is imaginary

$$\exp\left(-\frac{1}{k}\, i\int_a^x h(x)dx\right).$$

As we take the path from the left (positive x-a) to the right (negative x-a), we obtain

$$(x-a) \to -(x-a)$$

thus

$$\frac{1}{|x-a|^{1/4}} \rightarrow \frac{1}{|x-a|^{1/4}\,(-1)^{1/4}} = \frac{1}{|x-a|^{1/4}}\, e^{i\frac{\pi}{4}} \quad \text{as } e^{-i\pi} = \cos\pi - i\sin\pi = -1,$$

and the coefficient C_2 is given by $C_2 = \frac{1}{2}\,C\cdot e^{-\frac{\pi}{4}}$ and analogue $C_1 = \frac{1}{2}\cdot C\cdot e^{i\frac{\pi}{4}}$.

Applying these function theoretical methods, we transform

$$\frac{C}{\sqrt{h(x)}} \exp\left(-\frac{1}{k}\int_a^x h(x)dx\right) \rightarrow \frac{C}{\sqrt{h(x)}} \exp\left(-i\,\frac{1}{k}\int_a^x h(x)dx\right)\cdot\exp\left(i\,\frac{\pi}{4}\right) =$$

$$= \frac{C}{\sqrt{h(x)}}\cdot\exp\left(-i\left(\frac{1}{k}\int_a^x h(x)dx - \frac{\pi}{4}\right)\right),$$

and taking the real part, we have the transformed influence-wave

$$\frac{C}{\sqrt{h(x)}} \cos\left(\frac{1}{4}\int_a^x |h(x)|dx - \frac{\pi}{4}\right).$$

Now we consider the influence potential U of the form

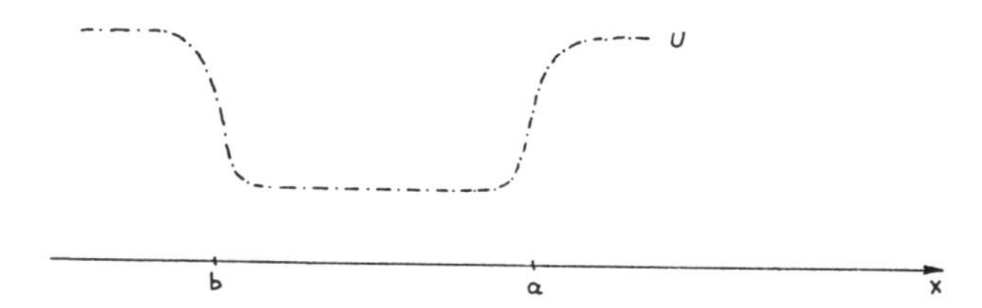

Applying our transformation method, we obtain that the influence wave to the left of the boundary point x = a, is given by

$$\psi = \frac{C}{\sqrt{h}} \cos\left(\frac{1}{k}\int_a^x h(x)dx - \frac{\pi}{4}\right),$$

and in the region to the right of the boundary point x = b

$$\psi = \frac{C^*}{\sqrt{h}} \cos\left(\frac{1}{4}\int_x^b h(x)dx - \frac{\pi}{4}\right).$$

IV. 5.3 "Behavioural features" of individuals

These expressions must be the same throughout the region $b \leq x \leq a$. Thus, the arguments of the cos-expressions are integral multiples of :

$$\left(\frac{1}{k}\int_a^x h(x)dx - \frac{\pi}{4}\right) + \left(\frac{1}{k}\int_x^b h(x)dx - \frac{\pi}{4}\right) = n\cdot\pi \;,\; n = \text{integer},$$

thus

$$\frac{1}{k}\int_a^b h(x)dx - \frac{\pi}{2} = n\cdot\pi.$$

Choosing the constants $C = (-1)^n C^*$, we obtain finally

$$\frac{1}{2\pi k}\int_a^b h(x)dx = n + \frac{1}{2}\,.$$

Now we give some possible applications in economics.
There is a shopping centre offering the latest summer fashion:

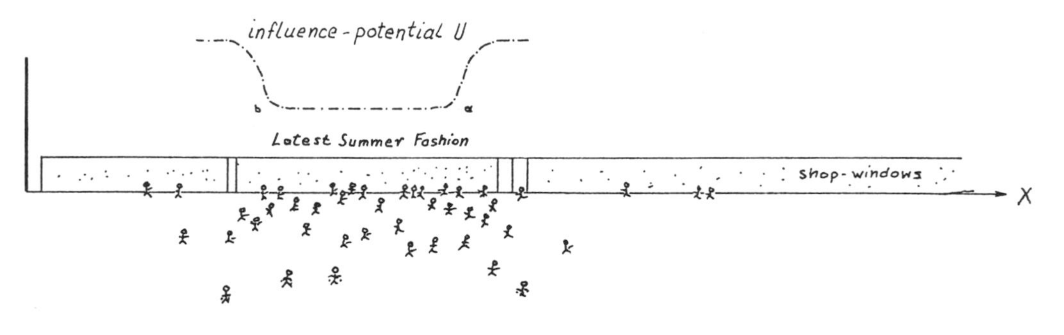

In the general case

$$\frac{1}{k}\int_a^b \sqrt{2q(I-U(x))}\;dx - \frac{\pi}{2} = n\cdot\pi,$$

as we interpret U = the advertising potential of the shopping centre, we can say that advertising generates possible "behavioural features".

The integer n characterizes possible "Verhaltenszustände", this means "behavioural features" of the individuals influenced by the advertising potential U. As

$$\psi \sim \cos\left(\frac{1}{k}\int_x^a h(x)dx - \frac{\pi}{4}\right)$$

it is easy to see that the integer n is equal to the number of zeros of the wave function ψ, for $x = a$ we obtain the argument $\frac{\pi}{4}$,

$$x = b \qquad \frac{1}{k}\int_a^b h(x)dx - \frac{\pi}{4} = n\cdot\pi + \frac{\pi}{2} - \frac{\pi}{4} = (n + \frac{1}{4})\pi.$$

Thus the cosine vanishes n times in the above range, but outside $b \leqq x \leqq a$ it decreases monotonically and has no zeros at a finite distance; hence n characterizes the number of possible "Verhaltenszustände" of the individuals. The number of the zeros of the influence-wave can be interpreted as the spectrum of possible impressions made upon the influenced individuals. These impressions generate groups of individuals of the same behaviour. The group-building process is nonlinear and governed by dispersion and nonlinear effects.

Now we consider the distribution of the "Verhaltenszustände" of the individuals under the influence of the spectrum of impressions generated by advertising.

IV. 5.4 Nonlinear phenomena and group-building processes

The crowd movement along the shop front can be described by density waves. The profile of the crowd density wave steepens in front of the shop window, where advertising is influencing the pedestrian. Let σ be a constant, given by

$$\sigma = \left(\frac{u_o}{\beta}\right)^{\frac{1}{2}} \cdot \ell$$

where u_o is the strength of advertising, β the dispersion of the individuals, and ℓ the front of the window.

Numerical computation indicates that, at values of σ exceeding a critical value σ_o, a soliton (solitary wave) appears in the frontal part of the solution.

For purposes of illustration, we present the results of numerical computation of the Korteweg-de Vries equation for an initial condition of the form

$$U(x,0) = - u_o \frac{\partial}{\partial x} \left(\frac{1}{\sqrt{\pi}} e^{-\frac{x^2}{\ell^2}}\right)$$

and σ is defined as $\sigma = \left(\frac{u_o}{\beta}\right)^{\frac{1}{2}} \cdot \ell$.

The numerical solutions are represented in the following diagrams for $\sigma = 2$ and at three points in time $t_1 = 0$, and t_2, t_3 (34).

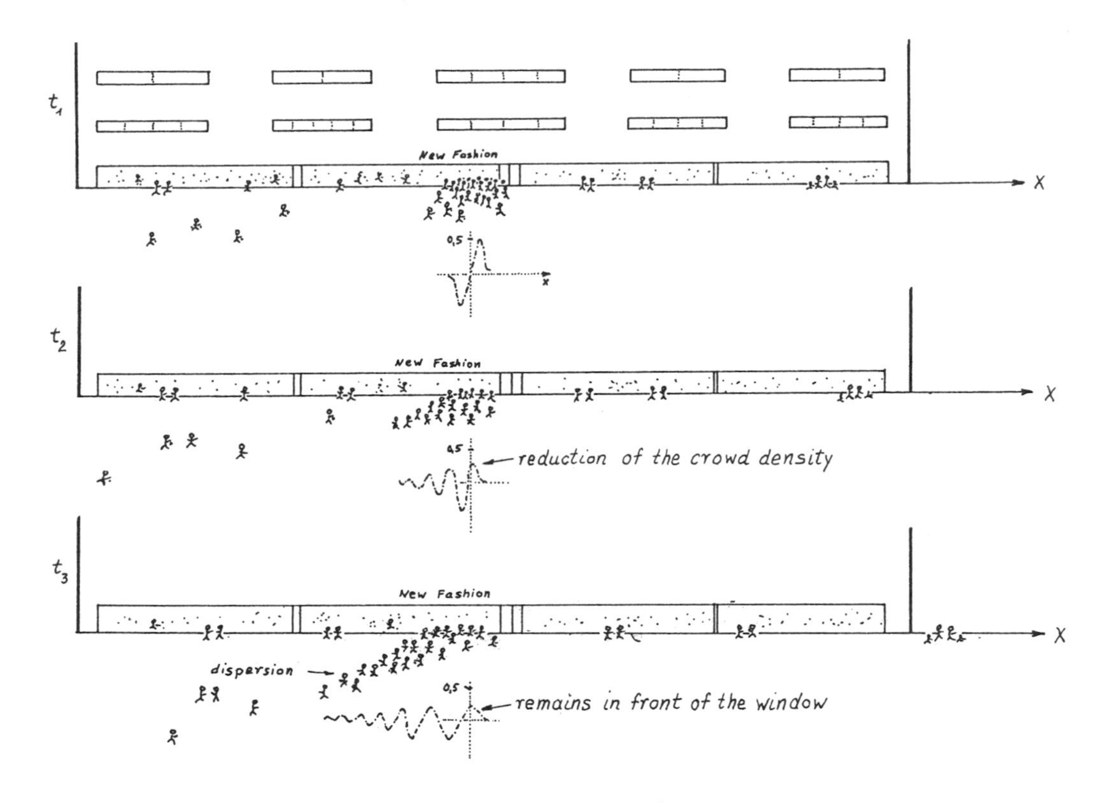

The maximum density to the right does not move (it remains under the influence of advertising), while the others move to the left. The amplitude of the density maxima decrease with increases in t. The third diagram shows that the front of the wave moves to the right and for $t \to \infty$ separates from the tail which spreads to the left.

At large values of t, the amplitude of the wave remains constant. The coefficient σ reads

$$\sigma = \left(\frac{u_0}{\beta}\right)^{\frac{1}{2}} \cdot \ell,$$

if β, the dispersion of the individuals is small, then σ becomes great; and if advertising is powerful, then u_0 is great and σ becomes great, too.

This phenomenon is treated in the following model. The diagram shows, the density wave of the solution for $\sigma = 40$. It can be seen that at least two solitons are formed in this case.

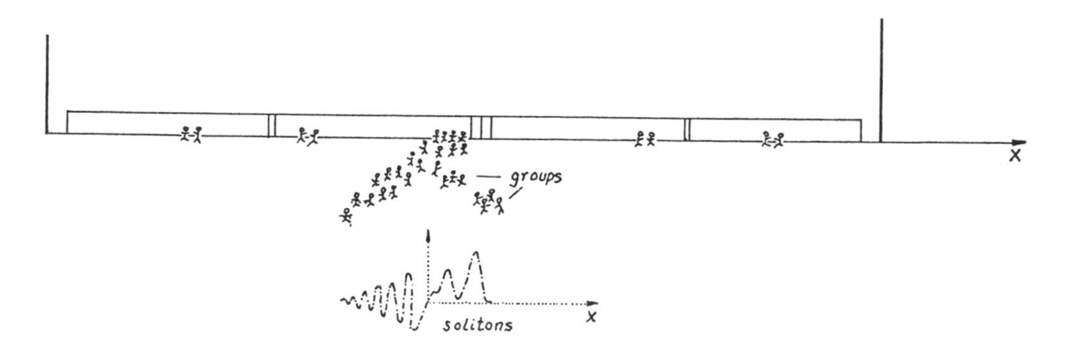

The formation of solitons, given as an analogy-model, can be interpreted as the formation of groups under the influence of advertising.

We have seen that the profile of the crowd density wave steepens in front of the shopping centre, where advertising influences the pedestrians. When the steepness is fairly large, dispersion becomes important, and as a result the number of pedestrians is split up into individual groups (solitary groups).

As shown by numerical computations of the Korteweg-de Vries equation ((360), (361), (331)) and by analytical theory (122), these groups are eventually transformed into solitary waves or into a nearly linear wave packet which eventually diffuses.

As an illustration of the above, we present the following diagrams showing the numerical solutions of the Korteweg-de Vries equations for an initial pulse of the form

$$u(0,\xi) = e^{-\xi^2},$$

and values of the parameter $\sigma_I = 1{,}9$ and $\sigma_{II} = 16{,}5$: (33)

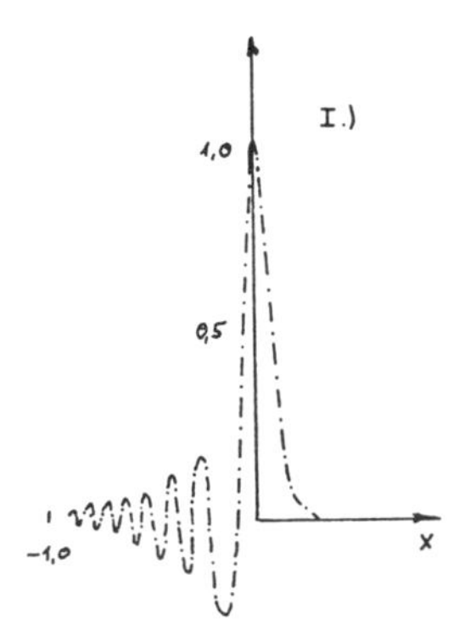

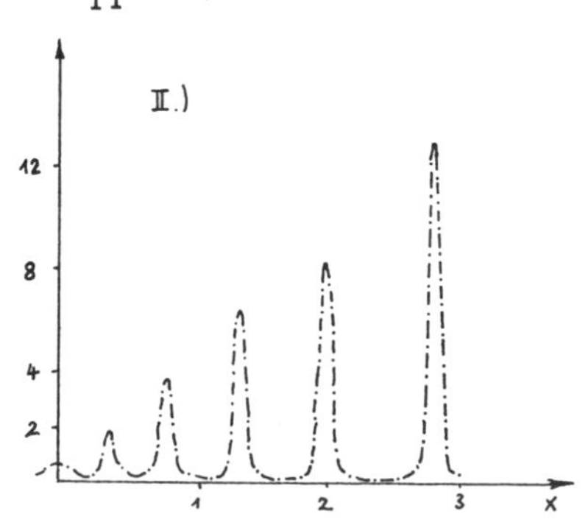

Possible applications as above. In the first diagram, advertising generates two group-phenomena within the population, described by the decomposition into the soliton and the wave packet. The amplitude of the generated soliton and the wave packet. The amplitude of the generated soliton is given by

$$a = -2u_o \cdot I$$

where u_o is the strength of advertising, and I the eigenvalue (or influence coefficient).

Individuals remaining for a long time under the influence of advertising u_o decompose into solitary groups influenced by $a = -2u_o I$. In the second case advertising generates six groups, described by six solitary waves, but the wave packet has a very small amplitude and is not shown in the diagram.

IV. 5.5 Analytic solution of the Korteweg-de Vries equation

In the pioneering paper (122) a profound analysis is given which provides an understanding of various general laws of the evolution of perturbations, described by the Korteweg-de Vries equation

$$1) \quad u_t - 6u \cdot u_x + u_{xxx} = 0,$$

Burger's equation

$$u_t + u \cdot u_x - yu_{xx} = 0$$

can be solved in closed form by the nonlinear transformation

$$u = - 2\nu \frac{\psi_x}{\psi} .$$

Gardner, Green, Kruskal, and Miura (122) took it further with the choice

$$u = \frac{\psi_{xx}}{\psi} + \lambda$$

and solved the Korteweg-de Vries equation.

Rewriting this nonlinear transformation we obtain the Schrödinger equation in a reduced form (186)

$$2) \quad \psi_{xx} + (\lambda - u)\psi = 0.$$

Now consider this equation as the associated scattering problem from which information on ψ could be used to diagnose properties of u; in this view, the wave profile u(x,t) provides the scattering potential.

To use (2), we have to find the equation for ψ from (1). Since values of λ would belong to the spectrum of the scattering problem (2) and the problem changes with t, we assume in the first instance that λ is a function of t.

Substituting (2) in (1), it is found (after certain ingenuity) that

$$3)\quad \frac{d\lambda}{dt}\psi^2 = 6\frac{\partial}{\partial x}\{(\psi\frac{\partial}{\partial x} - \psi_x)\cdot(\psi_t - \frac{3\psi_x\cdot\psi_{xx}}{\psi} + \psi_{xxx} - \lambda\psi_x)$$

$$= 6\frac{\partial}{\partial x}\{(\psi\frac{\partial}{\partial x} - \psi_x)\ (\psi_t - \frac{1}{2}\psi_x(u+\lambda) + \psi_{xx})\}.$$

Restricting our discussion to the solution of

$$U_t - 6\ uu_x + u_{xxx} = 0,$$

with $u \to 0$ as $|x| \to \infty$ and u integrable.

Under these conditions the spectrum of (2) is discrete in $\lambda > 0$, continuous in $\lambda > 0$. The point eigenvalues, $\lambda = -k_n^2$, the corresponding eigenfunctions ψ_n satisfy

$$|\psi_n| \to 0 \text{ for } |x| \to 0, \text{ and } 0 < \int_{-\infty}^{+\infty} \psi_n^2\, dt = \text{finite}.$$

Integrating (3) from $-\infty$ to $+\infty$, we deduce that λ_n is independent of t. For the continuous spectrum we may choose a $\lambda > 0$ independent of t and consider the behaviour of the solutions ψ with t.

As $\frac{d\lambda}{dt} = 0$, we deduce from (3) that

$$4)\quad \psi_t + \psi_{xxx} - 3(n+\lambda)\psi_x = C\cdot\psi,$$

where C is independent of x.

It is known, that any solution ψ of (2) (for λ fixed) will develop in time according to (4).

Considering the eigenvalue problem (2) and introducing $\lambda = \mu^2$, we have the function χ, which satisfies (2) and the condition

$$5)\quad \chi \sim e^{i\mu x} \text{ as } |x| \to \infty.$$

We require that this function may be an asymptotic solution of (4) for all t; this requires the choice

$$C = -4i\mu^3.$$

We obtain the equations

6) $\psi_{xx} + (\mu^2 - u)\psi = 0$

7) $\psi_t + \psi_{xxx} - 3(u+\mu^2)\psi_x + 4i\mu^3\psi = 0.$

Now, we formulate the direct scattering problem:

We consider the problem of finding $u(x,t)$ for $t > 0$, when $u(x,0)$ is given. For the given $u(x,0)$, we first solve the eigenvalue problem (6), and determine the point eigenvalues $\mu = ik_n$, the corresponding eigenfunctions ψ_n, and the reflection coefficient β for incoming waves.

Our choice for the eigenfunctions is

$$\psi_n(x) = \chi(ik_n, x),$$

where χ is given by (5), the normalization constants are

$$\gamma_n = \left(\int_{-\infty}^{+\infty} \psi_n^2 \, dx\right)^{-1}.$$

Let β be the reflection coefficient, we determine the solution Θ of (6) with respect to the initial condition $u(x,0)$, which has the form

$$\Theta(k,x) \sim \begin{cases} e^{-ikx} + \beta(k)e^{ikx} & \text{for } x \to +\infty \\ \alpha(k)e^{-ikx} & \text{for } x \to -\infty, \end{cases}$$

for k = real and positive.

This determines the reflection coefficient $\beta(k)$ and the transmission coefficient $\alpha(k)$.

For given $u(x,0)$ we have found the scattering parameter $k_n, \gamma_n, \beta(k)$.

Now the inverse scattering problem would be to determine $u(x,0)$ from the knowledge of k_n, γ_n, $\beta(k)$ as follows.

We turn to the development of these solutions in time. It is known that the k_n are unchanged.

From (6) and (7) follows

8) $\frac{d}{dt}\int_{-\infty}^{+\infty} \chi^2 dx = (2\chi\chi_{xx} + 4\chi_x^2 + 6\mu^2\chi^2)\Big|_{-\infty}^{+\infty} - 8i\mu^3 \int_{-\infty}^{+\infty} \chi^2 dx.$

For the eigenfunctions

$$\mu = ik_n, \text{ and } \psi_n(x,t) = \chi(x,t,ik_n) \to 0 \text{ as } x \to \pm\infty,$$

therefore the normalization constants are

$$9) \quad C_n(t) = \left(\int_{-\infty}^{+\infty} \psi_n^2 dx\right)^{-1} = \gamma_n e^{3k_n^3 t}.$$

The solution $\Theta(k,x,t)$ for scattered waves will have the behaviour

$$\Theta(k,x,t) \sim f(k,t)e^{-ikx} + g(k,t)e^{ikx} \text{ for } x \to \infty.$$

For $\mu = k$ this must be an asymptotic solution of (7). Substituting, we deduce that

$$f(k,t) = e^{-ik^3 8t}, \quad g(k,t) = \beta,$$

and the reflection coefficient

$$b(k,t) = \frac{g(k,t)}{f(k,t)} = \beta(k)e^{i8k^3 t},$$

thus the inverse scattering theory provides the construction of $u(x,t)$ from

$$K_n, C_n(t), \text{ and } b(k,t).$$

The solution of the inverse problem is provided by the famous Gelfand-Levitan paper in (1951 (130)). Their paper is phrased in terms of determining the scattering potential u from the spectral function $p(\lambda)$, which has jumps of magnitude C_n at the point eigenvalues $\lambda = -k_n^2$ and a continuous spectrum $0 < \lambda < \infty$ related to b (see Kay (187) and Moses (188), Marchenko (231), the review by Faddeyev (in 1959)).

The solution of the Korteweg-de Vries equation is given by

$$10) \quad u(x,t) = -2 \frac{d}{dt} K(x,y,t),$$

where $K(x,y,t)$ satisfies the famous linear integral equation

$$11) \quad K(x,y,t) + B(x+y,t) + \int_x^\infty K(x,z,t)B(z+y,t)dz = 0$$

for $y > x$, where

$$12)\quad B(x+y,t) = \sum_n C_n(t)\exp(-K_n(x+y)) + \frac{1}{2\pi}\int_{-\infty}^{+\infty} b(k,t)\exp(ik(x+y))dk$$

$$= \sum_n \gamma_n \exp(-k_n(x+y)+8k_n^3\cdot t) + \frac{1}{2\pi}\int_{-\infty}^{+\infty} \beta(k)\exp(ik(x+y)+8ik^3t)dk$$

by the normalization condition (7).

The initial function $u(x,0)$ provides $k_n, \gamma_n, \beta(k)$.

This procedure makes it possible to solve the Korteweg-de Vries equation (in principle) on the basis of the initial condition $u(x,0)$.

In particular, we can obtain information on the solitons formed from the initial perturbation $u(x,0)$ from the above equations (10) - (12).

Let us examine the asymptotic form of the solution at $t \to \infty$ and $x \sim 8k_1^2t$, where k_1 is the largest of k_n. In this case we can neglect the integral on the right-hand side of (12). We need retain in the sum only the main term with $n = 1$, thus

$$B(x,t) = C_1(0)\cdot\exp(-k_1x + 8k_1^3t).$$

The asymptotic solution of (11) then becomes (derived in (187))

$$K(x,y,t) = -C_1(0)\exp(-K_1(x-y)+8K_1^3t)\cdot\{1 + \frac{C_1(0)}{2K_1}\exp(-2K_1x+8K_1^3t)\}^{-1},$$

substituting this in (10)

$$u(x,t) = -2K_1^2 \frac{1}{\cosh^2(K_1(x-x_o)-4k_1^2t)} = -2K_1^2 \operatorname{sech}^2(K_1(x-x_o)-4K_1^2t)$$

where $C_1(0) = 2K_1\cdot\exp(2K_1x_o)$, (see Novikov et al. (268), p. 28).
This expression describes a solitary wave with amplitude $a = 2\cdot k_1^2$ and velocity $v = 4\cdot k_1^2$. Setting

$$\alpha\cdot x = k_1(x-x_o)-4k_1^2t, \text{ then } \alpha = k_1-k_1\frac{x_o}{x} - 4k_1^2\cdot t.\frac{1}{x}.$$

Analogously we can derive the asymptotic solution in the region $t\to\infty$ and $x \sim 8K_n^2\cdot t$. We can obviously use the expression (13) for $B(x,t)$, where we replace $c_1 \to c_n$ and $k_1 \to k_n$.

The solution of the Korteweg-de Vries equation will again have the form of a solitary wave, with the parameters k_n and c_n. Thus for $t \to \infty$ and large positive values of x, the solution is representable as a family of solitary waves moving in the positive direction with velocities v_n proportional to their amplitudes

$$u(x,t) = - \sum_n 2\,|\lambda_n| \cdot \mathrm{sech}^2((\frac{\lambda_n}{6})^{\frac{1}{2}} \cdot (x - x_o - v_n t)),$$

where $v_n = \frac{2}{3}\,|\lambda_n|$ and λ_n are the eigenvalues of the discrete spectrum of the Schrödinger equation (6) at $t = 0$.

Let us examine the solution of

$$\frac{\partial u}{\partial x} + u \cdot \frac{\partial u}{\partial x} + \beta \cdot \frac{\partial^3 u}{\partial x^3} = 0,$$

with respect to the initial condition (perturbation)

$$u(x,0) = u_o \cdot \varphi(\frac{x}{\ell}),$$

$\varphi(\xi)$ is a dimensionless function characterizing the initial profile of a simple wave and u_o the amplitude of the initial perturbation.

Changing to the dimensionless variables

$$\theta = \frac{u}{u_o}, \quad \xi = \frac{x}{\ell} \quad \eta = \frac{t \cdot u_o}{\ell},$$

the Korteweg-de Vries equation is reduced to

$$\theta_\eta + \theta \cdot \theta_\xi + \frac{1}{\sigma^2}\,\theta_{\xi\xi\xi} = 0,$$

with the initial condition

$$\theta(\xi,0) = \varphi(\xi)$$

and the dimensionless coefficient

$$\sigma = \ell \cdot (\frac{u_o}{\beta})^{\frac{1}{2}}.$$

The corresponding Schrödinger equation is given by

$$\psi''(\xi) + \frac{\sigma^2}{6} \cdot (\varphi(\xi) + \lambda_n)\psi(\xi) = 0,$$

where λ_n are the eigenvalues of the discrete spectrum. The above representations of the solution show that the amplitudes a_n of the generated solitary waves are related to the eigenvalues of the Schrödinger equation by

$$a_n = -2u_o \cdot \lambda_n,$$

where u_o is the amplitude of the initial perturbation.

Let us now consider the region of small negative values of x. The asymptotic form of the solution of this region is given by the integral term of (12) (at long times t). As the reflection coefficient b(k,t) is time dependent, the integral term

$$\frac{1}{2\pi} \int_{-\infty}^{+\infty} b(k,t) \cdot \exp(ik(x+y))dk$$

describes an oscillation (see page 310) in the form of a wave-group-packet (its detailed investigation is difficult).

This wave-group-packet can be estimated by invariants.

IV. 5.6 Invariants

Let the initial perturbation decompose into N solitons with amplitudes a_ν for $\nu = 1,\ldots,N$, where

$$a_\nu = -2u_o \cdot \lambda_\nu$$

and a wave packet.

After the solitons have separated from the wave packet and covered a fairly large distance, we can assume, that the value of the invariant $I_m(0)$ is equal to the sum of the invariants I_s for the individual solitary waves and the wave packet, thus

$$I_s + I_w = I_m(0),$$

where I_w is the invariant of the wave packet. These invariants are defined as follows:

The Korteweg-de Vries equation

$$\frac{\partial u}{\partial t} + u \cdot \frac{\partial u}{\partial x} + \beta \frac{\partial^3 u}{\partial x^3} = 0$$

can be written in the divergence form

$$\frac{\partial u}{\partial t} + \frac{\partial}{\partial x}\left(\frac{u^2}{2} + \beta u_{xx}\right) = 0,$$

thus, we obtain an invariant

$$I_1 = \int_{-\infty}^{+\infty} Q_1(x,t)dt,$$

where the density Q_1 satisfies

$$\frac{\partial Q_1(x,t)}{\partial t} + \frac{\partial}{\partial x} P_1(x,t) = 0,$$

and P_1 is a polynomial of β and the spatial derivatives of u. In our case, we have to choose

$$Q_1(x,t) = u(x,t) \text{ and } P_1(x,t) = \frac{u^2}{2} + \beta u_{xx}.$$

It was shown in (201), (248) that there exists an infinite number of invariants corresponding to the Korteweg-de Vries equation

$$I_m = \int_{-\infty}^{+\infty} Q_m(x,t)dx,$$

where the densities $Q_m(x,t)$ satisfy the conditions

$$\frac{\partial Q_m(x,t)}{\partial t} + \frac{\partial}{\partial x} P_m(x,t) = 0 \text{ for } m = 1,2,3,...$$

and Q_m, P_m are polynomials of β and the spatial derivatives of u. As an example we present the first densities of the invariants (360), (248), (201),

$$Q_1(u) = u,\ Q_2(u) = \frac{u^2}{2},\ Q_3(u) = \frac{u^3}{3} - \beta u_x^2,\ Q_4 = \frac{u^4}{4} - 3\beta u u_x^2 + \frac{2}{5}\beta^2 u_{xx}^2.$$

If the quantity β is small, we remark that the coefficient in $Q_m(n)$, containing β to the first power, we have

$$Q_m(n) = \frac{u^m}{m} - \beta\frac{(m-1)(m-2)}{2} u_x^2 u^{m-3} + O(\beta^2).$$

For solitary waves

$$u=a.\text{sech}^2 \left(\left(\frac{a}{12\beta}\right)^{\frac{1}{2}} \left(x - \frac{a}{3}t\right)\right),$$

the invariants of arbitrary order are given (184) by

$$I_s = I_m(n) = \int_{-\infty}^{+\infty} Q_m(n)dx = \sqrt{12\cdot\beta} \cdot \frac{2^m((m-1)!)^2}{(2m-1)!} a^{\frac{2m-1}{2}}.$$

The relation of the invariants $I_s + I_w = I_m(0)$ can be written as

$$\sqrt{12\cdot\beta}\; 2^m \frac{((m-1)!)^2}{(2m-1)!} \sum_{\nu=1}^{N} a_\nu^{\frac{2m-1}{2}} + I_w = I_m(0),$$

where

$$I_m(0) = \int_{-\infty}^{+\infty} Q_m(u(x,0))dx = u_o^{m-\frac{1}{2}}\; \beta^{\frac{1}{2}}\cdot\sigma\cdot \int_{-\infty}^{+\infty} \tilde{Q}_m(\varphi(\xi))d\xi,$$

we obtain the quantity $\tilde{Q}_m(\varphi(\xi))$ by the transformation $u \to \varphi(\xi)$

$$\beta \to \frac{1}{\sigma^2}\;.$$

For large values of σ we obtain (185)

$$\tilde{Q}_m(\varphi(\xi)) = \frac{\varphi^m(\xi)}{m} - \frac{1}{\sigma^2}\,\frac{(m-1)(m-2)}{2}\;\varphi_\xi^2\;\varphi^{m-3} + O(\sigma^{-4})$$

by the above asymptotic expression.

IV. 5.7 Application of the invariants: Advertising and group-building

Now we continue the models in economics given in this chapter. Let $\mathcal{D}(a)$ be the distribution function of M solitary waves with amplitudes within the interval (a,a+da) (see page 331), then

$$\mathcal{D}(a) = \frac{\sigma}{4\pi}\left(\frac{1}{3u_o}\right)^{\frac{1}{2}} \int_{2u_o\varphi(\xi)>a} \frac{1}{\sqrt{2u_o\cdot\varphi(\xi)-a)}}\; d\xi$$

is generated by the expression $u_o\cdot\varphi(\xi)$, where u_o can be interpreted as the strength of advertising.

We are interested in the magnitude of the wave groups behind the solitary waves. These wave groups are described by the invariants I_ω. According to the above condition of invariants

$$I_s + I_\omega = I_m(0),$$

the invariants I_s of the solitary waves are given by

$$I_s = (12\cdot\beta)^{\frac{1}{2}}\, \frac{2^m((m-1)!)^2}{(2m-1)!} \int_0^\infty a^{\frac{2m-1}{2}}\; \mathcal{D}(a)da + I_\omega = I_m(0)$$

and

$$I_m(0) = \frac{\ell u_o^m}{m} \int_{-\infty}^{+\infty} \varphi^m(\xi)d\xi = \frac{1}{m}\int_{-\infty}^{+\infty} u^m(x,0)dx,$$

substituting the distribution function $\mathcal{D}(a)$ in the condition of invariants, we have with (183):

$$I_\omega = \frac{1}{m}\int_{u(x,0)<0} u^m(x,0)dx \text{ for } m = 1,2,\dots .$$

Thus, I_ω has only contributions for regions, where $u(x,0) < 0$, i.e. the initial condition is negative, i.e. the influence of advertising is negative.

If the initial condition $u(x,0) \geqq 0$ is positive for all values of x, i.e. the influence of advertising is positive along the shop window front, it follows that the invariant $I_\omega = 0$.

In this case the initial condition completely separates into solitary waves. This means that the positive influence of advertising completely separates the individuals into solitary groups (stable groups seen as solitary waves).

The collision of two solitons is investigated in (268), p. 30): A large amplitude soliton, initially behind, overtakes a soliton with smaller amplitude. The perturbation appearing after their superposition again decomposes into two solitons, which coincide exactly with the first solitons, but now arranged in reverse order. This result is essentially an experimental proof of the stability of solitons: they retain their characteristics after interaction.

IV. 5.8 Influence potential

Now we consider the influence potential of the form:

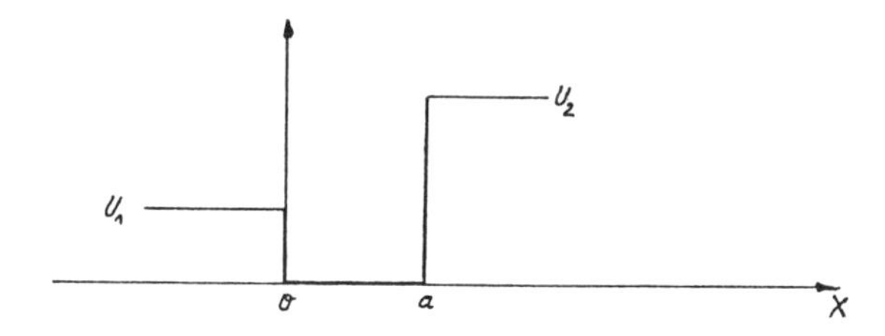

Let the influence coefficient I be $< U_1$, the wave equation in the region $x < 0$ can be written as

$$\frac{d^2\psi}{dx^2} + \frac{2q}{k^2}(I-U_1)\psi = 0,$$

and in the region $x > 0$

$$\frac{d^2\psi}{dx^2} + \frac{2q}{k^2}(I-U_2)\psi = 0;$$

the corresponding solutions, which vanish at infinity, are

$$\psi = \psi_1 = C_1 e^{\mu_1 x}, \text{ where } \mu_1 = \frac{1}{k}\sqrt{2q(U_1-I)}$$

and

$$\psi = \psi_2 = C_2 e^{-\mu_2 x} \text{ where } \mu_2 = \frac{1}{k}\cdot\sqrt{2q(U_2-I)}.$$

Within the region $0 \leqq x \leqq a$ the wave equation is given by

$$\frac{d^2\psi}{dx^2} + \frac{2q}{k^2} I\cdot\psi = 0$$

with its solution

$$\psi = C\cdot\sin(mx+\gamma), \text{ where } m = \frac{1}{k}\sqrt{2q\cdot I}, \ \gamma = \text{constant}.$$

The solution ψ and its derivative ψ' may be continuous at $x = 0$ and $x = a$, too, thus we require the continuity of ψ and its logarithmic derivative $\frac{\psi'}{\psi}$. In the region $x < 0$ we have the solution

$$\psi = C_1 e^{\mu_1 x}, \text{ then } \psi' = \mu_1 e^{\mu_1 x} \text{ and } \frac{\psi'}{\psi} = \mu_1;$$

and in the region $x > a$ there is $\frac{\psi'}{\psi} = -\mu_2$.

In the region $0 \leqq x \leqq a$, the solution can be written as

$$\psi = C\cdot\sin(mx+\gamma), \ \psi' = C\cdot m\cos(mx+\gamma),$$

hence

$$\frac{\psi'}{\psi} = m.\cot(m\cdot x+\gamma)\Big|_{x=0} = m\cdot\cot\gamma = \mu_1 \underset{\text{def.}}{=} \sqrt{\frac{2q}{k^2}U_1 - m^2} \text{ by continuity}$$

and quite analogously

$$m\cdot\cot(m\cdot a-\gamma) = -\mu_2 = -\sqrt{\frac{2q}{k^2} - m^2},$$

or

$$\sin\gamma = \frac{mk}{\sqrt{2qU_1}}\ , \ \sin(m\cdot a+\gamma) = -\frac{m\cdot k}{\sqrt{2q\cdot U_2}}\ .$$

Then we have

$$\gamma = \arcsin\left(\frac{m\cdot k}{\sqrt{2qU_1}}\right), \quad ma + \gamma = \arcsin\left(-\frac{m\,k}{\sqrt{2q\cdot U_2}}\right)$$

the eliminating γ, there follows a transcendental equation

$$ma = n\cdot\pi = \arcsin\frac{m\,k}{\sqrt{2qU_1}} - \arcsin\frac{m\,k}{\sqrt{2qU_2}} \quad \text{for } n = 1,2,3,\dots\ .$$

For each n there is in general one root, if the values of arc sin $(\cdot)$ are taken between 0 and $\frac{\pi}{2}$. Then the influence coefficient, or eigenvalue, can be written

$$I = \frac{m^2k^2}{2q}\ ,$$

and m is a solution of the transcendental equation.

As the argument of arc sin cannot exceed unity, then the m-values are in $\left(0, \sqrt{\frac{2q}{k^2}U_1}\right)$.

The left-hand side of the transcendental equation is monotone increasing with m and the right-hand side monotone decreasing.

Then for $m = \sqrt{\frac{2q}{k^2}U_1}$ it is necessary for a root to exist, that the right-hand side should be less than the left-hand side,

$$\frac{a}{k}\sqrt{2qU_1} = m\cdot a \geqq n\pi - \arcsin 1 - \arcsin\sqrt{\frac{U_1}{U_2}}\ ,$$

for n = 1 we obtain

$$\frac{a}{k}\sqrt{2qU_1} \geqq \frac{\pi}{2} - \arcsin\sqrt{\frac{U_1}{U_2}}\ .$$

If the inequality is satisfied, there exists at least one solution m_1; therefore, there exists at least one influence coefficient

$$I = \frac{m_1^2\cdot k^2}{2q}\ .$$

But for general influence potentials $U_1 \neq U_2$ there exist widths, denoted by a, so small that there is no solution of the transcendental equation, and no influence coefficient I.

There are some applications in economics. Supposing the influence potential of the form

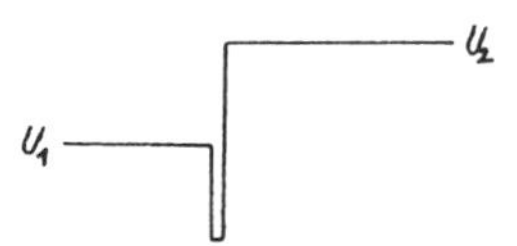

describing discontinuous phenomena, we can say, the individuals are not influenced. But for influence potentials like

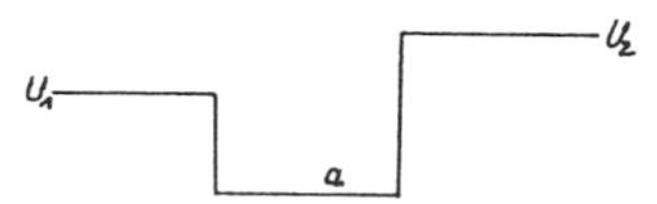

there exists at least one coefficient I influencing the individuals. Under the condition $U_1 = U_2$ the inequality is always satisfied and the transcendental equation reduces to

$$\text{arc sin } \frac{mk}{\sqrt{2qU}} = \frac{n\cdot\pi - m\cdot a}{2} \quad ,$$

then

$$\frac{mk}{\sqrt{2qU}} = \sin\left(\frac{n\pi}{2} - \frac{ma}{2}\right) = \sin\frac{n\cdot\pi}{2}\cos\frac{ma}{2} - \cos\frac{n\pi}{2}\sin\frac{m\cdot a}{2} = \begin{cases} \cos\frac{ma}{2} & \text{for odd } n \\ \sin\frac{ma}{2} & \text{for even } n. \end{cases}$$

We introduce $\theta = \frac{ma}{2}$ and obtain for odd n

$$\cos\theta = \frac{m\cdot k}{\sqrt{2qU}} = \pm\,\theta\xi, \text{ where } \xi = \frac{k}{a}\sqrt{\frac{2}{q\cdot U}} \quad .$$

The solutions of $\cos\theta = \pm\,\theta\xi$ have to satisfy $\tan\theta > 0$. For every n we have to solve

$$\sin\theta = \pm\,\theta\xi,$$

where those solutions must be taken for which $\tan\theta < 0$. With these solutions we obtain the influence coefficient, or eigenvalue, as

$I = \frac{m^2k^2}{2q}$ and $\theta = \frac{m\cdot a}{2}$, then $\theta^2 = \frac{m^2a^2}{4}$, thus $I = 2\,\frac{\theta^2k^2}{g\cdot a^2}$.

I can be interpreted also as influence-level.

For a shallow influence potential, which satisfies $U << \frac{k^2}{q \cdot a^2}$, we have $\xi >> 1$, as

$$\sqrt{q \cdot U} << \frac{k}{a} \text{ and } \sqrt{2}\,\frac{a}{k} \cdot \frac{1}{\sqrt{q \cdot U}} >> \frac{k}{a} \cdot \frac{a}{k}\sqrt{2} = \sqrt{2} > 1,$$

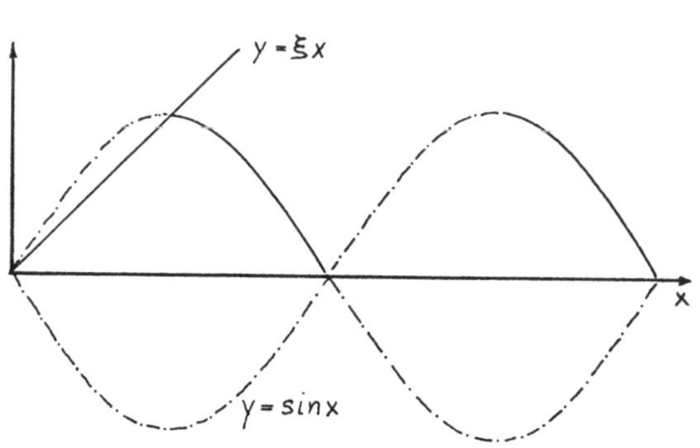

thus $\xi >> 1$ and $\sin\theta = \pm\,\theta\xi$ has no roots. The roots are given as follows: Solve

$$\sin x = \pm\,\xi x \text{ for } \tan x < 0,$$

as $\frac{\sin x}{\cos x} < 0$, if $\sin x > 0$, then $\cos x < 0$, thus we obtain the <u>regions</u> of the solutions (see the diagram).

The solution of $\cos\theta = \pm\,\theta\xi$, where $\tan\theta > 0$, is given approximately by

$$\theta = \frac{1}{\xi}\left(1 - \frac{1}{2\xi^2}\right),$$

and the eigenvalue can be derived

$$I = \frac{2\xi^2 k^2}{q \cdot a^2} = \frac{2k^2}{qa^2} \cdot \frac{\left(1 - \frac{1}{2\xi^2}\right)^2}{\frac{k^2}{a^2} \cdot \frac{2}{qU}} = U\left(1 - \frac{1}{2}\,\frac{qa^2}{2k^2}\,U\right)^2,$$

thus, the influence coefficient

$$I = U\left(1 - \frac{qa^2}{2k^2}\,U + \ldots\right),$$

hence

$$I \sim U - \frac{qa^2}{2k^2}\,U^2.$$

We have derived a very intersting phenomenon in economics, that only the reduced advertising potential is influencing the customers.

IV. 5.9 Influence level

We consider the influence potential of the form

$u(x,0) = u_o \cdot \varphi(x)$

where

$$\varphi(x) = \frac{1}{\cos^2 h(\alpha \cdot x)} .$$

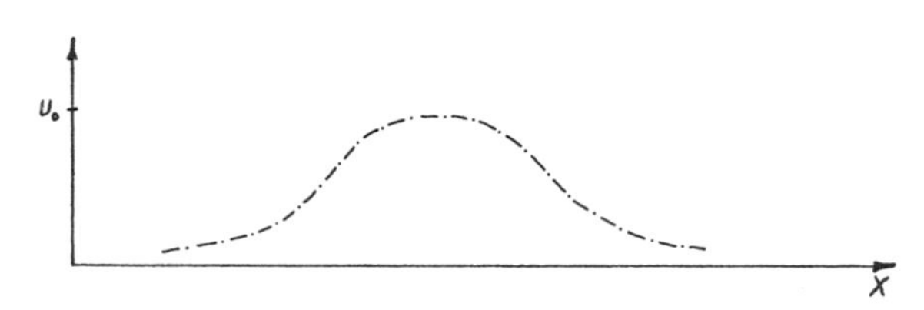

As the spectrum of positive eigenvalues is continuous, while that of negative values is discrete, we consider the negative eigenvalues.

The wave equation reads

$$\frac{d^2\psi}{dx^2} + \frac{\sigma^2}{6}(I + \varphi(x))\psi(x) = 0,$$

where $\sigma^2 = \frac{u_o}{\beta} \cdot \ell^2$, u_o = amplitude, ℓ = width of the impulse. Substituting $z = \tan h\ x$ and using the notations

$$x = \frac{\sigma}{\alpha} \cdot \sqrt{-I} \cdot \frac{1}{\sqrt{6}}, \quad \frac{\sigma^2}{\alpha^2 6} = \tau(\tau+1) \text{ and } \tau = \frac{1}{2}(-1 + \sqrt{1 + \frac{2}{3}(\frac{\sigma}{\alpha})^2}).$$

we obtain the equation

$$\frac{d}{dz}((1-z^2)\frac{d\psi}{dz}) + (\tau(\tau+1) - \frac{x^2}{1-z^2})\psi(z) = 0.$$

This is the differential equation of the generalized Legendre polynomials. Making the substitution

$$\psi = (1-z^2)^{\frac{x}{2}}\omega(z),$$

and changing the variable to $\frac{1}{2}(1-z) = u$, we obtain the associated hypergeometric differential equation

$$u(1-u)\omega'' + (x+1)\cdot(1-2u)\omega' - (x-\tau)(x+\tau+1)\omega = 0.$$

The solution of this equation is finite for $z = +1$ (i.e. $x = +\infty$) and given by

$$\psi = (1-z^2)^{\frac{x}{2}} F\{x-\tau, x+\tau+1, x+1, \frac{1}{2}(1-z)\},$$

where F is the hypergeometric function.

We must have

$$x - \tau = -n, \text{ for } n = 0,1,2,\ldots,$$

that ψ remains finite for $z = -1$ (i.e. $x = -\infty$).

Now F is a polynomial of degree n. The eigenvalues, or influence coefficients I_n, are determined by $x + n = \tau$, i.e.

$$\frac{\sigma}{\alpha} \frac{1}{\sqrt{6}} \sqrt{-I_n} + \frac{2}{2} \cdot n = \frac{1}{2}(1 - \sqrt{1 + \frac{2}{3}(\frac{\sigma}{\alpha})^2})$$

then

$$\frac{\sigma}{\alpha} \cdot \frac{1}{\sqrt{6}} \sqrt{-I_n} = \frac{1}{2} (-(1+2n) - \sqrt{1 + \frac{2}{3} (\frac{\sigma}{\alpha})^2} ,$$

and

$$I_n = - \frac{3}{2} (\frac{\alpha}{\sigma})^2 (-(1+2n) + \sqrt{1 + \frac{2}{3}(\frac{\sigma}{\alpha})^2})^2,$$

specialized for $\alpha = 1$

$$I = I_1 = - \frac{3}{2} \frac{1}{\sigma^2} (-(1+2n) + \sqrt{1 - \frac{2}{3} \sigma^2})^2.$$

There exist a finite number n of influence coefficients (levels), determined by the condition $x > 0$, as $\tau - n = x > 0$ we have $n < \tau$, i.e.

$$n < \frac{1}{2} (-1 + \sqrt{1 + \frac{2}{3}(\frac{\sigma}{\alpha})^2}).$$

As

$$x = \frac{\sigma}{\alpha} \cdot \frac{1}{\sqrt{6}} \cdot \sqrt{-I} = \frac{\ell}{\sqrt{\beta}} \frac{1}{\alpha \cdot \sqrt{6}} \sqrt{-I} \sqrt{u_o} > 0,$$

we obtain that $u_o > 0$.

IV. 5.10 Possible group-building processes

If there is positive advertising, $u_o > 0$, of the shopping centre, then we have $x > 0$ and there exist a finite number n of influence levels, such that (for $\alpha = 1$)

$$n < \frac{1}{2} (-1 + \sqrt{1 + \frac{2}{3} \cdot \sigma^2}).$$

The influence of advertising generates solitary groups (seen as solitary waves).

Now if

$\sigma^2 = 30$, there are generated $n = 2$ solitary groups

$\sigma^2 = 50$, there are generated $n = 3$ solitary groups

$\sigma^2 = 100$, there are generated $n = 4$ solitary groups.

As $\sigma^2 = \ell^2 \cdot \frac{u_o}{\beta}$, where u_o = influence potential of the shop-window

ℓ = width of the influence potential

and ℓ^2 may be interpreted as the advertising area of the shopping centre.

The amplitudes a_n of the solitary wave are determined by

$$a_n = -2u_o \cdot I_n,$$

hence (for $\alpha = 1$)

$$\frac{a_n}{u_o} = -\,2I_n = +\,\frac{3}{\sigma^2}\,((1+2n) + \sqrt{1 + \tfrac{2}{3}\,\sigma^2})^2 \;.$$

We can determine the invariants numerically and show that the invariants for a wave group (packet) decrease with increase in σ, even at $n = 2$ solitary waves, the contribution of the wave packet is slight.

We observe, that for large values of the parameter σ, the number of the formed solitary groups (two or three persons) is large, too. This may be given by a distribution function $\mathcal{D}(a)$, which is the number n of solitary waves with amplitudes within (a, a+da)

$$dn = \mathcal{D}(a)da.$$

IV. 5.11 Variational principle

The existence of this distribution function can be given by variational methods.

Group-building processes are governed by the nonlinear partial differential equation

$$u_t + \alpha u u_x + \beta u_{xxx} = 0.$$

By normalization, we obtain

$$\eta_t + 6\eta \cdot \eta_x + \eta_{xxx} = 0,$$

for this equation there exists no variational principle, but by choosing $\eta = \varphi_x$ the equation becomes

$$\varphi_{xt} + 6\varphi_x \varphi_{xx} + \varphi_{xxxx} = 0,$$

and the Lagrangian is

$$L = -\frac{1}{2}\varphi_t \varphi_x - \varphi_x^3 + \frac{1}{2}\varphi_{xx}^2.$$

We calculate the average Lagrangian

$$\mathcal{L} = k_1 \int_{-\infty}^{+\infty} L dx,$$

by applying the solitary wave solution

$$\eta = a_1 \cdot \operatorname{sech}^2((\frac{a_1}{2})^{\frac{1}{2}} (x - \frac{\omega_1}{k_1} \cdot t)),$$

we obtain

$$\mathcal{L} \approx \omega_1 a_1^{\frac{3}{2}} - \frac{5}{6} k_1 \cdot a_1^{\frac{5}{2}}.$$

We remember that the description of a slowly varying wave in $\mathbb{R}^3$ involves a phase $\varphi(x,t)$, where $x = (x_1, x_2, x_3)$. Then we define the frequency ω by $\omega = -\frac{\partial\varphi}{\partial t}$ and the wave number k by $k = \frac{\partial\varphi}{\partial x}$.

Eliminating φ by applying the theorem of H.A. Schwarz, we have

$$\frac{\partial k_i}{\partial t} + \frac{\partial\omega}{\partial x_i} = 0 \text{ and } \frac{\partial k_i}{\partial x_j} - \frac{\partial k_j}{\partial x_i} = 0.$$

For functions $a(x,t)$, $\varphi(x,t)$ we propose the average variational principle

$$\delta \iint L(-\varphi_t, \varphi_x, a) dx = 0.$$

Since derivatives of a do not occur, the Euler-Lagrange equations for variation in a are given by

$$\delta_a : L_a = 0,$$

and the equation for

$$\delta_\varphi : \frac{\partial}{\partial t} L_{\varphi t} + \frac{\partial}{\partial x_j} L_{\frac{\partial \varphi}{\partial x_j}} = 0,$$

with the consistency conditions

$$\frac{\partial k_i}{\partial t} + \frac{\partial \omega}{\partial x_i} = 0 \quad \text{and} \quad \frac{\partial k_i}{\partial x_j} - \frac{\partial k_j}{\partial x_i} = 0.$$

If we are working with (ω,k,a) instead of $(\varphi_t, \varphi_x, a)$, the associated variational equations are

$$L_a = 0$$

$$\frac{\partial L_\omega}{\partial t} + \frac{\partial}{\partial x_j} L_{k_j} = 0$$

$$\frac{\partial k_i}{\partial t} + \frac{\partial \omega}{\partial x_i} = 0, \quad \frac{\partial k_i}{\partial x_j} - \frac{\partial k_j}{\partial x_i} = 0.$$

Applying this to

$$L \sim \omega_1 a_1^{\frac{3}{2}} - \frac{6}{5} k_1 \cdot a^{\frac{5}{2}},$$

then

$$L_a = \frac{2}{3} \omega_1 a_1^{\frac{1}{2}} - \frac{6}{5}\frac{5}{2} k_1 a_1^{\frac{3}{2}} = 3a_1^{\frac{1}{2}} (\tfrac{1}{2}\omega_1 - k_1 a_1) = 0,$$

thus

$$\omega_1 = 2k_1 a_1.$$

With $L_\omega = a_1^{\frac{3}{2}}$ and $L_k = -\frac{6}{5} a_1^{\frac{5}{2}}$,

$$\frac{\partial L_\omega}{\partial t} - \frac{\partial}{\partial x} L_{k_1} = 0$$

can be written as

$$(a_1^{\frac{3}{2}})_t + (\tfrac{6}{5} a_1^{\frac{5}{2}})_x = 0.$$

Using $\omega_1 = 2k_1 \cdot a_1$ in the consistency relation $(k_1)_t + (\omega_1)_x = 0$, we obtain

$$(k_1)_t + (2a_1 k_1)_x = 0.$$

A particular solution is

$$a_1 = \frac{x}{2t} \quad \text{and} \quad k_1 = \frac{1}{2t}\cdot f\left(\frac{x}{2t}\right),$$

where f is an arbitrary function; this can be seen by direct verification. The arbitrary function f is interpretable as a sequence of solitary waves. The amplitude a_1 and f can be determined only from the initial conditions as follows.

IV. 5.12 Distribution function

We have the distribution function $\mathcal{D}(a)$, which is the number n of solitary waves with amplitudes within (a, a + da) given by

$$dn = \mathcal{D}(a)da.$$

On page 307 we derived the relation

$$\frac{1}{2\pi k}\int h(x)dx = n + \frac{1}{2},$$

where n characterizes possible "Verhaltenszustände" of the individuals. As n is a function of the influence coefficient I, we obtain by differentiating with respect to I

$$\frac{\partial n}{\partial I} = \frac{1}{2\pi k}\int \frac{\partial h}{\partial I}\,dx, \text{ as } h(x) = \sqrt{2q(\varphi(x)-I)},\ \frac{\partial h}{\partial I} = \frac{1}{h(x)}\,2q = \frac{\sqrt{2q}}{2}\cdot\frac{1}{\sqrt{\varphi(x)-I}},$$

then

$$\frac{\partial n}{\partial I} = \frac{\sqrt{2q}}{4\pi k}\int \frac{d\xi}{\sqrt{\varphi(\xi)-I}}\ .$$

Identifying the number of solitary waves with the "Verhaltenszustände" of the individuals, we have

$$\frac{\partial n}{\partial a} = \frac{\partial n}{\partial I}\cdot\frac{\partial I}{\partial a},$$

on page we derived $a = 2\cdot u_o I$, thus $\frac{\partial I}{\partial a} = \frac{1}{2u_o}$, as $\frac{\sigma^2}{6} = \frac{2q}{k^2}$,

then

$$\mathcal{D}(a) = \frac{\partial n}{\partial a} = \frac{\partial n}{\partial I}\frac{\partial I}{\partial a} = \frac{\partial n}{\partial I}\cdot\frac{1}{2u_o} = \frac{\sqrt{2q}}{4\pi k}\frac{1}{\sqrt{2u_o}\ \sqrt{2u_o}}\int\frac{d\xi}{\sqrt{\varphi(\xi)-I}} = \frac{\sigma}{4\pi\cdot\sqrt{6}}\cdot\frac{1}{\sqrt{\cdot}\ \sqrt{\cdot}}$$

$$\times\int\frac{d\xi}{\sqrt{\dfrac{2\cdot\varphi\cdot u_o}{2u_o}-\dfrac{a}{2u_o}}} = \frac{\sigma}{4\pi\cdot\sqrt{6}}\cdot\frac{1}{\sqrt{2u_o}}\int\frac{d\xi}{\sqrt{2\varphi u_o - a}}$$

we obtain finally

$$\mathcal{D}(a) = \frac{\sigma}{4\pi\cdot\sqrt{6}}\cdot\left(\frac{1}{2u_o}\right)^{\frac{1}{2}}\int\frac{d\xi}{\sqrt{2\varphi(\xi)u_o-a}}\ .$$

Integrating with respect to $2\varphi(\xi)u_o > a$, we thus obtain

$$\mathcal{D}(a) = 0 \text{ for } a > 2u_o\cdot\max\ \varphi(\xi).$$

This means that the amplitudes of the formed solitary waves do not exceed twice the maximum of the initial condition, thus

$$0 \leqq \begin{array}{c}\text{amplitudes of the}\\ \text{solitary waves}\end{array} < 2\cdot u_o\cdot\max(\varphi(\xi)).$$

The maximum strength of the influence on the individuals is given by $2\cdot u_o\cdot\max$ (influence of advertising).

On page 330 we introduced the amplitude $a_1 = \frac{x}{2t}$ of the solitary waves, and the arbitrary function f( ), hence the distribution of the amplitudes is given by

$$0 < \frac{x}{2t} < 2\cdot u_o\cdot\max(\varphi(\xi)),$$

and the solitary wave of amplitude a_1 moves with velocity $2\cdot a_1$ and is found at

$$x = 2a_1 t \text{ as } t \to \infty.$$

The number of waves $N(x,t)$ in the interval $(x,x+dx)$ can be given as

$$N dx = f(a)da,$$

hence

$$N(x,t) = \frac{1}{2t}\,f\left(\frac{x}{2t}\right),$$

where the arbitrary function f is given by the distribution function $\mathcal{D}(a)$. Thus, we have fixed the arbitrary function, obtained by the variational principle.

Integrating the distribution function $\mathcal{D}(a)$, we obtain the whole number of generated solitary waves

$$n = \int \mathcal{D}(a)da = (\frac{\sigma}{4\pi\cdot\sqrt{6}}) \int_{\varphi(\xi)>0} (\varphi(\xi))^{\frac{1}{2}} d\xi,$$

this number is generated by the initial condition

$$u(x,0) = u_0 \cdot \varphi(x).$$

Solitary waves are formed only for regions where the initial condition is positive.

If we interpret the initial condition $u(x,0) > 0$ as positive advertising, then we can say that stable, solitary groups of individuals are formed in front of the shop-windows.

On page 317 we introduced the invariants and have shown that the invariant of the wave packet behind the solitary waves are given by

$$I_\omega = \frac{1}{m} \int_{u(x,0)<0} u^m(x,0)dx;$$

thus we have only contributions for regions where the initial condition is negative. If the initial conditions are negative, then there is a negative influence of advertising on the individuals, thus, they are leaving the shopping centre. (See the diagram on page 310, the wave packet (leaving individuals) is spreading to the left.)

The dynamic process of advertising described in this chapter is only on a first level of approximation.

V Dynamic stochastic models

V.1 Some remarks on stochastic modelling

V.1.1 Stochastic differential equations and impulse control

The word "system" will be applied to any physical or economical or social phenomenon whose evolution changes with time. The dynamics of the system is characterized by its state at each instant t, denoted by $y(t)$. The mapping $t \to y(t)$ describes the evolution of the system. This evolution is provided by a model.

In this book, we shall only consider the deterministic and some stochastic models without delay, leading to a continuous function $y(t)$.

In the simplest deterministic case, $y(t)$ is the solution of a differential equation

$$\frac{dy}{dt} = g(y(t),t),$$

with initial condition $y(0) = y_0$.

The variation in the state $y(t)$ between the instants t and $t + \Delta t$ can be given by

$$\Delta y(t) = g(y(t),t) \cdot \Delta t.$$

In the stochastic case, the variation $\Delta y(t)$ is random. It is natural to add a random number, with zero mean, to the right-hand side. We assume that the random number is normally distributed and two small random numbers corresponding to two non overlapping time intervals are independent. We obtain the model

$$\Delta y(t) = g(y(t),t)\Delta t + \sigma(y(t),t) \cdot \Delta w(t),$$

where σ is a matrix and $w(t)$ is normally distributed with mean 0, it is sufficient to assume that $w(t)$ is an n-dimensional Wiener process

$$\Delta w = w(t + \Delta t) - w(t).$$

If Δt tends to 0, the above model is called a stochastic differential equation, and can be written in the differential form

$$dy(t) = g(y(t),t)dt + \sigma(y(t),t)dw(t),$$

where the Wiener process $w(t)$ is continuous and has bounded variation, but it is not differentiable.

The solution of the stochastic differential equation is a continuous Markoff process, where g is called the drift of the process and σ the diffusion term.

Now we describe the way in which control can influence the evolution of the state, we consider in this volume continuous control. In continuous control, the drift

$$g = g(y(t),v(t),t)$$

and the diffusion term

$$\sigma(y(t),v(t),t)$$

depend on the state $y(t)$, the control $v(t)$, and the time t. The control is a process $v(t)$ whose value can be decided at every instant t as a function of available information at that moment; in general, the value $v(t)$ satisfies certain constraints.

Now the controlled stochastic differential equation is written

$$dy(t) = g(y(t),v(t),t)dt + \sigma(y(t),v(t),t)dw(t).$$

In impulse control the state undergoes jumps (seen as impulses) at certain instants ((23)-(32)).

We denote

$\mathfrak{D}^1 \leqq \mathfrak{D}^2 \leqq \mathfrak{D}^3 \leqq \ldots$ a sequence of impulse instants, and
$\zeta^1, \zeta^2, \zeta^3, \ldots$ a sequence of intensities of the impulses (as jumps).

The evolution of the state is describable:

$$dy(t) = g(y(t),t)dt + \sigma(y(t),t)dw(t) \text{ for } \mathfrak{D}^i \leqq t \leqq \mathfrak{D}^{i+1}$$
$$y(\mathfrak{D}^i) = y(\mathfrak{D}^i - 0) + \zeta^i, \; y(0) = y_0,$$

where

$$y(\mathfrak{D}^i - 0) = y(\mathfrak{D}^{i-1}) + \int_{\mathfrak{D}^{i-1}}^{\mathfrak{D}_i} g(y(t),t)dt + \int_{\mathfrak{D}^{i-1}}^{\mathfrak{D}_i} \sigma(y(t),t)dw(t).$$

The instant $\mathfrak{D}^i$ (representable as stopping time) and the intensity ζ^i are random, and can be chosen as a function of the available information.

A particular case of impulse control is the problem of evolution of the state described by the controlled stochastic differential equation.

We decide to interrupt this evolution at an instant $\mathfrak{D}$, which is a stopping time. This instant can be seen as the decision variable.

V. 1.2 Model in economics

Now we consider some possibilities in applying the continuous - or impulse control models in economics.

Let f be the integral cost, g the final cost, T the horizon of the model ($T \to \pm\infty$) and the discount - or interest factor, then we consider the functional J, where E denotes the expectation

$$J = E(\int_0^T e^{-\alpha s} f(y(s),v(s),s)ds + g(y(T))e^{-\alpha \cdot T}).$$

Starting the control process at the instant t in the state x, we have to consider the evolution described by the controlled stochastic differential equation with respect to the initial condition $y(t) = x$.

Then the functional J is a function of the control v, the initial state $y(t) = x$ and the time t, thus

$$J_{x,t}(v) = E(\int_t^T e^{-\alpha(s-t)} f(y(s),v(s),s)ds + g(y(T))e^{-\alpha(T-t)}).$$

Under certain conditions there exists

$$u(x,t) = \inf_v \; J_{x,t}(v),$$

and the equivalent formulation

$$-u(x,t) = \sup_v \; (-J_{x,t}(v)).$$

Now we have some interesting economical interpretations: $-u(x,t)$ is the profit that results from the deposition at the instant t, from the level of the state x at t, and by applying the control v.

Then $\frac{\partial u(x,t)}{\partial x} = p(x,t)$ is the revenue corresponding to having an additional unit of each of the components of the set. We can say, that $p(x,t)$ represents the market price at the instant t, when the quality x is available.

Applying the formal argument, that is based on the optimal principle of dynamic programming (Bellman (22)), we obtain an analytical description of $u(x,t)$.

The above optimal principle tells us that if $v(s)$ is chosen on t, $t+\delta$ in such a way that we minimise the functional, then

$$u(x,t) = \inf_{v(t,t+d)} E \int_t^{t+d} e^{-\alpha(s-t)} f(y(s),v(s),s)ds + g(y(t+\delta),t+\delta)e^{-\alpha\delta}),$$

expanding the right-hand side up to first order in δ,

$$E(u(y(t+\delta),t+\delta)) = u(x,t) + \delta \frac{\partial u}{\partial x} g(x,v,t) + \delta \frac{\partial u}{\partial t} + \delta \sum_{i,j} a_{ij} \frac{\partial u}{\partial x_i \partial x_j} + O(\delta)$$

and remembering that the Wiener process satisfies

$$E(w(t+\delta)-w(t))^2 = \delta,$$

then we obtain in a formal way by Bellman (22):

$$-(\frac{\partial u}{\partial t} + \sum_{i,j} a_{ij} \frac{\partial^2 u}{\partial x_i \partial x_j} + \inf(\frac{\partial u}{\partial x} \cdot g(x,v,t) + f(x,v,t)) + \alpha u = 0.$$

We consider a firm or shopping centre which invests in the presence of uncertainty, this means in a random environment.

In the presence of uncertainties funds are of interest as a protection against risks. Denoting by

$y(t)$ = level of the total investments at the instant t,

$r(t)$ = disposable reserves at t,

$I(t)$ = investment rate at t,

$D(t)$ = rate of dividents at t.

Let y be the level of capital of the firm or the shopping centre, then the corresponding level during the interval dt is a random quantity

$$f(y(t),t)dt + \varepsilon dW(t),$$

where $W(t)$ is a Wiener process and $f(y(t),t)$ the mean revenue as a function of the level of the invested capital $y(t)$. It is clear, that the derivative of the level $y(t)$ of the total investments with respect to t is equal to the investment rate at t,

$$\frac{dy(t)}{dt} = I(t).$$

We assume that the investment rate + the rate of dividends cannot exceed the mean available revenue, thus

$$I(t) + D(t) \leqq f(y(t),t).$$

So we obtain the relation:

the disposable reserve = the available mean revenue + a Wiener process
- investment rate - rate of dividends,
and written in differential form

$$dy(t) = I(t)dt$$

$$dr(t) = f(y(t),t) + x\cdot\gamma dW(t) - I(t)dt - D(t)dt.$$

In our case we assume that the Wiener process does not depend on the level of the invested capital and γ the diffusion coefficient.

We can say that a firm is bankrupt if its funding $r(t) = 0$; so the moment of bankruptcy is representable as

$$t_b = \inf\,\{t|r(t) = 0\}.$$

In the above sense, we introduce the functional

$$J = E\int_0^{t_b} e^{-it}D(t)dt,$$

where i is interpretable as the rate of interest.

We have to maximise J with respect to the possible initial controls

$$y(t)|_{t=0} = x, \text{ and } r(t)|_{t=0} = r.$$

Denoting the maximum of J by $u(x,r)$, we obtain by applying the dynamic programming

$$\gamma \cdot \frac{\partial^2 u(x,r)}{\partial r^2} + \max(1, \frac{\partial u}{\partial x}, \frac{\partial u}{\partial r}) f(x) - i \cdot u = 0,$$

with respect to the initial condition $u(x,0) = 0$.

Thus $u(x,r)$ is the maximum of the realized profit and depends on r = the disposable reserves, and on x = the level of the capital. If $r(t) = 0$, the firm is bankrupt and thus the profit $u = 0$. Now we consider optimal strategies of firms or shopping centres.

V. 1.3 Optimal strategies

First we assume that the shopping centre distributes dividends without investing, so $I(t) = 0$ and $D = f(x)$, then we take

$$\max\ (1, \frac{\partial u}{\partial x}, \frac{\partial u}{\partial r}) = 1$$

and obtain the differential equation

$$\gamma^2 \frac{\partial^2 u}{\partial r^2} - iu + f(x) = 0,$$

with the condition $u(x,0) = 0$.

Considering the disposable reserve $r(t)$ in the interval $0 \leqq r \leqq \pi$, we are able to construct the general closed solution of the differential equation. It is known (193) that the equation

$$y'' + \lambda y = f(x) \text{ in } 0 \leqq x \leqq \pi$$

is solvable by the finite Laplace transform for $\lambda - n^2 \neq 0$, where n is a natural number, then

$$y(x) = \frac{1}{2} \int_{-\pi}^{+\pi} g(x-\xi) f(\xi) d\xi + y(0) \frac{\sin \sqrt{\lambda}(\pi - x)}{\sin \sqrt{\lambda}\ \pi} + y(\pi) \frac{\sin \sqrt{\lambda} \cdot x}{\sin \sqrt{\lambda}\ \pi},$$

where

$$g(x) = - \frac{\cos(\sqrt{\lambda}(\pi - x))}{\sqrt{\lambda}\ \sin \sqrt{\lambda}\ \pi}$$ is Green's function ((76)-(82)).

Applying this general solution by defining $\lambda = - \frac{i}{\gamma^2}$, we obtain the maximum of the realized profit

$$u(x,r) = \frac{1}{2} \frac{1}{\gamma^2} \int_{-\pi}^{+\pi} g(r-\xi) f(x,\xi) d\xi + u(x,\pi) \frac{\sin \sqrt{\lambda}\ r}{\sin \sqrt{\lambda}\ \pi},$$

the rate of interest is given by $i = \gamma^2 \frac{1}{\sqrt{2}}$, where γ is the diffusion coefficient of the Wiener process.

For a constant level x_0 of the invested capital the maximum of the realized profit $u(x_0,r)$ depends on

i) the available mean revenue f
ii) the constant λ
iii) the disposable reserve r
iv) the rate of interest
v) the coefficient γ of the Wiener process
vi) the given known profit at $r = \pi$.

Thus the shopping centre has influenced their customers by distributing the dividends.

Other optimal strategies of firms or shopping centres are:

$$\frac{\partial u}{\partial x} = \max\left(1, \frac{\partial u}{\partial x}, \frac{\partial u}{\partial r}\right).$$

If the firm makes maximum investment, then $D = 0$ and $I = f(x)$, and the maximum profit is a solution of the Hamilton-Jacobi-Bellman equation, that reduces to the heat equation which can be solved in closed form.

In the case

$$\frac{\partial u}{\partial r} = \max\left(1, \frac{\partial u}{\partial x}, \frac{\partial u}{\partial r}\right),$$

the firms or shopping centres neither invest nor pay dividends, and the Hamilton-Jacobi-Bellman equation can be written as an ordinary differential equation that is solvable.

V. 2 Some remarks on the Wiener process

In the following theorem p^t may be thought of as the probability distribution at time t of an individual starting a Brownian motion at $x = 0$, at $t = 0$.

Theorem:

Let p^t, $0 \leq t \leq \infty$, be a family of probability measures on the real line $\mathbb{R}$, denoting by "*" the convolution, we have

$$p^t * p^s = p^{t+s} \text{ for } 0 \leq t,s < \infty,$$

and for each $\varepsilon > 0$, let

$$p^t(\{x \mid |x| \geqq \varepsilon\}) = o(t) \text{ for } t \to 0,$$

and for each $t > 0$, p^t is invariant under the transformation $x \to -x$, then either

i) $p^t = \delta$ (Dirac) for all $t > 0$

or

ii) there is a $D > 0$, the diffusion coefficient, such that for all $t > 0$, p^t has the density

$$p(t,x) = \frac{1}{\sqrt{4\pi . Dt}} \exp(- \frac{x^2}{4Dt}),$$

with respect to the x-coordinate, and p satisfies the diffusion equation

$$\frac{\partial p}{\partial t} = D . \frac{\partial^2 p}{\partial x^2} \text{ for } t > 0.$$

The proof can be seen in Hunt (170), where he treats nonlocal processes on Lie groups as well. Gaussian processes are discussed in J.L. Doob ((83), p.71-78).

A Gaussian measure on $\mathbf{R}^n$ is a measure which is the transform of the measure with density

$$\frac{1}{(2\pi)^{n/2}} \exp(- \frac{1}{2}|x|^2)$$

under an affine transformation.

Two jointly Gaussian random variables are independent if and only if they are uncorrelated, this means their covariance

$$r(x,y) = E(x-Ex)(y-Ey)$$

is zero, where E denotes the expectation.

We define the mean m and the covariance r of a probability measure μ on $\mathbf{R}^n$ as follows

$$m_i = \int x_i \ \mu(dx), \quad r_{ij} = \int (x_i-m_i)(x_j-m_j)\mu(dx),$$

where x_j are the components of x, then the Gaussian measure with mean m and covariance r, if r is a nonsingular matrix, has the density

$$\frac{1}{(2\pi)^{n/2}} \cdot \frac{1}{(\det r)^{\frac{1}{2}}} \cdot \exp(-\frac{1}{2} \sum_{i,j} (\frac{1}{r})_{ij} (x_i - m_i) \cdot (x_j - m_j).$$

By the basic existence theorem for stochastic processes, we have the

Theorem:

Let I be a set, m a function on I, r a function of positive type on $I \times I$, then there exists a Gaussian stochastic process indexed by I with mean m and covariance r.

Now we can define the Wiener integral (see Doob's book (83); for a Hilbert space approach, I. Segal (316); and for an account of deeper facts the book of Ito and McKean (172)).

The difference of the Wiener process

$$W(t) - W(s) \text{ for } 0 \leqq s \leqq t < \infty$$

form a Guassian stochastic process, indexed by pairs of positive numbers s,t with $s \leqq t$. This process has mean 0 and covariance

$$E(W(t)-W(s) \cdot (W(t')-W(s')) = \sigma^2 \, L\{[s,t] \cap [s',t']\},$$

where L denotes the Lebesgue's measure, and σ^2 is the variance parameter of the Wiener process.

It is known, that the sample paths of the Wiener process are continuous, with probability one, but not differentiable.

For square-integrable f, the integrals of the form

$$\int_{-\infty}^{+\infty} f(t) \cdot d(W(t))$$

can be defined by the

Theorem:

Let Ω be the probability space of the differences of the Wiener process. Then there exists a unique isometric operator from $L^2(\mathbb{R},\sigma^2 dt)$ to $L^2(\Omega)$ denoted by

$$f \to \int_{-\infty}^{+\infty} f(t) d(W(t)),$$

such that

$$\int_{-\infty}^{+\infty} \varphi_{[a,b]}(t)d(W(t)) = W(b)-W(a) \text{ for all } -\infty < a \leqq b < \infty,$$

where $\varphi_A(t) = \begin{cases} 1 & t \in A \\ 0 & t \notin A \end{cases}$ is the characteristic function, and

$$\int_{-\infty}^{+\infty} f(t)d(W(t))$$

is Gaussian.

V. 2.1 Wiener integral

For the Wiener integral on the real line we have the

Theorem:

Let f be of bounded variation of the real line with compact support, and let W be a Wiener process, then

$$\int_{-\infty}^{+\infty} f(t)d(W(t)) = - \int_{-\infty}^{+\infty} d(f(t))W(t),$$

if f is absolutely continuous on [a,b], then

$$\int_a^b f(t)d(W(t)) = - \int_a^b f'(t)W(t)dt + f(b)\cdot W(b)-f(a)\cdot W(a).$$

Now we are able to consider a special class of stochastic differential equations.

V. 2.2 Infinitesimal generator

If X is a Banach space, we call $\{P^+\}_t$ a contraction semigroup on X ((367), p. 246f) if P^+ is a family of bounded linear operators from X to X, defined for $0 \leqq t, s < \infty$ and for all $f \in X$ (see page 365).

The infinitesimal generator A is defined by

$$\lim_{t \to 0^+} \frac{P^t f - f}{t} = Af$$

on the domain $\mathcal{D}(A)$ of all f for which this limit exists.

If X is a locally compact Hausdorff space, then C(X) denotes the Banach space of all continuous functions vanishing at infinity in the norm

$$\|f\| = \sup_{x \in X} |f(x)|.$$

A Markovian semigroup on C(X) is a contraction semigroup on C(X), such that

$$f \geqq 0 \text{ imples } P^t f \geqq 0 \text{ for } 0 \leqq t < \infty,$$

and such that for all $x \in X$ and $0 < t < \infty$

$$\sup_{\substack{0 \leqq f < 1 \\ f \in C(X)}} P^t(f(x)) = 1,$$

if X is compact, we have $p^t 1 = 1$.

V. 2.3 Existence of stochastic processes

A Markov process w on $\mathbb{R}^n$ with infinitesimal generator Q of the form

$$\sum_{i,j=1}^{n} q_{ij} \frac{\partial^2}{\partial x_i{}^i \partial x_j},$$

where q_{ij} is a constant real matrix of positive type, is a special Wiener process.

Then w(t) - w(s) are Gaussian with mean 0 and covariance matrix $2q_{ij}|t-s|$. This in mind, we obtain:

Theorem I

Let $b:\mathbb{R}^n \to \mathbb{R}^n$ be a function satisfying a global Lipschitz condition, so that

$$|b(x_0) - b(x_1)| < k|x_0 - x_0|$$

for some constant k, and all $x_0, x_1 \in \mathbb{R}^n$.

Let w be a Wiener process on $\mathbb{R}^n$ with infinitesimal generator Q given above. Then there exists a unique stochastic process x(t), for each $x_0 \in \mathbb{R}^n$, and $0 \leqq t < \infty$ such that

$$x(t) = x_0 + \int_0^t b(x(s))ds + W(t) - W(s) \text{ for all } t.$$

The process x(t) has continuous sample paths with probability one. Furthermore we define $P^t(f(x_0))$ for $0 \leqq t < \infty$, $x_0 \in \mathbb{R}^n$, $f \in C(\mathbb{R}^n)$ by

$$P^t(f(x_0)) = E(f(x(t))),$$

where E denotes the expectation on the probability space of the Wiener process, then P^t is a Markovian semigroup on $C(\mathbf{R}^n)$. Let A be the infinitesimal generator of P^t, then the functions of $C^2(\mathbf{R}^n)$ with compact support in $\mathbf{R}^n$ are in $\mathcal{D}(A)$, and A can be written as

$$Af = b \cdot \text{grad} f + Qf,$$

for all functions of $C^2(\mathbf{R}^n)$ with compact support in $\mathbf{R}^n$.

Proof:

It is known that the sample paths of a Wiener process are continuous, therefore we have only to prove the existence and uniqueness of x(t), where w is a fixed and continuous function of t. This may be proved by the fixed point theorem.

Let $\mathcal{H}$ be the Banach space of all continuous functions ζ from $[0,t]$ to $\mathbf{R}^n$, with norm

$$\|\xi\| = \sup_{0 \leqq s \leqq t} e^{-\lambda s} |\zeta(s)|,$$

where $t \geqq 0$, $\lambda > k$, and k is the Lipschitz constant.

We consider the non linear transformation $N : \mathcal{H} \to \mathcal{H}$ defined by

$$N\zeta(s) = \zeta(0) + \int_0^s b(x(r))dr + W(s) - W(0).$$

We have to show that N has a unique fixed point.

Applying the global Lipschitz condition (this is also necessary in general), we obtain

$$\|N\zeta - N\eta\| \leqq |\zeta(0)-\eta(0)| + \sup_{0 \leqq s \leqq t} e^{-\lambda s} \int_0^s \{b(\zeta(r))-b(\eta(r))\}dr$$

$$\leqq |\zeta(0)-\eta(0)| + \sup_{0 \leqq s \leqq t} e^{-\lambda s} \cdot k \cdot \int_0^s (\zeta(r)-\eta(r))dr,$$

and by the definition of the norm

I) $\|N\zeta - N\eta\| \leqq |\zeta(0)-\eta(0)| + \sup e^{-\lambda s}\, k \int_0^s e^{\lambda r} |\zeta-\eta| dr = |\zeta(0)-\eta(0)| + \beta \|\zeta-\eta\|,$

where $\beta = \dfrac{k}{\lambda} < 1$.

Defining $\mathcal{H}_{x_0} = \{\xi \in \mathcal{H} | \xi(0) = x_0\}$, then $\mathcal{H}_{x_0}$ is a complete metric space and N is a contraction on it, therefore, N has a unique fixed point x in $\mathcal{H}_{x_0}$. Now since t is arbitrary, then there is a unique continuous function x(t) from $[0,\alpha]$ to $\mathbb{R}^n$ satisfying

$$x(t) = x_0 + \int_0^t b(x(s))ds + W(t)-W(0).$$

Now we prove the second part of the theorem.
We consider

$$P^t f(x_0) = E(f(x(t))) \text{ for } 0 \leq t < \infty,\ x_0 \in \mathbb{R}^n,\ f \in C(\mathbb{R}^n)$$

and show that

$$P^t : C(\mathbb{R}^n) \to C(\mathbb{R}^n)$$

is a Markovian semigroup on $C(\mathbb{R}^n)$.

Indeed by induction on I) we obtain

$$\|N^n\zeta - N^n\eta\| \leq |\zeta(0)-\eta(0)| \cdot (1+\beta + \beta^2 + \ldots + \beta^{n-1}) + \beta^n \|\zeta-n\| .$$

This expression can be written as

$$\|x-y\| \leq \mathfrak{X} \|x_0-y_0\| \text{ for } \mathfrak{X} = \frac{1}{1-\beta},$$

if x,y are the fixed points of N with respect to x(0) = x, y(0) = y and

$$x = \lim_{n\to\infty} N^n x_0 \text{ and } y = \lim_{n\to\infty} N^n y_0.$$

By definition of the norm, we obtain

$$|x(t) - y(t)| \leq e^{\lambda t}\mathfrak{X}|x_0-y_0|.$$

Let f satisfy a global Lipschitz condition with constant k, then

$$|f(x(t))-f(y(t))| \leq k\cdot|x(t)-y(t)| \leq k\cdot e^{\lambda t}\mathfrak{X}|x_0-y_0|.$$

Taking the expectation on the left, the estimate remains true

$$|P^t f(x_0)-P^t f(y_0)| \leq k\cdot e^{\lambda t}\mathfrak{X}|x_0-y_0|,$$

where $Ef(x,t) = P^t(f(x_o))$.

It is known that the Liptschitz functions are dense in $C(\mathbb{R}^n)$ and if f is in $C(\mathbb{R}^n)$, then $P^t f$ is a bounded continuous function. We only need to show that $P^t f$ tends to 0 as x tends to infinity, it vanishes at infinity.

We obtain by the first part of the theorem

$$x(t) = x(s) + \int_s^t b(x(r))dr + W(t) - W(s),$$

using the Lipschitz condition, then

$$|x(t)-x(s)| \leqq \int_s^t |b(x(r))-b(x(t))|dr + (t-s)\cdot b(x(t)) + W(t)-W(s)$$

$$\leqq k \int_0^t |x(r)-x(t)|dr + t|b(x(t))| + |W(t)-W(s)|$$

$$\leqq k\cdot t \sup_{0\leqq r\leqq t} |x(r)-x(t)| + t|b(x(t))| + \sup_{0<s<t} |W(t)-W(s)|$$

for all s, $0 \leqq s < t$;

taking the $\sup_{0\leqq s\leqq t}$ on the left,

$$(1-kt)(\sup_{0\leqq s\leqq t} |x(t)-x(s)|) \leqq t\cdot|b(x(t))| + \sup_{0<s<t} |W(t)-W(s)|,$$

thus, there exists a constant $\mu = \frac{1}{1-kt}$ and if we assume that $k\cdot t < 1$, we obtain

$$\text{II)} \quad |x(t)-x_o| < \sup_{0\leqq s<t} |x(t)-x(s)| \leqq \mu(t|b(x(t))| + \sup_{0<s<t} |W(t)-W(s)|.$$

Let z_o be in the support of f and $\sup |b(z_o)| = \delta$, then for $f(x(t)) = 0$ we have

$$\inf_{z_o \in supp(f)} |z_o - x_o| \leqq \mu(t\cdot\delta + \sup_{0<s<t} |W(t)-W(s))).$$

The probability that w satisfies the inequality, will tend to 0, if x_o tends to infinity. This means that

$$Ef(x(t)) = P^t f(x_o)$$

tends to 0 as x_o tends to infinity.

Now we shall show that P^t is a Markovian semigroup. As we know that

$$x(t) = x(s) + \int_s^t b(x(r))dr + W(t)-W(s)$$

has a unique solution for $0 \leqq s < t$, the conditional distribution of $x(t)$, with given $x(r)$ for $0 \leqq r < s$, is a function of $x(s)$, thus $x(t)$ is a Markovian process, with

$$E\{f(x(t))|x(r) \quad 0 \leqq r < s\} = E\{f(x(t))|x(s)\} = P^{t-s}f(x(s)),$$

for all $f \in C(\mathbb{R}^n)$ and $0 \leqq s < t$. So we have

$$P^{t+s}f(x_0) = E(f(x(t+s)))=EE\{f(x(t+s))|x(r) \ 0 \leqq r < s\} = EP^t f(x(s))=P^s P^t f(x_0)$$

and

$$\sup_{0 \leqq f \leqq 1} P^t(f(x_0)) = 1 \text{ for all } x_0 \text{ and } t.$$

The functions $C^2(\mathbb{R}^n)$ with compact support on $\mathbb{R}^n$ are dense in $C(\mathbb{R}^n)$ and the operators P^t have norm one, this implies that

$$P^t f \to f, \text{ as } t \to 0 \text{ for } f \text{ in } C(\mathbb{R}^n),$$

and now P^t is a Markovian semigroup.

We consider the infinitesimal generator of P^t. Let f be a function in $C^2(\mathbb{R}^n)$ with compact support on $\mathbb{R}^n$ and choose a compact set C, such that the support of f is contained in the interior of C. We have to consider two cases, x_0 in C and x_0 in the complement of C.

Now let x_0 be in C and by definition of the operator P^t and by the representation of the Markoff process $x(t)$, we have

$$P^t(f(x_0)) = E(f(x(t))) = Ef(x_0 + \int_0^s b(x(s)ds + W(t)-W(0)).$$

Applying Taylor's formula to

$$f(x_0 + \int_0^t b(x(s))ds + W(t)-W(0)),$$

we obtain

$$f(x_0 + \int_0^t b(x(s))ds + (W(t)-W(0))) = f(x_0) + t\cdot b(x_0)\cdot \text{grad } f(x_0) +$$

$$+ (W(t)-W(0))\cdot \text{grad } f(x_0) + \frac{1}{2} \sum_{i,j} (W_i(t)-W_i(0))(W_j(t)-W_j(0))\frac{\partial^2}{\partial x_i \partial x_j} f(x_0)$$

$$+ R(t),$$

where the remainder can be written as

$$R(t) = O(|W(t)-W(0)|^2) + O(\int_0^t (b(x(s))-b(x(0)))ds).$$

The infinitesimal generator Q of the Wiener process is given by assumption

$$Q = \sum_{i,j} q_{ij} \frac{\partial^2}{\partial x_i \partial x_j}$$

and applying the expectation E, it becomes

$$\frac{P^t f(x_0)-f(x_0)}{t} = b(x_0)\cdot \text{grad } f(x_0) + Q\cdot f(x_0) + \frac{1}{t} \cdot E(R(t)).$$

E(R(t)) can be estimated in the following way. We know that

$$E(|W(t)-W(0)|^2) < \text{constant } t.$$

By substituting $b(x_0)$ instead of $b(x,t)$ in the inequality II, we can derive in a quite analogous way

$$|x(t)-x_0| \leq \mu(t|b(x_0)| + \sup_{0\leqq s<t} |W(0)-W(s)|);$$

and now

$$\sup_{x_0\in C} \frac{1}{t} \int_0^t |b(x(s))-b(x_0)|ds \leq \sup_{x_0\in C} \frac{1}{t} k\cdot\int_0^t |x(s)-x_0|ds \leq$$

$$\leq \sup_{x_0\in C} \frac{1}{t} k\cdot\mu \int_0^t (t\cdot|b(x_0)| + \sup_{0\leqq s<t} |W(0)-W(s)|)ds$$

taking the expectation E, then

$$E \sup_{x_0\in C} \frac{1}{t} \int_0^t |b(x(s))-b(x_0)|ds \leq E(\sup_{x_0\in C} \mu\cdot k(t\cdot|b(x_0)|+ \sup_{0\leqq s<t} |W(0)-W(s)|)),$$

the integrand is integrable and decreases to 0 as $t \to 0$. Thus we have finally, if A is the infintesimal generator of P^t,

$$Af = b \cdot \text{grad } f + Qf$$

for all functions of $C^2(\mathbb{R}^n)$ with compact support on $\mathbb{R}^n$.

Generalizations with respect to the matrix q_{ij} which depends on x and t, are given by Doob ((183), p. 273f) where he uses Ito's stochastic integrals; see also Nelson (260).

V. 2.4 Crowd movements as a dynamic stochastic process

Pedestrians sometimes meet friends in the shopping street. Each individual has some tendency to persist approximately in uniform rectilinear motion for small intervals of time. Let x be the coordinate of the individual at time t, then the motion is describable by a stochastic process. Let x(t) be a stochastic process; it is known that for many important processes x(t) is not differentiable (for example Wiener processes). Thus we need a substitute for the derivative, to discuss the dynamics of stochastic processes. In the sense of Doob we define the mean forward derivative. For precise definitions we need some remarks.

Let I be an interval, let x be an $\mathbb{R}^n$-valued stochastic process (Markov process) indexed by I. Let $\mathcal{B}_t$ for t in I be an increasing family of σ-algebras, such that each x(t) is $\mathcal{B}_t$-measurable, thus $\mathcal{B}_t$ contains the σ-algebra generated by the x(s) with $s \leq t$ and $s \in I$.

We assume that each stochastic process x(t) is in L^1 and the mapping $t \to x(t)$ is continuous from I to L^1.

$E\{\ldots \mid \mathcal{B}_t\}$ denotes (Doob (83)) the conditional expectation. The notation $\Delta t \to 0^+$ means that Δt tends to 0 through positive values. The random variable D(x(t)) is $\mathcal{B}_t$-measurable. For each t in I the expression

$$Dx(t) = \lim_{\Delta t \to 0^+} E\{\frac{x(t+\Delta t)-x(t)}{\Delta t} \mid \mathcal{B}_t\}$$

exists as a limit in L^1 and $t \to D(x(t))$ is continuous from I to L^1, and we call it the mean forward derivative.

In an analogous way we define the mean backward derivative. Let $\mathcal{A}_t$ be a decreasing family of σ-algebras, such that each x(t) is $\mathcal{A}_t$-measurable; we can say that $\mathcal{B}_t$ represents the past, and $\mathcal{A}_t$ the future. For each $t \in I$

$$D^*x(t) = \lim_{\Delta t \to 0^+} E\{\frac{x(t)-x(t-\Delta t)}{t} | \mathbb{A}_t\}$$

exists as limit in L^1 and $t \to D^*(x(t))$ is continuous from I to L^1 and is called the mean backward derivative.

The standard viewpoint of the theory of stochastic processes is that the past is known and that the future develops from the past by certain probabilistic laws. But in our models we assume that the past and the future are on equal level. Therefore we give a treatment of stochastic motion in which a complete symmetry between past and future is maintained. Let us assume that the past $\mathbb{B}_t$ and the future $\mathbb{A}_t$ are conditionally independent given the present $\mathbb{B}_t \cap \mathbb{A}_t$.

We assume that if f is an $\mathbb{A}_t$-measurable function in L^1, then

$$E\{f|\mathbb{B}_t\} = E\{f|\mathbb{A}_t \cap \mathbb{B}_t\},$$

and if f is any $\mathbb{B}_t$-measurable function in L^1, then

$$E\{f|\mathbb{A}_t\} = E\{f|\mathbb{B}_t \cap \mathbb{A}_t\}.$$

Let x be a Markov process, then $\mathbb{B}_t$ is generated by the x(s), where $s \leqq t$.
Let x be a Markov process, then $\mathbb{A}_t$ is generated by the x(s), where $s \geqq t$.
Let x be a stochastic process and with the above assumptions on $\mathbb{B}_t$ and $\mathbb{A}_t$

$$D(x,(t)),\ D^*(x(t)) \text{ are } \mathbb{B}_t \cap \mathbb{A}_t\text{-measurable},$$

so we obtain, if the expressions exist,

$$DD^*(x(t)),\ D^*D(x(t)),$$

and we can define

$$\frac{1}{2}\, DD^*(x(t)) + \frac{1}{2}\cdot D^*D(x(t))$$

as the mean second derivative.

V. 2.5 Dynamics of Markovian motion

Now we describe the dynamics of Markovian motion. Consider a Markov process x on $\mathbf{R}^n$ of the form

$$dx(t) = b(x(t),t)dt + dw(t),$$

where w(t) is a Wiener process on $\mathbf{R}^n$ with diffusion coefficient ν. The above stochastic differential equation means that

$$x(t)-x(s) = \int_s^t b(x(r)),r)dr + w(t)-w(s)$$

for all t and s.

We have shown in Theorem I that, if $b : \mathbf{R}^n \to \mathbf{R}^n$ satisfies a global Lipschitz condition, this means for some constant $\mathfrak{K}$, we have

$$|b(x_0)-b(x_1)| \leqq \mathfrak{K}|x_0-x_1|$$

for all x_0, x_1 in $\mathbf{R}^n$ and let w be a Wiener process on $\mathbf{R}^n$. For each $x_0 \in \mathbf{R}^n$ there is a unique stochastic process x(t) for $0 \leqq t < \infty$ such that for all t

$$x(t) = x_0 + \int_0^t b(x(r),r)dr + w(s)-w(0).$$

Now let w(t) - w(s) be independent of the x(r), whenever r < s and r < t, we obtain by the above remarks

$$Dx(t) = b(x(t),t).$$

As a Markov process with time reversed (see Doob (83), p. 83f.) is again a Markov process, we have the stochastic differential equation

$$dx(t) = b^*(x(t),t)dt + dw^*(t),$$

where $w^*(t)$ is a Wiener process with diffusion coefficient ν; hence we obtain, if the $dw^*(t)$ are independent of the x(s) with $s \leqq t$, then

$$D^*(x(t)) = b^*(x(t),t).$$

We consider an elementary example. The individuals in the shopping street may be influenced by an external field, generated, for example, by the shop-windows.

The individual performs a Markov process, so its stage at time t_o is given by a point $x(t_o)$ in the coordinate space of the plane.

However, to know the motion of the individuals we also need to know what the Markov process is. We denote by $x(t)$ the position of the individual at time t, by $v(t)$ its velocity, and by

$F = -\text{grad}\ V$ the influence field.

The influence is generated by a potential field V.

V. 2.6 System of stochastic differential equations

For some coefficient β, describing friction phenomena of the population movement (see page 272), the following stochastic differential equation system (in the sense of Langevin) can be written

$$dx(t) = v(t)dt$$

$$dv(t) = -\beta v(t)dt + F(x(t),t)dt + dw(t),$$

where $w(t)$ is a Wiener process representing the random meetings of individuals during their shopping trips. The $dw(t)$ are Gaussian with mean zero.

This is of the form

$$d\begin{pmatrix} x(t) \\ v(t) \end{pmatrix} = \begin{pmatrix} v(t) \\ F(x(t),t)-\beta v(t) \end{pmatrix} dt + d\begin{pmatrix} 0 \\ w(t) \end{pmatrix},$$

where the $dw(t)$ are independent of all the $x(s)$, $v(s)$ with $s \leqq t$. It we assume an asymmetry in time, the second equation of the stochastic differential equation system can be written

$$dv(t) = \beta v(t)dt + F(x(t),t)dt + dw^*(t),$$

where the $dw^*(t)$ are independent of $x(s)$, $v(s)$ with $s \geqq t$. Let $x(t)$ be differentiable, then $\frac{dx(t)}{dt} = v(t)$ and by applying the forward, $Dx(t)$, and backward derivative $D^*x(t)$, we obtain

$$Dx(t) = D^*x(t) = v(t).$$

But w(t) and v(t) are not differentiable, however, we have $Dw(t) = 0$. This is clear, because $w(t+\Delta t) - w(t)$ is independent of x(t) and v(t) and has expectation zero. As the operators D, D* are linear, we apply them to the above stochastic differential equation, so we find

$$Dv(t) = -\beta v(t) + F(x(t),t) \text{ and } D^*v(t) = \beta v(t) + F(x(t),t).$$

It follows immediately that

$$\frac{1}{2} DD^*x(t) + \frac{1}{2} D^*Dx(t) = F(x(t),t),$$

thus, we define the mean second derivative of the stochastic process x(t) by

$$\frac{1}{2} DD^*x(t) + \frac{1}{2} D^*Dx(t).$$

If in our example the individual is not influenced by the potential field, this means that the influence of the shop-windows $V = 0$, then $F = 0$ and we obtain

$$Dv(t) = - D^*v(t) = -\beta v(t),$$

where β describes friction phenomena of the population movement.

Thus, we have a damping effect; we can say that the velocity has a tendency toward zero.

The motion of individuals within the population movement, not influenced by external socio-economic fields, can be described by the Wiener process w(t).

We assume dw(t) are Gaussian with mean zero and

$$E(dw_i(t) \cdot dw_j(t)) = 2\delta_{ij}\, \nu dt,$$

where ν is the diffusion coefficient.

The motion satisfies the Markov process x(t) in the form

$$dx(t) = b(x(t),t)dt + dw(t),$$

where b and b* are vector-valued functions, and with the asymmetry in time t

$$dx(t) = b^*(x(t),t)dt + dw^*(t).$$

If the dw(t) are independent of the x(s) with $s \leq t$, then we obtain by applying the forward D and backward derivative D* to the stochastic differential equation

$$Dx(t) = b(x(t),t) \text{ and } D^*(x(t)) = b^*(x(t),t),$$

where dw*(t) is independent of x(s) with $s \geq t$.

V. 2.7 Characterization of Markov processes

Let $k(x(t),t)$ be a smooth function on $\mathbb{R}^{n+1}$; expanding it by Taylor, we can write

$$k(x(t+\Delta t),t+\Delta t)-k(x(t),t) = \frac{\partial}{\partial t} k(x(t),t)\cdot\Delta t + (x(t+\Delta t)-x(t))\cdot \text{grad } k(x(t),t)$$

$$+ \frac{1}{2} \sum_{i,j} (x_i(t+\Delta t)-x_i(t))\cdot(x_j(t+\Delta t)-x_j(t)).\frac{\partial^2 k(x(t),t)}{\partial x_i\, \partial x_j} + \text{higher order terms.}$$

Thus, up to terms of order two, we have

$$dk(x(t),t) = \frac{\partial k(x(t),t)}{\partial t} dt + dx(t)\cdot \text{grad } k(x(t),t) +$$

$$+ \frac{1}{2} \sum_{i,j} dx_i dx_j \frac{\partial^2 k(x(t),t)}{\partial x_i\, \partial x_j} .$$

We derived the stochastic differential equation

$$dx(t) = b(x(t),t)dt + dw(t),$$

and $Dx(t) = b(x(t),t)$ since dw(t) is independent of x(t) and has mean zero; written as the system

$$dx_i(t) = b(x_i(t),t) + dw_i(t)$$

$$dx_j(t) = b(x_j(t),t) + dw_j(t),$$

we can thus replace

$dx_i(t)\cdot dx_j(t)$ by $dw_i(t)\cdot dw_j(t)$.

If we take the expectation of the above expression, then

$$\frac{1}{2}\sum_{i,j} E(dw_i(t)\cdot dw_j(t))\,\frac{\partial^2 k(x(t),t)}{\partial x_i\,\partial x_j} = \frac{1}{2}\sum_{i,j} 2\delta_{ij}\,\nu dt\,\frac{\partial^2 k(x(t),t)}{\partial x_i\,\partial x_j},$$

and $dx(t)\cdot \mathrm{grad}\, k(x(t),t)$ is replaced by $b(x(t),t)\cdot \mathrm{grad}\, k(x(t),t)$, we obtain finally

$$Dk(x(t),t) = \left(\frac{\partial}{\partial t} + b.\mathrm{grad} + \nu.\Delta\right)(k(x(t),t));$$

in the same way we obtain the reverse direction

$$D^*k(x(t),t) = \left(\frac{\partial}{\partial t} + b^*\,\mathrm{grad} - \nu^*\Delta\right)(k(x(t),t)).$$

In the following we show that $\nu = \nu^*$. But first we have to show the

Lemma:

Let $x(t)$ and $y(t)$ be stochastic processes with respect to the family of σ-algebras $\mathbb{A}_t$ and $\mathbb{B}_t$. If we assume that $Dx(t)$, $Dy(t)$, $D^*x(t)$, $D^*y(t)$ exist and lie in L^2 and are continuous in L^2 with respect to t, then

$$\frac{d}{dt} E(x(t)\cdot y(t)) = EDx(t)\cdot y(t) + Ex(t)\cdot D^*y(t).$$

It remains to show that

$$E(x(b)\cdot y(b) - x(a)y(a)) = \int_a^b E(Dx(t)\cdot y(t) + x(t)D^*y(t))dt.$$

Dividing [a,b] into n equal parts $\tau_n = a + \frac{n(b-a)}{\nu}$ for $n = 0,1,\dots,\nu$, then we obtain

$$E(x(b)\cdot y(b)-x(a)\cdot y(a)) = \lim_{\nu\to\infty}\sum_{n=1}^{\nu-1} E(x(t_{n+1})\cdot y(t_n) - x(t_n)\cdot y(t_{n-1}))$$

$$= \lim_{\nu\to\infty}\sum_{n=1}^{\nu-1} E(x(t_{n-1}) - x(t_n))\frac{y(t_n)+y(t_{n-1})}{2} + \frac{x(t_{n+1})+x(t_n)}{2}(y(t_n)-y(t_{n-1}))$$

$$= \lim_{\nu\to\infty}\sum_{n=1}^{\nu-1} E(Dx(t_n)\cdot y(t_n) + x(t_n)\cdot D^*y(t_n))\,\frac{b-a}{\nu}$$

$$= \int_a^b E(Dx(t)\cdot y(t) + x(t)\cdot D^*y(t))dt.$$

Applying the Lemma, we have

$$\int_{-\infty}^{+\infty} EDf(x(t),t)\cdot g(x(t),t)dt = - \int_{-\infty}^{+\infty} Ef(x(t),t)\cdot D^*g(x(t),t)dt,$$

if f and g have compact support in time.

Let $\rho(x,t)$ be the probability density of $x(t)$ and substituting $Df(x(t),t)$ and $D^*f(x(t),t)$, we obtain

$$\int_{-\infty}^{+\infty} \int_{\mathbf{R}^n} (\frac{\partial}{\partial t} + b\cdot \text{grad} + \nu\Delta)f(x,t)\cdot g(x,t)\rho(x,t)dxdt =$$

$$= \int_{-\infty}^{+\infty} \int_{\mathbf{R}^n} f(x,t)\cdot(\frac{\partial}{\partial t} + b^*\cdot \text{grad} - \nu^*\Delta)g(x,t)\cdot\rho(x,t)dxdt,$$

let

$$A = \frac{\partial}{\partial t} + b\cdot \text{grad} + \nu\Delta$$

be the differential operator and let

$$A^* = -\frac{\partial}{\partial t} - b^* \text{grad} + \nu^*\Delta$$

be its adjoint with respect to ρ-times the Lebesgue measure, then the above equation can be written

$$\int_{\mathbf{R}^{n+1}} Af\ \rho\cdot gdxdt = \int_{\mathbf{R}^{n+1}} f\ A^*(g)\ \rho dxdt$$

for f,g with compact support with respect to the time t.

For another interpretation let A^L be the adjoint of A with respect to the Lebesgue measure on $\mathbf{R}^{n+1}$, we find by integration by parts

$$\int_{\mathbf{R}^{n+1}} Af\ g\cdot\rho dxdt = \int_{\mathbf{R}^{n+1}} f\cdot A^L(g\cdot\rho)dxdt.$$

By comparison we obtain $\rho A^*(g) = A^L(\rho\cdot g)$ thus $A^* = \rho^{-1}A^L\rho$.

If $T: u \to \text{grad } u$ is given as a linear operator with dense domain, then its adjoint T^* exists and is formally given by

$$T^* = -\text{div}(v).$$

In fact, we have by Kato ((186), p. 347)

$$(\text{grad } u, v) = -(u, \text{div}(v))$$

in L^2 at least if the vector-valued function v is continuous differentiable and has compact support.

As div(b...) = ... div b + b. grad..., the adjoint operator can be written as

$$\left(\frac{\partial}{\partial t} + b\cdot\text{grad}+\nu\Delta\right)^L = -\frac{\partial}{\partial t} - \text{div}(b \ \dots\ ) - \nu \text{ div grad} = -\frac{\partial}{\partial t} -$$

$$b \text{ grad} - \text{div } b + \nu \text{ div grad}.$$

As $A^* = \rho^{-1}A^L\rho$ we obtain by

$$\text{div grad } (\rho\cdot g) = g\cdot\Delta\rho+\rho\Delta g+2 \text{ grad } g\cdot\text{grad } \rho \quad \text{and}$$

$$b\cdot\text{grad}(\rho\cdot g) = b\cdot(\text{grad } \rho)g + b\cdot\rho\cdot\text{grad } g; \ -\frac{\partial}{\partial t} - b^*\text{grad} + \nu^*\Delta =$$

$$\rho^{-1}\left(-\frac{\partial}{\partial t} - b \text{ grad} - \text{div } b + \nu\cdot\text{div grad}\right)g\cdot\rho$$

$$= -\frac{\partial g}{\partial t} - \rho^{-1}\frac{\partial \rho}{\partial t}g - b \text{ grad} g - \underset{\dots\dots\dots}{\rho^{-1}b(\text{grad}\rho)g} - \underline{(\text{div} b)g} + \rho^{-1}\cdot\nu\Delta\rho \ g +$$

$$\frac{\nu}{\rho} \ 2 \text{ grad } \rho\cdot\text{grad } g + \nu\cdot \Delta g.$$

Now we recall the forward Kolmogoroff-Focker-Planck equation

$$\frac{\partial\rho}{\partial t} = -\text{div}(b\cdot\rho) + \nu\Delta\rho = -\rho\text{div } b - b \text{ grad } \rho + \nu\Delta\rho$$

and obtain the expression

$$-\frac{1}{\rho}\frac{\partial\rho}{\partial t} \ g = \underline{(\text{div } b)g} + \underset{\dots\dots\dots}{b\cdot\frac{1}{\rho} \text{ grad } \rho\cdot g} - \nu\frac{1}{\rho} \ \Delta\rho\cdot g$$

and obtain the above equation

$$-A^*g = -\frac{\partial g}{\partial t} - b \text{ grad } g + \frac{1}{\rho} \cdot \nu \cdot 2 \text{ grad } g \cdot \text{grad } \rho + \nu \cdot \Delta g$$

thus

$$(-\frac{\partial}{\partial t} - b^*\text{grad} + \nu^*\Delta) = (-\frac{\partial}{\partial t} - b \text{ grad} + 2\cdot\nu \frac{1}{\rho} \text{grad}\rho\cdot\text{grad} + \nu\Delta)g.$$

We find

$$-b^* = b-2\nu\cdot \frac{1}{\rho} \text{ grad } \rho$$

and by comparison

$$\nu=\nu^*, \text{ hence } \frac{1}{2}(b-b^*) = \nu\frac{1}{\rho} \cdot \text{grad } \rho.$$

If we define $u = \frac{b-b^*}{2}$, we obtain the explicit velocity $u = \nu\cdot\frac{1}{\rho}\cdot\text{grad } \rho$. One possible interpretation of this formula is given as follows. Let μ be the number of individuals moving in the shopping street. The external field F of influence may be generated by the potential field V of the shop window display. $\frac{F}{\beta}$ can be interpreted (see page 353) as the velocity of the individuals. Thus, the influence field F imparts to each individual a velocity of the form $\frac{F}{\beta}$. We have $\mu\cdot\frac{F}{\beta}$ individuals passing a region on the shopping street due to the action of the external influence field. If diffusion alone were acting, μ would satisfy the diffusion equation (see Theorem on page 339)

$$\frac{\partial\mu}{\partial t} = D\cdot\Delta\mu,$$

where D is the diffusion coefficient. So we have

$-D\cdot\text{grad } \mu$ individuals passing the region in the shopping street due to diffusion.

V. 2.8 Dynamic equilibrium

$$\mu\cdot\frac{F}{\beta} = D\cdot\text{grad } \mu$$

thus we obtain

$$\frac{F}{\beta} = D \cdot \frac{1}{\mu} \cdot \text{grad } \mu.$$

If we introduce the probability density ρ that is defined by the number density μ divided by the total number of individuals, the equation may be written as

$$\frac{F}{\beta} = D \cdot \frac{1}{\rho} \cdot \text{grad } \rho.$$

The left-hand side is interpretable as the velocity of the individuals influenced, for example, by the shop-windows or the video-display, and

$$D \cdot \frac{1}{\rho} \text{ grad } \rho$$

is the velocity required of the individuals to counteract diffusion effects (e.g. of the crowd movement).

We have defined $u = \frac{1}{2}(b+b^*)$ and obtained $u = \frac{\nu}{\rho}$ gradρ. Now, we define $v = \frac{1}{2}(b+b^*)$ and derive for u and v in the next section a system of nonlinear partial differential equations.

Recalling the Kolmogoroff-Focker-Planck equation for time reversed

$$\frac{\partial\rho}{\partial t} = - \text{div}(b^*\rho) - \nu\Delta\rho = - \rho \text{ div } b^* - b^* \text{ grad } \rho - \nu\Delta\rho$$

Defining $v = \frac{1}{2}(b+b^*)$ we have from the above equations

$$\frac{\partial\rho}{\partial t} = - (\tfrac{1}{2}(\text{div } b + \text{div } b^*)\rho - \text{grad } \rho.\ \tfrac{1}{2}(b + b^*),$$

thus we obtain an equation of continuity

$$\frac{\partial\rho}{\partial t} = - \text{div}(v \cdot \rho)$$

V. 2.9 System of coupled nonlinear partial differential equations

The velocity u can be written as

$$u = \nu \cdot \frac{1}{\rho} \cdot \text{grad}\rho = \nu \text{ grad } \ln \rho,$$

and by the equation of continuity

$$\frac{\partial u}{\partial t} = \nu \cdot \text{grad} \frac{\partial}{\partial t} \ln \rho = \nu \, \text{grad}\left(\frac{1}{\rho} \cdot \frac{\partial \rho}{\partial t}\right) = \nu \, \text{grad}\left(\frac{1}{\rho} \cdot (-\text{div} \nu \rho)\right) =$$

$$= - \nu \, \text{grad} \, (\text{div}(v) + v \frac{1}{\rho} \, \text{grad} \, \rho),$$

thus

$$\frac{\partial u}{\partial t} = - \nu \cdot \text{grad div}(v) - \text{grad}(u \cdot v).$$

The mean second derivative of the stochastic process $x(t)$ is defined by

$$d(t) = \frac{1}{2} D \cdot D^*(x(t)) + \frac{1}{2} D^*D(x(t)),$$

as $Dx(t) = b(x(t),t)$, $D^*(x(t)) = b^*(x(t),t)$, and on page 357 we have constructed

$$Db^*(x(t),t) = \frac{\partial}{\partial t} b^*(x(t),t) + b \cdot \text{grad } b^*(x(t),t) + \nu \Delta b^*(x(t),t)$$

$$D^*b(x(t),t) = \frac{\partial}{\partial t} b(x(t),t) + b^* \cdot \text{grad } b(x(t),t) - \nu \Delta b(x(t),t),$$

then, the mean second derivative of $x(t)$ becomes

$$d(t) = \frac{\partial}{\partial t} \left(\frac{1}{2}(b + b^*)\right) + \frac{1}{2} \cdot b \cdot \text{grad } b^* + \frac{1}{2} b^* \text{ grad } b + \nu \Delta \left(\frac{1}{2}(b-b^*)\right)$$

As we can show that

$$u \cdot \text{grad } u - v \cdot \text{grad } v = \frac{1}{2}(b-b^*)\text{grad}\left(\frac{1}{2}(b-b^*)\right) - \frac{1}{2}(b+b^*)\text{grad}\left(\frac{1}{2}(b+b^*)\right)$$

$$= \frac{1}{4}(b-b^*)(\text{grad}b-\text{grad}b^*) - \frac{1}{4}(b+b^*)(\text{grad}b+\text{grad}b^*)$$

$$= - \frac{1}{2} b \text{ grad } b^* - \frac{1}{2} b^* \text{ grad } b,$$

therefore

$$\frac{\partial v}{\partial t} = d(t) + u \cdot \text{grad } u - v.\text{grad } v + \nu \Delta u.$$

Thus we obtain a coupled nonlinear partial differential equation system

$$\frac{\partial u}{\partial t} = -\nu \cdot \text{grad div}(v) - \text{grad}(u \cdot v)$$

$$\frac{\partial v}{\partial t} = d(t) + u \cdot \text{grad } u - v \cdot \text{grad } v + \nu \Delta u.$$

We can solve the Cauchy problem of this nonlinear partial differential equation system with respect to given initial values $u(x,t_0)$ and $v(x,t_0)$. If the solutions u and v are described by $u = \frac{1}{2}(b-b^*)$, and $v = \frac{1}{2}(b+b^*)$, we obtain b* and b, and then the Markov process is completely known.

Thus we have characterized a Markov process by a Cauchy problem of a nonlinear partial differential equation system. In a special case we can reduce the Cauchy problem of the nonlinear partial differential equations to the Cauchy problem of a linear partial differential equation, which is solvable in closed form. We are able to show that the Markov process of the form

$$d(x(t)) = b(x(t),t)dt + dw(t),$$

where w(t) is a Wiener process in $\mathbf{R}^n$, which is independent of x(r), r < t, is formally equivalent to a partial differential equation system.

V. 2.10 Economic interpretations

We shall assume that each individual of a certain crowd movement performs a Markov process of the above form. We shall postulate that the diffusion coefficient, $\nu = \frac{a}{2\omega}$, is proportional to an action-parameter a and inversely proportional to ω = weight, the mean degree of importance or influence of an individual. We suppose that the individuals are subject to external fields, such as information -, innovation -, advertising - fields (along shopping streets) etc.

We take the special ansatz for the external field F by

$$F = \omega \cdot d,$$

where ω = the mean degree of importance or influence of an individual,

d = the mean second derivative of the stochastic process.

We consider the case where the external field is derived by a potential V (e.g. seen as the influence-potential of the shop-window display)

$$F(x,t) = - \text{grad } V(x,t).$$

In the above nonlinear partial differential equation system, we have to put

$$d = \frac{F}{\omega} = - \frac{1}{\omega} \text{grad } V,$$

hence

$$\frac{\partial u}{\partial t} = - \frac{a}{2\omega} \cdot \text{grad div}(v) - \text{grad } (u \cdot v)$$

$$\frac{\partial v}{\partial t} = - \frac{1}{\omega} \cdot \text{grad } V - v \text{ grad } v + u \cdot \text{grad } u + \frac{a}{2\omega} \Delta u.$$

By solving this system we merely find what stochastic process the individual obeys.

Now we are in a position to reduce the nonlinear differential equation system into a linear partial differential equation. On page 358 we derived the velocity $u = \nu \cdot \text{grad} \ln \rho$, where $\nu = \frac{a}{2\omega}$ is the diffusion coefficient. We obtain

$$\frac{\omega}{a} u = \text{grad } R, \text{ where } R = \frac{1}{2} \ln \rho,$$

and we choose the ansatz $\frac{\omega}{a} v = \text{grad } S$ (quite by analogy). Multiplying both sides of the above system by $\frac{\omega^2}{a^2}$, we have

$$\text{I.} \quad \frac{\omega}{a} \frac{\partial}{\partial t} \text{grad } R = - \frac{1}{2} \text{grad div(grad } S) - \text{grad (grad } R \cdot \text{grad } S)$$

$$\frac{\omega}{a} \frac{\partial}{\partial t} \text{grad } S = - \frac{1}{a^2 m} \text{grad } V - \text{grad } \{- \frac{1}{2} (\text{grad} R)^2 + \frac{1}{2} (\text{grad} S)^2 + \frac{1}{2} \Delta R\},$$

taking the grad (...), then

$$\frac{\omega}{a} \frac{\partial R}{\partial t} = - \frac{1}{2} \Delta S - (\text{grad } R \cdot \text{grad } S)$$

$$\frac{\omega}{a} \frac{\partial S}{\partial t} = - \frac{1}{2} \Delta R - \frac{1}{a^2 \omega} V + \frac{1}{2} (\text{grad } S)^2 - \frac{1}{2} (\text{grad } R)^2,$$

by adding the equations, we obtain

$$\frac{\partial R}{\partial t} + i\,\frac{\partial S}{\partial t} = i^2\,\frac{a}{2\omega}\Delta S + i^2\,\frac{a}{2\omega}\,2\cdot \mathrm{grad}R\cdot\mathrm{grad}S + i\,\frac{a}{2\omega}(\mathrm{grad}S)^2 -$$

II.

$$- i\,\frac{a}{2\omega}\,(\mathrm{grad}R)^2 + i\,\frac{a}{2\omega}\,\Delta R - \frac{1}{a^2\omega}\,\frac{a}{\omega}\,V$$

$$= i\,\frac{a}{2\omega}\,(\Delta R + i\Delta S) + (\mathrm{grad}\,(R+iS))^2 - i\,\frac{1}{a\omega^2}\,V.$$

Now we define the transformation

$$\psi = \exp(R + iS),$$

and by differentiating

$$\frac{\partial\psi}{\partial t} = \left(\frac{\partial R}{\partial t} + i\,\frac{\partial S}{\partial t}\right)\psi,$$

we obtain

$$\frac{\partial^2\psi}{\partial x^2} = \left(\frac{\partial^2 R}{\partial x^2} + i\,\frac{\partial^2 S}{\partial x^2}\right)\psi + \left(\frac{\partial R}{\partial x} + i\,\frac{\partial S}{\partial x}\right)^2\psi \quad \text{and} \quad \frac{\partial^2\psi}{\partial y^2} = \left(\frac{\partial^2 R}{\partial y^2} + i\,\frac{\partial^2 S}{\partial y^2}\right)\psi + $$

$$+ \left(\frac{\partial R}{\partial y} + i\,\frac{\partial S}{\partial y}\right)^2\psi$$

and

$$\Delta\psi = \frac{\partial^2\psi}{\partial x^2} + \frac{\partial^2\psi}{\partial y^2} = (\Delta R + i\Delta S)\psi + (\mathrm{grad}(R + iS))^2\psi.$$

The equation can be written as

$$\frac{\partial\psi}{\partial t} = \left(\frac{\partial R}{\partial t} + i\,\frac{\partial S}{\partial t}\right)\psi \;= i\,\frac{a}{2\omega}(\Delta R + i\Delta S)\psi + (\mathrm{grad}(R+iS))^2\psi - i\,\frac{1}{a\cdot\omega^2}\,V\cdot\psi$$

$$= i\,\frac{a}{2\omega}\,\Delta\psi - i\,\frac{1}{a\omega^2}\,V\cdot\psi.$$

We have reduced with help of the transformation $\psi = \exp(R + iS)$, the non-linear partial differential equation system to the linear partial differential equation

$$\frac{\partial\psi}{\partial t} = i\,\frac{a}{2\omega}\,\Delta\psi - i\,\frac{1}{a\omega^2}\,V\cdot\psi$$

(as we have taken grad(...) in I, we must add a term $-iC\psi$, where C is a constant).

In the time-dependent case we have $\frac{\partial\psi}{\partial t} = 0$ and obtain

$$\Delta\psi - \frac{2}{a^2\omega} V\cdot\psi = 0.$$

Conversely, if ψ satisfies the linear differential equation and if we define R and S by

$$\text{grad } R = \frac{\omega}{a}\cdot u, \quad \text{grad } S = \frac{\omega}{a}\cdot v \text{ and } \psi = \exp(R + iS),$$

then u and v satisfy the nonlinear partial differential equation system. This is clear by taking in equation II the gradients and separating the real and imaginary part.

Thus we have characterized the Markov process of the motion of individuals under the influence of an external potential field V.

V. 2.11 Probability density of a Markov process

We obtain the

Theorem:

The solution of

$$\frac{\partial\psi}{\partial t} = i\frac{a}{2\omega}\Delta\psi - i\frac{1}{a\omega^2} V\cdot\psi$$

may be written as $\psi = \exp(R + iS)$ and we obtain

$$u = \frac{a}{\omega} \text{ grad } R, \quad v = \frac{a}{\omega} \text{ grad } S.$$

Then the Markov process

$$dx(t) = b(x(t),t) + dw(t)$$

is completely known, as we can write

$$b = v + u \text{ and } b^* = v - u,$$

and w(t) is a Wiener process (representing e.g. random meetings of individuals).

Then there follows the remarkable

Lemma:

The probability density ρ of the Markov process is given by

$$\rho = \psi \cdot \bar{\psi} = |\psi|^2.$$

Proof:

As $\psi = \exp(R + iS)$ and $\bar{\psi} = \exp(R - iS)$, we have

$$\psi\bar{\psi} = e^{2R} = \rho, \text{ as } R = \frac{1}{2} \ln \rho$$ (see pages 359 , 362)

V. 3 Feller semigroups and the uncertainty relation

Let D be a bounded domain in $\mathbf{R}^n$ with smooth boundary ∂D and let $C(\bar{D})$ be the space of real valued continuous functions on $\bar{D} = D \cup \partial D$. Now we define the Feller semigroup.

Let M be a compact metric space and let C(M) be as above with norm

$$\|g\| = \sup_{x \in M} |g(x)|.$$

A family $\{T_t\}_{t \geq 0}$ of bounded linear operators on C(M) is called a Feller semigroup, if $\{T_t\}_{t \geq 0}$ satisfies the following conditions

i) $T_t \circ T_s = T_{t+s}$

ii) $\{T_t\}$ is strongly continuous in t on the interval $[0,\infty]$, i.e.,

$$\lim_{t \downarrow 0} \|T_{t+s}f - T_s f\| = 0 \text{ for f in } C(M) \text{ and } 0 \leq s < \infty,$$

iii) $\{T_t\}$is nonnegative and contractive on C(M), i.e.,

$$f \text{ in } C(M) \text{ and } \quad 0 \leq f \leq 1 \text{ on } M \text{ implies } 0 \leq T_t f \leq 1 \text{ on } M.$$

Let $\{T_t\}_{t \geq 0}$ be a Feller semigroup on $\bar{D}$. Its infinitesimal generator

$\mathbb{A}: C(\bar{D}) \to C(\bar{D})$ is defined by

$$\mathbb{A}f = \lim_{t \searrow 0} \frac{T_t f - f}{t} \text{ in } C(\bar{D}).$$

The domain $\mathcal{D}(\mathbb{A})$ of $\mathbb{A}$ consists of all f in $C(\bar{D})$ for which the above limit exists.

It is known (42, 91, 92, 356) that there corresponds to a Feller semigroup $\{T_t\}_{t \geq 0}$ on $\bar{D}$ a strong Markov process on $\bar{D}$ whose transition function $V(t,x,dy)$ satisfies

$$T_t f(x) = \int_{\bar{D}} f(y) \cdot V(t,x,dy) \text{ for f in } C(\bar{D}).$$

Under certain continuity hypotheses concerning the transition function $V(t,x,dy)$ such as

$$\lim_{t \searrow 0} \frac{1}{t} \int_{|y-x|>\varepsilon} V(t,x,dy) = 0 \text{ for } \varepsilon > 0 \text{ and } x \text{ in } \bar{D},$$

we can describe the infinitesimal generator $\mathbb{A}$ of $\{T_t\}_{t \geq 0}$ as follows: Let x be a fixed point of the interior of D. For a C^2-function u in the domain $\mathcal{D}(\mathbb{A})$ of the infinitesimal generator $\mathbb{A}$, we expand $u(y) - u(x)$ by Taylor and obtain from the representation of the Markov process on $\bar{D}$ and the continuity hypotheses

$$\begin{aligned}
\mathbb{A}u(x) &= \lim_{t \searrow 0} \frac{T_t u(x) - u(x)}{t} = \lim_{t \searrow 0} \frac{1}{t} \int_{\bar{D}} V(t,x,dy)u(y) - u(x) \\
&= \lim_{t \searrow 0} \left\{ \frac{1}{t} \int_{\bar{D}} V(t,x,dy)(u(y) - u(x)) + \frac{1}{t} \int_{\bar{D}} (V(t,x,dy) - 1)u(x) \right\} \\
&= \lim_{t \searrow 0} \Bigg[\frac{1}{t} \int_{|y-x|<\varepsilon} (V(t,x,dy) - 1)u(x) + \\
&\qquad + \sum_{i=1}^{N} \frac{1}{t} \int_{|y-x|<\varepsilon} (y_i - x_i) V(t,x,dy) \frac{\partial u(x)}{\partial x_i} + \\
&\qquad + \sum_{i,j=1}^{N} \frac{1}{t} \int_{|y-x|<\varepsilon} (y_i - x_i)(y_j - x_j) V(t,x,dy) \frac{\partial^2 u(x)}{\partial x_i \partial x_j} + \text{remainder terms} \Bigg].
\end{aligned}$$

Thus we obtain

$$\mathbb{A}u(x) = c(x)u(x) + \sum_{i=1}^{N} b_i(x) \frac{\partial}{\partial x_i} u(x) + \sum_{i,j=1}^{N} a_{ij}(x) \frac{\partial^2}{\partial x_i \partial x_j} u(x),$$

where the limits can be written

$$c(x) = \lim_{t \searrow 0} \frac{1}{t} \left(\int_{|y-x|<\varepsilon} V(t,x,dy) - 1 \right); \; b_i(x) = \lim_{t \searrow 0} \frac{1}{t} \int_{|y-x|<\varepsilon} (y_i - x_i) V(t,x,dy),$$

and

$$a_{ij}(x) = \lim_{t \searrow 0} \frac{1}{t} \int_{|y-x|<\varepsilon} (y_i - x_i)(y_j - x_j) V(t,x,dy),$$

this exists independently for sufficiently small $\varepsilon > 0$. The coefficients may satisfy

$$c(x) \leqq 0, \; a_{ij}(x) = a_{ji}(x) \text{ and } \sum_{i,j=1}^{N} a_{ij} \xi_i \xi_j \geqq 0 \text{ for } \xi = (\xi_1, \xi_2, \ldots, \xi_n) \in \mathbf{R}^n.$$

Let A be a linear partial differential operator of second order of the form

$$Au(x) = \sum_{i,j=1}^{N} a_{ij}(x) \frac{\partial^2}{\partial x_i \partial x_j} u(x) + \sum_{i=1}^{N} b_i(x) \frac{\partial}{\partial x_i} u(x) + c(x)u(x),$$

then we obtain

$$\mathbb{A}u(x) = Au(x) \text{ for } u \text{ in } \mathcal{D}(\mathbb{A}) \cap C^2(\bar{D}),$$

thus, Kolmogoroff's partial differential equation can be written in the form

$$\frac{\partial u(x)}{\partial t} = - \sum_{i,j=1}^{N} \frac{\partial^2}{\partial x_i \partial x_j} (a_{ij}(x) \, u(x)) - \sum_{j=1}^{N} \frac{\partial}{\partial x_j} (b_j(x) u(x)).$$

It is known that the probability density ρ satisfies the Focker equation given by

$$\frac{\partial \rho}{\partial t} = - \sum_j \frac{\partial}{\partial y_j} (b_j \, \rho) + \sum_{i,j} \frac{\partial^2}{\partial y_i \partial y_j} (a_{ij} \, \rho)$$

$$= - \sum_j \frac{\partial}{\partial y_j} (b_j \, \rho - \sum_i \frac{\partial}{\partial y_i} (a_{ij} \rho)).$$

On page 359 we obtained the relation div $\mathfrak{G} = -\frac{\partial\rho}{\partial t}$ where $\mathfrak{G}$ is interpretable as the current density, hence

$$\frac{\partial\rho}{\partial t} = -\sum_j \frac{\partial}{\partial y_j} \mathfrak{G}_j,$$

thus we attach to each Markov process the statistical current density

$$\mathfrak{G}_j = b_j\rho - \sum_k \frac{\partial}{\partial y_k}(a_{jk}\rho)$$

and the stochastic velocity is defined by

$$g_i = \frac{1}{\rho}\mathfrak{G}_i = b_i - \frac{1}{\rho}\sum_k \frac{\partial}{\partial y_k}(a_{ik}\rho).$$

Defining the operator

$$G_i = b_i - \sum_k \frac{\partial}{\partial y_k}(a_{ik}),$$

we obtain the stochastic velocity

$$g_i = \frac{1}{\rho} G_i\rho,$$

and show that the multiplication $y_\ell \cdot G_i$ is not commutative. In fact

$$(y_\ell \cdot G_i - G_i y_\ell)\rho = y_\ell b_i\rho - y_\ell \sum_k \frac{\partial}{\partial y_k}(a_{ik}\rho) - b_i y_\ell\rho + \sum_k \frac{\partial}{\partial y_k}(a_{ik} y_\ell \cdot \rho) =$$

$$= \rho \sum_k a_{ik} \frac{\partial y_\ell}{\partial y_k} = a_{i\ell}\rho,$$

hence

$$y_\ell \cdot G_i - G_i y_\ell = a_{i\ell}.$$

Defining the statistical relations:

$$\bar{y}_\ell = \int y_\ell \cdot \rho\, dy, \quad \overline{y_\ell^2} = \int y_\ell^2 \rho dy, \quad \bar{G}_i = \bar{g}_i = \int G_i\rho dy,$$

$$\overline{g_i^2} = \int \frac{1}{\rho}(G_i\rho)^2 dy, \text{ where } dy = dy_1 \cdot dy_2 \cdots dy_n,$$

we obtain

$$(\Delta y_\ell)^2 = \overline{(y_\ell-\bar{y}_\ell)^2} = \int (y_\ell-\bar{y}_\ell)^2\rho \, dy \text{ and } (\Delta g_i)^2 = \overline{(g_i-\bar{g}_i)^2} = \int \frac{1}{\rho}(G_i-\bar{g}_i)\rho)^2 dy.$$

We have

$$(y_\ell-\bar{y}_\ell)(G_i-\bar{g}_i)\rho = \frac{1}{\sqrt{\rho}}(y_\ell-\bar{y}_\ell)\rho \cdot \frac{1}{\sqrt{\rho}}(G_i-\bar{g}_i)\rho.$$

Now we apply the Cauchy-Schwarz inequality to

$$\frac{1}{\sqrt{\rho}}(y_\ell-\bar{y}_\ell)\rho \ , \quad \frac{1}{\sqrt{\rho}}(G_i-\bar{g}_i)\rho,$$

and obtain

$$\left|\int (y_\ell-\bar{y}_\ell)(G_i-\bar{g}_i)\rho dy\right| \leq \left(\int \left|\frac{1}{\sqrt{\rho}}(y_\ell-\bar{y}_\ell)\rho\right|^2 dy\right)^{\frac{1}{2}} \left(\int \left|\frac{1}{\sqrt{\rho}}(G_i-\bar{g}_i)\rho\right|^2 dy\right)^{\frac{1}{2}},$$

then

$$\left|\int (y_\ell-\bar{y}_\ell)(G_i-\bar{g}_i)\rho dy\right|^2 \leqq \int \frac{1}{\rho}\left|(y_\ell-\bar{y}_\ell)\rho\right|^2 dy \int \frac{1}{\rho}\left|(G_i-\bar{g}_i)\rho\right|^2 dy =$$

$$= \int (y_\ell-\bar{y}_\ell)^2\rho \, dy \cdot (\Delta G_i)^2 = (\Delta y_\ell)^2 \, (\Delta G_i)^2.$$

The integral on the left-hand side can be written

$$\int (y_\ell \cdot G_i - y_\ell \bar{g}_i - \bar{y}_\ell \cdot G_i + \bar{y}_\ell \cdot \bar{g}_i)\rho \, dy = \int y_\ell \cdot G_i \rho dy - \bar{g}_i \int y_\ell \cdot \rho \, dy - \bar{y}_\ell \int G_i \rho dy +$$

$$+ \bar{y}_\ell \cdot \bar{g}_i = \overline{y_\ell \cdot G_i} - \bar{y}_\ell \cdot \bar{G}_i$$

thus

$$\Delta y_\ell \cdot \Delta G_i \geqq |\overline{y_\ell \cdot G_i} - \bar{y}_\ell \cdot \bar{G}_i|.$$

As

$$y_\ell \cdot G_i = \frac{y_\ell \cdot G_i + G_i y_\ell}{2} + \frac{y_\ell G_i - G_i y_\ell}{2}$$

and by the commutation relation

$$\frac{y_\ell G_i - G_i y_\ell}{2} = \frac{a_{i\ell}}{2} ,$$

we find

$$\overline{y_\ell \cdot G_i} - \bar{y}_\ell \cdot \bar{G}_i = \frac{\overline{y_\ell G_i} + \overline{G_i y_\ell}}{2} - \bar{y}_\ell \cdot \bar{G}_i + \frac{\bar{a}_{i\ell}}{2} ,$$

then $\overline{y_\ell \cdot G_i}$ can be written as

$$\overline{y_\ell \cdot G_i} = \int y_\ell \cdot G_i \cdot \rho dy = \int y_\ell (b_i \rho - \sum_\kappa \frac{\partial}{\partial y_k} (a_{ik}\rho))dy =$$

$$= \overline{y_\ell \cdot b_i} - \sum_k \int y_\ell \cdot \frac{\partial}{\partial y_k} (a_{ik}\rho) dy_1 dy_2 \dots dy_n ;$$

$$\int y_\ell \cdot \frac{\partial}{\partial y_k} (a_{ik}\rho) dy_k = y (a_{ik}\rho) \Big|_{\dots}^{\dots} -$$

$$- \int a_{ik}\rho \frac{\partial y_\ell}{\partial y_k} = \begin{cases} 0 & \text{for } \ell \neq k \text{ and } \rho = 0 \text{ on the boundary} \\ -\bar{a}_{i\ell} & \text{for } \ell = k \text{ and } \rho = 0 \text{ on the boundary.} \end{cases}$$

Thus

$$\overline{y_\ell G_i} = \overline{y_\ell b_i} + \bar{a}_{i\ell} \text{ and } \frac{\overline{y_\ell \cdot G_i}}{2} = \frac{\overline{y_\ell \cdot b_i}}{2} + \frac{\bar{a}_{i\ell}}{2}$$

and the expression given above can be written

$$\overline{y_\ell \cdot G_i} - \bar{y}_\ell \cdot \overline{G_i} = \frac{\overline{y_\ell \cdot G_i}}{2} + \frac{\overline{G_i y_\ell}}{2} - \bar{y}_\ell \cdot \bar{G}_i + \frac{\overline{a_{i\ell}}}{2}$$

$$= \frac{\overline{Y_\ell \cdot b_i}}{2} + \frac{\overline{G_i y_\ell}}{2} - \bar{y}_\ell \cdot \bar{G}_i + \bar{a}_{i\ell},$$

then

$$\overline{y_\ell \cdot G_i} - \bar{y}_\ell \cdot \bar{G}_i \geqq \bar{a}_{i\ell} \text{ and } \Delta y_\ell \cdot \Delta G_i \geqq \quad |\overline{y_\ell \cdot G_i} - \bar{y} \cdot \bar{G}_i| \geqq \bar{a}_i ,$$

and finally we obtain

$$\Delta y_\ell \cdot \Delta G_i > \bar{a}_{i\ell} > 0.$$

The uncertainty relation reads

$$\Delta y_\ell \cdot \Delta G_i > \bar{a}_{i\ell} > 0, \text{ where } a_{i\ell} \neq 0.$$

Thus we have derived the uncertainty relation from Fokker's equation.

Now the processes of wave mechanics or quantum mechanics can be seen as special Markov processes and the uncertainty relation of quantum mechanics can be derived by setting

$$a_i = \frac{\hbar}{4\pi m} .$$

where $\hbar$ is called Planck's constant, then

$$\Delta y_\ell \cdot \Delta \sigma_i > \frac{\hbar}{4\pi}$$

is the well known uncertainty relation of W. Heisenberg.

A complete description of the state of a physical system in classical mechanics is effected by stating all its coordinates and velocities at a given instant; with these initial data, the equations of motion completely determine the behaviour of the system at all subsequent instants. In quantum mechanics such a description is, in principle, impossible, since the coordinates and the corresponding velocities cannot exist simultaneously (by the uncertainty principle of W. Heisenberg). A very important consequence follows from this regarding the nature of the predictions made in quantum mechanics. Whereas a classical description suffices to predict the future motion of a mechanical system with complete accuracy, the less detailed description given in quantum mechanics evidently cannot be enough to do this. This means that, even if an electron is in a state described in the most complete manner possible in quantum mechanics, its behaviour at subsequent instants is still in principle uncertain. Hence, quantum mechanics cannot make completely definite predictions concerning the future behaviour of the electron. For a given initial state of the electron, a subsequent measurment can give various results. The problem in quantum mechanics consists of determining the probability of obtaining various results on performing this measurement.

There exist noteworthy applications in economics, motivated by quantum mechanics. In the sense of quantum mechanics, our influence-models, too, cannot make complete predictions concerning the future behaviour of the individuals in front of the shop-windows. Our knowledge with respect to the future is bounded by the uncertainty relation.

VI Nonlinear interaction

On model-building A. Kuipers said in ((205), p. 128)

"The equations of Maxwell offer a further specification of the wave-theory, representing them as (periodical) variations of electric and magnetic field-strengths. Thus the model of light has become a mathematical equation which will serve as a starting-point from which we derive the various properties of light. The quantities occurring in this equation are of a physical character, so that they cannot vary at will and in solving the equations we are tied to the actual situation which has induced us to form these equations".

On page 129 of the same reference:

"The physical model is a link between the direct observations and the abstract mathematical theory, which knits these experiences logically together and thus explains them."

In the following we try to give an example in opto-electronics considering the nonlinear interaction of optical waves. Possible applications in economics, in the sense of James Clerk Maxwell, are optical advertising processes.

We start with the Maxwell equations

$$\text{curl } H = \mathcal{J} + \frac{\partial}{\partial t} \mathcal{D}$$

$$\text{curl } E = - \frac{\partial}{\partial t} \mu_o H,$$

where $\mathcal{D} = \varepsilon_o \cdot E + P$, E is the electric, H the magnetic field with the constants ε_o and μ_o, P is the polarization given by

$$P = \varepsilon_o \chi E + P_{NL}, \; (P_{NL})_i = 2d_{ijk} E_j E_k, \; \chi = \text{constant},$$

$$d_{ijk} = \text{const. (group-symmetry)}.$$

Thus the first equation can be written

$$\text{curl } H = \sigma \cdot E + \frac{\partial}{\partial t} \varepsilon E + \frac{\partial}{\partial t} P_{NL},$$

σ-represents the conductivity or the loss. The second Maxwell equation reads

$$\text{curl } E = - \frac{\partial}{\partial t} \mu_o H,$$

taking the curl on both sides

$$\text{curl curl } E = - \frac{\partial}{\partial t} \mu_o \text{curl } H = - \frac{\partial}{\partial t} \mu_o \sigma E - \frac{\partial}{\partial t} \mu_o \frac{\partial}{\partial t} \varepsilon E - \frac{\partial}{\partial t} \mu_o \frac{\partial}{\partial t} P_{NL},$$

as

$$\text{curl curl } E = - \Delta E \text{ as div } E = 0,$$

we obtain finally

$$\Delta E = \mu_o \sigma \frac{\partial E}{\partial t} + \mu_o \varepsilon \frac{\partial^2 E}{\partial t^2} + \mu_o \frac{\partial^2 P_{NL}}{\partial t^2} .$$

Specializing this problem to one dimension only by taking $\frac{\partial}{\partial x} = 0$, $\frac{\partial}{\partial y} = 0$, such that the waves are propagating only in z-direction, we have

$$\frac{\partial^2 E}{\partial z^2} = \mu_o \sigma \frac{\partial E}{\partial t} + \mu_o \varepsilon \frac{\partial^2 E}{\partial t^2} + \mu_o \frac{\partial^2 P_{NL}}{\partial t^2} .$$

The corresponding field has the form of travelling plane waves with frequencies $\omega_1, \omega_2, \omega_3$:

$$E_1(z,t) = \frac{1}{2} \{C_1(z) \cdot e^{i(\omega_1 t - k_1 z)} + \text{conjugate complex}\}$$

$$E_2(z,t) = \frac{1}{2} \{C_2(z) e^{i(\omega_2 t - k_2 z)} + \text{c.c.}\}$$

$$E_3(z,t) = \frac{1}{2} \{C_3(z) e^{i(\omega_3 t - k_3 z)} + \text{c.c.}\} .$$

With these "ansätzen" the wave equation reduces to

$$\Delta E \rightarrow \frac{\partial^2 E}{\partial z^2} = \frac{\partial^2}{\partial z^2}(E_1(z,t)) = \frac{1}{2}\frac{\partial}{\partial z}\left(\frac{1}{2}\frac{\partial C_1(z)}{\partial z}\cdot e^{i(\omega_1 t - k_1 z)} + C_1(z)(-1)ik_1 e^{i(\omega_1 t - k_1 z)} + c.c.\right)$$

$$= \frac{1}{2}\left(\frac{\partial^2 C_1}{\partial z^2} e^{i(.)} + \frac{\partial C_1}{\partial z}(-1)ik_1 e^{i(\cdot)} + (-1)\frac{\partial C_1}{\partial z} ik_1 e^{i(\cdot)} + C_1(z) i^2 k_1^2 \cdot e^{i(\cdot)} + c.c.\right).$$

Let the variation of the amplitude be small $\frac{\partial^2 C_1}{\partial z^2} << k_1 \frac{\partial C_1}{\partial z}$, then

$$\frac{\partial^2 E}{\partial z^2} = -\frac{1}{2}\left(K_1^2 C_1(t) + 2k_1 i \frac{\partial C_1}{\partial z}\right)e^{i(\omega_1 t - k_1 z)}$$

and

$$\mu_o \sigma \frac{\partial E}{\partial t} \rightarrow \mu_o \sigma \frac{\partial}{\partial t} E_1(z,t) = -i\omega_1 \mu_o \cdot \sigma \cdot \frac{1}{2} C_1(z) e^{i(\omega_1 t - k_1 z)} + c.c;$$

recognizing that $k_1^2 = \omega_1^2 \mu_o \varepsilon$, we obtain

$$ik_1 \frac{\partial C_1}{\partial z} e^{-ikz} = -i \frac{\omega_1 \sigma \cdot \mu_o}{2} C_1 e^{-ikz} + \mu_o \cdot \frac{\partial^2 P_{NL}}{\partial t^2}.$$

Thus the reduced wave equation can be written as

$$-\frac{1}{2}\left(K_1^2 C_1(z) + 2ik_1 \frac{\partial C_1(x)}{\partial z}\right)e^{i(\omega_1 t - k_1 z)} + c.c$$

$$= \left\{\frac{1}{2}(-i\omega_1 \mu_o \sigma + \omega_1^2 \mu_o \varepsilon) C_1(z) e^{i(\omega_1 t - k_1 z)} + c.c.\right\} + \mu_o \frac{\partial^2 P_{NL}}{\partial t^2}.$$

VI. 1.1 Polarization phenomena

We consider phenomena which describe the nonlinear response of individuals to the field (e.g. the optical field of advertising). In any real system, polarization induced in the medium (population Haken (148), p. 315f.) is not only proportional to the field E.

We express the polarization in a Taylor series expansion

$$P_i = \lambda_{ij}E_j + 2d_{ijk}E_{jk} + 4\chi_{ijkl}E_jE_kE_l,$$

where the P_i is the i-th component of the instantaneous polarization and E_i is the i-th component of the field, d_{ijk} is responsible for second order harmonic generation (114) (frequency doubling, parameter amplification and oscillation); χ_{ijkl} for third order harmonic generation (230), Raman and Brillouin scattering (364) self-focusing and optical phase conjugation (109).

The four wave interaction can be expressed by the polarization

$$P_i = 4\chi_{ijkl}E_jE_kE_l,$$

the form of χ_{ijkl} is given by the group-symmetry of the medium (population). We consider four waves E_1, E_2, E_3, E_4 coupled by means of the nonlinear polarization P_i. The waves are taken as

$$E_1 = C_1(z)e^{i(\omega t-k_1z)}, \quad E_2 = C_2(z)e^{i(\omega t-k_2z)},$$

where C_1, C_2 are their complex amplitudes.

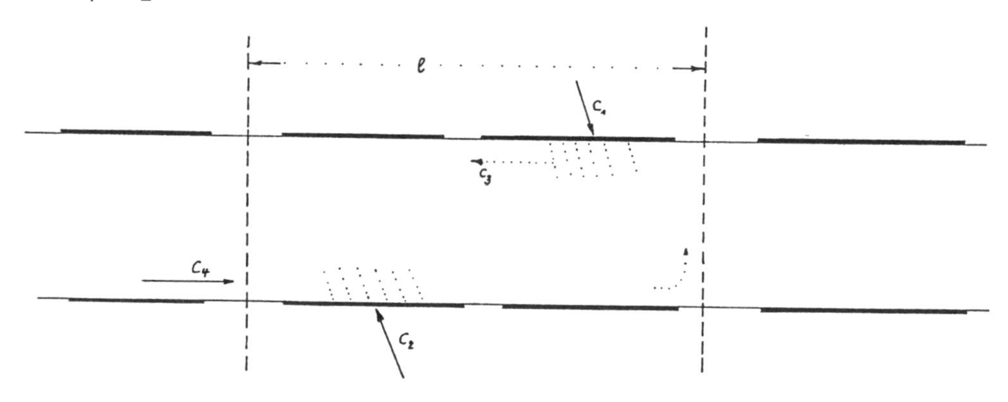

E_4 is some kind of input wave (with respect to a shopping-street of the length ℓ) given by

$$E_4 = C_4(z)e^{i(\omega t-kz)}.$$

If the frequency ω_4 of the wave E_4 is $\omega_4 = \omega$, this means the same frequency as that of the waves E_1 and E_2 (e.g. generated by the optical field of advertising). Then there will be generated a wave E_3 with frequency $\omega_3 = \omega$. It will be shown that the amplitude of E_3 will be the complex conjugate of the input wave E_4. The generation of E_3 can be described by the nonlinear polarization term P_i as follows.

The interaction of the advertising waves E_1, E_2 and E_4 is given by

$$P_i = \frac{1}{2}\chi\cdot E_1\cdot E_2\cdot \bar{E}_4,$$

where $\bar{E}_4$ is the complex conjugate of E_4, thus

$$\bar{E}_4 = C_4(z)e^{i(kz-\omega t)}$$

and

$$P_i = \frac{1}{2}\cdot\chi\cdot C_1\cdot C_2\cdot\bar{C}_4 e^{i(\omega+\omega-\omega)t-(k_1+k_2)z+k\cdot z)} = \frac{1}{2}\chi\cdot C_1\cdot C_2\cdot\bar{C}_4 e^{i(\omega t+k\cdot z)},$$

as $k_1 + k_2 = 0$ by the given direction of C_1 and C_2, thus the excited wave E_3 has the form

$$E_3 = C_3(z)\cdot e^{i(\omega t+kz)}.$$

Now the unknown amplitude will be derived.

The generated field E_3 satisfies

$$\Delta E_3 = \mu_o\cdot\sigma\frac{\partial E_2}{\partial t} + \mu_o\cdot\varepsilon\frac{\partial^2 E_3}{\partial t^2} + \mu_o\frac{\partial^2 P_i}{\partial t^2}$$

where P_i is the above interaction term.

VI. 1.2 System of differential equations for interacting waves

By the above reduction process to one dimension, we obtain

$$ik_1\frac{\partial C_3}{\partial t}e^{i(\omega t-k_1 z)} = -i\frac{\omega_1\sigma.\mu_o}{2}e^{i(\omega t-k_1 z)} + \mu_o\frac{\partial^2}{\partial t^2}P_i(z,t).$$

Let us take the loss $\sigma = 0$ and replace k_1 by $-k$, we finally obtain, remembering that $k^2 = \omega^2\cdot\mu\cdot\varepsilon$,

$$\frac{\partial C_3}{\partial z} = i\,\frac{\omega}{2}\,\frac{\sqrt{\mu}}{\sqrt{\varepsilon}}\cdot\chi\cdot C_1 C_2\cdot\bar{C}_4.$$

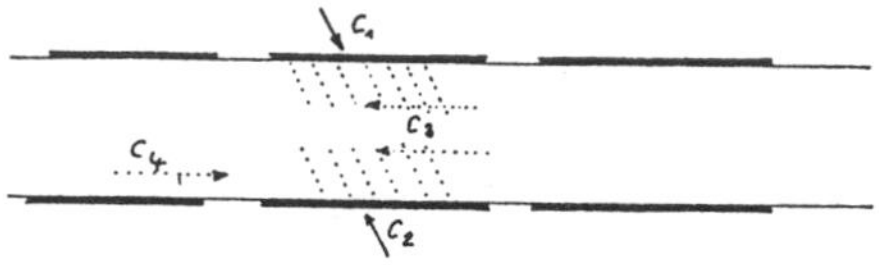

Once generated, this new wave E_3 will interact with the advertising waves C_1, C_2 and generate a polarization

$$P = \frac{1}{2}\cdot\chi\cdot C_1 C_2 \bar{C}_3 e^{i(\omega t - k\cdot z)}.$$

P has the same frequency and propagation vector as E_4. This P will interact strongly with the input wave E_4. Now we can describe the interaction and possible exchange of the waves, where E_3 and E_4 are mediated by the advertising waves E_1 and E_2.

Quite similarly the influenced field of the input wave E_4 is given by

$$\frac{d\bar{C}_4}{\partial z} = i\,\frac{\omega}{2}\cdot\frac{\sqrt{\mu}}{\sqrt{\varepsilon}}\;\chi\bar{C}_1\bar{C}_2 C_3$$

where we have used $k = \frac{\omega}{\mu\cdot\varepsilon}$.

The above derivation of the interaction of the four waves can be written as a differential equation system

$$\frac{dC_3}{dz} = i\cdot\varkappa^*.\bar{C}_4$$

$$\frac{d\bar{C}_4}{dz} = i\cdot\varkappa\cdot C_3,$$

where $\varkappa^* = \frac{\omega}{2}\,\frac{\sqrt{\mu}}{\sqrt{\varepsilon}}\;\chi C_1\cdot C_2.$

Specifying the complex amplitudes $C_3(z)$, $C_4(z)$ at the boundary $z = 0$ and $z = \ell$ and applying the Laplace transform, we obtain the closed solution

$$C_3(z) = \frac{\cos|\varkappa|\cdot z}{\cos|\varkappa|\cdot\ell}\cdot C_3(\ell) + i\,\frac{\varkappa^*}{|\varkappa|}\cdot\frac{\sin|\varkappa|\cdot(z-\ell)}{\cos|\varkappa|\cdot\ell}\;\bar{C}_4(0)$$

$$\bar{C}_4(z) = i\,\frac{|\varkappa|}{\varkappa^*}\cdot\frac{\sin|\varkappa|\cdot z}{\cos|\varkappa|\cdot\ell}\;C_3(\ell) + \frac{\cos|\varkappa|(z-\ell)}{\cos|\varkappa|\ell}\cdot\bar{C}_4(0).$$

Proof:

Introducing the Laplace transform with respect to z, denoted by

$$L\{c\} = \int_0^\infty e^{-nz}C(z,s)dz$$

and applied to the equation system it follows that

$$L\,\{\frac{dC_3}{dz}\} - i\varkappa^*\, L\{\bar{C}_4\} = 0$$

$$L\,\{\frac{d\bar{C}_4}{dz}\} - i\varkappa\;\; L\{C_3\} = 0.$$

The Laplace transform explicitly defined by

$$\mathfrak{C}(n,s) = \int_0^\infty e^{-nz}\; C(z,s)dz.$$

and applied to the differential equation system, we obtain ((76)-(82), (46)-(48))

$$n\cdot\mathfrak{C}_3(n,s) - C_3(0,s) - i\varkappa^*\;\bar{\mathfrak{C}}_4(n,s) = 0$$

$$n\cdot\bar{\mathfrak{C}}_4(n,s) - \bar{C}_4(0,s) - i\varkappa\;\;\mathfrak{C}_3(n,s) = 0,$$

where $C_3(0,s)$ and $\bar{C}_4(0,s)$ are the initial conditions. The solution can be written

$$\mathfrak{C}_3(n,s) = \frac{\begin{vmatrix} \frac{C_3}{n} & -i\,\frac{\varkappa^*}{n} \\ \bar{C}_4 & n \end{vmatrix}}{\begin{vmatrix} 1 & -i\,\frac{\varkappa^*}{n} \\ i\varkappa & n \end{vmatrix}} = \frac{C_3(0,s)\cdot n - \bar{C}_4(0,s)\cdot(-i\varkappa^*)}{n^2-(-i\varkappa^*)(-i\varkappa)}$$

and

$$\bar{\mathfrak{C}}_4(n,s) = \frac{\bar{C}_4(0,s)\cdot n - C_3(0,s)\cdot(-i\,\varkappa)}{n^2-(-i\varkappa^*)\cdot(-i\varkappa)}$$

Defining

$$\tau^2(s) = (-i\varkappa^*)\cdot(-i\varkappa) = i^2\varkappa^2\varkappa = -1\cdot\varkappa^*\varkappa,$$

thus

$$\tau(s) = i\cdot\sqrt{\varkappa^*\varkappa} = i\sqrt{|\varkappa|^2} = i\,|\varkappa|\quad,$$

and we obtain

$$\mathcal{E}_3(n,s) = C_3(0,s)\ \frac{n}{n^2-\tau^2(s)} - \frac{\bar{C}_4(0,s)(-i\varkappa^*)}{\tau(s)}\ \frac{\tau(s)}{n^2-\tau^2(s)}\ :$$

taking the inverse transform, we have (82)

$$1)\quad C_3(z,s) = C_3(0,s)\cdot\cosh(\tau(s)z)-\bar{C}_4(0,s)\ \frac{-i\varkappa^*}{\tau(s)}\ \sinh(\tau(s)\cdot z)$$

and in a quite analogous way

$$2)\quad \bar{C}_4(z,s) = \bar{C}_4(0,s)\cdot\cosh(\tau(s)\cdot z) - C_3(0,s)\cdot\ \frac{-i\varkappa}{\tau(s)}\ \sinh(\tau(s)dz).$$

The first equation can be written

$$C_3(0,s)\cdot\cosh(\tau(s)\cdot z) = C_3(z,s) + \bar{C}_4(0,s)\ \frac{-i\varkappa^*}{\tau(s)}\ \sinh(\tau(s)\cdot z),$$

setting $z = \ell$

$$C_3(0,s) = C_3(\ell,s)\ \frac{1}{\cosh(\tau(s)\cdot z)} - \bar{C}_4(0,s)\cdot\frac{i\varkappa^*}{\tau(s)}\cdot\frac{\sinh(\tau(s)\cdot\ell)}{\cosh(\tau(s)\cdot\ell)}\ ,$$

using this in 2), we have

$$\bar{C}_4(z,s) = \bar{C}_4(0,s)\{\cosh(\tau(s)\cdot z) - \frac{-i\varkappa^*}{\tau(s)}\cdot\frac{\sinh(\tau(s)\cdot\ell)}{\cosh(\tau(s)\cdot\ell)}\cdot\frac{-i\varkappa}{\tau(s)}\cdot\sinh(\tau(s)\cdot z)\}-$$

$$- C_3(\ell,s)\ \frac{\frac{-i\varkappa}{\tau(s)}}{\cosh(\tau(s)\cdot z)}\ \sinh(\tau(s)\cdot z)$$

the expression in {.....} can be written as

$$\frac{\cosh(\tau(s)\cdot z)\cdot\cosh(\tau(s)\cdot\ell)-\sinh(\tau(s)\cdot z)\cdot\sinh(\tau(s)\cdot\ell)}{\cosh(\tau(s)\cdot\ell)} =$$

$$= \frac{\cosh(\tau(s)(z-\ell)}{\cosh(\tau(s)\cdot\ell)}\ ,\ \text{as}\ \frac{(-i\varkappa^*)(-i\varkappa)}{\tau(s)^2} = 1$$

and finally

$$\bar{C}_4(z,s) = \frac{\varkappa}{|\varkappa|}\cdot\frac{\sinh(\tau(s)\cdot z)}{\cosh(\tau(s)\cdot\ell)}\ C_3(\ell,s) + \frac{\cosh(\tau(s)\cdot(z-\ell)}{\cosh(\tau(s)\cdot\ell)}\cdot\bar{C}_4(0,s).$$

Remembering that

$$\cos(|\varkappa|\cdot z) = \cosh(i|\varkappa|z),\ \sin(|\varkappa|\cdot z) = \frac{1}{i}\sinh(i|\varkappa|\cdot z) \text{ and } \tau(s) = i|\varkappa|,$$

then

$$\bar{C}_4(z,s) = i\,\frac{\varkappa}{|\varkappa|}\cdot\frac{\sin(|\varkappa|z)}{\cos(|\varkappa|\cdot z)}\,C_3(\ell,s) + \frac{\cos(|\varkappa|\cdot(z-\ell)}{\cos(|\varkappa|\cdot\ell)}\cdot\bar{C}_4(0,s)$$

and analogously

$$C_3(z,s) = \frac{\cos(|\varkappa|\cdot z)}{\cos(|\varkappa|\cdot\ell)}\cdot C_3(\ell,s) + i\,\frac{\varkappa^*}{|\varkappa|}\cdot\frac{\sin(|\varkappa|\cdot(z-\ell)}{\cos(|\varkappa|\ell)}\cdot\bar{C}_4(0,s).$$

VI. 1.2.1 The solution

The solution of the differential equation system can be given in closed form

$$C_3(z) = \frac{\cos(|\varkappa|\cdot z)}{\cos|\varkappa|\ell}\cdot C_3(\ell) + i\,\frac{\varkappa^*}{|\varkappa|}\cdot\frac{\sin|\varkappa|\cdot(z-\ell)}{\cos|\varkappa|\ell}\cdot\bar{C}_4(0)$$

$$\bar{C}_4(z) = i\,\frac{|\varkappa|}{\varkappa^*}\cdot\frac{\sin|\varkappa|\cdot z}{\cosh|\varkappa|\ell}\cdot C_3(\ell) + \frac{\cos|\varkappa|\cdot(z-\ell)}{\cos|\varkappa|\cdot\ell}\,\bar{C}_4(0).$$

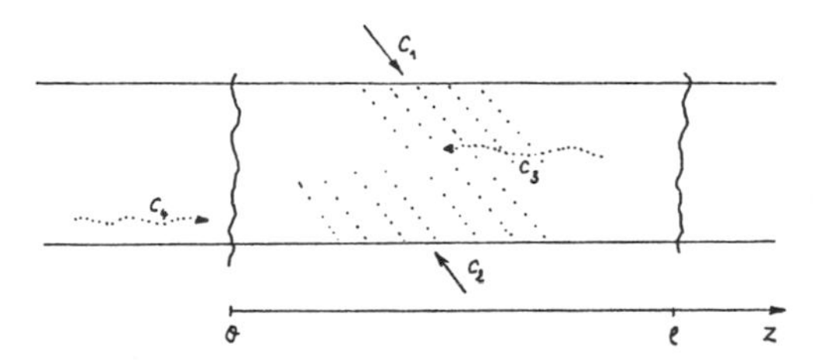

The boundary condition of the reflected wave C_3 may be $C_3(\ell) = 0$. The reflected wave at the input $z = 0$ can be written as

$$C_3(0) = i\,\frac{\varkappa^*}{|\varkappa|}\cdot\tan|\varkappa|\cdot\ell\ \bar{C}_4(0),\ \text{where } \varkappa^* = \frac{\omega}{2}\sqrt{\frac{\mu}{\varepsilon}}\ \chi\cdot C_1\cdot C_2.$$

This means that the reflected wave C_3 is proporational to the complex conjugate of the input wave C_4. Considering the input wave C_4 at the output $z = \ell$, we obtain

$$\bar{C}_4(\ell) = \frac{1}{\cos|\varkappa|\ell}\cdot\bar{C}_4(0);$$

thus, for a single input wave C_4 the transmitted wave is given by

$$C_4(\ell) = \frac{1}{\cos|\varkappa|\cdot\ell} \cdot C_4(0).$$

This can be interpreted to mean there is an amplification of the wave C_4 generated by the wave-produce $C_1 \cdot C_2$, hence

$$\frac{C_4(\ell)}{C_4(0)} = \frac{1}{\cos|\varkappa|\cdot\ell} \to \infty \text{ for } |\varkappa| \cdot \ell \to \frac{\pi}{2},$$

and

$$\frac{C_4(\ell)}{C_4(0)} \to \frac{1}{0{,}7} \quad \text{for } |\varkappa| \cdot \ell \to \frac{\pi}{4}.$$

This shows a situation where a vanishingly small input $C_4(0)$ gives rise to a finite output $C_4(\ell)$.

We have by construction that

$$\varkappa^{*\prime} = \frac{\omega}{2}\sqrt{\frac{\mu}{\varepsilon}}\ \chi \cdot C_1 \cdot C_2,$$

denoting

$$B = \frac{\omega}{2}\sqrt{\frac{\mu}{\varepsilon}}\ \chi,$$

then

$$\varkappa^* = B \cdot C_1 \cdot C_2, \text{ then } \varkappa = B \cdot (C_1 \cdot C_2)^* \text{ and}$$

$$|\varkappa|^2 = \varkappa \cdot \varkappa^* = B^2 C_1 \cdot C_1^* \cdot C_2 C_2^* = B^2 . |C_1|^2 \cdot |C_2|^2,$$

and finally

$$\varkappa = B \cdot |C_1| \cdot |C_2|.$$

We have oscillatory phenomena for $|\varkappa|\ell = B \cdot |C_1| \cdot |C_2|\ell \to \frac{\pi}{2}$.

We obtain only small amplification for $|\varkappa| \cdot \ell = B \cdot |C_1| \cdot |C_2|\ell \to \frac{\pi}{4}$.

VI. 1.3 Application in economics (analogue models)

We will attempt one possible application in economics using analogue models motivated by Kuipers (205), p. 131:

"From the macrophysical world we know laws and models whose applicability and fertility had already been proven. We have discovered them by extrapolating our ordinary experiences to the field that our senses cannot survey.

The success of this passage from our own experience into the domain of the cosmos via the model causes us to suppose that the same laws will hold good in microcosm. Whether this supposition is correct will have to appear from its results.

In constructing modern physics we notice a shifting from what may be concretely pictured to what is mathematically abstract."

Let C_1 and C_2 be optical advertising waves influencing the population movement

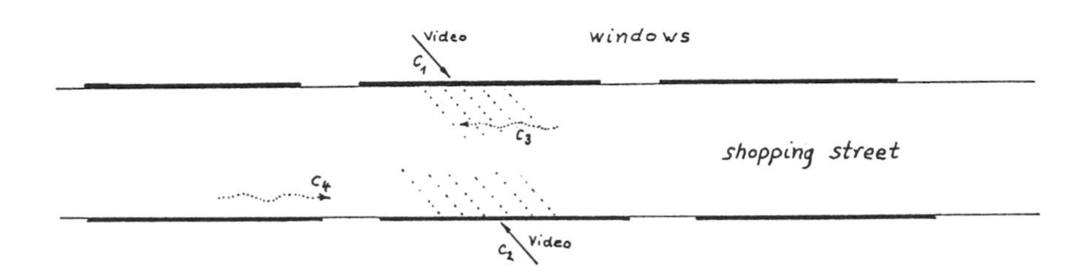

We assume that

$$|\varkappa| \cdot \ell = \ell \cdot B \cdot |C_1| \cdot |C_2| = \frac{\pi}{2} ,$$

let the advertising intensities C_1, C_2 on both sides of the shopping street be

$$|C_1| = |C_2| = A, \text{ then } \ell B A^2 = \frac{\pi}{2} , \text{ as } B = \frac{\omega}{2} \sqrt{\frac{\mu}{\varepsilon}} \chi.$$

We have in (141), p. 62) that $\varepsilon \cdot \mu = \frac{1}{v_0^2}$ where v_0 is interpretable as the innovation velocity and μ, ε are certain indices, then

$$\sqrt{\frac{\mu}{\varepsilon}} = \frac{1}{v_0 \cdot \varepsilon} \quad \text{and} \quad B = \frac{\omega}{2} \chi \cdot \frac{1}{v_0 \cdot \varepsilon} .$$

The innovation velocity follows to

$$v_0 = 2 \cdot \frac{\omega}{\pi \cdot \varepsilon} \chi \cdot A^2 \cdot \ell.$$

The innovation velocity depends on the square of the advertising intensities A, and on the group symmetry of the pedestrian χ (seen as solitary groups on page 308f.), and on the length of the shopping street.

Let the boundary condition $C_3(\ell)$ for the reflected wave C_3 be $C_3(\ell) = 0$, then the reflected wave at the input z = 0 can be written as

$$C_3(0) = -i(\frac{\varkappa^*}{|\varkappa|} \cdot \tan|\varkappa|\ell) \cdot C_4^*(0),$$

then

$$(\frac{C_3(0)}{C_4^*(0)})^* = i \cdot \frac{\varkappa}{|\varkappa|} \cdot \tan|\varkappa| \cdot \ell,$$

and

$$\left|\frac{C_3(0)}{C_4^*(0)}\right|^2 = \frac{C_3(0)}{C_4^*(0)} \cdot \frac{C_3^*(0)}{C_4^{**}(0)} = -i^2 \frac{\varkappa \cdot \varkappa^*}{|\varkappa|^2} \cdot \tan^2|\varkappa|\ \ell =$$

$$= \frac{\varkappa \cdot \varkappa^*}{|\varkappa|^2} \tan^2|\varkappa| \cdot \ell = \frac{\varkappa \cdot \varkappa^*}{|\varkappa|^2} \tan^2|\varkappa|\ell = \tan^2|\varkappa| \cdot \ell,$$

thus

$$\left|\frac{C_3(0)}{C_4^*(0)}\right| = \tan|\varkappa|\ell.$$

As $|\varkappa| \cdot \ell$ is proportional to $|C_1| \cdot |C_2|$ and increases to $\frac{\pi}{2}$, we obtain a very high amplification.

Let the optical advertising intensity be increased, given by the product $|C_1| \cdot |C_2|$ and approaching $|\varkappa|\ell = \frac{\pi}{2}$ then

$$\left|\frac{C_3(0)}{C_4^*(0)}\right| \to \infty$$

this mean that there is high amplification of the influence of advertising on the passers-by :

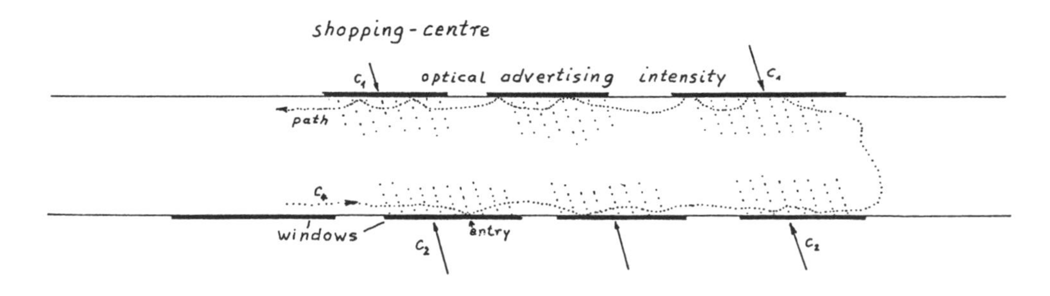

The figure shows a possible path (trajectory) of a pedestrian on his shopping trip. At the beginning the influence-level $C_4^*(0)$ is low. Now the individuals are moving to the influence region in front of the shop-windows. Some individuals are escaping from the influence region (see page 295f.). If the frequency of escape is bounded by the advertising field of the shopping centre, we have shown on page 297f that there remains "bleibender Einfluß", a "lasting influence" or "lasting impression".

The above consideration says that the "lasting influence" on the individuals can be amplified under suitable conditions.

VI. 2 Nonlinear polarization phenomena

VI. 2.1 Image transmission

The interaction of four optical waves can be described by the differential equation system

$$\frac{dC_3}{dz} = i \cdot \varkappa^* \bar{C}_4 \qquad \text{where } \varkappa^* = \frac{\omega}{2}\sqrt{\frac{\mu}{\varepsilon}}\, \chi \cdot C_1 \cdot C_2 .$$

$$\frac{d\bar{C}_4}{dz} = -i\varkappa\, C_3 .$$

Specifying the complex amplitudes $C_3(z)$, $C_4(z)$ at the boundary value $z = 0$ and $z = \ell$, and applying the Laplace transform, we obtain the closed solution

$$C_3(z) = \frac{\cos|\varkappa|\cdot z}{\cos|\varkappa|\cdot \ell} \cdot C_3(\ell) + i\,\frac{\varkappa^*}{|\varkappa|} \cdot \frac{\sin|\varkappa|\cdot(z-\ell)}{\cos|\varkappa|\ell} \cdot \bar{C}_4(0)$$

$$\bar{C}_4(z) = i\,\frac{|\varkappa|}{\varkappa^*} \cdot \frac{\sin|\varkappa|\cdot z}{\cos|\varkappa|\cdot \ell} \cdot C_3(\ell) + \frac{\cos|\varkappa|\cdot(z-\ell)}{\cos|\varkappa|\cdot \ell} \cdot \bar{C}_4(0).$$

Let the boundary condition $C_3(\ell)$ for the reflected wave C_3 be $C_3(\ell) = 0$, then the reflected wave at the point $z = 0$ is given by

$$C_3(0) = -i\left(\frac{\varkappa^*}{|\varkappa|} \cdot \tan|\varkappa|\cdot \ell\right)\bar{C}_4(0)$$

and the input wave reads

$$\bar{C}_4(\ell) = \frac{1}{\cos|\varkappa|\cdot \ell} \cdot \bar{C}_4(0).$$

Thus the reflected wave $C_3(0)$ is proportional to $\bar{C}_4(0)$, that is, the complex conjugate of the input wave $C_4(0)$. By the linearity of the differential equation system we obtain the following image transmission problem.

Let the incident incoming field $C_4(0)$ at $z = 0$ be a complex wavefront

$$C_4(0) = \text{Real } (\varphi(r)e^{i(\omega t-kz)})$$

then the reflected wave $C_3(0)$ is

$$C_3(0) = -i(\frac{\varkappa^*}{|\varkappa|} \cdot \tan|\varkappa|\ell) \cdot \bar{C}_4(0),$$

thus the reflected field is given by

$$E_3(r)_{z<0} = \text{Real}(-i(\frac{\varkappa^*}{|\varkappa|} \cdot \tan|\varkappa|\ell)\overline{\varphi(r)}e^{i(\omega t+k\cdot z)}).$$

This can be interpreted by the propagation of wavefronts through distorting media.

The fundamental principle of phase conjugation can be applied for long distance image transmission in optical wave-guides: (364, 365, 366):

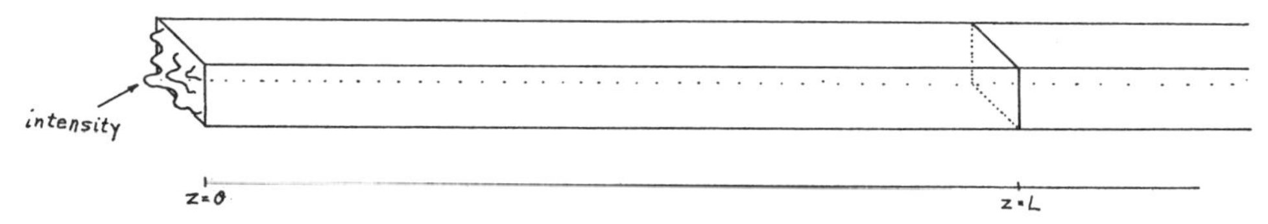

Let $E_{nm}(x,y)$ be the spatial mode functions (intensity, density)

$$f(x,y,z = 0,t) = \sum_{n,m} A_{nm}E_{nm}(x,y)e^{i\omega t}$$

propagating each mode a distance L, and summing up, then there follows the output field at $z = L$

$$f(x,y,z = L,t) = \sum_{n,m} A_{nm}E_{nm}(x,y)e^{i(\omega t-\beta_{nm}\cdot L)},$$

where β_{nm} are the propagation constants of mode n,m by neglecting the mode dependent losses. The output field is a scrambling of spatial information.

Generating the complex conjugate of the outpute field, we obtain

$$f_{out}(x,y,z = L,t) = \sum_{n,m} \bar{A}_{nm} \bar{E}_{nm}(x,y)e^{i(\omega t+\beta_{nm}L)}$$

by leaving the factor $e^{i\omega t}$ intact.

Let the way the fields are propagating be identical in length L and all other characteristics, the output field of the conjugate complex input field is given by

$$f_{out}(x,y,z = 2L,t) = \sum_{n,m} \bar{A}_{nm}\bar{E}_{nm}(x,y)e^{i(\omega t+\beta_{nm}L-\beta_{nm}L)} =$$

$$= \sum_{n,m} \bar{A}_{nm}\bar{E}_{nm}(x,y)e^{i\omega t}.$$

Thus, except for the complex conjugate, the field at z = 2L is the same as the input field (365, 366).

Now the spatial information carried by the wave has been recovered.

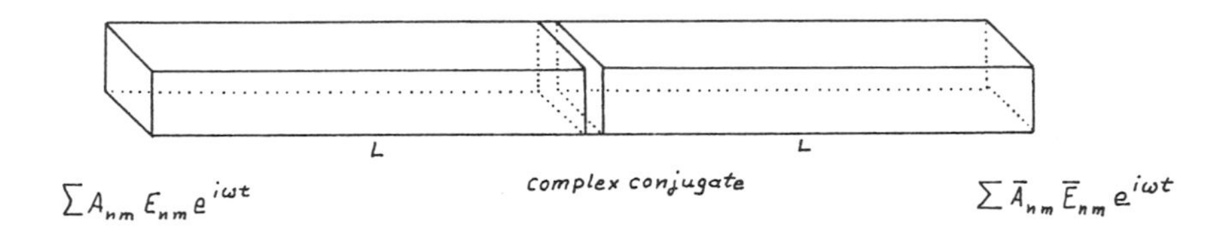

This result gives us the reconstruction of the input image. First demonstrations of the image transmission through a multimode fibre are given by Dunning and Lind (88):

"The results of the far-field experiments are shown in Fig. 3. [Their figure 3]. The input beam is monitored at image plane 2 and is typical of an apodized Gaussian beam (Fig. 3a). The output from the fibre, monitored at image plane 3, is shown in Fig. 3b. It is seen to be very irregular in intensity, indicating the severe aberrations introduced by the fibre.

This output is then conjugated and retraverses the fibre. The reconstructed beam is monitored at image plane 1 and is seen very nearly to reconstruct the initial input beam (Fig. 3c).

The corresponding intensity cross sections 3d, 3e, and 3f were taken at the position of the horizontal line in Figs. 3a, 3b, and 3c."

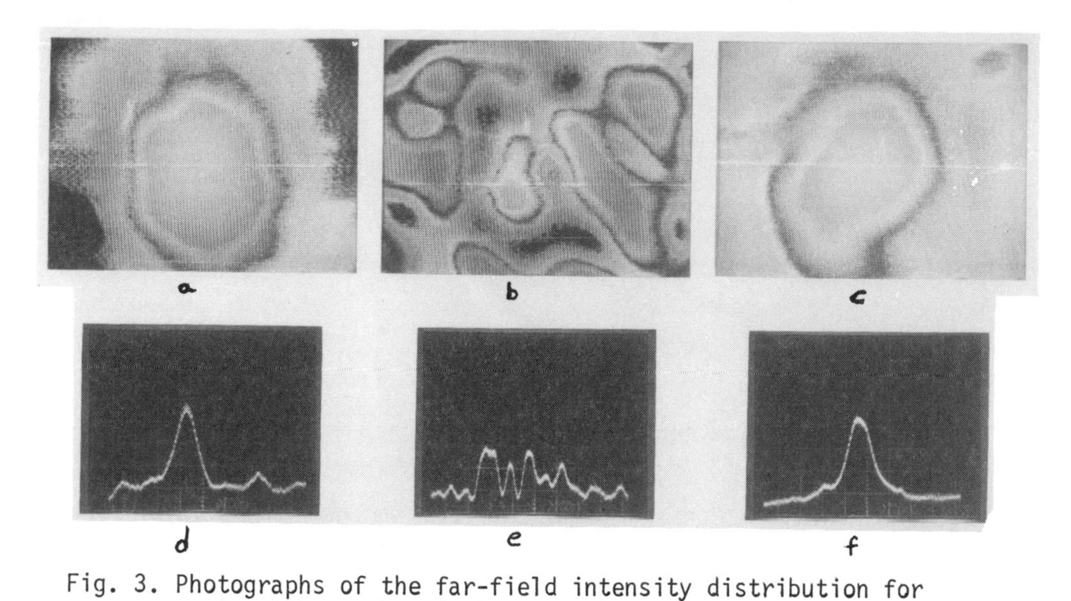

Fig. 3. Photographs of the far-field intensity distribution for a, input beam; b, single pass through fibre; c, corrected beam. Oscilloscope traces d, e, and f are the corresponding cross sections.

(Figure from (88) page 560, by kind permission of The Publishers.)

Applying to nonlinear media external fields, it is possible to generate the complex conjugate. These phenomena are derived in the following section.

VI. 2.2 Solution of the nonlinear wave equation

Nonlinear polarization phenomena in space are governed by

$$\frac{\partial^2 E}{\partial x^2} + \frac{\partial^2 E}{\partial y^2} + \frac{\partial^2 E}{\partial z^2} - \mu\cdot\varepsilon\,\frac{\partial^2 E}{\partial t^2} = \frac{\partial^2}{\partial t^2} P_{NL}(r,t) \quad (364,\ 365,\ 366).$$

We try the ansatz

$$E_2(r,t) = \frac{1}{2}\cdot\sum_m B_m(z)\,\mathcal{E}_2^m(x,y)\,e^{i(\omega_2 t-\beta m_2 . z)} .$$

The z-dependent elements read

$$B_m(z)\cdot e^{-i\beta m_2 z},$$

differentiating

$$\frac{\partial}{\partial z}(B_m(z)\cdot e^{-i\beta m_2 z}) = \frac{\partial B_m}{\partial z}\,e^{-i\beta m_2 z} + B_m(z)(-i)\beta m_2 e^{-i\beta m_2 z},$$

and

$$\frac{\partial^2}{\partial z^2}(B_m(z)\cdot e^{-i\beta m_2 z}) = (\frac{\partial^2 B_m(z)}{\partial z^2} + 2(-i)\beta m_2\cdot\frac{\partial B_m(z)}{\partial z} - \beta^2 m_2^2 B_m(z))e^{-i\beta m_2 z}$$

thus

$$\frac{\partial^2}{\partial z} E_2(r,t) = \sum_m (\frac{1}{2}\frac{\partial^2 B_m(z)}{\partial z^2} - i\beta m_2 \frac{\partial B_m(z)}{\partial z})\mathcal{E}_2^m(x,y)e^{-i\beta m_2 z}e^{i\omega_2 t} -$$

$$- \sum_m \beta^2 m_2^2 B_m(z)\cdot\mathcal{E}_2^m(x,y)e^{-i\beta m_2 z}e^{i\omega_2 t}.$$

The other expressions can be written as

$$\frac{\partial^2}{\partial x^2} E_2(r,t) = \frac{1}{2}\sum_m B_m(z)e^{-i\beta m_2 z}\frac{\partial^2 \mathcal{E}_2^m(x,y)}{\partial x^2} e^{i\omega_2 t}, \quad \frac{\partial^2 E_2(r,t)}{\partial y^2} =$$

$$= \frac{1}{2}\sum_m B_m(z)e^{-i\beta m_2 z}\frac{\partial^2 \mathcal{E}_2^m(x,y)}{\partial y^2}\cdot e^{i\omega_2 t}$$

and

$$\frac{\partial^2 E_2(r,t)}{\partial t^2} = \frac{1}{2}\sum_m B_m(z)e^{-i\beta m_2 z}\mathcal{E}_2^m(x,y)i^2\omega_2^2 e^{i\omega_2 t},$$

then

$$-\mu\cdot\varepsilon\frac{\partial^2 E_2(r,t)}{\partial t^2} = \mu\cdot\varepsilon\,\omega_2^2\cdot\frac{1}{2}\sum_m B_m(z)e^{-i\beta m_2 z}\mathcal{E}_2^m(x,y)e^{i\omega_2 t}.$$

Using the above expression, we obtain

$$\frac{\partial^2}{\partial x^2} E_2(r,t) + \frac{\partial^2}{\partial y^2} E_2(r,t) - \mu\cdot\varepsilon\frac{\partial^2 E_2(r,t)}{\partial t^2} =$$

$$= \sum_m (\frac{\partial^2}{\partial x^2}\mathcal{E}_2^m(x,y) + \frac{\partial^2}{\partial y^2}\mathcal{E}_2^m(x,y) +$$

$$+ \mu\cdot\varepsilon\cdot\omega_2^2\mathcal{E}_2^m(x,y))\frac{1}{2}\cdot B_m(z)e^{-i\beta m_2 z}\cdot e^{i\omega_2 t},$$

$$\frac{\partial^2}{\partial z^2} E_2(r,t) = \sum_m (\ \ldots\ -\beta^2 m_2^2\mathcal{E}_2^m(x,y))\cdot B_m(z)e^{-i\beta m_2 z}e^{i\omega_2 t}.$$

Let $\mathcal{E}_2^m(x,y)$ be a fundamental solution of the equation

$$\left(\frac{\partial^2}{\partial x^2} + \frac{\partial^2}{\partial y^2} - \beta^2 m_2^2 + \mu\cdot\varepsilon\cdot\omega^2\right)E(x,y) = 0,$$

by choosing the normalization constants of $\mathcal{E}_2^m(x,y)$ according to

$$\int \mathcal{E}_2^m \cdot \mathcal{E}_2^n \, dxdy = \delta_{n,m},$$

there remains

$$\sum_m \left(\frac{1}{2}\frac{\partial^2 B_m(z)}{\partial z^2} - i\beta m_2 \frac{\partial B_m(z)}{\partial z}\right) \mathcal{E}_2^m(x,y)e^{-i\beta m_2 z} e^{i\omega_2 t} + \text{c.c.} = \mu \frac{\partial^2}{\partial t^2} P_{NL}(r,t)$$

(see the right-hand side of the above expression). Assuming that

$$\left|\frac{d^2 B_m(z)}{dz^2}\right| < \beta\cdot m_2 \left|\frac{dB_m(z)}{dz}\right| ,$$

the equation reduces to

$$\sum_m \left(-i\beta m_2 \frac{\partial B_m(z)}{\partial z} \mathcal{E}_2^m(x,y)e^{-i\beta m_2 z} e^{i\omega_2 t}\right. + \text{c.c.} = \mu \frac{\partial^2}{\partial t^2} P_{NL}(r,t).$$

We have

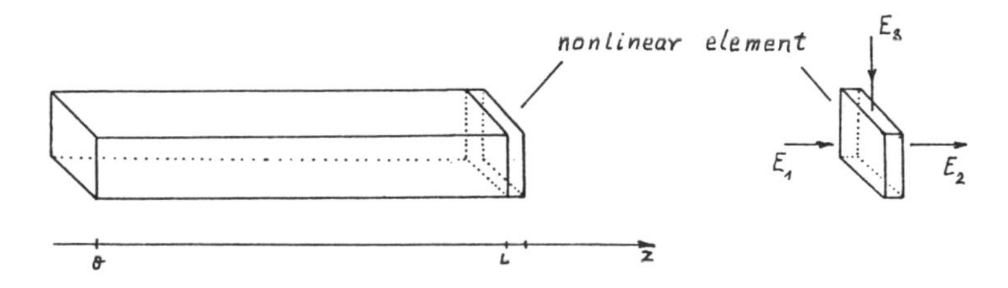

Let the influence field be given by $E_3(z,t) = \frac{1}{2} E_3 e^{i(\omega_3 t - \beta_3 z)}$. The picture field E_1 may be given by

$$E_1(r,t) = \frac{1}{2} \sum_\ell A_\ell(z)\cdot\mathcal{E}_1^\ell(x,y)e^{i(\omega_1 t - \beta_{\ell 1}\cdot z)},$$

by taking the complex conjugate

$$\overline{E_1(r,t)} = \frac{1}{2} \sum_\ell \bar{A}_\ell(z)\mathcal{E}_1^\ell(x,y)e^{-i(\omega_1 t - \beta_{\ell 1} z)}.$$

Now the polarization is defined by

$$P(..) = \chi\, E_3\cdot\bar{E}_1 \text{ at } \omega_2, \text{ where } \omega_2 = \omega_3 - \omega_1;$$

thus we obtain

$$P(\cdot) = \frac{1}{2}\chi \cdot \sum_{\ell} E_3(z,t)\bar{A}_{\ell}(z)\mathcal{E}_1^{\ell}(x,y)e^{-i(\omega_1 t-\beta_{\ell 1}z)}$$

$$= \frac{1}{2} \cdot \frac{1}{2}\chi \sum_{\ell} E_3\ \bar{A}_{\ell}(z)\mathcal{E}_1^{\ell}(x,y)e^{i(\omega_3 t-\beta_3 z)}\, e^{-i(\omega_1 t-\beta_{\ell 1}z)}$$

and by construction

$$e^{i(\omega_3 t-\omega_1 t)}\, e^{i(\beta_{\ell 1} - \beta_3)z} = e^{i\omega_2 t}\ e^{-i(\beta_3-\beta_{\ell_1})z}$$

The polarization can be written as

$$P(\cdot) = \frac{1}{4}\cdot\chi \sum_{\ell} E_3\bar{A}_{\ell}(z)\ \mathcal{E}_1^{\ell}(x,y)e^{i\omega_2 t}\ e^{-i(\beta_3-\beta_{\ell_1})z},$$

differentiating

$$\mu \frac{\partial^2}{\partial t^2} P(\cdot) = \mu\chi\ E_3\cdot\frac{\omega_2^2}{4} \sum_{\ell}\ \bar{A}_{\ell}(z)\mathcal{E}_1^{\ell}(x,y)e^{-i(\beta_3-\beta_{\ell 1})z}\, e^{i\omega_2 t};$$

this is used in the above expression

$$\sum_m -i\beta m_2 \frac{\partial B_m(z)}{\partial z}\ \mathcal{E}_2^m(x,y)e^{-i\beta m_2 z}\, e^{i\omega_2 t} + c.c =$$

$$= \chi\ E_3\ \mu \frac{\omega_2^2}{4} \sum_{\ell} \bar{A}_{\ell}(z)\mathcal{E}_1^{\ell}(x,y)\cdot e^{i\omega_2 t}\, e^{-i(\beta_3-\beta_{\ell_1})z},$$

integrating

$$\int\int \sum_m\ .\ .\ .\ .\ \mathcal{E}_2^m(x,y)\cdot\mathcal{E}_2^s(x,y)...dxdy =$$

$$= \chi\ E_3\mu \frac{\omega_2^2}{4} \int\int \sum_{\ell}\ .\ .\ .\ \mathcal{E}_1^{\ell}(x,y)\cdot\mathcal{E}_2^s(x,y)\ ...\ dxdy,$$

by normalization

$$\int\int \mathcal{E}_2^m(x,y)\cdot\mathcal{E}_2^s(x,y)dxdy = \delta_{m,s};$$

for m = s we obtain

$$-i\beta s_2 \frac{\partial B_s(z)}{\partial z} = -\chi\, E_3\, \mu \cdot \frac{\omega_2^2}{4} \sum_{\ell} \bar{A}_\ell(z) \iint \mathcal{E}_1^\ell(x,y)\mathcal{E}_2^s(x,y)dxdy\; e^{-i(\beta_3-\beta_{\ell 1})z}\; e^{i\beta s_2 z}$$

and for $e^{-i(\beta_3-\beta_{\ell 1}-\beta s_2)z}$ we define $\beta_3-\beta_{\ell_1}-\beta s_2 = \Delta\beta$.

We ignore the indices 1 and 2, as ω_1, ω_2 are sufficiently close together for $\ell = s$ we have $\Delta\beta = \beta_3 - \beta s_2 - \beta_{\ell_1}$ and

$$\iint \mathcal{E}_1^\ell(x,y)\cdot\mathcal{E}_2^s(x,y)dxdy = \delta_{\ell,s},$$

we obtain finally

$$-i\beta s_2 \frac{\partial B_s(z)}{\partial z} = -\chi^\mu \cdot E_3 \cdot \frac{\omega_2^2}{4} \bar{A}_s(z)e^{-i(\Delta\beta)z}.$$

Hence by defining the coefficient

$$\gamma = \frac{\chi\cdot\mu}{2\beta s_2} \cdot E_3\, \omega_2^2,$$

we obtain

$$\frac{\partial B_s(z)}{\partial z} = -i\frac{\gamma}{2}\bar{A}_s(z)e^{-i\Delta\beta z}.$$

Analogously we derived

$$\frac{\partial \bar{A}_s}{\partial z} = +i\frac{\chi\cdot\mu}{4\beta s_2}\cdot\omega_1^2\, E_3\cdot B_s(z)e^{i\Delta\beta z}$$

and

$$\frac{\partial \bar{A}_s(z)}{\partial z} = i\cdot\frac{\gamma'}{2} B_s(z)e^{i\Delta\beta z}.$$

Finally we obtain the differential equation system

$$\frac{\partial B_s(z)}{\partial z} = -i\frac{\gamma}{2}\bar{A}_s(z)e^{-i\Delta\beta z}$$

$$\frac{\partial \bar{A}_s(z)}{\partial z} = i\frac{\gamma'}{2} B_s(z)e^{i\Delta\beta z}.$$

The solution of this system with respect to the boundary condition $B_s(L) = 0$ can be given (by analogous methods derived above)

$$B_s(z) = -i\,\frac{\gamma}{2}\cdot\frac{1}{K}\,\bar{A}_s(0)e^{i\beta s_1 L}\sinh(K(z-L))e^{i\frac{\Delta\beta s}{2}(z-L)}, \quad \text{where}$$

$$K = \frac{1}{2}\,[\gamma^2-(\Delta\beta_s)^2]^{\frac{1}{2}}.$$

VI. 2.3 Image reconstruction

For $\gamma >> \Delta\beta$ we can approximate the solution by

$$B_s(z) = -i\bar{A}_s(0)e^{i\beta s_1 L}\cdot\sinh(\tfrac{1}{2}\gamma(z-L))e^{i\frac{1}{2}\Delta\beta(z-L)}.$$

The solution of the nonlinear wave equation (with respect to polarization phenomena) can be written as

$$E_2(x,y,z,t) = \frac{1}{2}\sum_m B_m(z)\mathcal{E}_2^m(x,y)e^{i(\omega_2 t-\beta m_2 . z)},$$

using the explicit solution $B_s(z)$, we obtain

$$E_2(x,y,z,t) = -\frac{1}{2}\,i\sum_m \bar{A}_m(0)e^{i\beta m_1 L}\sinh(\tfrac{1}{2}\gamma(z-L)e^{i\frac{1}{2}\Delta\beta(z-L)}\cdot\mathcal{E}_2^m(x,y)e^{i(\omega_2 t-\beta m_2 z)}.$$

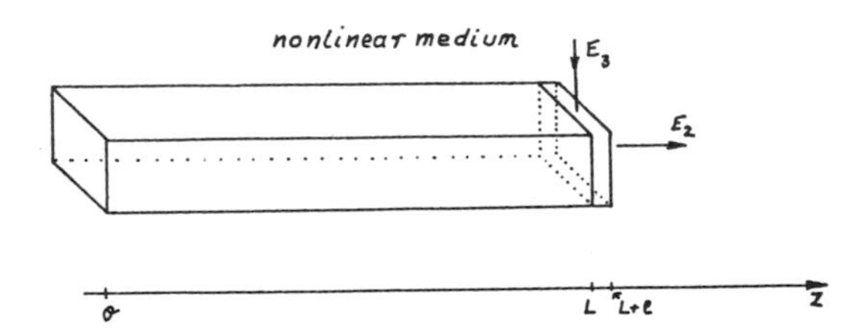

The nonlinear wave at the output of the nonlinear medium can be written

$$E_2(x,y,z = L+\ell,t) = -\frac{1}{2}\,i\sum_m \bar{A}_m(0)e^{i\beta m_1 L}\sin(\tfrac{1}{2}\gamma\cdot\ell)e^{i\frac{1}{2}\Delta\beta\ell}\mathcal{E}_2^m(x,y)e^{i(\omega_2 t-\beta m_2(L+\ell))}.$$

The input wave at the transition of the linear to the nonlinear medium is given on page 390 as

$$E_1(x,y,z,t) = \frac{1}{2}\sum_m A_m(z)\cdot\mathcal{E}_1^m(x,y)e^{-i\beta m_1 z}e^{i\omega_1 t},$$

thus at z = L we obtain

$$E_1(x,y,z = L,t) = \frac{1}{2}\sum_m A_m(L)\mathcal{E}_1^m(x,y)e^{-i\beta m_1 L}e^{i\omega_1 t},$$

as the input wave $A_m(0)$ at the output z = L is given by the relation (see on page 381)

$$A_m(L) = \frac{1}{\cos \varkappa|L|} \cdot A_m(0),$$

we have

$$E_1(x,y,z=L,t) = \frac{1}{2} \sum_m A_m(0) e^{-\beta m_1 L} \mathcal{E}_1^m(x,y) \frac{1}{\cos|\varkappa| \cdot L} e^{i\omega_1 t},$$

then

$$\bar{E}_1 = \frac{1}{2} \sum_m \bar{A}_m(0) e^{\overline{-i\beta m_1 L}} \mathcal{E}_1^m(x,y) \frac{1}{\cos|\varkappa|L} e^{i\omega_1 t}$$

leaving $e^{i\omega_1 t}$ intact, the nonlinear wave can be written as

$$\bar{E}_1(x,y,z = L,t) = \frac{1}{2} \sum_m \bar{A}_m(0) e^{i\beta m_1 L} \mathcal{E}_1^m(x,y) \frac{1}{\cos|\varkappa| \cdot L} e^{i\omega_1 t}$$

Now we are comparing the output field E_2 and the conjugate complex input field E_1 and obtain:
the amplitude of the output field is proportional to the complex conjugate amplitude of the input field.

Thus the problem of the reconstruction of the input image is solved. The conjugate wave could be generated by using nonlinear effects and an influence field E_3. In ((109), p. XV): "We know that almost any nonlinear optical effect can convert an incoming light beam into one with these remarkable image-restoring properties".

Instructive demonstrations are given by Dunning and Lind in (88): "The results are shown in Fig. 2. [Their figure 2]. The first photograph shows the input image of a portion of a U.S. Air Force resolution chart. The second photograph shows the output at the end of the fibre.

The output is typical of a highly multimode fibre in which a large number of modes are excited. From the output, it is impossible to discern that any spatial information has been transmitted through the fibre. The last photograph shows the image that has been reconstructed after double passing the fibre (approximately 25- μm spacings are shown)."

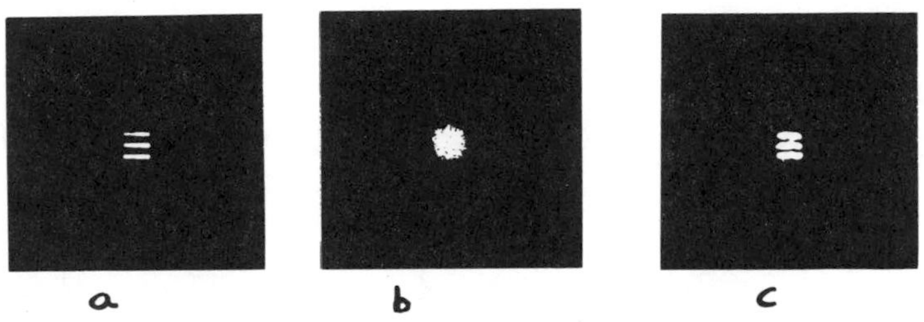

Fig. 2. Demonstration of image transmission through an optical fibre: a, input image resolution pattern; b, output after a single pass through the fibre; c, reconstruction of resolution pattern after output has been phase conjugated and retraverses fibre.

(Figure from (88), by kind permission of The Publishers.)

VI. 2.4 Information reconstruction in multiconnected domains

In the following only one possible application of modelling spatial reaction-diffusion phenomena by methods constructed in this section is given.

We assume that reaction-diffusion phenomena in a multiconnected domain D are governed by partial differential equations (see Chapter III)

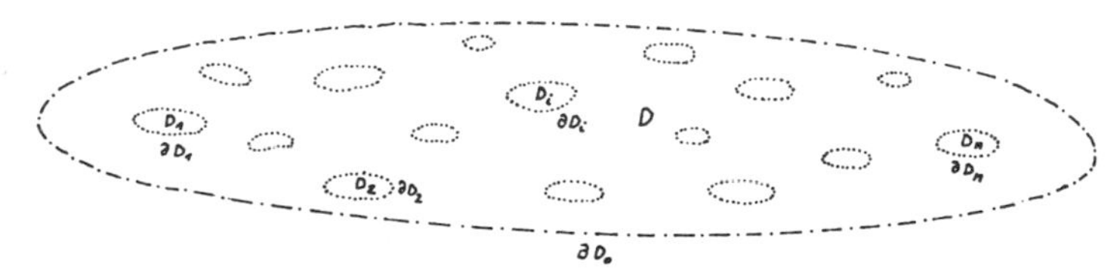

A reaction-diffusion process in D influences (see p. 19f., 42f., 54f., 119f., 134f., 149f., 239) the density distributions in the interior of the domains $D_1, D_2, \ldots, D_n$. If the boundary conditions on $\partial D_1, \partial D_2, \ldots, \partial D_n$ and the densities at prescribed internal points of D_i are given, the density distribution in the domains D_i for $i = 1,\ldots,n$ can be constructed by the methods of interpolating, 2-dimensional Lg-splines in (139).

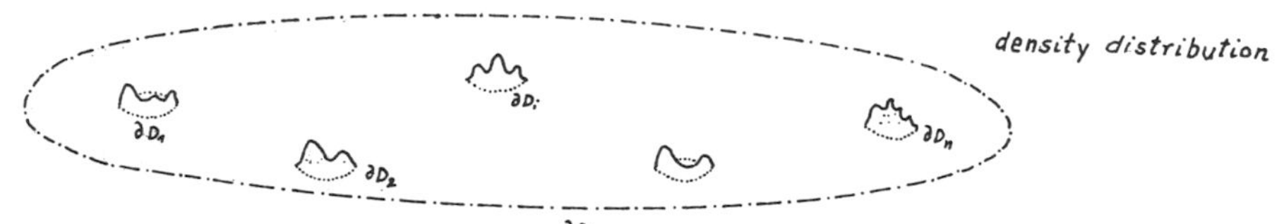

The generated density distribution in the domains D_i are now influencing the reaction-diffusion process in the multiconnected domain (e.g. by the asymptotic methods of Chapters I, II).

The density structure described by interpolating, 2-dimensional Lg-splines can be reproduced by the methods given in this section. We need a nonlinear medium and an external field E_3 both generating the complex conjugate of the solution of the nonlinear wave equation.

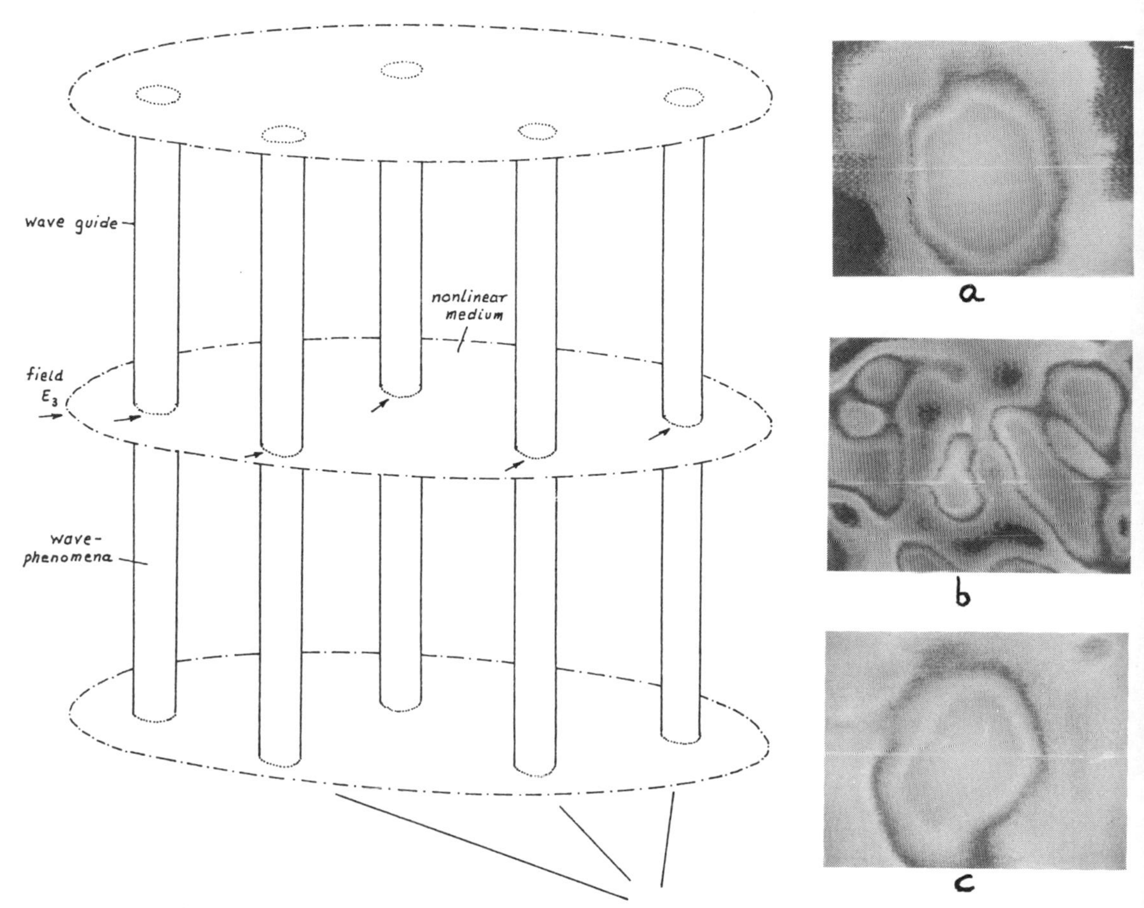

The density distributions in the domains D_i, reconstructed by nonlinear methods, generate diffusion-reaction processes in the multiconnected domain D. These new methods of model-building can be applied in morphogenesis, physics, chemistry, biology and medicine.
(Note: right-hand side of figure from (88) page 560, by kind permission of The Publishers.)

VII Vortical processes in space

VII. 1 Methods for modelling the genesis of cardiac arrhythmias

VII. 1.1 Review on the myocard contraction

Each contraction of the heart is preceded by an electrical wave of excitation, which follows a definite route over the heart cells, triggering this biochemical contractile apparatus and providing the required sequence of contractions (287):

ICV - inferior vena cava
SCV - superior vena cava
Ao - aorta
SN - sinus node
RAA - right atrial appendage

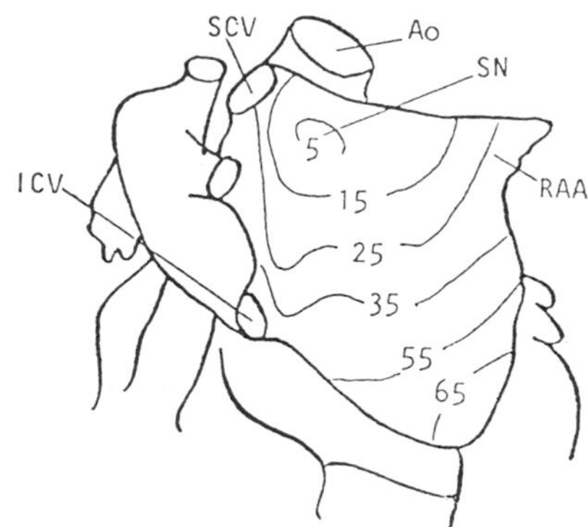

Figure VII.1 from (287)

The thin lines show a succession of the wavefront positions, the numbers denoting time in ms.

"The electrical current resulting from propagation of the wave causes an alteration of the potential on the surface of the body, which is recorded as ECG and is used in medical diagnostics

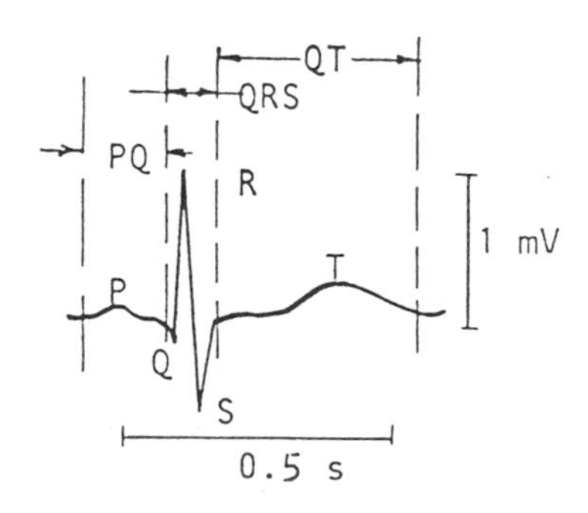

Figure VII.2 from (287)

where ECG records the change in the potential on the surface of the body.

The P, Q, R, S, T deflections characterize the rate of the propagation of excitation in the different heart chambers: the P deflection is associated with the contraction of the atrial, the QRS complex is caused by the ventricular activity; the QT interval and the T deflection characterize the recovery time and the shape of the back wavefront, respectively (287)".

The equations describing the propagation of the excitation wave in the heart are

$$C \frac{\partial V}{\partial t} = \frac{1}{r} \frac{\partial^2 V}{\partial x^2} + I(V,g)$$

$$\frac{\partial g}{\partial t} = \varphi(V,g)$$

where C is the specific capacitance of the cell membrane, V the potential of the membrane, r the longitudinal resistance of the heart fibres, I the total current across the cell membrane; g stands for the kinetic variables describing the dynamics of the membrane conductivity for specific ions.

The kinetic equations

$$C\cdot \frac{\partial V}{\partial t} = I(V,g) \quad \text{and} \quad \frac{\partial g}{\partial t} = \varphi(V,g)$$

can be reduced to

$$C\cdot \frac{dV}{dt} = f(V)-g \quad \text{and} \quad \frac{dg}{dt} = \frac{1}{\tau(V)}\,(g(V)-g).$$

As we are treating the process of propagation, the function f is derived from the current-voltage membrane characteristics, g(V) and $\tau(V)$ are the established value and the relaxtion time constant of the variable g.

The excitation wave propagating over the heart is a typical autowave (195). It shows no damping when propagating; it retains its shape; it does not reflect from the boundaries. There is no interference of autowaves in the heart.

The waves of excitation in the heart are generated by the specific cells of the sinus node (see figure VII.1). They exhibit an auto-oscillatory nature, the period of oscillations of the membrane potential being about 1 sec. under normal conditions. On the other hand, the cells of the working myocard have a single stable stationary state, corresponding to the resting potential. They respond to externally applied current with single pulses.

However, excitation sources which are not associated with the transition of the cells to the auto-oscillatory mode may also occur in the working heart under some pathological conditions. These sources are due to the generation of rotating waves, or "reverberators" (vortices in the heart), "leading cycles" in the physiological context. Once appeared, such a vortex governs the rhythm of the sinus node cell, thus resulting in a sharp increase in the heart-rate-paroxysmal tachycardia, a disease which ranks, along with fibrillation, among the most severe heart disorders.

Cardiac tissue is essentially three dimensional (the left ventricule wall thickness exceeds 1 cm). However, little is known of the structure of reverberators in the thickness of the heart wall. In a three-dimensional medium, the reverberator can be observed only in certain cross-sections intersecting its filament.

VII. 1.2 Generation of a vortex ring

Figure VII.3a shows schematically a mechanism of wave generation on the myocard surface (238):

A scroll ring in the heart wall:

a) a scheme of retracting wave pattern formation during the scroll rotation, the inner boundary of the ring of excitation progressively shrinks to the centre. This boundary value problem can be described by the methods given in Chapter I.

b,c,d) a record of the wave pattern observed experimentally on the surface of myocard during arrythmias. The region of the initial excitation is blackened. Dashed is the region involved in excitation.

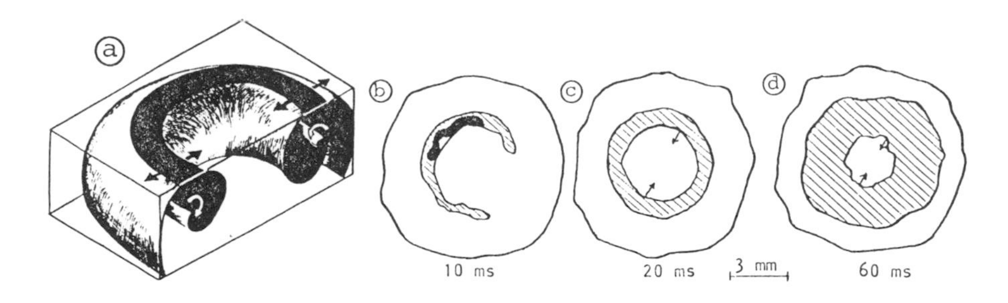

Figure VII.3 (Figure from (238) page 198 by kind permission of Springer-Verlag)

We try a hypothetical mechanism of generation of a scroll ring in the thick of the heart wall during arrhytmias. In principle, it does not differ from the mechanisms of vortex generation in two dimensions (195), (287), (200), (363). The mechanism of generation given in (238):

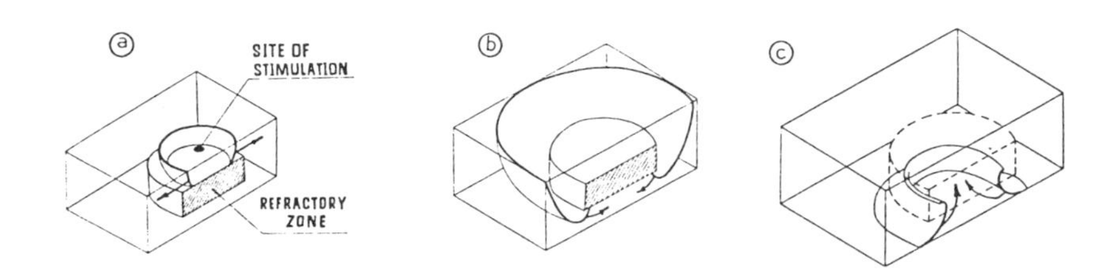

Figure VII.4

a,b) wave propagation initiated by a stimulus applied at the phase of relative refractoriness;

c) invasion of the excitation wave into the region of increased refractoriness following the recovery of excitability (the birth of a scroll ring). The shape of the ring thread is determined by the boundary of the cross section of the region of increased refractoriness and is close to a circumference;

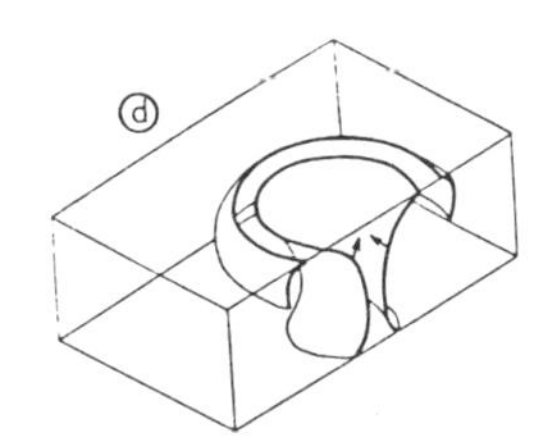

d) emergence on the myocard surface of a wave generated by the vortex ring and formation of retracting and outward-spreading waves. [Author's note: This can be described by our methods in (140, 141).]

Figure VII.4

(Figure from (238) page 198 by kind permission of The Publishers.)

Pertsov and Grenadier in ((287), p. 187) said:

"No analytical approach to solve the problem is, as yet, available.".

By Krinsky (195):

"The most important property of rotating vortices is their ability for reproduction (196, 197). In an inhomogeneous medium, the waves emitted by a rotating reverberator can produce new reverberators when breaking at inhomogeneities, and so on".

We show that the cross section of the generated vortex ring is multi-connected, thus, under suitable conditions the vortex ring breaks into new

vortices.
"The chain reaction of the reproduction of reverberators leads to a decrease in the characteristic scales of the wave pattern, and the whole medium turns out to be filled with fragments of rotating vortices. The resulting chaotic regime resembles a small-scale turbulence." (195) (see our considerations on weak turbulence in (141)).

"The properties of vortices in active media (their ability to occur at inhomogeneities during wave propagation, to synchronize other wave sources and to reproduce themselves) gives insight into fundamental mechanisms of a number of serious heart disorders (197, 198). A vortex-like wave source (reverberators) occurring in heart muscle suppresses the normal heart pacemaker (SA-node) with a resulting abrupt increase in the caridiac rate and dramatic impairment of homodynamics. This is just the case with paroxysmal tachycardia (severe cardiac arrhytmia), as shown in experiments on animals (7).

The chain reaction of reverberator production by inhomogeneities in heart tissue is the cause of another serious heart disease, ventricular fibrillation.

During fibrillation, the myocardial cells contract asynchronously and the living pump turns out to be a chaotically pulsing muscle bag uncapable of pumping blood.

This is the dramatic way a turbulent autowave regime manifests itself in the heart." (195)

Processes such as fibrillation can be described by high-frequency wave trains.

VII. 1.3 Methods of model-building

In the following we touch only one class of problems related to vortical processes.

Pertsov and Grenadier ask for: "The key parameters that control the birth of vortices" (287), p. 189).

Medvinsky, Panfilov and Pertsov gave in ((238), p. 198) a hypothetical mechanism of generation of a scroll ring in the middle of the heart wall during arrhythmias. In figure VII.5 a), b) the wave propagation is initiated by a stimulus applied at the phase of relative refractoriness;

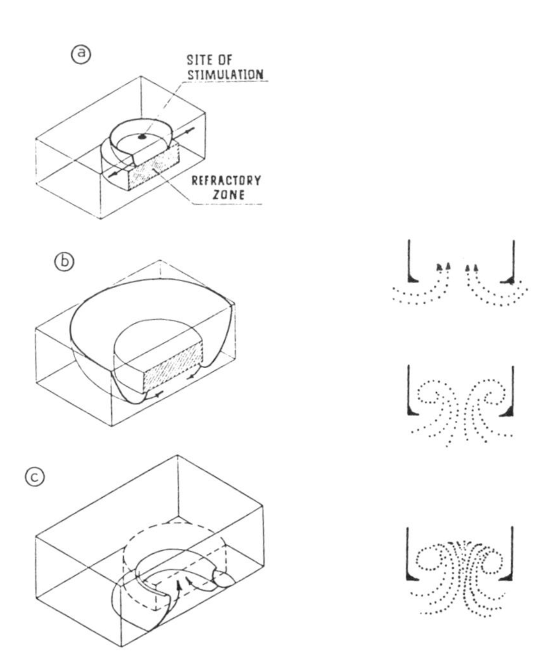

In c) is shown the invasion of the excitation wave into the region of increased refractoriness following the recovery of excitability (the birth of the vortex ring). The slope of the vortex ring is determined by the boundary of the cross-section of the region (for analytical treatment see (140, 141, 142)).

Figure VII.5
(From (238) page 198 by kind permission of The Publishers.)

VII. 1.4 The cross-section as a free boundary-value problem

The generation of the cross-section of the vortex ring can be formulated as a free boundary value problem.

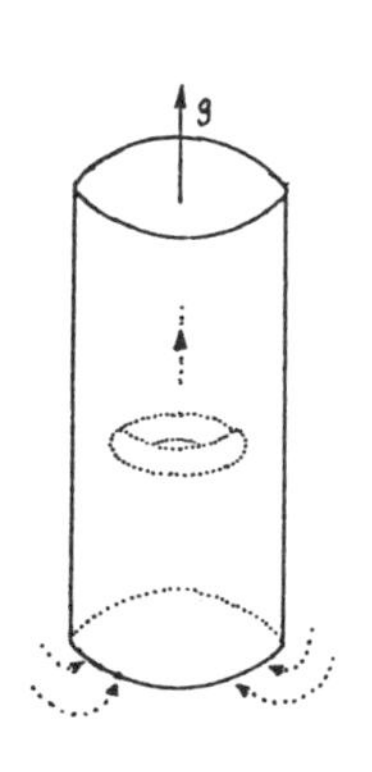

Introducing the three-dimensional real Euclidean space $\mathbb{R}^3$ with cylinder coordinates r, θ, z. Let Σ be the Stokes stream function, and the vorticity is given by

$$\omega = \text{curl } v,$$

where v is the velocity, then

$$\omega = -\Delta(\Sigma) = -\text{div grad } \Sigma,$$

where Δ = Laplace operator.

Now we reduce the generation problem of vortices from three to two dimensions (see the modelling-methods in (141)). To simplify the problem, we take the plane of the cross section of the vortex ring, then

$$-\Delta(\Sigma) = \omega$$

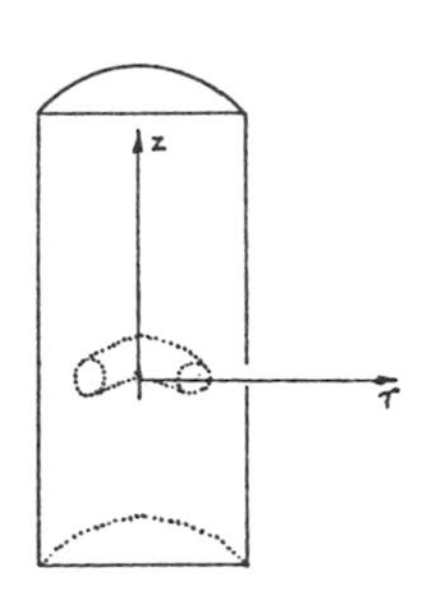

reduces to

$$-r \frac{\partial}{\partial r} \left(\frac{1}{r} \frac{\partial}{\partial r} (\Sigma)\right) + \frac{\partial^2}{\partial z^2} \Sigma = \omega,$$

where ω is the magnitude of the vorticity and r the distance to the symmetry axis. We assume that $\frac{\omega}{r}$ is constant on each stream surface, so we define

$$\frac{\omega}{r} = \lambda \cdot f(\Sigma),$$

where $f(\Sigma)$ is a nonlinear vorticity function and λ the parameter of the vortex-strength. Let D be the Domain

$$D = \{(r,z) | 0 < r < a \text{ and } |z| < b\},$$

with the boundary conditions:

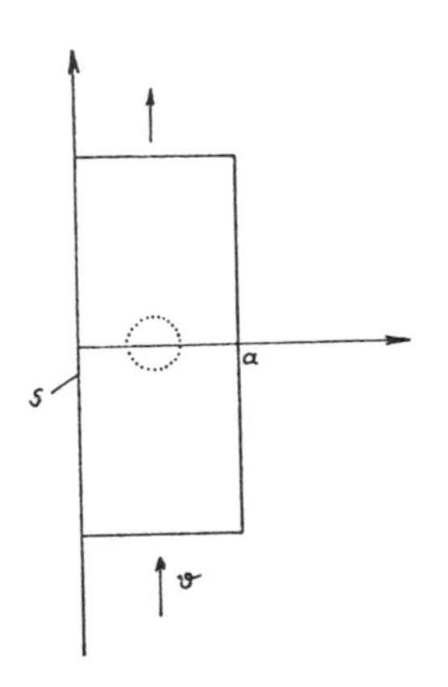

On the symmetry axis there is prescribed a flux constant s interpreted as a control or strategy parameter which influences the generation of the vortex.

The vortex moves with velocity v only in z-direction, and the radial velocity component is zero at the cylinder wall r = a.

Thus, the stream function can be written as

$$\Sigma(r,z) = \varphi(r,z) - \frac{1}{2} r^2 \cdot v - s,$$

where $\varphi(r,z)$ is generated by the vortex ring, hence, the vortex stream function $\varphi(r,z)$ satisfies

$$r \cdot \frac{\partial}{\partial r}\left(\frac{1}{r} \frac{\partial}{\partial r} \varphi(r,z)\right) + \frac{\partial^2 \varphi(r,z)}{\partial z^2} = -\lambda \cdot r^2 \cdot f(\Sigma) \text{ in } D,$$

$$\varphi = 0 \text{ on the boundary } \partial D,$$

and $\frac{1}{2} r^2 \cdot v + s$ is a fundamental solution of the partial differential equation.

Let $C_0^\infty(D)$ be the set of real valued functions with compact support in D and derivatives of arbitrary order.

Let $\mathcal{H}$ be the completion of $C_0^\infty(D)$ with respect to the norm, given by the inner product (e.g. Lions (216), Adams (2))

$$(\varphi,\varepsilon) = \iint_D \frac{1}{r^2} \left(\frac{\partial\varphi}{\partial r} \cdot \frac{\partial\varepsilon}{\partial r} + \frac{\partial\varphi}{\partial z} \cdot \frac{\partial\varepsilon}{\partial z}\right) dr \cdot r \cdot dz.$$

If $\varphi \in C^2(\bar{D})$ then integration by parts gives

$$(\varphi,\varepsilon) = \iint_D \frac{1}{r^2} \left(\frac{\partial\varphi}{\partial r} \frac{\partial\varepsilon}{\partial r} + \frac{\partial\varphi}{\partial z} \cdot \frac{\partial\varepsilon}{\partial z}\right) r dr dz =$$

$$= \int_{\partial D} \frac{1}{r^2} \left(\frac{\partial\varphi}{\partial r} \cdot \varepsilon + \frac{\partial\varphi}{\partial z} \cdot \varepsilon\right) d(\partial D) - \iint_D \left(\frac{\partial}{\partial r}\left(\frac{1}{r^2}\left(\frac{\partial\varphi}{\partial r}\right)\right)\varepsilon + \frac{1}{r^2} \frac{\partial^2\varphi}{\partial z^2} \varepsilon\right) r dr dz$$

for $\varepsilon \in \mathcal{H}$, then the trace $(\varepsilon) = 0$, i.e., $\varepsilon|_{\partial D} = 0$,

$$(\varphi,\varepsilon) = \lambda \cdot \iint_D \varepsilon \cdot f(\Sigma) r dr dz.$$

The generalized solution of the above problem can be written

$$(\varphi,\varepsilon) = \lambda \iint_D \varepsilon \cdot f(\Sigma) \cdot r dr dz \text{ for } \varepsilon \in \mathcal{H}.$$

VII. 1.5 Variational principle and generation of a vortex

We show that the solution $\varphi(r,z)$ of the partial differential equation characterizes a critical point of a variational problem. Let μ be a constant, then consider

$$K(\mu) = \{u \in \mathcal{H} \mid \|u\|^2 = \iint_D \left(\frac{1}{r^2} \left(\left(\frac{\partial u}{\partial r}\right)^2 + \left(\frac{\partial u}{\partial z}\right)^2\right) r dr dz = \mu\right\},$$

and the functional is given as

$$J(u) = \iint_D \Xi\left(u(x) - \frac{1}{2} v \cdot r^2 - s\right) dr \cdot dz,$$

where

$$\Xi(t) = \int_0^t f(s)ds.$$

For

$$u(x) - \frac{1}{2} v \cdot r^2 - s > 0 \text{ we have } u(x) > \frac{1}{2} \cdot v \cdot r^2 + s,$$

thus instead of D we take the subdomain

$$\{x \in D | u(x) > \frac{1}{2} \cdot v \cdot r^2 + s\}$$

denoted by D_n, which becomes the cross-section of the vortex ring if $u = \varphi$.

Theorem:

The generation of a vortex is governed by the variational principle

$$\sup_{u \in K(\mu)} J(u).$$

There exists a solution φ satisfying $J(\varphi) > 0$ and $\varphi \geqq 0$ almost everywhere in D.

Proof:

We show that J is continuous with respect to the weak convergence in $\mathcal{H}$ and bounded in $K(\mu)$ (thus there exists $\sup_{u \in K(\mu)} J(u)$.)

Let u be in $\mathcal{H}$, we define

$$\psi = u(x) - \frac{1}{2} v \cdot r^2 - s.$$

Because the functional depends only on the non-negative part ψ^+ of ψ, where $\psi^+ = \max(\psi, 0)$ (Ladyzhenskaya (206), p. 50), as $\Xi(\psi) = \Xi(\psi^+)$, we obtain the following estimation by assuming that Ξ is locally Hölder continuous, i.e.

$$|\Xi(\psi^+) - \Xi(\chi^+)| \leq (1 + (C \cdot \psi^+)^m + (C \cdot \chi^+)^m) \cdot (\psi^+ - \chi^+),$$

with C = constant. Applying this we have

$$|J(u)-J(v)| = \left|\iint_D (\Xi(\psi^+)-\Xi(\chi^+))drdz\right| \leqq \iint_D (1+(C\psi^+)^m + + (C\cdot\chi^+)^m)\cdot(\psi^+-\chi^+)drdz.$$

By definition

$$|\psi^+-\chi^+| \leqq |u-v| \text{ and } |\psi^+| \leqq u,$$

thus

$$|J(u)-J(v)| \leqq \|u-v\| + C^m(\|u\|^m_{2m} + \|v\|^m_{2m})\cdot \|u-v\| \quad u,v \in \mathfrak{H}.$$

Let $L^p(D)$ denote the completion of the set $C_0^\infty(D)$ with respect to the norm

$$\left(\iint_D |u|^p drdz\right)^{1/p} = \|u\|_{p,D},$$

and let $W^{1,2}(D)$ be the completion of the set $C_0^\infty(D)$, with norm

$$\left(\iint_D \left(\left(\frac{\partial u}{\partial r}\right)^2 + \left(\frac{\partial u}{\partial z}\right)^2\right)drdz\right)^{1/2}$$

then by Sobolev's famous theorem (Sobolev (323), p. 84) $W^{1,2}(D)$ is embedded compactly in $L^p(D)$ for $p \geqq 1$, thus

$$\mathfrak{H} \subset W^{1,2}(D) \subset L^p(D).$$

This means that the weak convergence in $W^{1,2}(D)$ implies strong convergence in L^p (that is to say, the weak topology of $W^{1,2}$ is stronger than the strong topology of L^p) and also, that a bounded domain in $W^{1,2}$ is a precompact set in L^p.

If $u_n \to u$ weakly in $\mathfrak{H}$, then $u_n \to u$ strongly in L^p for each $p \geqq 1$, then

$$|J(u_n)-J(u)| \leqq \|u_n-u\| + C^m(\|u_n\|^m_{2m} + \|u\|^m_{2m})\|u_n-u\|,$$

hence

$$J(u_n) \to J(u).$$

Setting above $v = 0$ and $J(0) = 0$, we have

$$|J(u)| \leqq \|u\| + C^m(\|u\|^m_{2m}) \cdot \|u\| = \mu^{\frac{1}{2}} + C^m (\mu^{m/2}) \mu^{\frac{1}{2}},$$

thus $J(u)$ is bounded above on $K(\mu)$.

Denoting

$$\sup_{u \in K_\mu} J(u) = \tau,$$

and $\tau > 0$, let u_n be a sequence in $K(\mu)$ such that $J(u_n) \to \tau$. By construction $u_n \in K(\mu)$, i.e., $\|u_n\|^2 = \mu$, u_n is bounded, the sequence u_n is weakly compact in $\mathcal{H}$ and by a known lemma, there exists a subsequence φ_n and an element $\varphi \in \mathcal{H}$ such that $\varphi_n \to \varphi$ weakly in $\mathcal{H}$.

Applying the consideration above, then

$$\lim_{n\to\infty} J(\varphi_n) = J(\varphi) = \tau.$$

We have only to show that $\varphi \in K(\mu)$, too.

By the concept of weak convergence in Hilbert spaces, we have

$$\|\varphi\| \leqq \liminf_n \|\varphi_n\|$$

and by assumption

$$\|\varphi\|^2 \leqq \mu.$$

As $J(\varphi) = \tau > 0$ we have $\|\varphi\| \geqq 0$.

We assume that $\|\varphi\|^2 \lneqq \mu$ defining $\hat{\varphi} = \mu^{\frac{1}{2}} \cdot \frac{\varphi}{\|\varphi\|}$,

we obtain

$$\hat{\varphi} \in K(\mu) \text{ and } \hat{\varphi}(x) > \varphi(x) \text{ for } \varphi(x) > 0.$$

Let us assume that $\Xi(t)$ is strictly increasing for $t > 0$, then

$$\hat{\varphi}(x) > \varphi(x) \Rightarrow J(\hat{\varphi}) > J(\varphi) = \tau > 0.$$

As $\tau = \sup_{u \in K(\mu)} J(u)$, we have a contradiction, hence $\|\varphi\|^2 = \mu$, thus $\varphi \in K(\mu)$.

We must assume that Ξ is strictly increasing and locally Hölder continuous.

We have the functional

$$J(u) = \iint_D \Xi(\psi)drdz,$$

where

$$\Xi(t) = \int_0^t f(s)ds,$$

then the Fréchet derivative can be defined by

$$(J'(\omega),\varepsilon) = \iint_D \varepsilon\cdot f(\psi)drdz = \iint_D \varepsilon\cdot f(u-\tfrac{1}{2}vr^2-s)drds,$$

this is clear by Hölder's inequality

$$|J(u+\varepsilon)-J(u) - \iint_D \varepsilon\cdot f(u-\tfrac{1}{2}\cdot vr^2-s)drdz| \leq C_1+C_1(\|u\|^{m-1}-\|\varepsilon\|^{m-1})\|\varepsilon\|^{1+\mu}$$

$$u,\varepsilon \in H$$

where C_i = constant.

A known result (see Vainberg (344), p. 96, or Chow and Hale (58), p.116) that the maximizer φ of the functional J is a critical point of the restriction of J to $K(\mu)$, such that

φ and $J'(\varphi)$ are parallel vectors,

this means, there exist constants C_3, $C_4 \neq 0$, such that

$$C_3(\varphi,\varepsilon) = C_4 \iint_D \varepsilon\cdot f(\Sigma)drdz \qquad \varepsilon \in H,$$

hence

$$(\varphi,\varepsilon) = \frac{C_4}{C_3}\iint_D \varepsilon\cdot f(\Sigma)drdz \text{ for all } \varepsilon \in H.$$

The general solution of the partial differential equation satisfies

$$(\varphi,\varepsilon) = \lambda\iint_D \varepsilon\cdot f(\Sigma)drdz \text{ for } \varepsilon \in H,$$

by choosing $\varphi = \varepsilon$ we obtain

$$(\varphi,\varphi) = \frac{C_4}{C_3}\iint_D \varphi f(\Sigma)drdz, \text{ as } \varphi \in K(\mu),\ \|\varphi\|^2 = \mu,$$

by comparison we obtain λ, the parameter of the vortex strength

$$\lambda = \mu \cdot \frac{1}{\int_D\int \varphi \cdot f(\Sigma)drdz} .$$

Now we show that λ is bounded above. By definition

$$\Sigma = \varphi(x) - \frac{1}{2} v \cdot r^2 - s,$$

thus

$$0 < \Sigma(x) < \varphi(x)$$

in the subdomain D_p and by assumption let f be locally Hölder continuous and strictly increasing, then

$$f(\Sigma(x)) < f(\varphi(x))$$

and

$$\int_{D_n}\int \varphi \cdot f(\Sigma)drdz \leqq \int_D\int \varphi(1 + (C \cdot \varphi)^m)drdz \leq C_5 \mu^{\frac{1}{2}} + C_6 \mu^{\frac{m+1}{2}}$$

by inequalities of Nirenberg ((264), p. 128). As

$$\Xi(t) = \int_0^t f(\tau)d\tau$$

and f be increasing, then

$$\Xi(t) \leqq t \cdot f(t),$$

hence

$$J(u) < \int_{D_n}\int \Xi(\Sigma)drdz \leqq \int_{D_n}\int \Sigma \cdot f(\Sigma)drdz < \int_{D_n}\int \varphi \cdot f(\Sigma)drdz,$$

as $\Sigma(x) < \varphi(x)$, for any $u \in K(\mu)$ the integral is bounded

$$J(u) \leqq \int_{D_n}\int \varphi \cdot f(\Sigma)drdz$$

and finally

$$\lambda = \mu \cdot \frac{1}{\int_{D_n}\int \varphi \cdot f(\Sigma)drdz} .$$

Remark 1

Let

$$D^+ = \{x \in D | z > 0\}$$

and φ be the maximizer of the theorem, i.e.,

$$\sup_{u \in K(\mu)} J(u) = J(\varphi),$$

then $\frac{\partial \varphi}{\partial z} \leq 0$ almost everywhere in D^+ and φ can be symmetrized by

$$\varphi(r,z) = \varphi(r,-z).$$

Remark 2

By the generalized maximum principle (224) we are able to show that

$$\varphi(r,z) = \varphi(r,-z) \text{ on } \bar{D} \text{ and } \frac{\partial \varphi}{\partial z} \lneqq 0 \text{ in } D^+ \text{ and } \varphi > 0 \text{ in } D.$$

VII. 1.6 A free boundary value problem

The generation of the cross-section of the vortex can be formulated as a free boundary value problem.

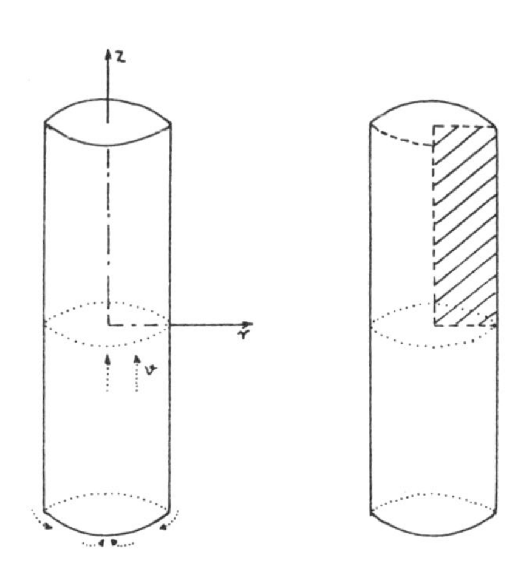

The stream function can be written as

$$\Sigma(r,z) = \varphi(r,z) - \frac{1}{2} v \cdot r^2 - s,$$

where v is the velocity and s the flux constant seen as a control or strategy parameter, where the vortex stream function satisfies

$$r \frac{\partial}{\partial r} \left(\frac{1}{r} \frac{\partial}{\partial r} (\varphi(r,z))\right) + \frac{\partial^2 \varphi}{\partial z^2} = -\lambda r^2 f(\Sigma) \text{ in } D$$

and $\varphi = 0$ on the boundary ∂D.

φ is the maximizer of the variational problem

$$\sup_{u \in K(\mu)} J(u) = J(\varphi),$$

where the functional

$$J(u) = \iint_{\{x \in D | u > \frac{1}{2} v \cdot r^2 + s\}} \Xi(u(x) - \frac{1}{2} v \cdot r^2 - s) dr dz.$$

In the beginning the vortex, located interior of the domain D, may not influence the boundary conditions. Let the control or strategy parameter s be > 0. This parameter is applied to the symmetry line, the boundary r = 0 of the domain D, by construction

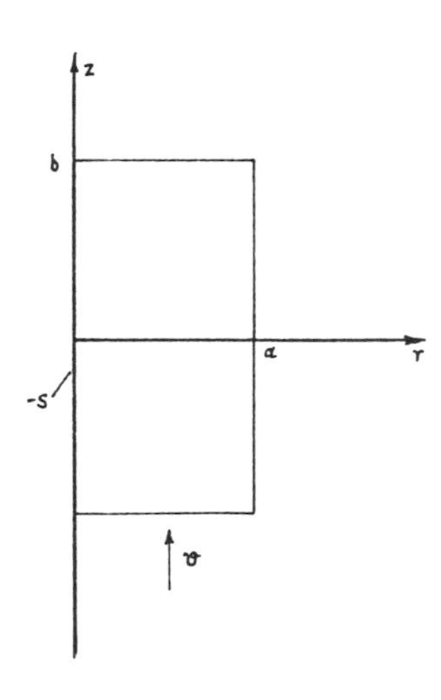

$$\Sigma(0,0) = -s < 0.$$

As

$$\Sigma(r,z) = \varphi(r,z) - \frac{1}{2} \cdot r^2 \cdot v - s$$

the boundary value at r = a follows

$$\Sigma(a,0) = -\frac{1}{2} a^2 \cdot v - s < 0$$

and for a point r_i

$$\Sigma(r_i, b) < 0.$$

The variational problem says that within the domain

$$D = \{(r,z) | 0 \leqq r \leqq a, |z| < b\}$$

there exists a subdomain

$$D_\varphi = \{(x \in D | \varphi(x) > \frac{1}{2} v \cdot r^2 + s\}$$

where the vortex is located.
We have

$$D_\varphi = \{(r,z) \in D | \Sigma(z,r) \geqq 0\}$$

and especially the subset

$$I = \{r | \Sigma(r,0) > 0\}.$$

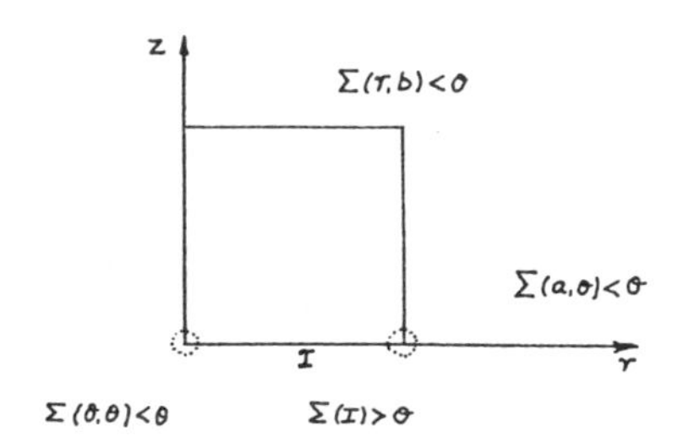

Let r_i be a point in the boundary interval I, then

$$\Sigma(r_i,0) > 0,$$

as the boundary values are all < 0, there must exist a point interior D, where the stream function satisfies

$$\Sigma(r_i,.) = 0,$$

mathematically we have to apply the implicit function theorem. Let r_i be in I, then there exists a unique point $(r_i,g(r_i))$ in D^+ at which

$$\Sigma(r_i,g(r_i)) = 0,$$

because the boundary value

$$\Sigma(r_i,b) < 0,$$

and the functional determinant $\neq 0$, thus in our case

$$\frac{\partial\Sigma}{\partial z} = \frac{\partial\varphi}{\partial z} < 0$$

is satisfied, then

$$g \in C^{2+\mu}$$

in an open interval containing r_i.

VII. 1.7 Localizing the cross-section of the vortex ring

The space $C^{2+\mu}$, where $0 < \mu < 1$, consists of functions having derivatives of order 2 that are uniformly Hölder continuous with exponent μ.

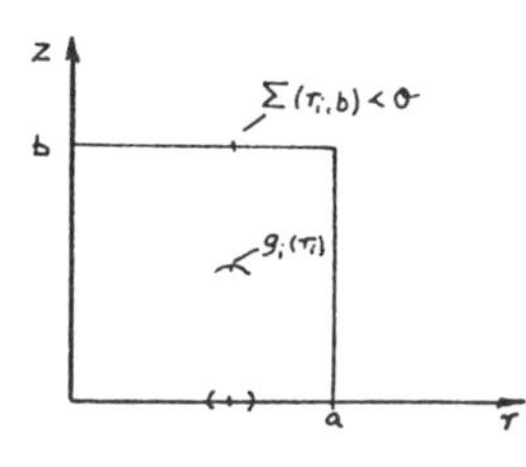

That $g \in C^{2+\mu}$ can be shown by regularity results of partial differential equations. We find that there exists a denumerable union of disjoint simple connected subsets of the form

$$\gamma_i = \{(r,z)|x_i < r_i < y_i \quad \text{and} \quad -g(r_i) < z < g(r_i)\}$$

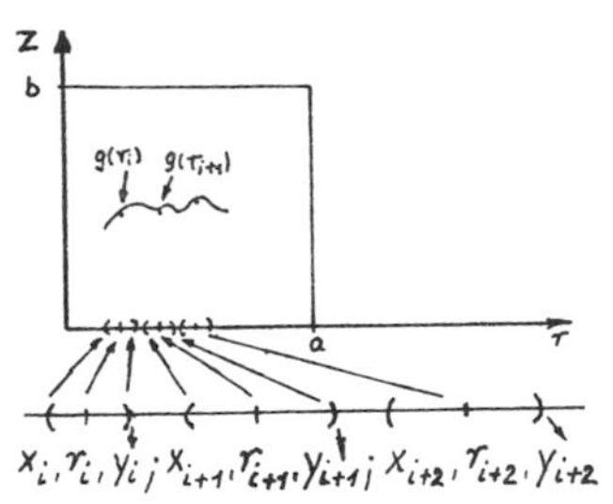

with $0 < x_i, y_i < a$ and $g \in C^{2+\mu}(x_i,y_i)$ for each i describing the cross-section D_φ of the vortex. In the case where $r \notin \bigcup_i (x_i,y_i)$ we define $g(r)$ = constant and obtain in a global sense $g \in C[0,a]$.

Now we obtain the

Theorem:

If the vortex function f is Lipschitz continuous with constant M, then the cross-section of the vortex ring contains only a finite number of γ_i.

Proof:

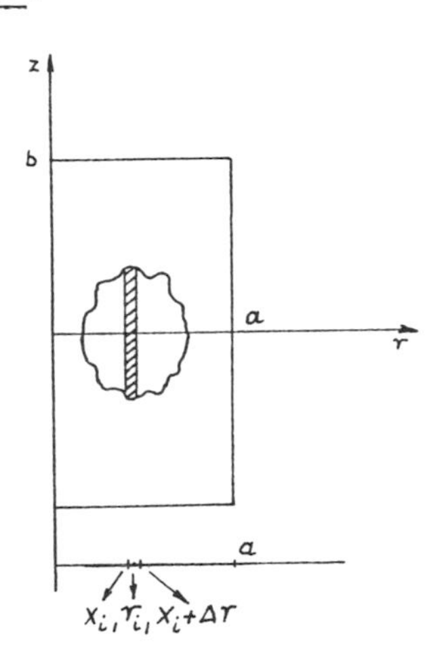

Let r_i be a point of the section γ_i the corresponding neighbourhood may be given by the interval $(x_i, x_i + \Delta r)$, and the z-coordinate can be written as $g(r_i)$, general as $z = g(r)$. We define a function ε by

$$\varepsilon(r,z) = \Sigma(r,z) \text{ in } \gamma$$

and $\varepsilon(x) = 0$ elsewhere, hence $\varepsilon \in \mathfrak{H}$. The generalized solution of the partial differential equation of the vortex stream can be written

$$(\varphi,\varepsilon) = \lambda \iint \varepsilon \cdot f(\Sigma) drdz \quad \varepsilon \in \mathfrak{H},$$

as

$$\Sigma(r,z) = \varphi(r,z) - \frac{1}{2} r^2 v - s \text{ and } \Sigma(r,z) - \varphi(r,z) = -\frac{1}{2} r^2 v - s,$$

then

$$\frac{\partial}{\partial r} (\varphi(r,z) - \Sigma(r,z)) = -r \cdot v, \quad \frac{\partial}{\partial z}(\varphi - \Sigma) = \frac{\partial}{\partial z} (\frac{1}{2} \cdot r^2 v + s) = 0;$$

with the representation of the generalized solution

$$(\varepsilon,\varphi-\Sigma) = \iint_D \frac{1}{r^2} \left(\frac{\partial\varepsilon}{\partial r} \cdot \frac{\partial}{\partial r}(\varphi-\Sigma) + \frac{\partial}{\partial z} (\varphi-\Sigma)\cdot\frac{\partial\varepsilon}{\partial z}\right) r \, drdz$$

we obtain

$$(\varepsilon,\varphi-\Sigma) = \iint_D \frac{1}{r^2} \left((-r\cdot v)\cdot\frac{\partial\varepsilon}{\partial r} + 0\right) r\cdot drdz = v \iint_{\gamma_i} \varepsilon_r drdz$$

As

$$\varepsilon(r,z) = \Sigma(r,z),$$

we have

$$(\varepsilon,\varphi-\varepsilon) = (\varepsilon,\varphi) - (\varepsilon,\varepsilon) = v \iint_{\gamma_i} \varepsilon_r drdz = 0,$$

thus

$$(\varepsilon,\varepsilon) = (\varepsilon,\varphi),$$

and we are able to estimate

$$\|\varepsilon\|^2 = (\varepsilon,\varepsilon) = (\varepsilon,\varphi) = \lambda \iint_D \varepsilon f(\Sigma) drdz \leqq \lambda \cdot M \iint_D \varepsilon\cdot\Sigma \, drdz =$$

$$= \lambda\cdot M \iint_{\gamma_i} \varepsilon^2 drdz,$$

as f is Lipschitz continuous and $\varepsilon = \Sigma$ on γ_i.

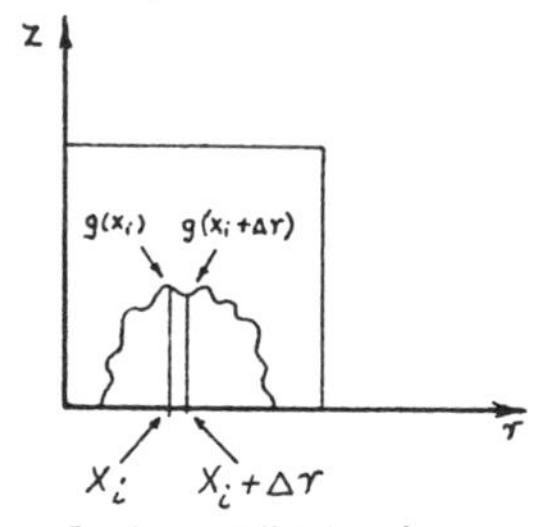

Let $v_0 \in (x_i, x_i + \Delta r)$ then

$$\int_{x_i}^{r_0} \frac{\partial}{\partial r} \varepsilon(r,z) dr = \varepsilon(r_0,z) - \varepsilon(x_i,z),$$

and applying Hölder's inequality

$$\left|\int_{x_i}^{r_0} 1\cdot\varepsilon_r dr\right| \leq \left(\int_{x_i}^{r_0} |1| dr\right)^{\frac{1}{2}} \cdot \left(\int_{x_i}^{r_0} |\varepsilon_r|^2 dr\right)^{\frac{1}{2}} = |r_0 - x_i|^{\frac{1}{2}} \left(\int_{x_i}^{r_0} |\varepsilon_r|^2 dr\right)^{\frac{1}{2}},$$

let $\varepsilon(x_i,z) = 0$, then

$$\varepsilon^2(r_o,z) = \left(\int_{x_i}^{r_o} \frac{\partial\varepsilon}{\partial r}\, dr\right)^2 \leqq |r_o - x_i| \int_{x_i}^{r_o} |\varepsilon_r|^2 dr;$$

quite analogously we have

$$\int_0^{g(x_i)} \frac{\partial\varepsilon(r,z)}{\partial z}\, dz = \varepsilon(r,g(x_i)) - \varepsilon(r,0)$$

and

$$\left|\int_0^{g(x_i)} 1.\ \frac{\partial\varepsilon}{\partial z}\, dz\right| \leqq \left(\int_0^{g(x_i)} |1|^2 dz\right)^{\frac{1}{2}} \left(\int_0^{g(x_i)} \left|\frac{\partial\varepsilon}{\partial x}\right|^2 dz\right)^{\frac{1}{2}};$$

let $\varepsilon(r,0) = 0$, then

$$\varepsilon^2(r,g(x_i)) = \left(\int_0^{g(x_i)} \left|\frac{\partial\varepsilon}{\partial z}\right| dz\right)^2 \leqq |g(x_i) - 0| \int_0^{g(x_i)} \left|\frac{\partial\varepsilon}{\partial z}\right|^2 dz,$$

so we obtain

$$\int_0^{g(x_i)} \varepsilon^2(r_o,z)dz \leqq |g(x_i)-0| \cdot |r_o - x_i| \int_0^{g(x_i)} \int_{x_i}^{r_o} \left|\frac{\partial\varepsilon}{\partial r}\right| drdz$$

and

$$\int_{x_i}^{r_o} \varepsilon^2(r,g(x_i))dr \leqq |g(x_i)-0|\ |r_o - x_i| \int_{x_i}^{r_o} \int_0^{g(x_i)} \left|\frac{\partial\varepsilon}{\partial z}\right|^2 drdz.$$

We have $|x_i - x_o| < \Delta r$ and $|g(x_i) - 0| < \frac{b}{2}$ and by definition

$$\|\varepsilon\|^2 = \iint \frac{1}{r^2} \left(\left(\frac{\partial\varepsilon}{\partial r}\right)^2 + \left(\frac{\partial\varepsilon}{\partial z}\right)^2\right) rdrdz,$$

we obtain

$$\iint \left(\left(\frac{\partial\varepsilon}{\partial r}\right)^2 + \left(\frac{\partial\varepsilon}{\partial z}\right)^2\right) drdz \leqq \int\int \frac{r_o}{r} \left(\left(\frac{\partial\varepsilon}{\partial r}\right)^2 + \left(\frac{\partial\varepsilon}{\partial z}\right)^2\right) drdz \leqq$$

$$\leqq \sqrt{r_o} \iint \frac{1}{r^2} \left(\left(\frac{\partial\varepsilon}{\partial r}\right)^2 + \left(\frac{\partial\varepsilon}{\partial z}\right)^2\right) r\cdot drdz$$

using this in the above inequality

$$\int_{x_i}^{r_o} \int_0^{g(x_i)} \varepsilon^2 \, drdz \leq (\Delta r \cdot \frac{b}{2}) \int_0^{g(x_i)} \int_{x_i}^{r_o} \frac{1}{r^2} ((\frac{\partial \varepsilon}{\partial r})^2 + (\frac{\partial \varepsilon}{\partial z})^2) rdrdz =$$

$$= \Delta r \cdot \frac{b}{2} \ \|\varepsilon\|^2 .$$

Now we obtain finally

$$\|\varepsilon\|^2 \leqq \lambda \cdot M \int\int_{\gamma_i} \varepsilon^2 drdz \leqq \lambda \cdot M \cdot \Delta r \cdot \frac{b}{2} \cdot \|\varepsilon\|^2 ,$$

thus

$$1 \leqq \lambda M \cdot \Delta r \cdot \frac{b}{2} \quad \text{and} \quad \Delta r > \frac{2}{b \cdot \lambda \cdot M} ,$$

so Δr is finite and the cross section of the vortex ring contains only a finite number of disjoint simple connected subsets. We suppose that the cross-section contains

$$n \cdot \Delta r \text{ parts (where } n = \text{integer}),$$

and let us assume that $a = 3 \cdot n \cdot \Delta r$, then there follows the inequality

$$a = 3n\Delta r > \frac{2}{b\lambda \cdot M} 3n .$$

The partition of the cross section can be estimated by

$$n < \frac{1}{6} \cdot a \cdot b \cdot C \cdot M ,$$

where C is the upper bound of the parameter λ of the vortex strength and M the Lipschitz constant of the vortex function f. In the case $b = 3 \cdot a$ and by taking $a = 1$ and $C \cdot M = 4$, then

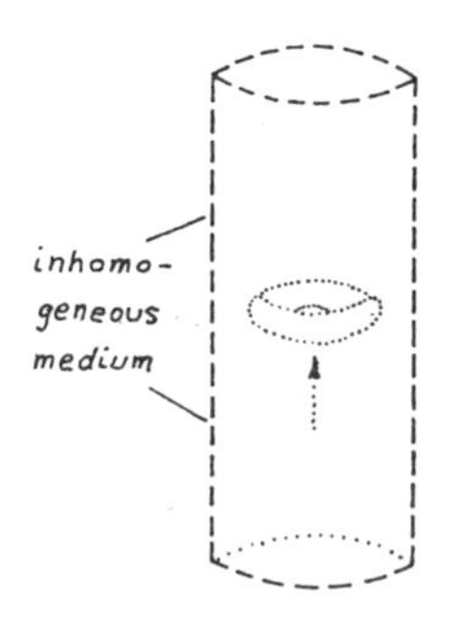

$$n < \frac{1}{6} a \cdot b \cdot C \cdot M = \frac{1}{6} 3a \cdot 4 = \frac{12}{6} = 2 ,$$

thus $n = 1$ and the cross-section is a simple connected domain. Under suitable conditions the cross-section of the vortex ring is multi-connected, decomposed into disjoint subsets. It is known that vortices are unstable if they are passing through inhomogeneous media.

The most important property of rotating vortices is their ability to reproduce themselves. In an inhomogeneous medium the waves emitted by a rotating vortex can produce new vortices when they meet up with inhomogeneties. Thus, the multiconnected cross section of the vortex ring breaks into the disjoint subsets and new vortices appear.

Another application is based on the model of Krinsky (195), where he proposed a theory of fibrillation.

"According to the theory, fibrillation is a result of the reproduction of vortices due to gradients in the refractory period. The theory shows that the vortices (reverberators) in inhomogeneous media are unstable and have a finite lifetime determined by parameter gradients. However, they emit high-frequency trains of waves which give birth to new reverberators by the above described mechanism. At a certain critical relationship between the medium parameters, the number of the nascent reverberators may exceed the number of those which died away." (287)

This results in a sustained chaotic wave activity, i.e. fibrillation ((287), p. 189). Figure VII.3a above (from Medvinsky et al. (238)) shows schematically a mechanism of wave generation on the myocard surface, following the occurrence of a scroll ring with its thread lying in the plane parallel to the surface. The real picture, as recorded from the surface of rabbit heart atrium, is shown in figure VII.3b,c,d. The thread plane is somewhat tilted with respect to the surface, and a horseshoe shape of the wave initially results; figures VII.3b,c,d show a record of the wave pattern observed experimentally on the surface of myocard during arrythmias. The region of the initial excitation is blackened. The region involved in excitation is shown by dashes.

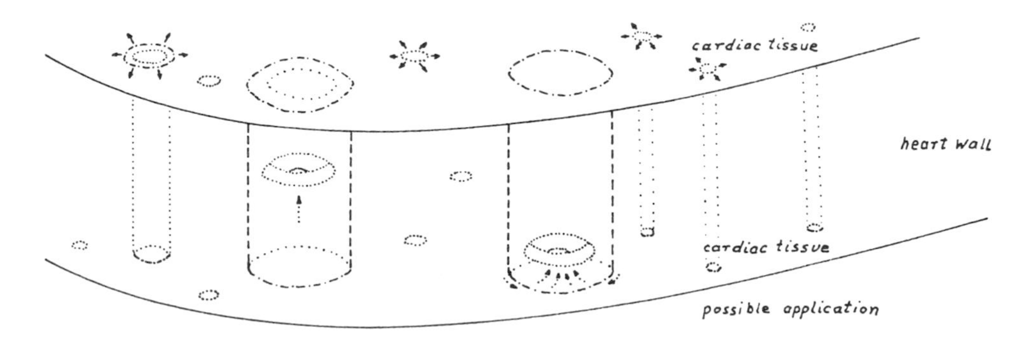

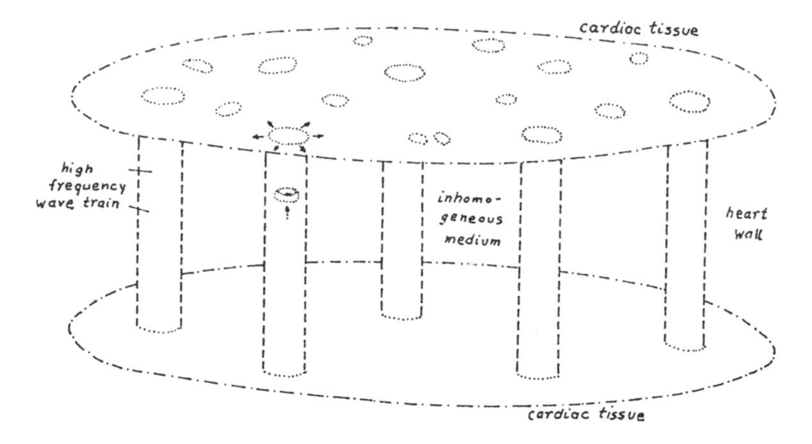

The vortex distribution on the surface of the heart and the interdependence of the vortices with respect to diffusion and certain turbulence phenomena can be given in explicit, analytic form (see page 148 and (140, 141, 142).

The process of the emergence of the wave on the myocard surface generated by the vortex ring and the process of formation of retracting and outward spreading waves can be treated by the modelling methods.

Phenomena of local processes occurring in a two-connected domain are described by

$$\frac{\partial u_o}{\partial t} = \Delta u_o + f_o(u_o) \text{ in } G_o$$

$$\varepsilon\frac{\partial u_1}{\partial t} = \Delta u_1 + f_1(u_1) \text{ in } G_1$$

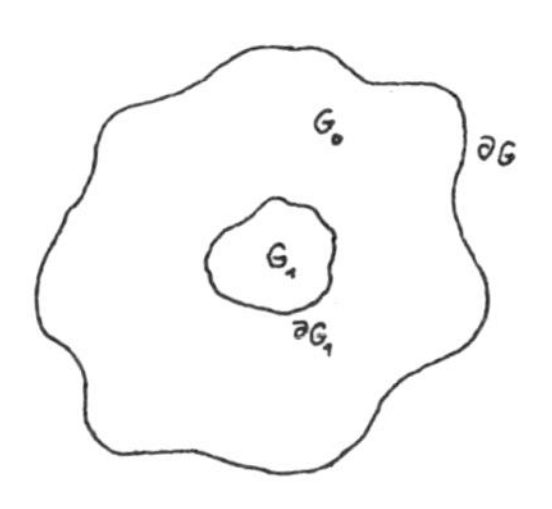

with the boundary conditions

$$u_o = 0 \text{ on } \partial G \text{ and } u_o = u_1, \; \frac{\partial u_o}{\partial n} = \frac{\partial u_1}{\partial n} \text{ on } \partial G_1,$$

and the initial conditions

$$u_o = u_{oo} \text{ in } G_o \text{ and } u_1 = u_{o1} \text{ in } G_1.$$

The solution, constructed by asymptotic methods (see page 27f), describes the evolution process of retracting and outward spreading waves. Controlled processes in three-connected domains are given on page 149ff .

VII. 1.8 Vortex ring

Lemma:

The cross-section of the vortex ring depends on the velocity v, the nonlinear vorticity function f, the control or strategy parameter s and the constant μ, where $\|\varphi\|^2 = \mu$. Thus, there exist curves

$$z = \pm \frac{\mu}{v(v \cdot r^2 + 2s)}$$

containing the vortex ring.

The stream function in D can be written as

$$\theta(r,z) = \varphi(r,z) - \frac{1}{2} r^2 \cdot v - s,$$

where $\varphi(r,z)$ satisfies

$$\Delta\varphi(r,z) = -\lambda \cdot r^2 f(\theta)$$

especially

$$r \cdot \frac{\partial}{\partial r} \left(\frac{1}{r} \frac{\partial}{\partial r} (\varphi(r,z)\right) + \frac{\partial^2 \varphi}{\partial z^2} = -\lambda r^2 f(\theta) \text{ in } D$$

$$\varphi = 0 \text{ on } \partial D.$$

$\varphi(r,z)$ describes the influence of the vortex.

The vortex can be found at the subset of D (see the diagram on page 402)

$$D_r = \{x \in D | \varphi(x) > \frac{1}{2} \cdot r.^2 v + s\}.$$

We derive that

$$\|\varphi\|^2 - \|\theta\|^2 = v^2 \text{ measure } (D_\varphi).$$

Proof:

$$\|\varphi\|^2 - \|\theta\|^2 = \int_{D_\varphi} \frac{1}{r^2} \left(\left(\frac{\partial\varphi}{\partial r}\right)^2 + \left(\frac{\partial\varphi}{\partial z}\right)^2\right) r dr dz -$$

$$- \int_{D_\varphi} \frac{1}{r^2} \left(\left(\frac{\partial\theta}{\partial r}\right)^2 + \left(\frac{\partial\theta}{\partial z}\right)^2\right) r dr dz,$$

the second integral can be written as

$$\int_{D_\varphi} \frac{1}{r^2}\left(\left(\frac{\partial\varphi}{\partial r} - v\cdot r\right)^2 + \left(\frac{\partial\varphi}{\partial z}\right)^2\right) r\,dr\,dz = \int_{D_\varphi} \frac{1}{r^2}\left(\left(\left(\frac{\partial\varphi}{\partial r}\right)^2 - 2\cdot v\cdot r\,\frac{\partial\varphi}{\partial r} + v^2\cdot r^2\right) + \left(\frac{\partial\varphi}{\partial z}\right)^2\right) r\,dr\,dz,$$

and

$$\|\varphi\|^2 - \|\theta\|^2 = \int_{D_\varphi} \frac{1}{r^2}\left(2r\cdot v\,\frac{\partial\varphi}{\partial r} - v^2 r^2\right) r\,dr\,dz = v \int_{D_\varphi} \left(\frac{2}{r}\frac{\partial\varphi}{\partial r} - v\right) r\,dr\,dz;$$

the integrand can be expressed by

$$\theta = \varphi - \frac{1}{2} v\cdot r^2 - s,$$

then

$$\frac{\partial\theta}{\partial r} = \frac{\partial\varphi}{\partial r} - v\cdot r, \text{ thus } \frac{2}{r}\frac{\partial\varphi}{\partial r} = \frac{2}{r}\frac{\partial\theta}{\partial r} + 2\cdot v \text{ and } \left(\frac{2}{r}\frac{\partial\varphi}{\partial r} - v\right) = \frac{2}{r}\frac{\partial\theta}{\partial r} + v,$$

hence

$$\|\varphi\|^2 - \|\theta\|^2 = v\int_{D_\varphi}\left(\frac{2}{r}\frac{\partial\theta}{\partial r} + v\right) r\,dr\,dz = 2\cdot v\int_{D_\varphi} \frac{\partial\theta}{\partial r}\,dr\,dz + v^2 \text{ measure } (D_\varphi).$$

We are using the fact that θ^+ belongs to the Sobolev space $W^{1,2}(D)$ (see Ladyzhenskaya (206), p. 50), as φ vanishes on the boundary ∂D, than θ^+ vanishes on ∂D_φ, thus we obtain

$$\int_{D_\varphi} \frac{\partial\theta^+}{\partial r}\, 1\,dr\,dz = \int_{\partial D_\varphi} \theta^+\, d(\partial D_\varphi) - 0 = 0, \text{ as } \theta^+ = 0 \text{ on } \partial D_\varphi;$$

the above equation follows.

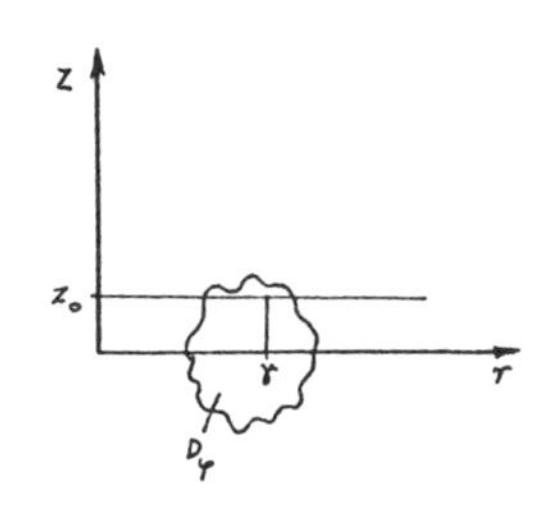

We take all points to the left of D_φ, thus

$D_{left} = \{(r,z) \mid$ there exists $(\gamma, z_0) \in D_\varphi$ such that $0 < r \leq \gamma$ and $z = z_0\}$.

The boundary of D_{left} contains a part IV of the z-axis, two straight lines I and III and the boundary D_φ, thus

$$\partial D_{left} \cap \partial D_\varphi.$$

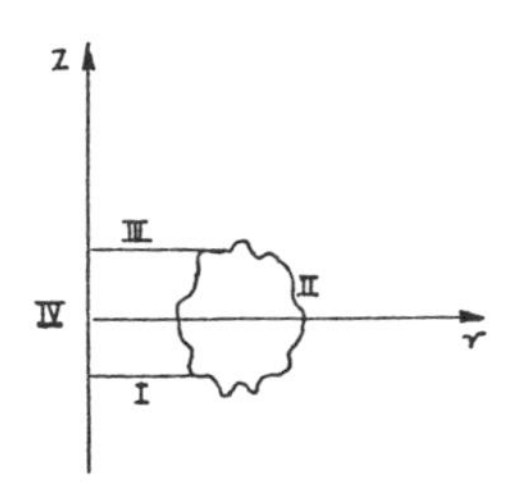

$$\partial D_{left} = I \cup \{D_\varphi \cap \partial D_{left}\} \cup III \cup IV.$$

Let

$$\theta(r,z) = \varphi(r,z) - \frac{1}{2} r^2 \cdot v - s$$

be the stream function, then we obtain as derived above

$$\|\varphi\|^2 - \|\theta\|^2 = v \cdot \int_{D_{left}} \frac{2}{r} \left(\frac{\partial\varphi}{\partial r} - v\right) r\,dr\,dz = 2 \cdot v \int_{D_{left}} \frac{\partial\varphi}{\partial r}\, dr\,dz - v^2 \cdot \text{measure}\,(D_{left}).$$

We have

$$\int_{D_{left}} \frac{\partial\varphi}{\partial r}\, dr\,dz = \int_{\partial D_{left}} \varphi\, dz - 0 \text{ and } \int_{\partial D_{left}} \varphi\, dz = \int_I \varphi\, dz + \int_{\partial D_\varphi \cap \partial D_{left}} \varphi\, dz + \int_{III} \varphi\, dz + \int_{IV} \varphi\, dz,$$

φ satisfies the partial differential equation and $\varphi = 0$ on ∂D, thus

$\int_{IV} \varphi\, dz = 0$; as I and III are perpendicular to the z-direction,

$$\int_{I,III} \varphi\, dz = 0,$$

hence, as $D_\varphi = \{x | \varphi(x) > \frac{1}{2} v \cdot r^2 + s\}$, we have

$$\int_{D_{left}} \varphi\, dz = \int_{\partial D_\varphi \cap D_{left}} \varphi\, dz = \int_{\ldots\cap\ldots} \left(\frac{1}{2} v \cdot r^2 + s\right) ds = \frac{1}{2} v \int r^2 dz + s \int dz;$$

introducing the projection on the z-direction by

$$P_\varphi = \{z | (r,z) \in D_\varphi \text{ for some } r\} \quad \text{and} \quad |P_\varphi| = \int_{P_\varphi} dz$$

then

$$\int_{D_{left}} \varphi \, dz = \frac{1}{2} v . r^2 |P_\varphi| + s \cdot |P_\varphi|,$$

and finally

$$\|\varphi\|^2 - \|\theta\|^2 = 2 \cdot v \int_{D_{left}} \frac{\partial\varphi}{\partial r} \, drdz - v^2 \cdot \text{measure } (D_{left})$$

$$= v^2 \cdot r^2 |P_\varphi| + 2v \cdot s \cdot |P_\varphi| - v^2 \cdot \text{measure } (D_{left})$$

$$= v(v \cdot r^2 + 2s) \cdot |P_\varphi| - v^2 \cdot \text{measure } (D_{left}).$$

Consider the boundary point $(r_i, z_i) \in \partial D_\varphi$, then we have that all points (r_i, z) where $|z| < z_i$ belong to D_φ. Thus, z_i is the projection of the boundary point on the z-direction, hence we obtain

$$\|\varphi\|^2 - \|\theta\|^2 = v(vr_i^2 + 2s)z_i$$

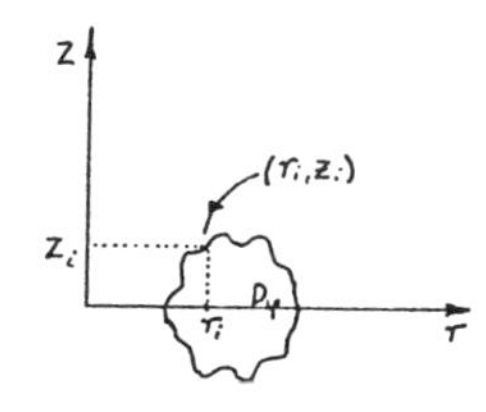

by construction $\varphi \in K(\mu)$, thus $\|\varphi\|^2 = \mu$ and then

$$v(v \cdot r_i^2 + 2s)z_i < \mu,$$

so we have

$$z_i < \frac{\mu}{v(v \cdot r_i^2 + 2s)} \quad .$$

There exist curves

$$z = \pm \frac{\mu}{v(v \cdot r^2 + 2s)}$$

containing the vortex ring.

Thus we have characterized the local behaviour of the vortex ring.

Bibliography

1. M. Abramowitz, I.A. Stegun: Handbook of mathematical functions, Dover, London, 1972.
2. R.A. Adams: Sobolev Spaces, Academic Press, New York, 1975.
3. S. Agmon: Lectures on elliptic boundary value problems, Van Nostrand Math. Studies, Vol. 2, Princeton, N.J., 1965.
4. S. Agmon, A. Douglis, L. Nirenberg: Estimates near the boundary for solutions of elliptic partial differential equations satisfying general boundary conditions, J. Comm. Pure Appl. Math., 12 (1959), 623-727.
5. N. Alikakos: An application of the invariance principle to reaction-diffusion equations. J. Diff. Equs., 33 (1979), 201-225.
6. N. Alikakos: Remarks on invariance in reaction-diffusion equations. Nonlinear Anal., 5 (1981), 593-614.
7. M. Allessie, F. Bonke, S. Shopman: Circ. Res. 33 (1973), 54.
8. H. Amann: On the existence of positive solutions of nonlinear elliptic boundary value problems. Ind. Univ. Math. J., 21 (1971), 125-146.
9. H. Amann: Invariant sets and existence theorems for semilinear parabolic and elliptic systems. J. Math. Anal. Appl., 65 (1978), 432-467.
10. N.R. Amundson: Nonlinear problems in chemical reactor theory, in Mathematical Aspects of chemical and biochemical problems in quantum chemistry, SIAM - AMS Proc. 8. Amer. Math. Soc., Providence, (1974), 59-84.
11. R. Aris: Prolegomena to the rational analysis of systems of chemical reactions, Arch. Rat. Mech. Analysis, Vol. 19 (1965), 81-99.
12. R. Aris: The mathematical theory of diffusion and reaction in permeable Catalysts I and II, Clarendon Press, Oxford, 1975.
13. R. Aris: Mathematical Modelling techniques, Research Notes in Math., Vol. 24, Pitman, London, 1978.
14. D.G. Aronson, H.F. Weinberger: Nonlinear diffusion in population genetics, combustion and nerve propagation, in Proceedings of the Tulane Program in Partial Differential Equations and related Topics, Lecture Notes in Mathematics, Vol. 446, Springer-Verlag, 1975, 5-49.

15. D.G. Aronson, H.F. Weinberger: Multidimensional nonlinear diffusion arising in population genetics, Adv. Math., 30 (1978), 33-76.
16. C.B. Bardos: A regularity theorem for parabolic equations. J. of Funct. Analysis 7 (1971), 311-322.
17. C.B. Bardos: Existence et unicité de la solution de l'équation d'Euler en dimension deux. J. Math. Anal. Appl. 40 (3), (1972), 769-790.
18. C. Bardos, J. Smoller: Instabilité des solutions stationaires pour des systèmes de réaction-diffusion. C.R. Acad. Sci. Ser. A, 285 (1977), 249-253.
19. C. Bardos, C.H. Matano, J. Smoller: Some results on the instability of solutions of systems of reaction-diffusion equations. 1er Colloque AFCET - SMF de Math. Appl., Ecole Polytechnique, t. II, (1979), 297-304.
20. N. Bebernes, K.N. Chueh, E. Fulks: Some applications of invariance for parabolic systems. Ind. Univ. Math. J., 28 (1979), 269-277.
21. M.J. Beckmann, T. Puu: Spatial economics: density, potential, and flow, North. Holland, Amsterdam, 1985.
22. R. Bellman: Dynamic Programming, Princeton University Press, 1957.
23. A. Bensoussan, J. Lesourne: Optimal growth of a self-financing firm in an uncertain environment, in Applied stochastic control in Econometrics and Management Science, ed. A. Bensoussan et al , North Holland, Amsterdam, 1980.
24. A. Bensoussan, J.L. Lions: Nouvelle formulation de problèmes de contrôles impulsionnel et applications (New Formulation of impulse control problems and applications). C.R. Acad. Sci. Paris, t. 276 (1973), 1189-1192.
25. A. Bensoussan, J.L. Lions: Contrôle impulsionnel et I.Q.V. d'évolution (impulse control and evolutionary Q.V.I.'s). C.R. Acad. Sci. Paris, t. 276 (1973), 1333-1338.
26. A. Bensoussan, J.L. Lions: Contrôle impulsionnel et contrôle continu (impulse control and continuous control). C.R. Acad. Sci. Paris, t. 278 (1974), 675-679.
27. A. Bensoussan, J.L. Lions: Contrôle impulsionnel et systèmes d'I.Q.V. (impulse control and systems of Q.V.I.'s). C.R. Acad. Sci. Paris, t. 278 (1974), 747-751.

28. A. Bensoussan, J.L. Lions: Sur de nouveaux problèmes aux limites pour des opérateurs hyperboliques (On new boundary problems for hyperbolic operators). C.R. Acad. Sci. Paris, t. 278 (1974), 1345-1349.
29. A. Bensoussan, J.L. Lions: Sur le contrôle impulsionnel et les I.Q.V. d'evolution (on impulse control and evolutionary Q.V.I.'s). C.R. Académie Sci. Paris, t. 280 (1975), 1049-1053.
30. A. Bensoussan, J.L. Lions: Sur l'approximation numerique d'I.Q.V. stationaires. Proc. Int. Symp. Computing Methods in Applied Sciences and Engineering, Versailles, 1973, in Lecture Notes in Computer Science, Vol. 11, Springer-Verlag, (1974), 325-338.
31. A. Bensoussan, J.L. Lions: Diffusion processes in bounded domains and singular perturbation problems for V.I.'s with Neumann boundary conditions, in Probabilistic Methods in Differential Equations, Lecture Notes in Mathematics, Vol. 451, Springer-Verlag, 1974.
32. A. Bensoussan, J.L. Lions: Applications des inéquations variationelles en contrôle stochastique, Dunod, Paris, 1978, Applications of variational inequalities in stochastic control, North Holland, Amsterdam, 1982.
33. Yu. A. Berezin, V.I. Karpman: Soviet Phys. JETP 24 (1966), 1043.
34. Yu. A. Berezin: Soviet Phys. Tech. Phys. 13 (1968), 16
35. M.S. Berger: Nonlinearity and functional analysis, Academic Press, New York, 1977.
36. M.S. Berger: Mathematical aspects of vorticity, in Nonlinear partial differential equations and their applications. College de France, Vol. 1, ed. by H. Brezis, J.L. Lions, Research Notes in Mathematics, Vol. 53, Pitman, Boston London, (1976), 52-75.
37. M.S. Berger, L.E. Fraenkel: Nonlinear desingularization in certain free-boundary problems, Communications in Mathematical Physics 77 (1980), 149-172.
38. M.S. Berger: The diagonalization of nonlinear differential operators. A new approach to complete integrability, in Nonlinear systems of partial differential equations in applied Mathematics, Lectures in Applied Math., Vol. 23, Part I, II, ed. by B. Nicolaenko et al ., Amer. Math. Soc., Providence, Rhode Island (1986), 223-239.
39. L. Bers, F. John, M. Schechter: Partial Differential Equations, Wiley-Interscience, New York, 1964.
40. N. Bloembergen: Nonlinear Optics, Benjamin, New York, 1965.

41. J.A. Boa: Multiple steady states in a model-biochemical reaction. Studies in Appl. Math. 54 (1975), 9-15.
42. J.-M. Bonny, P. Courrege, P. Priouret: Semigroups de Feller sur une variété a bord compacte et problèmes aux limites intégro-différentiels du second ordre donnant lieu au principe du maximum, Ann. Inst. Fourier (Grenoble) 18 (1968), 369-521.
43. J.R. Bowen, A. Acrivos, A.K. Oppenheim: Singular perturbation refinement to quasi-steady state approximations in chemical kinetics. Chem. Eng. Sci., Vol. 18 (1963), 177-188.
44. M. Bramson: Convergence of solutions of the Kolmogorov equation to travelling waves. Memoirs of the Am. Math. Soc., Providence, Rhode Island, Vol. 44, Number 285, 1983.
45. M. Bramson: Maximal Displacement of Branching Brownian Motion, Ph.D. Thesis, Cornell University, Comm. on Pure and Appl. Math., Vol. XXXI (1978), 531-581.
46. H. Brown: The resolution of boundary value problems by menas of the finite Fourier transformation: General vibration of a string. J. appl. Physics 14 (1943), 609-618.
47. H. Brown: General vibration of a hinged beam. J. appl. Physics 15 (1944), 410-413.
48. H. Brown: Resolution of temperature problems by the use of finite Fourier transformations. Bull. Amer. math. Soc., 50 (1944), 376-385.
49. P.N. Brown: Reaction-Diffusion-Equations in Population Biology, Ph.D. dissertation, Tulane University, 1978.
50. A.G. Butkovskii: Theory of optimal control of distributed parameter systems, Moscow, 1965, in Russian. English translation by American Elsevier, 1969.
51. A.P. Calderon: Lebesgue spaces of differentiable functions. Proc. Symp. Pure Math., Vol. 4, Am. Math. Soc., Providence, R.I., 1961.
52. F.J.L. von Capelle, M.J. Janse: Cardiac Arrhythmias During Acute Myocardial Ischemia, in Self-Organization, ed. by V.I. Krinksy, Synergetics, Springer-Verlag, (1984), 191-194.
53. G.F. Carrier, H.P. Greenspan: Water waves of finite amplitude on a sloping beach, Jour. Fluid Mech. 4 (1958), 97-109.
54. R.G. Casten, C.J. Holland: Stability properties of solutions to systems of reaction-diffusion equations, SIAM. J. Appl. Math., Vol. 33, 1977.

55. R.G. Casten, C.J. Holland: Instability results for reaction-diffusion equations with Neumann boundary conditions, J. Diff. Equations 27 (1978), 266-273.

56. S. Chandrasekhar: Hydrodynamic and Hydromagnetic Stability, Clarendon Press, Oxford, 1961.

57. S. Chandrasekhar: Stochastic problems in physics and astronomy, Reviews of Modern Physics, 15 (1943), 1-89.

58. S.N. Chow, J.K. Hale: Mathematics of bifurcation theory. Grundlehren der mathematischen Wissenschaften, Vol. 251, Springer-Verlag, Berlin Heidelberg New York, 1982.

59. K.N. Chueh: On the asymptotic behavior of solutions of semilinear parabolic partial differential equations, Ph.D. Thesis, Univ. of Wisconsin, 1975.

60. K.N. Chueh, C.C. Conley, J.A. Smoller: Positively invariant regions for systems of nonlinear diffusion equations, Indiana Univ. Math. J., 26 (1977), 373-391.

61. R. Churchill: Operational Mathematics, Sec. Ed. Mc-Graw-Hill, New York, 1958, p. 291.

62. C. Conley: Isolated Invariant Sets and the Morse Index. Conf. Board Math. Sci., No. 38, Amer. Math. Soc., Providence, 1978.

63. C. Conley, J. Smoller: Remarks on travelling wave solutions of non-linear diffusion equations. Proc. Battelle Sympos. Catastrophe Theory, in Lecture Notes in Mathematics, Vol. 525, (1975), 77-89.

64. C. Conley, J. Smoller: Isolated invariant sets of parametrised systems of differential equations. Lecture Notes in Mathematics, Vol. 668 (1978), 30-47.

65. C. Conley, P. Fife: Critical manifolds, traveling waves and an example from population genetics. J. Math. Biol., 14 (1982), 159-176.

66. C. Conley, J. Smoller: Remarks on the stability of steady-state solutions of reaction-diffusion equations, in Bifurcation phenomena in math. Physics and related phenomena, ed. by C. Bardos, Reidel, Dodrecht, 1980, 47-56,

67. C. Conley, J. Smoller: Topological techniques in reaction-diffusion equations. Biololgical Growth and Spread, Lecture Notes in Biomathematics, Vol. 38, Springer-Verlag, 1980, 473-483.

68. E. Conway, D. Hoff, J. Smoller: Large time behavior of solutions of systems of nonlinear reaction-diffusion equations, SIAM J. Appl. Math., 35 (1978), 1-16.

69. R. Courant, D. Hilbert: Methods of mathematical physics, Vol. 1, Wiley-Interscience, New York, 1953.

70. M.G. Crandall, P.H. Rabinowitz: Bifurcation from simple eigenvalues J. Funct. Analysis, 8 (1971), 321-340.

71. M.G. Crandall, P.H. Rabinowitz: Bifurcation, Perturbation of simple eigenvalues and linearized stability, Arch. Rat. Mech. Anal., 52 (1973), 161-180.

72. A. Damlamian, Li Ta-tsien: Comportement limite des solutions de certains problemes mixtes pour equations parabolique. C.R. Acad. Sci. Paris, t. 290, Serie A, (1980), 957-960.

73. G. Da Prato: Equations d'évolution dans des algebrès d'operateurs et applications. J. Math. Pures Appl., 48 (1969), 59-107.

74. G. De Prato: Somme d'applications non linéaires, Rome Symposium, May 1970.

75. G. Da Prato: Quelques resultats d'existence et regularité pour un problème non linéaire de la théorie du contrôle, Bordeaux, May 1971.

76. G. Doetsch: Handbuch der Laplace-Transformation Band I: Theorie der Laplace-Transformation, Birkhäuser, Basel, 1950.

77. G. Doetsch: Band II: Anwendungen der Laplace-Transformation, 1. Abteilung, Birkhäuser, Basel Stuttgart, 1955.

78. G. Doetsch: Band III: Anwendungen der Laplace-Transformation, 2. Abteilung, Birkhäuser, Basel Stuttgart, 1956.

79. G. Doetsch: Einführung in Theorie und Anwendung der Laplace-Transformation, Birkhäuser, Basel Stuttgart, 1958.

80. G. Doetsch: Anleitung zum praktischen Gebrauch der Laplace-Transformation, 2. Auflage, Oldenbourg, München, 1961.

81. G. Doetsch: Integration von Differentialgleichungen vermittels der endlichen Fourier-Transformation, Math. Ann., 112 (1935), 52-68.

82. G. Doetsch, H. Kniess, D. Voelker: Tabellen zur Laplace-Transformation und Anleitung zum Gebrauch, Springer-Verlag, Berlin Göttingen, 1947.

83. J.L. Doob: Stochastic Processes, John Wiley & Sons, Inc., New York, 1953.

84. J.L. Doob: The Brownian movement and stochastic equations, Annals of Mathematics, 43 (1942), 351-369.
85. N. Dunford, J.T. Schwartz: Linear Operators, Part I, Wiley-Interscience, New York, 1958.
86. N. Dunford, J.T. Schwartz: Linear Operators, Part II, Wiley-Interscience, New York, 1963.
87. N. Dunford, J.T. Schwartz: Linear Operators, Part III, Wiley-Interscience, New York, 1971.
88. C.J. Dunning, R.C. Lind: Demonstration of image transmission through fibers by optical phase conjugation, Opt. Lett., 7 (11) (1982), 558-560.
89. A. Duschek: Höhere Mathematik, III. Band, 2. Auflage, Springer-Verlag, Wien, 1960.
90. E.B. Dynkin: Infinitesimal operators of Markov processes, Teor. Veroj. i Primen. 1. (1956), 38-60 (in Russian).
91. E.B. Dynkin: Markov processes I, Grundlehren der mathematischen Wissenschaften, Vol. 121, Springer-Verlag, 1965.
92. E.B. Dynkin: Markov processes II, Grundlehren der mathematischen Wissenschaften, Vol. 122, Springer-Verlag, 1965.
93. W. Ebeling, R. Feistel: Physik der Selbstorganisation und Evolution, Akademie-Verlag, Berlin, 1982.
94. W. Ebeling, R. Feistel: Stochastic Models of Evolutionary Processes, in A. I. Zotin (Ed.): Thermodynamics and Regulation in Biological Processes, Nauka, Moscow, 1984.
95. W. Ebeling, Yu. Klimontovich: Selforganization and turbulence in liquids, Teubner, Berlin, 1984.
96. M. Eigen, P. Schuster: The Hypercycle - a Principle of Natural Self-Organization, Springer-Verlag, Berlin Heidelberg New York, 1979.
97. I. Ekeland, R. Temam: Analyse convexe et problemes variationnels, Dunod, Gauthier Villars, 1974. English translation, North Holland, 1979.
98. M. Feinberg, F.J.M. Horn: Dynamics of open chemical systems and the algebraic structure of the underlying reaction network. Chem. Eng. Sci., Vol. 29 (1974), 775-787.
99. M. Feinberg, F.J.M. Horn: Chemical mechanism structure and the coincidence of the stoichiometric and kinetic subspaces. Arch. Rat. Mech. Anal., 66 (1977), 83-97.

100. M. Feinberg: Mathematical aspects of mass action kinetics, Chap. 1., R.H. Wilhelm Memorial Vol. on Chemical Reaction Theory, 1978.
101. R. Feistel: Conservation Quantities in Selection Processes, Studia biophysica, 95 (1983), 107.
102. R. Reistel: Extremel Principles in Selection Processes, Studia biophysica, 96 (1983), 133.
103. W. Feller: The parabolic differential equations and the associated semi-groups of transformations. Ann. of Math., 55 (1952), 468-519.
104. W. Feller: On second order elliptic differential equations, Ann. of Math., 61 (1955), 90-105.
105. P.C. Fife, J.B. McLeod: The approach of solutions of nonlinear diffusion equations to travelling front solutions. Arch. Rat. Mech. Anal., 65 (1977), 335-361. Bull. Amer. Math. Soc., 81 (1975), 1075-1078.
106. P.C. Fife: Mathematical Aspects of Reacting and Diffusing systems, Lecture Notes in Biomathematics, Vol. 28, Springer-Verlag, Berlin Heidelberg New York, 1979.
107. R.A. Fisher: The Advance of advantageous Genes, Ann. Eugenics, Vol. 7 (1937), 355-369.
108. R.A. Fisher: Gene Frequencies in a Cline determined by selection and diffusion, Biometrics, 6 (1950), 353-361.
109. R. Fisher: Phase conjugate optics, Academic Press, New York, 1983.
110. R. Fitzhugh: Impulses and physiological states in theoretical methods of nerve membrane. Biophys. J., 1 (1961), 445-466.
111. L.E. Fraenkel: On steady vortex rings of small cross-section in an ideal fluid. Proc. Roy. Soc. London, A 316 (1970), 29-62.
112. L.E. Fraenkel: Examples of steady vortex rings of small cross-section in an ideal fluid, J. Fluid Mech., 51 (1972), 119-135.
113. L.E. Fraenkel, M.S. Berger: A global theory of steady vortex rings in an ideal fluid, Acta Math., 132 (1974), 14-51.
114. P.A. Franken, A.E. Hill, C.W. Peters: Second harmonic generation (frequency doubling), Phys. Rev. Lett., 7 (1961), 118.
115. A. Friedman: Partial Differential Equations of parabolic Type. Prentice Hall, Englewood Cliffs, N.J., 1964.
116. A. Friedman: Partial Differential Equations, Holt Rinehart and Winston, New York, 1969.

117. A. Friedman: Stochastic Differential Equations and Applications, Vol. I and II. Academic Press, New York, 1976.
118. A. Friedman: Variational principles and free-boundary problems. John Wiley & Sons, 1982.
119. K.O. Friedrichs: Water waves on a shallow sloping beach. Comm. Pure and Appl. Math., 1 (1948), 109-134.
120. H. Fujii, M. Mimura, Y. Nishiura: A picture of the global bifurcation diagram in ecological interacting and diffusing systems, Phys. D 5 (Nr. 1) (1982), 1-42.
121. C.J. Galvin: Waves breaking in shallow water, in Waves on beach, ed. by R.E. Meyer, Academic Press, New York, 1972.
122. C.S. Gardner, J.M. Green, M.D. Kruskal, R.M. Miura: Method for solving the Korteweg-de Vries equation. Phys. Rev. Lett., 19 (1967), 1095-1097.
123. R. Gardner: Asymptotic behavior of semilinear reaction-diffusion systems with Dirichlet boundary conditions. Ind. Univ. Math. J., 29 (1980), 161-190.
124. R. Gardner: Global stability of stationary solutions of reaction-diffusion systems. J. Diff. Equs., 37 (1980), 60-69.
125. R. Gardner: Comparison and stability theorems for reaction-diffusion systems. SIAM J. Math. Analy., 12 (1980), 603-616.
126. R. Gardner: Existence and stability of travelling wave solutions of competition models, a degree theoretic approach. J. Diff. Equs., 44 (1982), 343-364.
127. R. Gardner, J. Smoller: The existence of periodic travelling waves for singularly perturbed predator prey equations via the Conley index. J. Diff. Equs., 45 (1982).
128. G.R. Gavalas: Nonlinear Differential Equations of Chemically Reacting Systems, Springer Tracts in Natural Philosophy, Springer-Verlag, New York, 1968.
129. I.M. Gel-fand: Some problems in the theory of quasilinear equations. Usp. Mat. Nauk., 14 (1959), 87-158. English transl. in Amer. Math. Soc. Trans. Ser 2, 29 (1963), 295-381.
130. I.M. Gel'fand, B.M. Levitan: Izv. Akad. nauk. SSSR, Ser. Mat., 15 (1951), 309.
131. J.D. Gibbon, G. Zambotti: The interaction of n-dimensional Soliton wave fronts. Nuovo Cimento, 28 B, No. 1 (1975), 1.

132. J.W. Gibbs: The Scientific Papers of J. Willard Gibbs (1876-1878). New York, Longmans Green, 1906, and Dover, New York, 1961.
133. A. Gierer, H. Meinhardt: Biological pattern formation involving lateral inhibition. Lectures on Mathematics in the Life Science, 7 (1964), 163 ff.
134. A. Gierer, H. Meinhardt: Applications of a theory of biological pattern formation based on lateral inhibition. J. Cell. Sci., 15 (1974), 321 ff.
135. D. Gilbarg, F. Trudinger: Elliptic Partial Differential Equations of second order. Grundlehren der math. Wissenschaften, Vol. 224, Springer-Verlag, 1977.
136. P. Glansdorff, I. Prigogine: Thermodynamics of Structure, Stability, and Fluctuations. Wiley-Interscience, New York, 1971.
137. N.S. Goel et al : On the Volterra and other nonlinear models of interacting populations. Rev. Mod. Phys., 43 (1971), 231-276.
138. W. Greup: Linear Algebra. Grundlehren der math. Wissenschaften, Vol. 97, Springer-Verlag, Berlin Heidelberg New York, 1967.
139. K.-U. Grusa: Two-dimensional, interpolating Lg-splines and their applications. Lecture Notes in Mathematics, Vol. 916, ed. by A. Dold and B. Eckmann, Springer-Verlag, Berlin Heidelberg New York, 1982.
140. K.-U. Grusa: Two-dimensional differential-splines, unpublished manuscripts, 1981-1983.
141. K.-U. Grusa: Dynamic global models and two-dimensional splines I. Interdisciplinary Systems Research, Vol. 80, ed. by S. Klaczko-Ryndziun et al ., Verlag TÜV-Rheinland, Köln, 1984.
142. K.-U. Grusa: On nonlinear, dynamic processes, unpublished manuscripts, 1983-1987.
143. H. Haken: Laser Theory. Encyclopedia of Physics, Vol. XXV/2c, Springer-Verlag, Berlin Heidelberg New York, 1970.
144. H. Haken: Generalized Ginzburg-Landau Equations. Z. Phys., B 21 (1975), 105-114.
145. H. Haken: Generalized G.-L. Equations, Z. Phys., B 22 (1975), 69-72.
146. H. Haken: Generalized G.-L. Equations, Z. Phys., B 23 (1975), 388.
147. H. Haken, H. Ohno: Opt. Commun., 16 (1976), 205.
148. H. Haken: Synergetics, An Introduction, Springer-Verlag, Berlin Heidelberg New York, 3rd edition 1983.

149. H. Haken: Advanced Synergetics. Instability Hierarchies of Self-organizing Systems and Devices, Springer-Verlag, Berlin Heidelberg New York, 1983.

150. A. Haraux: Nonlinear Evolution-Equations-Global Behavior of solutions. Lecture Notes in Mathematics, Vol. 841, Springer-Verlag, Berlin Heidelberg New York, 1981.

151. B. Hassard. N. Kazarinoff, Y. Wan: Theory and Application of Hopf Bifurcation, Cambridge University Press, Cambridge, 1981.

152. S. Hastings: Some mathematical problems from neurobiology. AMS Monthly, 82 (1975), 881-895.

153. S. Hastings, J. Murray: The existence of oscillatory solutions in the Field-Noyes model for the Belousov-Zhabotinskii reaction. SIAM J. Appl. Math., 28 (1975), 678-688.

154. F.G. Heineken, H.M. Tsuchiya, R. Aris: On the mathematical status of the pseudo-steady state hypothesis of biochemical kinetics, Math. Biosci., 1 (1967), 95-113.

155. D. Henry: Geometric theory of semilinear parabolic equations. Lecture Notes in Mathematics, Vol. 840, Springer-Verlag, Berlin Heidelberg, New York, 1981.

156. M. Hestenes: Calculus of variations and optimal control theory. John Wiley & Sons, 1966.

157. E. Hewitt, K. Stromberg: Real and Abstract Analysis, Springer-Verlag, Berlin, Heidelberg New York, 1969.

158. E. Hille, R.S. Phillips: Functional Analysis and Semi-Groups, rev. edition, Amer. Math. Soc., Colloquium Publ., Vol. XXXI, 1957.

159. L. Hörmander: The Analysis of linear partial differential operators Vol. I, Distribution theory and Fourier Analysis.

160. L. Hörmander: Vol. II, Differential Operators with constant Coefficients.

161. L. Hörmander: Vol. III, Pseudo Differential Operators.

162. L. Hörmander: Vol. IV, Fourier Integral Operators, Grundlehren der mathematischen Wissenschaften, Volumes 256, 257, 274, 275, Springer-Verlag, Berlin Heidelberg New York, 1983, 1985.

163. A.L. Hodkin, A.F. Huxley: A quantitative description of membrane current and its applications to conduction and excitation in nerves. J. Physiol., 117 (1952), 500-544.

164. R.J. Holloway (ed.): Consumer behavior, Houghton Mifflin Company, Boston, 1971.

165. E. Hopf: Elementare Bemerkungen über die Lösungen partieller Differentialgleichungen zweiter Ordnung vom elliptischen Typus. Sitzb. preuss. Akad. d. Wiss., Berlin, 19 (1927), 141-152.

166. E. Hopf: The partial differential equation $u_t + u \cdot u_x = \mu \cdot u_{xx}$. Comm. Pure and Appl. Math., 3 (1950), 201-230.

167. F. Hoppensteadt: Mathematical Theories of populations: Demographics, Genetics, and Epidemics, Regional conference Series in Appl. Math. 20, Soc. for Indust. and Appl. Math. Philadelphia, 1975.

168. L. Howard, N. Kopell: Plane wave solutions to reaction-diffusion equations. Stud. Appl. Math., 52 (1973), 291-328.

169. P. Humbel: Preispolitische Gewinndifferenzierung im Einzelhandel, Zürich 1958, 78-79, "loss leader" (Zugartikel).

170. G.A. Hunt: Semi-groups of measures on Lie groups, Trans. of Amer. Math. Soc., 81 (1956), 264-293.

171. K. Itô: On Stochastic Differential Equations, Memoirs of the Am. Math. Soc., No. 4 (1951).

172. K. Itô, H.P. McKean Jr.: Diffusion processes and their sample paths. Die Grundlehren der mathematischen Wissenschaften, Vol. 125, Springer-Verlag, Berlin Heidelberg New York, 1965.

173. E.T. Jaynes: Gibbs vs. Boltzmann Entropies. Am. J. Phys., 33 (1965), 391-398.

174. E.T. Jaynes: Phys. Rev. A., 4 (1971), 747-750.

175. E.T. Jaynes: Where do we stand on Maximum Entropy? In the Maximum Entropy Formalism, ed. by R.O. Levine, M. Fribus, MIT Press, Cambridge, 1978, 15-118.

176. E.T. Jaynes: The minimum entropy production principle. Annual Rev. Phys. Chem., 31 (1980), 579-601.

177. E.T. Jaynes: Macroscopic Prediction, in Complex-Systems-Operational Approaches in Neurobiology, Physics, and Computers, Springer Series in Synergetics, Vol. 31, ed. by H. Haken, Springer-Verlag, 1985.

178. F. John: Plane waves and spherical means applied to partial differential equations, New York, Intersc. Publ., 1955.

179. M. Kac: A stochastic model related to the telegrapher's equation. Rocky Mountain Journal of Math., Vol. 4 (1974), 497-509.

180. Y. Kametaka: On the nonlinear diffusion equation of Kolmogorov-Petrovskii-Piskunov type. Osaka J. Math., 13 (1976), 11-66.

181. Ya. I. Kanel': On the stabilization of solutions of the Cauchy problem for the equations arising in the theory of combustion. Math. Sbornik, 59 (1962), 245-288.

182. Ya. I. Kanel': On the stabilization of the solutions of the equations of combustion theory with initial data of compact support. Math. Sbornik, 65 (1964), 398-413.

183. V.I. Karpman: Some asymptotic relations for solution of the Korteweg-de Vries equation, Phys. Lett., 26 A (1968), 619.

184. V.I. Karpman: Phys. Lett., 25 A (1967), 708.

185. V.I. Karpman, V.P. Sokolov: On solitons and the eigenvalues of the Schrödinger equation, Soviet. Phys. JETP, 27 (1968), 839-845.

186. T. Kato: Perturbation Theory of Linear Operators. Die Grundlehren der mathematischen Wissenschaften, Vol. 132, Springer-Verlag, Berlin Heidelberg New York, 1966.

187. I. Kay, H.E. Moses: The Determination of the scattering potential from the spectral measure function. Calculation of the scattering operator for the one-dimensional Schrödinger Equation. Nuovo Cim., 3 (1956), 276-304.

188. I. Kay, H.E. Moses: Reflectionless transmission through dielectrics and scattering potential. J. Appl. Physics, 27 (1956), 1503-1508.

189. H.B. Keller, D.A. Levine, G.B. Whitham: Motion of a bore over a sloping beach. Jour. Fluid Mech., 7 (1960), 302-316.

190. K. Kishimota: The diffusive Lotka-Volterra System with three species can have a stable non-constant equilibrium solution. J. Math. Biology, 16 (1982), 103-112.

191. Yu. L. Klimontovich: The kinetic theory of electromagnetic processes, Nauka, Moscow, 1980; im Springer-Verlag, Berlin Heidelberg New York, 1983.

192. Yu. L. Klimontovich: Statistical Physics. Nauka, Moscow, 1982; in Gordon and Breach, Harwood Academic publishers, New York, 1985.

193. H. Kniess: Lösung von Randwertaufgaben bei Systemen gewöhnlicher Differentialgleichungen vermittels endlicher Fourier-Transformation. Math. Zeitsch., 44 (1938), 266-292.

194. A. Kolmogoroff, I. Petrovsky, N. Piscounoff: Etude de l'equations de la Diffusion avec Croissance de la Quantite de Matiere et son Application a un Probleme Biologique. Bull. Univ. Moscow, Ser. Internat., Sec. A, 1 ≠ 6, (1937), 1-25.
195. V.I. Krinsky: Autowaves: Results, Problems, Outlook, in Self-Organization, ed. by V.I. Krinsky, Synergetics, Springer-Verlag, Berlin Heidelberg New York, 1984, 9-19.
196. V.I. Krinsky: Biofizika, 11 (1966), 676.
197. V.I. Krinksy: Problemy kibernetiki, 20 (1968), 59.
198. V.I. Krinsky: Internat. Encycl. of Pharma. and Therapeutics, Pergamon Press, London, 1981, 105.
199. V.I. Krinsky: Autowaves. Znanie Publ. Moskow, 1984 with A.S. Mikhailov.
200. V.I. Krinsky, A.M. Zhabotinsky: Autowave Processes in Systems with Diffusion, Gorky, 1981.
201. M.D. Kruskal, R.M. Miura, C.S. Gardner, N.J. Zabusky: Korteweg-de Vries Equation and Generalization V. Uniqueness and nonexistence of polynomial Conservation Laws. J. Math. Physics, 11 (1970), 952-960.
202. H.J. Kuiper: Existence and comparison theorems for nonlinear diffusion systems. J. Math. Anal. Appl., 60 (1977), 166-181.
203. H.J. Kuiper: Invariant sets for nonlinear elliptic and parabolic systems, Tech. Rep., Uni. of Wisconsin, Math. Research Center, 1978.
204. H.J. Kuiper: Invariant sets for nonlinear elliptic and parabolic systems, SIAM J. Math. Anal., 11 (1980), 1075-1103.
205. A. Kuipers: Model and insight, in the concept and the role of the model in Mathematics and Natural and Social Sciences, ed. by H. Freudenthal, D. Reidel publishing Co., Holland, 1961.
206. O.A. Ladyzhenskaya, N.N. Ural'tseva: Linear and quasilinear elliptic equations, Academic Press, 1968.
207. D.A. Larson: On the existence and stability of bifurcated solitary wave solutions to nonlinear diffusion equations. J. Math. Anal. Appl., 1977.
208. D.A. Larson: On models for two-dimensional spatially structured chemodiffusional signal propagation, preprint, 1977.
209. D.A. Larson: Transient bounds and time asymptotic behavior of solutions to nonlinear equations of Fisher type. SIAM J. Appl. Math., 34 (1978), 93-103.

210. P.D. Lax, N. Milgram: Parabolic equations. Contributions to the theory of partial differential equations. Annals of Math. Studies, No. 33, Princeton, 1954, 167-190.

211. S. Levin: Spatial Patterning and the Structure of Ecological communities, in some Mathematical Questions in Biology, VII, Lectures on Mathematics in the Life Sciences, Vol. 8, Amer. Math. Soc., Providence, R.I., 1976.

212. S.A. Levin, L.A. Segel: Hypothesis for origin of planktonic Patchiness, Nature 259, 1976, 659.

213. C.C. Lin, L.A. Segel: Deterministic problems in the natural sciences. MacMillan Publishing Co., 1974, New York.

214. J.L. Lions, G. Stampacchia: Variational inequalities. Comm. Pure and Appl. Math., XX (1967), 493-519.

215. J.L. Lions, G. Stampacchia: Inéquations variationnelles non coercives. C.R. Acad. Sci. Paris, t. 261 (1965), 27.

216. J.L. Lions, E. Magenes: Problèmes aux limites non homogènes et application. Dunod, Gauthier Villars, Vol. 1 and 2, 1968, Vol. 3, 1970, Traduction anglaise par P. Kenneth in Grundlehren der mathematischen Wissenschaften, Springer-Verlag, Berlin Heidelberg New York, Vol. 1,2 1972 and Vol. 3 1973.

217. J.L. Lions: Optimisation pour certaines classes d'évolution non lineaires. Ann. Math. Pura Appl., LXXII (1966), 275-294.

218. J.L. Lions: Equations différentielles et problèmes aux limites. Grundlehren der mathematischen Wissenschaften, Vol. 111, Springer-Verlag, Berlin Heidelberg New York, 1961.

219. J.L. Lions: Quelque methodes de resolution des problemes aux limites non lineaires. Dunod, Gauthier Villars, Paris, 1969.

220a. J.L. Lions: Sur le contrôle optimal des systèmes gouvernés par des équations aux derivées partielles. Dunod, Gauthier Villars, Paris, 1968.

220b. J.L. Lions: Some aspects of the optimal control of distributed parameter systems. Regional Conference Series in Applied Mathematics, SJAM 6,1972.

220c. J.L. Lions: Some methods in the mathematical analysis of systems and their control. Science Press, Beijing, China, 1981, Gordon and Breach Science Publishers Inc, New York 1981.

220d. J.L. Lions: Perturbations singulières dans les problèmes aux limites et en contrôle optimal. Lecture Notes in Mathematics, Volume 323, Springer-Verlag, 1973.

221. Li Ta-tsien: Propriétés d'espace fonctionnels intervénant en contrôle optimal. C.R. Acad. Sci. Paris, t. 289, serie 1, (1979), 315.

222. Li Ta-tsien: Propriétés d'espace fonctionnels et problèmes de contrôle optimal des systèmes gouvernés par des équations parabolique. C.A. Acad. Sci. Paris, t. 290, serie A (1980), 697-700.

223. Li Ta-tsien: Comportements limites des solutions de certaines équations paraboliques d'ordre supérieur et leurs applications au contrôle optimal. C.R. Acad. Sci. Paris, t. 293, serie I (1981), 205-208.

224. W. Littman: Generalized subharmonic functions: monotonic approximations and an improved maximum principle. Ann. Scuola Norm. Sup. Pisa, (3), 17 (1963), 207-222.

225. A.J. Lotka: Undamped oscillations derived from the law of mass action J. Amer. Chem. Soc., 42 (1920), 1595.

226. A.J. Lotka: J. Phys. Chem., 14 (1910), 271.

227. A.J. Lotka: Proc. Nat. Acad. Sci. US (Wash.), 6 (1920), 410.

228. A.J. Lotka: Elements of Mathematical Biology, Dover, New York, 1956.

229. K. Maginu: Stability of spatially homogeneous periodic solutions of reaction diffusion equations. J. Diff. Equations, 31 (1979), 130-138.

230. P.D. Maker, R.N. Terhune, C.M. Savage: Proceedings third conference on Quantum Electronics, Paris, 1963, ed. by P. Grivet, N. Bloembergen, Columb. Univ., 1964, 1559.

231. V.A. Marchenko: Dokl. Akad. nauk. SSSR, 104 (1955), 695.

232. J. Marsden, M. McCracken: The Hopf Bifurcation and Its Applications, Springer-Verlag, New York, 1976.

233. R.M. May: Stability and Complexity in Model Ecosystems. Princeton Univ. Press (2nd ed.), Princeton, N.J., 1974.

234. R.M. May: Theoretical Ecology: Principles and Applications. W.B. Saunders Co., Philadelphia, Toronto, 1976.

235. J. Maynard-Smith: Models in Ecology, Cambridge Uni. Press, Cambridge, 1974.

236. H.P. McKean: Application of Brownian motion to the equation of Kolmogorov-Petrovski-Piskunov, Comm. Pure Appl. Math., 28 (1975), 323-331.

237. J.B. McLeod, P.C. Fife: A phase plane discussion of convergence to travelling fronts. Arch. Rat. Mech. Anal., 75 (1981), 281-314.

238. A.B. Medvinsky, A.V. Panfilov, A.M. Pertsov: Properties of rotating waves in three dimensions. Scroll rings in myocard., in Self-Organization ed. by V.I. Krinsky, Synergetics, Springer-Verlag, Berlin Heidelberg New York, 1984 (28).

239. H. Meinhardt: The Formation of Morphogenetic Gradients and Fields Ber. Deutsch. Bot. Ges., 87 (1974), 101 ff.
240. H. Meinhardt: Models of Biological Pattern Formation, Academic Press, London, 1982.
241. M. Mimura: Spatial structures in nonlinear interaction-diffusion-systems. Kokyuroku 317, RIM, Kyoto Univ. Kyoto, Japan, 1978, 17-29.
242. M. Mimura, J.D. Murray: On a diffusive prey-predator model which exhibits Patchiness. J. Theoret. Biology, 75 (1979), 249-262.
243. M. Mimura, T. Nishida: On a certain semilinear parabolic system related to Lotka-Volterra's ecological model, Publ. Research Inst. Math. Sci., Kyoto Univ., 14 (1978), 269-282.
244. M. Mimura, Y. Nishiura: Spatial patterns for interaction-diffusion equations in Biology, Proc. International Symposium on Mathematical Topics in Biology, 1978, 136-146.
245. M. Mimura, Y. Nishiura, M. Yamaguti: Some diffusion prey and predator systems and their bifurcation problem, Ann. Y.Y. Acad. Sci., 316 (1976), 490-510.
246. M. Mimura, M. Tabata, Y. Hosono: Multiple solutions of two-point boundary value problems of Neumann type with small parameter. SIAM J. Math. Anal. Vol. II (1980), 613-631.
247. R.M. Miura: Korteweg-de Vries Equation and Generalizations I. J. Math. Physics, 9 (1968), 1202-1204.
248. R.M. Miura, C.S. Gardner, M.D. Kruskal: Korteweg-de Vries Equation and Generalization II. J. Math. Phys., 9 (1968), 1204-1209.
249. H.J.K. Moet: A note on the asymptotic behavior of solutions of the KPP-equation. SIAM J. on Math. Anal.
250. D. Mollison: Spatial contact models for ecological and epidemiological spread. J. Roy. Stat. Soc., B 39 (1977), 283-326.
251. X. Mora: Semilinear parabolic problems define semiflows on C^k spaces, Trans. Amer. Math. Soc. 1983.
252. W.H. Munk, H. Wimbush: Oceanology, 9 (1969), 56-59.
253. J.D. Murray: Lectures on nonlinear differential-equation Models in Biology, Clarendon Press, Oxford, 1977.
254. J. Nagumo, S. Arimoto, S. Yoshizawa: An active pulse transmission line simulating nerve axon. Proc. IRE, 50 (1964), 2061-2070.
255. T. Nagylaki: Selection in one-and two-locus Systems. Lecture Notes in Biomathematics, Vol. 15, Springer-Verlag, Berlin, 1977.

256. T. Nagylaki: Clines with asymmetric migration. Genetics, 88 (1978), 813-827.

257. T. Nagylaki: Random genetic drift in a cline. Proc. Nat. Acad. Sci., 75 (1978), 423-426.

258. T. Nagylaki: The geographical structure of populations. MAA Study in Math. Biology, Vol. II: Populations and communities, ed. by S.A. Levin, Mathematical Ass. of American, 1978, 588-624.

259. T. Nagylaki: A diffusion model for geographically structured populations J. Math. Biol., 6 (1978), 375-382.

260. E. Nelson: Les écoulements in compressibles d'energie finie. Colloques internationaux du Centre national de la recherche scientifique No. 117, Les équations aux derivées partielles. C.R. Acad. Sci.,Paris, 1962.

261. E. Nelson: Regular probability measures on function space. Annals of Math., 69 (1959), 630-643.

262. E. Nelson: Feynman integrals and the Schrödinger equation. Jour. of Math. Physics, 5 (1964), 332-343.

263. G. Nicolis, I. Prigogine: Self-Organization in Nonequilibrium Systems. J. Wiley & Sons, New York, London, Toronto, 1977.

264. L. Nirenberg: On elliptic partial differential equations. Ann. Scuola Norm. Sup. Pisa, (3), 13 (1959), 115-162.

265. L. Nirenberg: Topics in nonlinear functional analysis. Courant Inst. Math. Sci. Notes 1972/1973.

266. L. Nirenberg: Variational and topological methods in nonlinear problems. Bull. Amer. Math. Soc. (NS), 4 (1981), 267-302.

267. Y. Nishiura: Global Structure of bifurcating solutions of some reaction-diffusion-systems. SIAM J. Math. Analy., 13 (1982), 555-593.

268. S. Novikov, S.V. Manakov, L.P. Pitaevskii, V.E. Zakharov: Theory of Solitons. The inverse scattering methods. Consultants Bureau- New York and London, 1982.

269. F. Oberhettinger: Tabellen zur Fourier-Transformation, Springer-Verlag, Berlin Göttingen, Heidelberg, 1957.

270. H. Ohno, H. Haken: Transient ultrashort laser pulses. Opt. Commun., 16 (1976), 205.

271. A. Okubo: Diffusion and Ecological Problems: Mathematical Models. Springer-Verlag, Berlin Heidelberg New York, 1980.

272. O. Oleinik: Discontinuous solutions of nonlinear differential equations. Usp. Mat. Nauk. (N.S.), 12 (1957), 3-73. English transl. in Amer. Math. Soc. Transl. Ser. 2, 26, 95-172.

273. O. Oleinik: On the uniqueness of the generalized solution of the Cauchy problem for a nonlinear system of equations occurring in mechanics. Usp. Mat. Nauk. (N.S.), 12 (1957), 169-176.

274. O. Oleinik: Uniqueness and a stability of the generalized solution of the Cauchy problem for a quasilinear equation. Usp. Mat. Nauk., 14 (1959), 165-170. English transl. in Amer. Math. Soc. Transl. Ser. 2, 33 (1964), 285-290.

275. L. Onsager: Phys. Rev., 37 (1931), 405-426.

276. L. Onsager: Phys. Rev., 38 (1931), 2265-2279.

277. H.G. Othmre, L.E. Scriven: Instability and dynamic pattern in cellular networks. J. Theor. Biol., 32 (1971), 507-537.

278. D.M. Pepper: Nonlinear optical phase conjugation. Optical Engineering 21 (1982), 156.

279. A.S. Peters: Water waves over sloping beaches and the solution of a mixed boundary value problem for $\Delta^2\phi - k\phi = 0$ in a sector. Comm. Pure a. Appl. Math., 5 (1952), 87-108.

280. J.R. Philip: Sociality and sparce populations, Ecology, 38 (1957), 107-111.

281. L.S. Polak: Self-Organization in Nonequilibrium Physico-Chemical Systems (in Russian). M. Nauka Publ., 1983.

282. L.S. Pontryagin, V.S. Boltyanskii, R.V. Gamkrelidze, E.F. Mischenko: The mathematical theory of otpimal processes, Interscience, 1962.

283. I. Prigogine: Non-equilibrium statistical mechanics. J. Wiley & Sons, New York, 1962.

284. I. Prigogine: From being to becoming. Freeman, San Francisco, 1980.

285. I. Prigogine, I. Stengers: Order out of chaos. Bantham Books, Toronto, New York, London, 1984.

286. I. Prigogine, G. Nicolis: On symmetry-breaking instabilities in dissipative systems. J. Chem. Physics, 46 (1967), 3542-3550.

286a. M. Protter, H. Weinberger: Maximum Principles in Differential Equations. Prentice Hall, Englewood Cliffs, N.J., 1967.

287. A.M. Pertsov, A.K. Grenadier: The Autowave Nature of Cardiac Arrhythmias. Self-Organization, Synergetics, Springer-Verlag, 1984, 184-190, (28).

288. P. Rabinowitz: Some global results for nonlinear eigenvalue problems. J. Funct. Analysis, 1 (1971), 487-513.

289. J. Rauch: Stability of motion for semilinear equations. Boundary value problems of linear evolution partial differential equations, ed. by H. Garnir, Reidel, Dordrecht, 1977, 319-349.

290. J. Rauch, J. Smoller: Qualitative theory of the Fitz-Hugh-Nagumo equations. Adv. in Math., 27 (1978), 12-44.

291. Th. S. Robertson: Innovative behavior and communication. Holt, Rinehart and Winston, Inc., 1971.

292. T.R. Rockaffelar: Duality and stability in extremum problems involving convex functions. Pac. J. Math., 21 (1967), 167-187.

293. F. Rothe: Über das asymptotische Verhalten der Lösungen einer nichtlinearen parabolischen Differentialgleichung aus der Populationsgenetik, Dissertation 1975, Univ. Tübingen.

294. F. Rothe: Some analytical results about a simple reaction-diffusion system for morphogenesis. J. Math. Biol., 7 (1979), 375-384.

295. R. Rothe: Convergence to travelling fronts in semilinear parabolic equations. Proceedings of the Royal Society of Edinburgh, 80 A (1978), 213-234.

296. F. Rothe: Global solutions of Reaction-Diffusion Systems. Lecture Notes in Mathematics, Vol. 1072, Springer-Verlag, Berlin Heidelberg New York, 1984.

297. D. Sattinger: Stability of bifurcating solutions by Learay-Schauder degree. Arch. Rat. Mech. Anal., 43 (1971), 154-166.

298. D. Sattinger: Monotone methods in nonlinear elliptic and parabolic equations. Ind. Univ. Math. J., 21 (1972), 979-1000.

299. D. Sattinger: Topics in stability and Bifurcation Theory. Univ. of Minnesota, Lecture Notes 1971-1972.

300. S. Sawyer: Asymptotic properties of the equilibrium probability of identity in a geographically structured population. Adv. Appl. Prob., 9 (1977), 268-282.

301. S. Sawyer: Rates of consolidation in a selectively neutral migration model. Ann. of Probability, 5 (1977), 486-493.

302. S. Sawyer: A continuous migration model with stable demography. J. Math. Biol., 11 (1981), 193-205.

303. S. Sawyer: Isolation by distance in a hierarchically clustered population. J. Appl. Probab., 20 (1983), 1-10.

304. D. Sattinger: A nonlinear parabolic system in the theory of combustion, Quart. Appl. Math., 1975, 47-61.

305. D. Sattinger: On the stability of waves of nonlinear parabolic systems. Advances in Math., 22 (1976), 312-355.

306. D. Sattinger: Weighted norms for the stability of travelling waves. J. Diff. Equations, 25 (1977), 130-144.

307. J. Schauder: Der Fixpunktsatz in Funktionalräumen. Studia Math., 2 (1936), 171-180.

308. L. Schwartz: Theorie des distributions, T. 1, Hermann, Paris, 1950.

309. L. Schwartz: Theorie des distributions, T. 2, Hermann, Paris, 1951.

310. L. Schwartz: Theorie des distributions a valeurs vectorielles I. Annales Institut Fourier, 7 (1957), 1-141.

311. L. Schwartz: Theorie des distributions a valeurs vectorielles II. Annales Institut Fourier, 8 (1959), 1-209.

312. L. Schwartz: Theorie des noyaux. Proceedings of the international congress of mathematicians, 1 (1950), 220-230.

313. J. Schwartz: Nonlinear functional analysis. Courant Inst. Math. Sci., Notes (1963-64).

314. L.E. Scriven: Instability and dynamic pattern in cellular networks. J. Theor. Biol., 32 (1971), 507-537.

315. L.E. Scriven: Interactions of reaction and diffusion in open systems, Industrial and Engineering Chemistry Fund., 8 (1969), 302-313.

316. I. Segal: Algebraic integration theory. Bulletin Amer. Math. Soc., 71 (1965), 419-489.

317. L.A. Segel: Dissipative Structure: An explanation and an ecological example. J. Theor. Biology, 37 (1972), 545-559.

318. L.A. Segel, S.A. Levin: Application of nonlinear stability theory to the study of the effects of diffusion on predator-prey interactions, in: Topics in statistical Mechanics and Biophysics. A Memorial to Julius L. Jackson, AIP Conf. Proc. No. 27, ed. by R.A. Piccirelli, 1976, 123-152.

319. Shih Shu-Chung, J. Simon: Sur un espace intervenant en contrôle punctuel. C.R. Acad. Sci. Paris, t. 290, Serie A, 1980, 693-696.

320. J. Smoller, A. Tromba, A. Wasserman: Non-degenerate solutions of boundary value problems. J. Nonlinear Anal., 4 (1980), 207-216.

321. J. Smoller, H. Wasserman: Global bifurcation of steady-state-solution. J. Diff. Equations, 39 (1981), 269-290.

322. J. Smoller: Shock waves and reaction-diffusion equations. Grundlehren der mathematischen Wissenschaften, Vol. 258, Springer-Verlag, New York Heidelberg Berlin, 1983.

323. S.L. Sobolev: Applications of functional analysis in mathematical physics. Transl. Math. Monog., No. 7, Amer. Math. Soc., Providence, R.I., 1963.

324. H. von Stackelberg: Grundlagen der theoretischen Volkswirtschaftslehre, 2. Auflage, Tübingen, Mohr Bern. Franke, 1951, Kapitel IV.3.

325. G. Stampacchia: Formes bilinéaires coercitives sur les ensembles convexes. C.R. Acad. Sci. Paris, t. 258 (1964), 4413-4416.

326. G. Stampacchia: Le probleme de dirichlet pour les e'quations elliptiques du second ordre a coefficients discontinus. Ann. Inst. Fourier, 15 (1965), 189-258.

327. E.M. Stein: Singular Integrals and Differentiability Properties of Functions. Princeton Univ. Press, Princeton, New Jersey, 1970.

328. J.J. Stoker: Water Waves. Interscience, New York, 1957.

329. A.N. Stokes: On two types of moving front in quasilinear diffusion. Math. Biosci., 31 (1976), 307-315.

330. R. Temam: Sur l'équation de Riccati associée a des operateurs non bornes, en dimension infinie. J. Functional Analysis, 7 (1971), 85-115.

331. R. Temam, C. Foias: Some analytic and geometric Properties of the Solution of the Navier-Stokes Equations. J. Math. Pures Appl., 58 (1979), 339-368.

332. R. Teman, C. Foias: Determination of the solutions of the Navier-Stokes Equations by a Set of nodal Values. Math. Comput., 43 No. 167 (1984), 117-133.

333. R. Teman: Attractors for the Navier-Stokes Equations, in Nonlinear partial differential equations and their applications, Research Notes in Math., Vol. 122 (1985), 272-292.

334. R. Thom: Structural stability and Morphogenesis, W.A. Benjamin, Reading, Mass., 1975.

335. F. Treves: Relations de domination entre opérateurs differentiels. Acta Math., 101 (1959), 1-139.

336. F. Treves: Basic Linear Partial Differential Equations. Academic Press, New York, 1975.

337. F. Treves: Introduction to pseudodifferential and Fourier integral operators. Vol. 1: Pseudodifferential operators. Vol. 2: Fourier integral operators. Plenum Press, New York and London, 1980.
338. A.M. Turing: The Chemical Basis of Morphogenesis, Phil. Trans. Royal Soc., B 237, (37), 1952, 37-72.
339. R.J. Tykodi: Thermodynamics of steady states. Macmillan, New York, 1967.
340. K. Uchiyama: The behavior of solutions of the equation of Kolmogorov-Petrovsky-Piskunov. Proc. Japan Acad., Ser. A 53 (1977), 225-228.
341. K. Uchiyama: The behavior of solutions of some nonlinear diffusion equations for large time. J. Math. Kyoto Univ., 18 (1978), 453-508.
342. F. Ursell: Edge waves on a sloping beach. Proc. Royal Soc., A 214 (1952), 79-97.
343. G.E. Uhlenbeck: On the theory of Brownian motion I, in Physical Review, 36 (1930), 823-841.
344. M.M. Vainberg: Variational methods for the study of nonlinear operators. Holden-Day, 1964.
345. S.R.S. Varadhan: Lectures on diffusion problems and partial differential equations. Tata Institute of Fundamental Research, Springer-Verlag, Berlin Heidelberg New York, 1980.
346. I.M. Visik: On strongly elliptic differential equations. Mat. Sbornik, 29 (1951), 615-676.
347. I.M. Visik: Resolution of boundary value problems for quasi-linear parabolic equations of arbitrary order. Mat. Sbornik, 59 (1962), 289-335.
348. D. Voelker, G. Doetsch: Die zweidimensionale Laplace-Transformation, Birkhäuser, 1950.
349. A.I. Vol'pert, S.I. Hudyaev: Analysis in Classes of discontinuous functions and equations of the Mathematical Physics, Izd. Nauka, Moscow, 1975.
350. V. Volterra: Lecons sur la théorie mathematiques de la lutte pour la vie. Gauthier-Villars, Paris, 1931.
351. V. Volterra: Acta Biotheoret., 3 (1937), 1
352. M. Ch. Wang: On the theory of Brownian motion II, in Reviews of G.E. Uhlenbeck, Modern Physics, 17 (1945), 323-342.
353. P.K.C. Wang: Control of distributed parameter systems, in Advances in Control Systems, ed. by D.T. Leondes, Academic Press, 1 (1964), 75-172.

354. P.K.C. Wang: Optimal Control of a class of linear symmetric hyperbolic systems with applications to plasma confinement. J. of Math. Analysis and Appl., 28 (1969), 594-608.

355. H.F. Weinberger: Invariant sets for weakly coupled parabolic and elliptic systems. Rend. Mat. Univ. Roma, VI 8 (1975), 295-310.

356. A.D. Wentzell: On boundary conditions for multidimensional diffusion processes. Theor. Prob. and Appl., 4 (1959), 164-177.

357. S. Williams, P.L. Chow: Nonlinear reaction-diffusion models for interacting populations. J. Math. Anal. Appl., 62 (1978), 157-169.

358. A.T. Winfree: Geometry of Biological Time. Springer-Verlag, Berlin Heidelberg New York, 1980.

359a. G.B. Whitham: Lectures on wave propagation. TATA Institute of Fundamental Research, Narosa Publishing House, New Delhi, 1979, and Springer-Verlag, Berlin Heidelberg New York, 1980.

359b. G.B. Whitham: Linear and nonlinear waves, John Wiley & Sons, 1974.

360. N.J. Zabusky: Proceedings of the Symposium on Nonlinear Partial Differential Equations (ed. by W. Ames), Academic Press, New York, 1967, 223.

361. N.J. Zabusky, M.D. Kruskal: Interaction of "solitons" in a collisionsless plasma and the recurrence of initial states. Phys. Rev. Lett., 15 (1965), 240-243.

362. J. Zagrodzinski: The solution of the two-dimensional Sine-Gordon Equation, Physics Letters, 57 A, No. 3, (1976), 213-214.

363. A.M. Zhabotinsky: Concentrational Auto-Oscillations (in Russian) M. Nauka Publ., 1974.

364. A. Yariv: Quantum Electronics, second ed., J. Wiley & Sons, New York, 1975.

365. A. Yariv: Three-dimensional pictorial transmission in optical fibers. Appl. Phys. Lett., 28 (1976), 88.

366. A. Yariv: On transmission and recovery of three-dimensional image information in optical wave guides. J. Opt. Soc. Amer., 66 (1976), 301.

367. K. Yosida: Functional Analysis. Grundlehren der mathematischen Wissenschaften, Vol. 123, Springer-Verlag, Berlin Heidelberg New York, 1971.

142a. K.U. Grusa: On dynamic stochastic processes, preprint. Centre for Interdisciplinary Research (Zif), University of Bielefeld, 1988.

Index